T0259926

Springer Proceedings in Mathematics & Statistics

Volume 99

Springer Proceedings in Mathematics & Statistics

This book series features volumes composed of select contributions from workshops and conferences in all areas of current research in mathematics and statistics, including OR and optimization. In addition to an overall evaluation of the interest, scientific quality, and timeliness of each proposal at the hands of the publisher, individual contributions are all refereed to the high quality standards of leading journals in the field. Thus, this series provides the research community with well-edited, authoritative reports on developments in the most exciting areas of mathematical and statistical research today.

More information about this series at http://www.springer.com/series/10533

Kathrin Glau · Matthias Scherer
Rudi Zagst
Editors

Innovations in Quantitative Risk Management

TU München, September 2013

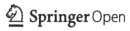

Editors
Kathrin Glau
Chair of Mathematical Finance
Technische Universität München
Garching
Germany

Rudi Zagst
Chair of Mathematical Finance
Technische Universität München
Garching
Germany

Matthias Scherer
Chair of Mathematical Finance
Technische Universität München
Garching
Germany

ISSN 2194-1009   ISSN 2194-1017   (electronic)
Springer Proceedings in Mathematics & Statistics
ISBN 978-3-319-35861-1   ISBN 978-3-319-09114-3   (eBook)
DOI 10.1007/978-3-319-09114-3

Mathematics Subject Classification (2010): 91B30, 91B82, 91B25, 91B24

Springer Cham Heidelberg New York Dordrecht London
© The Editor(s) (if applicable) and the Author(s) 2015. The book is published with open access at
SpringerLink.com.
Softcover reprint of the hardcover 1st edition 2015
Open Access This book is distributed under the terms of the Creative Commons Attribution
Noncommercial License which permits any noncommercial use, distribution, and reproduction in any
medium, provided the original author(s) and source are credited.
All commercial rights are reserved by the Publisher, whether the whole or part of the material is
concerned, specifically the rights of translation, reprinting, reuse of illustrations, recitation, broad-
casting, reproduction on microfilms or in any other physical way, and transmission or information
storage and retrieval, electronic adaptation, computer software, or by similar or dissimilar methodology
now known or hereafter developed.
The use of general descriptive names, registered names, trademarks, service marks, etc. in this
publication does not imply, even in the absence of a specific statement, that such names are exempt
from the relevant protective laws and regulations and therefore free for general use.
The publisher, the authors and the editors are safe to assume that the advice and information in this
book are believed to be true and accurate at the date of publication. Neither the publisher nor the
authors or the editors give a warranty, express or implied, with respect to the material contained
herein or for any errors or omissions that may have been made.

Printed on acid-free paper

Springer International Publishing AG Switzerland is part of Springer Science+Business Media
(www.springer.com)

Preface I

Dear Reader,

We would like to thank you very much for studying the proceedings volume of the conference "Risk Management Reloaded", which took place in Garching-Hochbrück, during September 9–13, 2013. This conference was organized by the KPMG Center of Excellence in Risk Management and the Chair of Mathematical Finance at Technische Universität München. The scientific committee consisted of Prof. Claudia Klüppelberg, Prof. Matthias Scherer, Prof. Wim Schoutens, and Prof. Rudi Zagst. Selected speakers were approached to contribute with a manuscript to this proceedings volume. We are grateful for the large number of high-quality submissions and would like to especially thank the many referees that helped to control and even improve the quality of the presented papers.

The objective of the conference was to bring together leading researchers and practitioners from all areas of quantitative risk management to take advantage of the presented methodologies and practical applications. With more than 200 registered participants (about 40 % practitioners) and 80 presentations we outnumbered our own expectations for this inaugural event. The broad variety of topics is also reflected in the long list of keynote speakers and their presentations: Prof. Hansjörg Albrecher (risk management in insurance), Dr. Christian Bluhm (credit-risk modeling in risk management), Prof. Fabrizio Durante (dependence modeling in risk management), Dr. Michael Kemmer (regulatory developments in risk management), Prof. Rüdiger Kiesel (model risk for energy markets), Prof. Ralf Korn (new mathematical developments in risk management), Prof. Alfred Müller (new risk measures), Prof. Wim Schoutens (model, calibration, and parameter risk), and Prof. Josef Zechner (risk management in asset management). Besides many invited and contributed talks, the conference participants especially enjoyed a vivid panel discussion titled "Quo vadis quantitative risk management?" with Dr. Christopher Lotz, Dr. Matthias Mayer, Vassilios Pappas, Prof. Luis Seco, and Dr. Daniel Sommer as participants and Markus Zydra serving as anchorman. Moreover, we had a special workshop on copulas (organized by Prof. Fabrizio Durante and Prof. Matthias Scherer), a DGVFM workshop on "Alternative interest guarantees in life insurance" (organized by Prof. Ralf Korn and Prof. Matthias Scherer), a workshop on "Advances in LIBOR modeling" (organized by Prof. Kathrin Glau), and a workshop on "Algorithmic differentiation" (organized by Victor Mosenkis and Jacques du Toit). Finally, the last day of the conference was dedicated to young researchers, serving as a platform to present results from ongoing Ph.D. projects. It is clearly worth mentioning, however, that there was enough time reserved for social events like a conference dinner at "Braustüberl Weihenstephan," a "Night watch man tour" in Munich, and a goodbye reception in Garching-Hochbrück. The editors of this volume would like to thank again all participants of the conference, all speakers, all members of the organizing committee (Kathrin Glau, Bettina Haas, Asma Khedher, Mirco Mahlstedt, Matthias Scherer, Anika Schmidt, Thorsten Schulz, and Rudi Zagst), all contributors to this volume, the referees, and finally our generous sponsor KPMG AG Wirtschaftsprüfungsgesellschaft.

Kathrin Glau
Matthias Scherer
Rudi Zagst

Preface II

The conference "Risk Management Reloaded" was held on the campus of Technische Universität München in Garching-Hochbrück (Munich) during September 9–13, 2013. Thanks to the great efforts of the organizers, the scientific committee, the keynote speakers, contributors, and all other participants, the conference was a great success, motivating academics and practitioners to learn and discuss within the broad field of financial risk management.

The conference "Risk Management Reloaded" and this book are part of an initiative called KPMG Center of Excellence in Risk Management that was founded in 2012 as a very promising cooperation between the Chair of Mathematical Finance at the Technische Universität München and KPMG AG Wirtschaftsprüfungsgesellschaft. This collaboration aims at bringing together practitioners from the financial industry in the areas of trading, treasury, financial engineering, risk management, and risk controlling, with academic researchers in order to supply trendsetting and realizable improvements in the effective management of financial risks. It is based on three pillars, consisting of the further development of a practical and scientifically challenging education of students, the support of research with particular focus on young researchers as well as the encouragement of exchange within the scientific community and between science and the financial industry.

The topic of financial risk management is a subject of great importance for banks, insurance companies, asset managers, and the treasury departments of industrial corporations that are exposed to financial risk. It has been of even greater attention ever since the financial crisis in 2008. Though regulatory focus rose and the requirements on internal risk models have become more pronounced and comprehensive, confidence in risk models and the financial industry itself has been damaged to some extent. We intended to discuss several questions concerning these doubts, for example, whether we need more or fewer quantitative risk models, and how to adequately use and manage risk models. We think that quantitative risk models are an important tool to understand and manage the risks of what continues to be a complex business. However, comprehensive regulation for internal models is necessary. It is important that models can be explained to internal and external stakeholders and are used in a suitable way.

The campus of the university in Garching-Hochbrück was a great place for the conference. The 200 participants, 55 % of whom were academics, 40 % practitioners, and 5 % students, had many fruitful discussions and exchanges during five days of workshops, talks, and great social events. Participants came from more than 20 countries, which made the conference truly international. Due to the broadness of the main theme and the many different backgrounds of the participants, the topics presented during the conference covered a large spectrum, ranging from regulatory developments to theoretical advances in financial mathematics and including speakers from both academia and the industry.

The first day of the conference was dedicated to workshops on copulas, algorithmic differentiation, guaranteed interest payments in life insurance contracts, and LIBOR modeling. During the following days, several keynote speeches and contributed talks treated various aspects of risk management, including market specific (insurance, credit, energy) challenges, and tailor-made methods (model building, calibration). The panel discussion on Wednesday brought together the views of prestigious representatives from academia, industry, and regulation on the necessity, reasonableness, and limitations of quantitative risk methods for the measurement and evaluation of risk. The conference was completed by a "Young Researchers Day" giving junior researchers the opportunity to present and discuss their results in front of a broad audience.

We would like to thank all the participants of the conference for making this event a great success. In particular, we express our gratitude to the scientific committee, namely Claudia Klüppelberg, Matthias Scherer, Wim Schoutens, and Rudi Zagst, the organizational team, namely Kathrin Glau, Bettina Haas, Asma Khedher, Mirco Mahlstedt, Matthias Scherer, Anika Schmidt, Thorsten Schulz, and Rudi Zagst, the keynote speakers, the participants of the panel discussion, namely Christopher Lotz, Luis Seco, and Vasilios Pappas, all speakers within the workshops, contributed talks, and the young researchers day, and, last but not least, all participants that attended the conference.

Dr. Matthias Mayer
KPMG AG Wirtschaftsprüfungsgesellschaft

Dr. Daniel Sommer
KPMG AG Wirtschaftsprüfungsgesellschaft

Contents

Contents

Part I
Markets, Regulation, and Model Risk

A Random Holding Period Approach for Liquidity-Inclusive Risk Management

Damiano Brigo and Claudio Nordio

Abstract Within the context of risk integration, we introduce risk measurement stochastic holding period (SHP) models. This is done in order to obtain a 'liquidity-adjusted risk measure' characterized by the absence of a fixed time horizon. The underlying assumption is that—due to changes in market liquidity conditions—one operates along an 'operational time' to which the P&L process of liquidating a market portfolio is referred. This framework leads to a mixture of distributions for the portfolio returns, potentially allowing for skewness, heavy tails, and extreme scenarios. We analyze the impact of possible distributional choices for the SHP. In a multivariate setting, we hint at the possible introduction of dependent SHP processes, which potentially lead to nonlinear dependence among the P&L processes and therefore to tail dependence across assets in the portfolio, although this may require drastic choices on the SHP distributions. We also find that increasing dependence as measured by Kendall's tau through common SHPs appears to be unfeasible. We finally discuss potential developments following future availability of market data. This chapter is a refined version of the original working paper by Brigo and Nordio (2010) [14].

1 Introduction

According to the Interaction between Market and Credit Risk (IMCR) research group of the Basel Committee on Banking Supervision (BCBS) [5], *liquidity conditions interact with market risk and credit risk through the horizon over which assets can be liquidated*. To face the impact of market liquidity risk, risk managers agree in adopting a longer holding period to calculate the market VaR, for instance 10 business

D. Brigo (✉)
Department of Mathematics, Imperial College London, 180 Queen's Gate,
London SW7 2AZ, UK
e-mail: damiano.brigo@imperial.ac.uk

C. Nordio
Risk Management, Banco Popolare, Milan, Italy
e-mail: claudio.nordio@bancopopolare.it

© The Author(s) 2015   3
K. Glau et al. (eds.), *Innovations in Quantitative Risk Management*,
Springer Proceedings in Mathematics & Statistics 99,
DOI 10.1007/978-3-319-09114-3_1

days instead of 1; recently, BCBS has prudentially stretched such liquidity horizon to 3 months [6]. However, even the IMCR group pointed out that *the liquidity of traded products can vary substantially over time and in unpredictable ways*, and moreover, *IMCR studies suggest that banks' exposures to market risk and credit risk vary with liquidity conditions in the market*. The former statement suggests a stochastic description of the time horizon over which a portfolio can be liquidated, and the latter highlights a dependence issue.

We can start by saying that probably the holding period of a risky portfolio is neither 10 business days nor 3 months; it could, for instance, be 10 business days with probability 99% and 3 months with probability 1%. This is a very simple assumption but it may have already interesting consequences. Indeed, given the FSA (now Bank of England) requirement to justify liquidity horizon assumptions for the Incremental Risk Charge modeling, a simple example with the two-points liquidity horizon distribution that we develop below could be interpreted as a mixture of the distribution under normal conditions and of the distribution under stressed and rare conditions. In the following we will assume no transaction costs, in order to fully represent the liquidity risk through the holding period variability. Indeed, if we introduce a process describing the dynamics of such liquidity conditions, for instance,

- the process of time horizons over which the risky portfolio can be fully bought or liquidated,

then the P&L is better defined by the returns calculated over such stochastic time horizons instead of a fixed horizon (say daily, weekly or monthly basis). We will use the "stochastic holding period" (SHP) acronym for that process, which belongs to the class of *positive processes* largely used in mathematical finance. We define the liquidity-adjusted VaR or Expexted Shortfall (ES) of a risky portfolio as the VaR or ES of portfolio returns calculated over the horizon defined by the SHP process, which is the 'operational time' along which the portfolio manager must operate, in contrast to the 'calendar time' over which the risk manager usually measures VaR.

1.1 Earlier Literature

Earlier literature on extending risk measures to liquidity includes several studies. Jarrow and Subramanian [17], Bangia et al. [4], Angelidis and Benos [3], Jarrow and Protter [18], Stange and Kaserer [25], Ernst, Stange and Kaserer [15], among others, propose different methods of extending risk measures to account for liquidity risk. Bangia et al. [4] classify market liquidity risk into two categories: (a) the exogenous illiquidity that depends on general market conditions is common to all market players and is unaffected by the actions of any one participant and (b) the endogenous illiquidity, which is specific to one's position in the market, varies across different market players and is mainly related to the impact of the trade size on the bid-ask spread. Bangia et al. [4] and Ernst et al. [15] only consider the exogenous illiquidity risk and propose a liquidity adjusted VaR measure built using the distribution of the bid-ask spreads. The other mentioned studies model and account for endogenous risk in the calculation of liquidity adjusted risk measures. In the context of the coherent risk measures literature, the general axioms a liquidity measure should satisfy are discussed in [1]. In that work coherent risk measures are defined on the vector space of portfolios (rather than on portfolio values). A key observation is that the portfolio value can be a nonlinear map on the space of portfolios, motivating the introduction of a nonlinear value function depending on a notion of liquidity policy based on a general description of the microstructure of illiquid markets.

As mentioned earlier, bid-ask spreads have been used to assess liquidity risk. While bid-ask spreads are certainly an important measure of liquidity, they are not the only one. In the Credit Default Swap (CDS) space, for example, Predescu et al. [22] have built a statistical model that associates an ordinal liquidity score with each CDS reference entity. The liquidity score is built using well-known liquidity indicators such as the already mentioned bid-ask spreads but also using other less accessible predictors of market liquidity such as number of active dealers quoting a reference entity, staleness of quotes of individual dealers, and dispersion in mid-quotes across market dealers. The bid-ask spread is used essentially as an indicator of market breadth; the presence of orders on both sides of the trading book corresponds to tighter bid-ask spreads. Dispersion of mid-quotes across dealers is a measure of price uncertainty about the actual CDS price. Less liquid names are generally associated with more price uncertainty and thus large dispersion. The third liquidity measure that is used in Predescu et al. [22] aggregates the number of active dealers and the individual dealers' quote staleness into an (in)activity measure, which is meant to be a proxy for CDS market depth. Illiquidity increases if any of the liquidity predictors increases, keeping everything else constant. Therefore, liquid (less liquid) names are associated with smaller (larger) liquidity scores. CDS liquidity scores are now offered commercially by Fitch Solutions and as of 2009 provided a comparison of relative liquidity of over 2,400 reference entities in the CDS market globally, mainly concentrated in North America, Europe, and Asia. The model estimation and the model generated liquidity scores are based upon the Fitch CDS Pricing Service database, which includes single-name CDS quotes on over 3,000 entities, corporates, and sovereigns across about two dozen broker-dealers back to 2000. This approach and the related results, further highlighting the connection between liquidity and credit quality/rating, are summarized in [14], who further review previous research on liquidity components in the pricing space for CDS.

Given the above indicators of liquidity risk, the SHP process seems to be naturally associated with the staleness/inactivity measure. However, one may argue that the random holding period also embeds market impact and bid-ask spreads. Indeed, traders will consider closing a position or a portfolio also in terms of cost. If bid-ask spreads cause the immediate closure of a position to be too expensive, market operators might wait for bid-asks to move. This will impact the holding period for the relevant position. If we take for granted that the risk manager will not try to model the detailed behavior of traders, then the stochastic holding period becomes a reduced form process for the risk manager, which will possibly incapsulate a number of aspects on liquidity risk. Ideally, as our understanding of liquidity risk progresses, we can move to a more structural model where the dynamics of the SHP is explained in terms of market prices and liquidity proxies, including market impact, bid-ask spreads, and asset prices. However, in this work we sketch the features the resulting model could have in a reduced form spirit.

This prompts us to highlight a further feature that we should include in future developments of the model introduced here: we should explicitly include dependence between price levels and holding periods, since liquidity is certainly related to the level of prices in the market.

1.2 Different Risk Horizons Are Acknowledged by BCBS

The Basel Committee came out with a recommendation on multiple holding periods for different risk factors in 2012 in [7]. This document states that

> The Committee is proposing that varying liquidity horizons be incorporated in the market risk metric under the assumption that banks are able to shed their risk at the end of the liquidity horizon.[...]. This proposed liquidation approach recognises the dynamic nature of banks trading portfolios but, at the same time, it also recognises that not all risks can be unwound over a short time period, which was a major flaw of the 1996 framework.

Further on, in Annex 4, the document details a sketch of a possible solution: assign a different liquidity horizon to risk factors of different types. While this is a step forward, it can be insufficient. How is one to decide the horizon for each risk factor, and especially how is one to combine the different estimates for different horizons for assets in the same portfolio into a consistent and logically sound way? Our random holding period approach allows one to answer the second question, but more generally none of the above works focuses specifically on our setup with random holding period, which represents a simple but powerful idea to include liquidity in traditional risk measures such as Value at Risk or Expected Shortfall. Our idea was first proposed in 2010 in [13].

When analyzing multiple positions, holding periods can be taken to be strongly dependent, in line with the first classification (a) of Bangia et al. [4] above, or independent, so as to fit the second category (b). We will discuss whether adding dependent holding periods to different positions can actually add dependence to the position returns.

The paper is organized as follows. In order to illustrate the SHP model, first in a univariate case (Sect. 2) and then in a bivariate one (Sect. 3), it is considerably easier to focus on examples on (log)normal processes. A brief colloquial hint at positive processes is presented in Sect. 2, to deepen the intuition of the impact on risk measures of introducing a SHP process. Across Sects. 3 and 4, where we try to address the issue of calibration, we outline a possible multivariate model which could be adopted, in line of principle, in a top-down approach to risk integration in order to include the liquidity risk and its dependence on other risks.

Table 1 Simplified discrete SHP

Holding period	Probability
10	0.99
75	0.01

Finally, we point out that this paper is meant as a proposal to open a research effort in stochastic holding period models for risk measures. This paper contains several suggestions on future developments, depending on an increased availability of market data. The core ideas on the SHP framework, however, are presented in this opening paper.

2 The Univariate Case

Let us suppose that we have to calculate the VaR of a market portfolio whose value at time t is V_t. We call $X_t = \ln V_t$, so that the log return on the portfolio value at time t over a period h is

$$X_{t+h} - X_t = \ln(V_{t+h}/V_t) \approx \frac{V_{t+h} - V_t}{V_t}.$$

In order to include liquidity risk, the risk manager decides that a realistic, simplified statistics of the holding period in the future will be the one given in Table 1. To estimate liquidity-adjusted VaR say at time 0, the risk manager will perform a number of simulations of $V_{0+H_0} - V_0$ with H_0 randomly chosen by the statistics above, and finally will calculate the desired risk measure from the resulting distribution. If the log-return $X_T - X_0$ is normally distributed with zero mean and variance T for deterministic T (e.g., a Brownian motion, i.e., a Random walk), then the risk manager could simplify the simulation using $X_{0+H_0} - X_0|_{H_0} \overset{d}{\sim} \sqrt{H_0}\,(X_1 - X_0)$ where $|_{H_0}$ denotes "conditional on H_0". With this practical exercise in mind, let us generalize this example to a generic t.

2.1 A Brief Review on the Stochastic Holding Period Framework

A process for the risk horizon at time t, i.e., $t \mapsto H_t$, is a positive stochastic process modeling the risk horizon over time. We have that the risk measure at time t will be taken on the change in value of the portfolio over this random horizon. If X_t is the log-value of the portfolio at time t, we have that the risk measure at time t is to be taken on the log-return

$$X_{t+H_t} - X_t.$$

For example, if one uses a 99 % Value at Risk (VaR) measure, this will be the 1st percentile of $X_{t+H_t} - X_t$. The request that H_t be just positive means that the horizon at future times can both increase and decrease, meaning that liquidity can vary in both directions.

There are a large number of choices for positive processes: one can take lognormal processes with or without mean reversion, mean reverting square root processes, squared Gaussian processes, all with or without jumps. This allows one to model the holding period dynamics as mean reverting or not, continuous or with jumps, and with thinner or fatter tails. Other examples are possible, such as Variance Gamma or mixture processes, or Levy processes. See for example [11, 12].

2.2 Semi-analytic Solutions and Simulations

Going back to the previous example, let us suppose that

Assumption 1 The increments $X_{t+1y} - X_t$ are logarithmic returns of an equity index, normally distributed with annual mean and standard deviation, respectively, $\mu_{1y} = -1.5\%$ and $\sigma_{1y} = 30\%$.

We suppose an exposure of 100 in domestic currency.

Before running the simulation, we recall some basic notation and formulas.

The portfolio log-returns under random holding period at time 0 can be written as

$$\mathbb{P}\left[X_{H_0} - X_0 < x\right] = \int_0^\infty \mathbb{P}\left[X_h - X_0 < x\right] \mathrm{d}F_{H,0}(h)$$

i.e., as a mixture of Gaussian returns, weighted by the holding period distribution. Here $F_{H,t}$ denotes the cumulative distribution function of the holding period at time t, i.e., of H_t.

Remark 1 (*Mixtures for heavy-tailed and skewed distributions*). Mixtures of distributions have been used for a long time in statistics and may lead to heavy tails, allowing for modeling of skewed distributions and of extreme events. Given the fact that mixtures lead, in the distributions space, to linear (convex) combinations of possibly simple and well-understood distributions, they are tractable and easy to interpret. The literature on mixtures is enormous and it is impossible to do justice to all this literature here. We just hint at the fact that static mixtures of distributions had been postulated in the past to fit option prices for a given maturity, see for example [24], where a mixture of normal densities for the density of the asset log-returns under the pricing measure is assumed, and subsequently [8, 16, 20]. In the last decade [2, 9, 10] have extended the mixture distributions to fully dynamic arbitrage-free stochastic processes for asset prices.

Table 2 SHP distributions and market risk

Holding period	VaR 99.96%	(Analytic)	ES 99.96%	(Analytic)
Constant 10 b.d.	20.1	(20.18)	21.7	(21.74)
Constant 75 b.d.	55.7	(55.54)	60.0	(59.81)
SHP (Bernoulli 10/75, $p_{10} = 0.99$)	29.6	(29.23)	36.1	(35.47)

Going back to our notation, $\text{VaR}_{t,h,c}$ and $\text{ES}_{t,h,c}$ are the value at risk and expected shortfall, respectively, for a horizon h at confidence level c at time t, namely

$$\mathbb{P}\left\{X_{t+h} - X_t > -\text{VaR}_{t,h,c}\right\} = c, \, \text{ES}_{t,h,c} = -\mathbb{E}\left[X_{t+h} - X_t | X_{t+h} - X_t \le -\text{VaR}_{t,h,c}\right].$$

We now recall the standard result on VaR and ES under Gaussian returns in deterministic calendar time.

Proposition 1 (VaR and ES with Gaussian log-returns on a deterministic risk horizon h) *In the Gaussian log-returns case where*

$$X_{t+h} - X_t \text{ is normally distributed with mean } \mu_{t,h} \text{ and standard deviation } \sigma_{t,h} \quad (1)$$

we obtain

$$\text{VaR}_{t,h,c} = -\mu_{t,h} + \Phi^{-1}(c)\sigma_{t,h}, \quad \text{ES}_{t,h,c} = -\mu_{t,h} + \sigma_{t,h} p\left(\Phi^{-1}(c)\right)/(1-c)$$

where p is the standard normal probability density function and Φ is the standard normal cumulative distribution function.

In the following we will calculate VaR and Expected Shortfall referred to a confidence level of 99.96%, calculated over the fixed time horizons of 10 and 75 business days, and under SHP process with statistics given by Table 1, using Monte Carlo simulations. Each year has 250 (working) days. The results are presented in Table 2.

More generally, we may derive the VaR and ES formulas for the case where H_t is distributed according to a general distribution

$$\mathbb{P}(H_t \le x) = F_{H,t}(x), \quad x \ge 0$$

and

$$\mathbb{P}(X_{t+h} - X_t \le x) = F_{X,t,h}(x).$$

Definition 1 (*VaR and ES under Stochastic Holding Period*) We define VaR and ES under a random horizon H_t at time t and for a confidence level c as

$$\mathbb{P}\left\{X_{t+H_t} - X_t > -\text{VaR}_{H,t,c}\right\} = c, \, \text{ES}_{H,t,c} = -\mathbb{E}\left[X_{t+H_t} - X_t | X_{t+H_t} - X_t \le -\text{VaR}_{H,t,c}\right].$$

We point out that the order of time/confidence/horizon arguments in the VaR and ES definitions is different in the Stochastic Holding Period case. This is to stress the different setting with respect to the fixed holding period case.

We have immediately the following:

Proposition 2 (VaR and ES for SHP independent of returns in deterministic calendar time) *Assume that H is independent of the log returns of X in deterministic calendar time. Using the tower property of conditional expectation it is immediate to prove that such a case VaR$_{H,t,c}$ obeys the following equation:*

$$\int_0^\infty \left(1 - F_{X,t,h}\left(-VaR_{H,t,c}\right)\right) dF_{H,t}(h) = c$$

whereas ES$_{H,t,c}$ is given by

$$ES_{H,t,c} = -\frac{1}{1-c}\int_0^\infty \mathbb{E}\left[X_{t+h} - X_t | X_{t+h} - X_t \le -VaR_{H,t,c}\right] Prob\left(X_{t+h} - X_t \le -VaR_{H,t,c}\right) dF_{H,t}(h)$$

For the specific Gaussian case (1) we have

$$\int_0^\infty \Phi\left(\frac{\mu_{t,h} + VaR_{H,t,c}}{\sigma_{t,h}}\right) dF_{H,t}(h) = c$$

$$ES_{H,t,c} = \frac{1}{1-c}\int_0^\infty \left[-\mu_{t,h}\Phi\left(\frac{-\mu_{t,h} - VaR_{H,t,c}}{\sigma_{t,h}}\right) + \sigma_{t,h}p\left(\frac{-\mu_{t,h} - VaR_{H,t,c}}{\sigma_{t,h}}\right)\right] dF_{H,t}(h)$$

Notice that in general one can try and obtain the quantile VaR$_{H,t,c}$ for the random horizon case by using a root search, and subsequently compute also the expected shortfall. Careful numerical integration is needed to apply these formulas for general distributions of H_t. The case of Table 2 is somewhat trivial, since in the case where H_0 is as in Table 1 integrals reduce to summations of two terms.

We note also that the maximum difference, both in relative and absolute terms, between ES and VaR is reached by the model under random holding period H_0. Under this model the change in portfolio value shows heavier tails than under a single deterministic holding period. In order to explore the impact of SHP's distribution tails on the liquidity-adjusted risk, in the following we will simulate SHP models with H_0 distributed as an Exponential, an Inverse Gamma distribution[1] and a Generalized

[1] Obtained by rescaling a distribution IG$(\frac{\nu}{2}, \frac{\nu}{2})$ with $\nu = 3$. Before rescaling, setting $\alpha = \nu/2$, the inverse gamma density is $f(x) = (1/\Gamma(\alpha))(\alpha)^\alpha x^{-\alpha-1}e^{-\alpha/x}$, $x > 0$, $\alpha > 0$, with expected value $\alpha/(\alpha - 1)$. We rescale this distribution by $k = 8.66/(\alpha/(\alpha - 1))$ and take for H_0 the random variable with density $f(x/k)/k$.

Pareto distribution[2] having parameters calibrated in order to obtain a sample with the same 99 %-quantile of 75 business days. The results are in Table 3.

The SHP process changes the statistical nature of the P&L process: the heavier the tails of the SHP distribution, the heavier the tails of P&L distribution. Notice that our Pareto distribution has tails going to 0 at infinity with exponent around 3, as one can see immediately by differentiation of the cumulative distribution function, whereas our inverse gamma has tails going to 0 at infinity with exponent about 2.5. In this example we have that the tails of the inverse gamma are heavier, and indeed for that distribution VaR and ES are larger and differ from each other more. This can change of course if we take different parameters in the two distributions.

3 Dependence Modeling: A Bivariate Case

Within multivariate modeling, using a common SHP for many normally distributed risks leads to dynamical versions of the so-called *normal mixtures* and *normal mean-variance mixtures* [19].

Assumption 2 In this section we assume that different assets have the same random holding period, thus testing an extreme liquidity dependence scenario. We will briefly discuss relaxing this assumption at the end of this section. We further assume that the stochastic holding period process is independent of the log returns of assets in deterministic calendar time.

Let the log returns (recall $X_t^i = \ln V_t^i$, with V_t^i the value at time t of the ith asset)

$$X_{t+h}^1 - X_t^1, \ldots, X_{t+h}^m - X_t^m$$

be normals with means $\mu_{t,h}^1, \ldots, \mu_{t,h}^m$ and covariance matrix $Q_{t,h}$.
Then

$$\mathbb{P}\left[X_{t+H_t}^1 - X_t^1 < x_1, \ldots, \ X_{t+H_t}^m - X_t^m < x_m \right]$$

$$= \int_0^\infty \mathbb{P}\left[X_{t+h}^1 - X_t^1 < x_1, \ldots, \ X_{t+h}^m - X_t^m < x_m \right] dF_{H,t}(h)$$

is distributed as a mixture of multivariate normals, and a portfolio V_t of the assets $1, 2, \ldots, m$ whose log-returns $X_{t+h} - X_t$ ($X_t = \ln V_t$) are a linear weighted combination $w_1, \ldots, w_m$ of the single asset log-returns $X_{t+h}^i - X_t^i$ would be distributed as

---

[2] With scale parameter $k = 9$ and shape parameter $\alpha = 2.0651$, with cumulative distribution function $F(x) = 1 - \left(\frac{k}{k+x}\right)^\alpha$, $x \geq 0$, this distribution has moments up to order α. So the smaller α, the fatter the tails. The mean is, if $\alpha > 1$, $\mathbb{E}[H_0] = k/(\alpha - 1)$.

Table 3 SHP statistics and resulting market risk

Distribution	Mean	Median	99%-q	VaR 99.96% simulation	VaR 99.96% root search	ES 99.96% simulation	ES 99.96% root search	ES/VaR-1 (%)
Exponential	16.30	11.3	75	39.0	39.2	44.7	44.7	14
Pareto	8.45	3.7	75	41.9	41.9	57.1	56.9	36
Inverse gamma	8.60	3.7	75	46.0	46.7	73.5	73.0	55

$$\mathbb{P}\left[X_{t+H_t} - X_t < z\right] = \int_0^\infty \mathbb{P}\left[w_1\left(X_{t+h}^1 - X_t^1\right) + \cdots + w_m\left(X_{t+h}^m - X_t^m\right) < z\right] \mathrm{d}F_{H,t}(h)$$

In particular, in analogy with the unidimensional case, the mixture may potentially generate skewed and fat-tailed distributions, but when working with more than one asset this has the further implication that VaR is not guaranteed to be subadditive on the portfolio. Then the risk manager who wants to take into account SHP in such a setting should adopt a coherent measure like Expected Shortfall.

A natural question at this stage is whether the adoption of a common SHP can add dependence to returns that are jointly Gaussian under deterministic calendar time, perhaps to the point of making extreme scenarios on the joint values of the random variables possible.

Before answering this question, one needs to distinguish extreme behavior in the single variables and in their joint action in a multivariate setting. Extreme behavior on the single variables is modeled, for example, by heavy tails in the marginal distributions of the single variables. Extreme behavior in the dependence structure of, say, two random variables is achieved when the two random variables tend to take extreme values in the same direction together. This is called tail dependence, and one can have both upper tail dependence and lower tail dependence. More precisely, but still loosely speaking, tail dependence expresses the limiting proportion according to which the first variable exceeds a certain level given that the second variable has already exceeded that level. Tail dependence is technically defined through a limit, so that it is an asymptotic notion of dependence. For a formal definition we refer, for example, to [19]. "Finite" dependence, as opposed to tail, between two random variables is best expressed by rank correlation measures such as Kendall's tau or Spearman's rho.

We discuss tail dependence first. In case the returns of the portfolio assets are jointly Gaussian with correlations smaller than one, the adoption of a common random holding period for all assets does not add tail dependence, *unless the commonly adopted random holding period has a distribution with power tails*. Hence, if we want to rely on one of the random holding period distributions in our examples above to introduce upper and lower tail dependence in a multivariate distribution for the assets returns, we need to adopt a common random holding period for all assets that is Pareto or Inverse Gamma distributed. Exponentials, Lognormals, or discrete Bernoulli distributions would not work. This can be seen to follow from properties of the normal variance-mixture model, see for example [19], p. 212 and also Sect. 7.3.3.

A more specific theorem that fits our setup is Theorem 5.3.8 in [23]. We can write it as follows with our notation.

Proposition 3 (A common random holding period with less than power tails does not add tail dependence to jointly Gaussian returns) *Assume the log returns to be* $W_t^i = \ln V_t^i$, *with* V_t^i *the value at time* t *of the ith asset,* $i = 1, 2$, *where*

$$W_{t+h}^1 - W_t^1, W_{t+h}^2 - W_t^2$$

*are two correlated Brownian motions, i.e., normals with zero means, variances h,
and instantaneous correlation less than 1 in absolute value:*

$$d\left(W^1, W^2\right)_t = dW_t^1 \, dW_t^2 = \rho_{1,2}dt, \quad |\rho_{1,2}| < 1.$$

*Then adding a common nonnegative random holding period H_0 independent of W's
leads to tail dependence in the returns*

$$W_{H_0}^1, \, W_{H_0}^2$$

if and only if $\sqrt{H_0}$ is regularly varying at ∞ with index $\alpha > 0$.

Theorem 5.3.8 in [23] also reports an expression for the tail dependence coefficients as functions of α and of the survival function of the student t distribution with $\alpha + 1$ degrees of freedom.

Summarizing, if we work with power tails, the heavier the tails of the common holding period process H, the more one may expect *tail dependence* to emerge for the multivariate distribution: by adopting a common SHP for all risks, dependence could potentially appear in the whole dynamics, in agreement with the fact that liquidity risk is a systemic risk.

We now turn to finite dependence, as opposed to tail dependence. First, we note the well-known elementary but important fact that one can have two random variables with very high dependence but without tail dependence. Or one can have two random variables with tail dependence but small finite dependence. For example, if we take two jointly Gaussian Random variables with correlation 0.999999, they are clearly quite dependent on each other but they will not have tail dependence, even if a rank correlation measure such as Kendall's τ would be 0.999, still very close to 1, characteristic of the co-monotonic case. This is a case with zero tail dependence but very high finite dependence. On the other hand, take a bivariate student t distribution with few degrees of freedom and correlation parameter $\rho = 0.1$. In this case the two random variables have positive tail dependence and it is known that Kendall's tau for the two random variables is

$$\tau = \frac{2}{\pi} \arcsin(\rho) \approx 0.1$$

which is the same tau one would get for two standard jointly Gaussian random variables with correlation ρ. This tau is quite low, showing that one can have positive tail dependence while having very small finite dependence.

The above examples point out that one has to be careful in distinguishing large finite dependence and tail dependence.

A further point of interest in the above examples comes from the fact that the multivariate student t distribution can be obtained by the multivariate Gaussian distribution when adopting a random holding period given by an inverse gamma distribution (power tails). We deduce the important fact that in this case *a common random holding period with power tails adds positive tail dependence but not finite dependence.*

In fact, one can prove a more general result easily by resorting to the tower property of conditional expectation and from the definition of tau based on independent copies of the bivariate random vector whose dependence is being measured. One has the following "no go" theorem for increasing Kendall's tau of jointly Gaussian returns through common random holding periods, regardless of the tail's power.

Proposition 4 (A common random holding period does not alter Kendall's tau for jointly Gaussian returns) *Assumptions as in Proposition 3 above. Then adding a common nonnegative random holding period H_0 independent of W's leads to the same Kendall's tau for*

$$W_{H_0}^1, W_{H_0}^2$$

as for the two returns

$$W_t^1, W_t^2$$

for a given deterministic time horizon t.

Summing up, this result points out that adding further finite dependence through common SHPs, at least as measured by Kendall's tau, can be impossible if we start from Gaussian returns. A different popular rank correlation measure, Spearman's rho, does not coincide for the bivariate t and Gaussian cases though, so that it is not excluded that dependence could be added in principle though dependent holding periods, at least if we measured dependence with Spearman's ρ. This is under investigation.

More generally, at least from a theoretical point of view, it could be interesting to model other kinds of dependence than the one stemming purely from a common holding period (with power tails). In the bivariate case, for example, one could have two different holding periods that are themselves dependent on each other in a less simplistic way, for example through a common factor structure, rather than being just identical. In this case it would be interesting to study the tail dependence implications and also finite dependence as measured by Spearman's rho.

We will investigate this aspect in further research, but increasing dependence may require, besides the adoption of power tail laws for the random holding periods, abandoning the Gaussian distribution for the basic assets under deterministic calendar time.

A further aspect worth investigating is the possibility to calculate semi-closed form risk contributions to VaR and ES under SHP along the lines suggested in [26], and to investigate the Euler principle as in [27, 28].

4 Calibration with Liquidity Data

We are aware that multivariate SHP modeling is a purely theoretical exercise and that we just hinted at possible initial developments above. Nonetheless, a lot of financial data is being collected by regulators, providers, and rating agencies, together with a consistent effort on theoretical and statistical studies. This will possibly result in available synthetic indices of liquidity risk grouped by region, market, instrument type, etc. For instance, Fitch already calculates market liquidity indices on CDS markets worldwide, on the basis of a scoring proprietary model [14].

4.1 Dependencies Between Liquidity, Credit, and Market Risk

It could be an interesting exercise to calibrate the dependence structure (e.g., copula function) between a liquidity index (like the Fitch's one), a credit index (like iTRAXX), and a market index (for instance Eurostoxx50) in order to measure the possible (nonlinear) dependence between the three. The risk manager of a bank could use the resulting dependence structure within the context of risk integration, in order to simulate a joint dynamics as a first step, to estimate later on the whole liquidity-adjusted VaR/ES by assuming co-monotonicity between the variations of the liquidity index and of the SHP processes.

4.2 Marginal Distributions of SHPs

A lot of information on SHP 'extreme' statistics of an OTC derivatives portfolio could be collected from the statistics, across Lehman's counterparties, of the time lags between the Lehman's Default Event Date and the trade dates of any replacement transaction. The data could give information on the marginal distribution of the SHP of a portfolio, in a stressed scenario, by assuming a statistical equivalence between data collected 'through the space' (across Lehman's counterparties) and 'through the time' under i.i.d. hypothesis.[3] The risk manager of a bank could examine a more specific and non-distressed dataset by collecting information on the ordinary operations of the business units.

5 Conclusions

Within the context of risk integration, in order to include liquidity risk in the whole portfolio risk measures, a stochastic holding period (SHP) model can be useful, being versatile, easy to simulate, and easy to understand in its inputs and outputs. In a single-portfolio framework, as a consequence of introducing an SHP model, the statistical distribution of P&L moves to possibly heavier tailed and skewed mixture distributions. In a multivariate setting, the dependence among the SHP processes to which marginal P&L are subordinated, may lead to dependence on the latter under drastic choices of the SHP distribution, and in general to heavier tails on the total

---

[3] A similar approach is adopted in [21] within the context of operational risk modeling.

P&L distribution. At present, lack of synthetic and consensually representative data forces to a qualitative top-down approach, but it is straightforward to assume that this limit will be overcome in the near future.

Acknowledgments This paper reflects the authors' opinions and not necessarily those of their current and past employers. The authors are grateful to Dirk Tasche for helpful suggestions and correspondence on a first version, to Giacomo Scandolo for further comments and correspondence, and to an anonymous referee and the Editors for important suggestions that helped in improving the paper.

Open Access This chapter is distributed under the terms of the Creative Commons Attribution Noncommercial License, which permits any noncommercial use, distribution, and reproduction in any medium, provided the original author(s) and source are credited.

References

1. Acerbi, C., Scandolo, G.: Liquidity risk theory and coherent measures of risk. Quant. Financ. **8**(7), 681–692 (2008)
2. Alexander, C.: Normal mixture diffusion with uncertain volatility: modelling short and long-term smile effects. J. Bank. Financ. **28**(12), 2957–2980 (2004)
3. Angelidis, T., Benos, A.: Liquidity adjusted value-at-risk based on the components of the bid-ask spread. Working paper, available at http://ssrn.com/abstract=661281 (2005)
4. Bangia, A., Diebold, F.X., Schuermann, T., Stroughair, J.D.: Modeling liquidity risk with implications for traditional market risk measurement and management. Working paper, Financial Institutions Center at The Wharton School (1999)
5. Basel Committee on Banking Supervision. Findings on the interaction of market and credit risk, BCBS working paper No 16 May 2009, available at http://www.bis.org (2009)
6. Basel Committee on Banking Supervision. Guidelines for computing capital for incremental risk in the trading book, BCBS guidelines, July 2009, available at http://www.bis.org (2009)
7. Basel Committee on Banking Supervision. Consultative document: fundamental review of the trading book, BIS, May 2012, available at http://www.bis.org/publ/bcbs219.pdf (2012)
8. Bhupinder, B.: Implied risk-neutral probability density functions from option prices: a central bank perspective. In: Knight, J., Satchell, S. (eds.) Forecasting Volatility in the Financial Markets, pp. 137–167. Butterworth Heinemann, Oxford (1998)
9. Brigo, D., Mercurio, F.: Displaced and mixture diffusions for analytically-tractable smile models. In: Geman, H., Madan, D.B., Pliska, S.R., Vorst, A.C.F. (eds.) Mathematical Finance-Bachelier Congress 2000. Springer, Berlin (2001)
10. Brigo, D., Mercurio, F., Rapisarda, F.: Smile at uncertainty, risk, May issue (2004)
11. Brigo, D., Dalessandro, A., Neugebauer, M., Triki, F.: A stochastic processes toolkit for risk management: geometric Brownian motion, jumps, GARCH and variance gamma models. J. Risk Manag. Financ. Inst. **2**(4), 365–393 (2009)
12. Brigo, D., Dalessandro, A., Neugebauer, M., Triki, F.: A stochastic processes toolkit for risk management. Mean reversing processes and jumps. J. Risk Manag. Financ. Inst. **3**, 1 (2009)
13. Brigo, D., Nordio, C.: Liquidity-adjusted market risk measures with stochastic holding period. Available at http://arxiv.org/abs/1009.3760 and http://ssrn.com/abstract=1679698 (2010)
14. Brigo, D., Predescu, M., Capponi, A.: Liquidity modeling for credit default swaps: an overview. In: Bielecki, Brigo, Patras (eds.) Credit Risk Frontiers: Sub prime crisis, Pricing and Hedging, CVA, MBS, Ratings and Liquidity, pp. 587–617. Wiley/Bloomberg Press, See also http://ssrn.com/abstract=1564327 (2010)

15. Ernst, C., Stange, S., Kaserer, C.: Accounting for non-normality in liquidity risk. Available at http://ssrn.com/abstract=1316769 (2009)
16. Guo, C.: Option pricing with heterogeneous expectations. Financ. Rev. **33**, 81–92 (1998)
17. Jarrow, R., Subramanian, A.: Mopping up liquidity. RISK **10**(10), 170–173 (1997)
18. Jarrow, R., Protter, P.: Liquidity Risk and Risk Measure Computation. Working paper, Cornell University (2005)
19. McNeil, A.J., Frey, R., Embrechts, P.: Quantitative Risk Management. Princeton University Press, New Jersey (2005)
20. Melick, W.R., Thomas, C.P.: Recovering an asset's implied PDF from option prices: an application to crude oil during the Gulf crisis. J. Financ. Quant. Anal. **32**, 91–115 (1997)
21. Moscadelli, M.: The modelling od operational risk: experience with the analysis of the data collected by the Base Committee, Banca d'Italia, Temi di discussione del Servizio Studi, Number 517 (July 2004), available at http://www.bancaditalia.it/pubblicazioni/econo/temidi (2004)
22. Predescu, M., Thanawalla, R., Gupton, G., Liu, W., Kocagil, A., Reyngold, A.: Measuring CDS liquidity. Fitch Solutions Presentation at the Bowles Symposium, Georgia State University, 12(2009)
23. Prestele, C.: Credit portfolio modelling with elliptically contoured distributions. Approximation, Pricing, Dynamisation. Doctoral dissertation, Ulm University. Available at http://vts.uni-ulm.de/docs/2007/6093/vts_6093_8209.pdf (2007)
24. Ritchey, R.J.: Call option valuation for discrete normal mixtures. J. Financ. Res. **13**, 285–296 (1990)
25. Stange, S., Kaserer, C.: Why and how to integrate liquidity risk into a VaR-framework. CEFS working paper 2008 No. 10, available at http://ssrn.com/abstract=1292289 (2008)
26. Tasche, D.: Measuring sectoral diversification in an asymptotic multi-factor framework. J. Credit Risk **2**(3), 33–55 (2006)
27. Tasche, D.: Capital allocation to business units and sub-portfolios: the euler principle. In: Resti, A., (ed.) Pillar II in the New Basel Accord: The Challenge of Economic Capital, Risk Books, pp. 423–453 (2008)
28. Tasche, D.: Capital allocation for credit portfolios with kernel estimators. Quant. Financ. **9**(5), 581–595 (2009)

Regulatory Developments in Risk Management: Restoring Confidence in Internal Models

Uwe Gaumert and Michael Kemmer

Abstract The paper deals with the question of how to restore lost confidence in the results of internal models (especially market risk models). This is an important prerequisite for continuing to use these models as a basis for calculating risk-sensitive prudential capital requirements. The authors argue that restoring confidence is feasible. Contributions to this end will be made both by the reform of regulatory requirements under Basel 2.5 and the Trading Book Review and by refinements of these models by the banks themselves. By contrast, capital requirements calculated on the basis of a leverage ratio and prudential standardised approaches will not be sufficient, even from a regulatory perspective, owing to their substantial weaknesses. Specific proposals include standardising models with a view to reducing complexity and enhancing comparability, significantly improving model validation and increasing transparency as to how model results are determined, also over time. The article reflects the personal views of the authors.

1 Introduction

Since 1997 ("Basel 1.5"), banks in Germany have been allowed to calculate their capital requirements for the trading book using internal value-at-risk (VaR) models that have passed a comprehensive and stringent supervisory vetting and approval process. Basel II and Basel III saw the introduction of further internal models complementing the standardised approaches already available—take, for example, the internal ratings-based (IRB) approach for credit risk under Basel II and the advanced credit valuation adjustment (CVA) approach for counterparty risk under Basel III. During the financial crisis, particular criticism was directed at internal market risk models, the design of which supervisors largely left to the banks themselves. This article therefore confines itself to examining these models, which are a good starting

U. Gaumert (✉) · M. Kemmer
Association of German Banks, Burgstr. 28, 10170 Berlin, Germany
e-mail: uwe.gaumert@bdb.de

M. Kemmer
e-mail: michael.kemmer@bdb.de

© The Author(s) 2015   19
K. Glau et al. (eds.), *Innovations in Quantitative Risk Management*,
Springer Proceedings in Mathematics & Statistics 99,
DOI 10.1007/978-3-319-09114-3_2

point for explaining and commenting on the current debate. Much of the following applies to other types of internal models as well.

Banks and supervisors learned many lessons from the sometimes unsatisfactory performance of VaR models in the crisis—one of the root causes of the loss of confidence by investors in model results. This led, at bank level, to a range of improvements in methodology, and also to the realisation that not all products and portfolios lend themselves to internal modelling. At supervisory level, Basel 2.5 ushered in an initial reform with rules that were much better at capturing extreme risks (tail risks) and that increased capital requirements at least threefold. Work on a fundamental trading book review (Basel 3.5), which will bring further methodological improvements to regulatory requirements, is also underway.

Nevertheless, models are still criticised as being

- too error-prone,
- suitable only for use in "fair-weather" conditions,
- too variable in their results when analysing identical risks,
- insufficiently transparent for investors and
- manipulated by banks, with the tacit acceptance of supervisors, with the aim of reducing their capital requirements.

As a result, the credibility of model results and thus their suitability for use as a basis for calculating capital requirements have been challenged. This culminated in, for example, the following statement by the academic advisory board at the German Ministry for Economic Affairs: "Behind these flaws (in risk modelling)[1] lie fundamental problems that call into question the system of model-based capital regulation as a whole."[2] It therefore makes good sense to explore the suitability of possible alternatives. The authors nevertheless conclude that model-based capital charges should be retained. But extensive efforts are needed to restore confidence in model results.

2 Loss of Confidence in Internal Models—How Did It Happen?

2.1 An Example from the First Years of the Crisis

The market disruption which accompanied the start of the financial crisis in the second half of 2007 took the form in banks' trading units of sharply falling prices with a corresponding impact on their daily P&Ls after a prolonged phase of low volatility. Uncertainty grew rapidly about the accuracy of estimated probabilities of default, default correlations of the underlying loans and the scale of loss in the event of default,

---

[1] Wording in brackets inserted by the authors.

[2] [31], p. 19.

and thus also about the probabilities of default and recovery rates of the securitisation instruments. This in turn caused spreads to widen, volatility to increase and market liquidity for securitisation products to dry up. A major exacerbating factor was that many market participants responded in the same way ("flight to simplicity", "flight to quality"). Later on, there were also jump events such as downgrades. Calibrating the above parameters proved especially problematic since there was often a lack of historical default or market data. Unlike in the period before the crisis, even AAA-rated senior or super senior tranches of securitisation instruments, which only start to absorb loss much later than their riskier counterparts, suffered considerably in value as the protective cushion of more junior tranches melted away, necessitating substantial write-downs.[3]

The performance of internal market-risk models was not always satisfactory, especially in the second half of 2007 and in the "Lehman year" of 2008. In this period, a number of banks found that the daily loss limits forecast by their models were sometimes significantly exceeded (backtesting outliers).[4] The performance results of Deutsche Bank, for instance, show that losses on some sub-portfolios were evidently serious enough to have an impact on the overall performance of the bank's trading unit. This demonstrates the extremely strong market disruption which can follow an external shock. When backtesting a model's performance, the current clean P&L—$P\&L_t$—is compared with the previous day's VaR forecast VaR_{t-1}.[5] At a confidence level of 99 %, an average of two to three outliers a year may be anticipated over the long term (representing 1 % of 250–260 trading days a year). In the years between 2007 and 2013, Deutsche Bank had 12, 35, 1, 2, 3, 2 and 2 outliers.[6] Although the models' performance for 2007 and 2008 looks bad at first sight, the question nevertheless arises as to whether or not these outliers are really the models' "fault", so to speak. By their very nature, models can only do what they have been designed to do: "If you're in trouble, don't blame your model." To function properly, the models needed liquid markets, adequate historical market data and total coverage of all market risks, particularly migration and default risk. These prerequisites were not always met by markets and banks. Anyone using a model has to be aware of its limitations and exercise caution when working with its results.

Even Germany's Federal Financial Supervisory Authority BaFin pointed out that, given the extreme combination of circumstances on the market in connection with the financial crisis, the figures do not automatically lead to the conclusion that the predicative quality of the models is inadequate.[7] The example could indicate that,

---

[3] Cf. [18], p. 128.

[4] [10], p. 8.

[5] Between 2007 and 2009, only so-called "dirty" P&L results were published in chart form, while outliers are based on "clean" P&L data. This inconsistency was eliminated in 2010. Dirty and clean P&L figures may differ. This is because clean P&L simply shows end-of-day positions revalued using prices at the end of the following trading day, whereas dirty P&L also includes income from intraday trading, fees and commissions and interest accrued.

[6] [11], Management Report, 2007, p. 88, 2008, p. 98, 2009, p. 85, 2010, p. 95, 2011, p. 104, 2012, p. 167, 2013, p. 170.

[7] [16], p. 133.

since 2009, the bank has been successful in eliminating its models weaknesses, at least at the highest portfolio level. It should nevertheless be borne in mind that market phases analysed after 2008 were sometimes quieter and that there has also been some reduction in risk. The increasing shift in the nature of the financial crisis from 2010 towards a crisis concerning the creditworthiness of peripheral European countries, which created new market disruption, is most certainly reflected at the highest level of the backtesting time series. Particularly large losses were incurred in March and May 2010, which only in May 2010 led to the two outliers realised that year. These outliers may be explained by the fears brewing at the time about the situation of the PIIGS states. Possibly, the scale of the corresponding trading activities was such that any problems with the models for these sub-portfolios made themselves felt at the highest portfolio level. The weaknesses outlined below were, by the banks own testimony, identified and rapidly addressed.[8] As mentioned above, two to three outliers per year represent the number to be expected and are not sufficient, in themselves, to call the quality of modelling into question.

The flaws banks identified in their models following the outbreak of the crisis revealed that a variety of areas needed work and improvement. These improvements have since been carried out. Some examples of model weaknesses which banks have now resolved are[9]:

1. *No coverage of default-risk driven "jump events", such as rating changes and issuer defaults.* At the outbreak of the crisis, models often failed to cover the growing amount of default risk in the trading book. The introduction of IRC models[10] to cover migration and default risk helped to overcome this.
2. *Insufficient coverage of market liquidity risk.* It was often not possible to liquidate or hedge positions within the ten-day holding period assumed under Basel 1.5. This led to risks being underestimated. Basel 2.5 takes account of market liquidity risk explicitly and in a differentiated way, at least for IRC models. Full coverage will be achieved under Basel 3.5.
3. *Slow response to external shocks (outlier clustering).* The introduction of stress VaR under Basel 2.5 went a long way towards eliminating the problem of under-estimating risks in benign market conditions. Historical market data for "normal VaR" are now adjusted daily, while monthly or quarterly adjustments were the norm before the crisis.
4. *Insufficient consideration of the risk factors involved in securitisation.* As a result, models designed for securitisation portfolios may no longer be used to calculate capital charges (with the exception of the correlation trading portfolio). Even before the rule change, some banks had already decided themselves to stop using these models.
5. *Flawed proxy approaches.* Prior to the crisis, it was often possible to assign a newly introduced product to an existing one and assume the market risk would

---

[8] Cf. [11], Management Report, 2010, p. 91.

[9] [30], pp. 13–17.

[10] IRC stands for incremental risk charge. This refers to risks such as migration and default risk, which were not covered by traditional market risk models before the crisis.

behave in the same way. During the crisis, this assumption proved to be flawed.[11] The supervisory treatment of such approaches is now much more restrictive.

6. *The approximation of changes in the price of financial instruments cannot accommodate large price movements (delta-gamma approximations).* Full revaluation of instruments is now standard practice.

7. *No and/or flawed scaling to longer time horizons.* Scaling practices of this kind, such as square-root-of-time scaling, are now subject to prudential requirements to ensure their suitability.

These problems were the basis of the review of market risk rules under Basel 2.5 and, as described above, were able to be eliminated both by banks themselves and by new supervisory requirements.[12] Despite this large-scale and appropriate response, distrust of internal model results and their use for prudential purposes persisted, leading to further fundamental discussions.[13]

2.2 Divergence of Model Results

This continuing distrust at the most senior level of the Basel Committee[14] led to the commissioning of the Standards Implementation Group for Market Risk (SIG-TB) to compare the results generated by the internal models of various banks when applied to the same hypothetical trading portfolios (hypothetical portfolio exercise). A major point of criticism has always been that internal model results are too variable even if the risks involved are the same. In January 2013, the SIG-TB published its analysis.[15] The following factors were identified as the key drivers of variation:

- *The legal framework*: some of the banks in the sample did not have to apply Basel 2.5. This means the US banks, for instance, supplied data from models that had neither been implemented nor approved. Analysis showed that some of these banks had significantly overestimated risk, though this did not, in practice, translate into higher capital requirements.
- *National supervisory rules for calculating capital requirements*: differences were noted, for example, in the multipliers set by supervisors for converting model results into capital requirements. In addition, some supervisors had already imposed restrictions on the type of model that could be used and/or set specific capital add-ons.
- *Legitimate modelling decisions taken by the banks*: among the most important of these was the choice of model (spread-based, transition matrix-based) in the absence of a market standard for modelling rating migration and default risk (IRC

---

[11] [18], p. 133.

[12] [21], pp. 59 ff., [25], p. 39.

[13] Cf. Sect. 3.

[14] The precise reasons for this distrust at senior level are not known.

[15] Cf. [6].

models). Different assumptions about default correlations also led to different results. In VaR and stressed VaR models, major factors were the length of data histories (at least one year, no maximum limit), the weighting system, the aggregation of asset classes and of general and specific market risk, and the decision whether to scale a one-day VaR up to ten days or estimate a 10-day VaR directly. The choice of stress period for the stressed VaR also played an important role.[16]

In summary, the differences noted were the result of legitimate decisions taken by banks with the approval of supervisors and of variations between supervisory approval procedures. There is no evidence to suggest manipulation with the aim of reducing capital requirements. Differences can also be explained by variations in the applicable legal framework and in the market phase on which the study was based. An issue related to the market phase is the length of the observation period used. Observation periods of differing lengths will have an impact if, for instance, the volatility of market data has changed from high (during a period over one year ago) to low (last year). In this example, a bank using a one-year data history will not capture the phase of higher volatility. This volatility will, by contrast, most certainly be captured by any bank using a longer data history (with the extent also depending on the weighting system applied to historical data).

It is also important to note that the study was based on a hypothetical portfolio approach at the lowest portfolio level and not on real portfolios. The study does not address the inherent weakness of this method. One major weakness is that the test portfolios used do not reflect portfolio structures in the real world. Portfolios for which banks calculate VaR are normally located at a far higher level in the portfolio "tree" and are consequently more diversified. If the portfolios analysed had been more realistic, variations would probably have been significantly less marked.[17]

Even if the variation between results can be readily explained and cannot be "blamed" on the banks, it is nevertheless difficult to communicate differences of, for instance, around 13–29 million euros in the results for portfolio 25, the most highly aggregated portfolio.[18] Efforts are most certainly needed to reduce the amount of variation by means of further standardisation, even if complete alignment would not make good sense (see Sect. 4.2). At first sight, the differences could also be interpreted as a quantitative measure of the uncertainty surrounding model results and thus as an expression of model risk. Section 4.7 will explore to what extent this is a reasonable analysis and whether banks should try to capitalise model risk themselves as things stand. As the next section shows, dispensing with internal models for prudential purposes would not, by contrast, be the correct response.

---

[16] Cf. [6], p. 10.

[17] The study by the SIG has now been expanded to cover more complex portfolios, cf. [7]. The results are nevertheless comparable. Variation increases with the complexity of the portfolios. In the first analysis, this was found to be particularly the case with IRC modelling compared to "normal" market risk modelling.

[18] Cf. portfolio 25, [6], p. 27.

3 Alternatives to Internal Models

3.1 Overview

Given the difficulties associated with modelling and the variation in results, it is legitimate to ask whether model-based, risk-sensitive capital charges should be dropped altogether. Such a step would, moreover, significantly simplify regulation. But it could also be asked whether it would not make more sense to address the undoubted weaknesses of internal models by means of the reforms already in place or in the pipeline without "throwing the baby out with the bath water", i.e. should we not try to learn from past mistakes instead of just giving up. These questions can best be answered systematically by examining to what extent the existing regulatory proposals could, together or on their own, replace model-based capital charges. There are essentially two alternatives under discussion:

- dropping risk-sensitive capital charges and introducing a leverage ratio as the sole "risk metric";
- regulatory standardised approaches: applying risk-sensitive capital charges while abandoning model-based ones.

3.2 The Leverage Ratio

An exclusively applicable, binding leverage ratio—defined as the ratio of tier 1 capital to total assets including off-balance-sheet and derivative positions[19]—is only a logical response if it must be assumed that neither banks nor supervisors are capable of measuring the risks involved in banking. Advocates of this approach talk of the "illusion of the measurability of risk."[20] They argue that we are in a situation of "uncertainty", not "risk". Uncertainty in decision theory is characterised by two things: neither are all conceivable results known, nor is it possible to assign probabilities to the results or estimate a probability density function. In this case, it would not, for example, be possible to calculate a VaR defined as a quantile of a portfolio loss distribution. This is only possible under "risk".

The concepts of "uncertainty" and "risk" are, however, abstract, theoretical extremes, while the various situations observed in reality usually lie somewhere in between. The answer to the question of whether it is more appropriate to assume a risk situation or an uncertainty situation is determined above all by the availability of the data needed for the model estimate (such as market data or historical default data). If, in addition, the risk factors associated with the financial instruments are known and taken into account, and if the potential changes in the value of a trading

---

[19] The most recent revision of the Basel Committee's definition of the leverage ratio can be found in draft form in [4] and, in its final form, in [8].

[20] Cf. [31], p. 19.

portfolio can be satisfactorily measured, (quality of the stochastic model, no normal loss distribution as a rule), determining a VaR of portfolio losses is likely to be appropriate.[21] This may be assumed for the vast majority of trading portfolios. Should this nevertheless not be the case, regulatory standardised approaches, which normally require less data to be available, could then be used. Reviewing and adjusting models is a never-ending task for banks. The model risks which undoubtedly exist (e.g. estimation errors) are also a focus of supervisors' attention. An awareness of the limits of a model and of such model risks does not, however, make the use of models obsolete.[22] Although modelling by its very nature always involve simplification of reality, quantitative and qualitative model validation is crucial. Supervisors set and enforce stringent rules for such validation.[23]

Advocates of the "uncertainty approach" propose a so-called heuristic as a "rule of thumb" and as a risk metric, at least for supervisors. Leverage ratios with widely differing minimum levels have been suggested as a heuristic for ensuring the solvency of banks. The levels called for range from 3 to 30 %.[24] As is generally recognised, it is not possible to infer a specific minimum level from theory.

The question of whether a leverage ratio is actually a suitable heuristic for ensuring solvency has not been satisfactorily answered, however. Empirical studies to determine to what extent the leverage ratio is a statistical, univariate risk factor that can distinguish between banks that survive and those that fail come to different conclusions.[25] Often, no such distinguishing ability can be demonstrated. This may have an economic explanation since the leverage ratio, as a vertical metric on the liabilities side of the balance sheet, cannot act as a horizontal metric of a bank's risk-bearing capacity by means of which sources of loss (causes of insolvency), which are mainly to be found on the assets side of the bank's balance sheet, are compared with a loss-absorbing indicator (capital). This can, by contrast, be accomplished by ratios such as the "core tier 1" or "tier 1" capital ratio. If, moreover, a leverage ratio were a measure capable of predicting the insolvency of certain types of banks, it would probably swiftly cease to be a good measure once it became a binding target (Goodhart's Law).

What is more, the leverage ratio has a very long—and already widely discussed—list of drawbacks.[26] These are the points of most relevance here:

- Perverse incentives and the potential for arbitrage: there are strong incentives to make business models more risky. Because assets are measured on a non-risk-weighted basis, an AAA investment, for instance, ties up just as much capital as does a B investment.

---

[21] Cf. [19], p. 36.

[22] See footnote 21

[23] See also Sect. 4.6.

[24] Cf., for example, [31], p. 23 (15 % capital ratio), cf. [26], p. 182 (20–30 % capital ratio). Leverage ratios set at this level would override risk-based standards, thus rendering them obsolete.

[25] Cf., for example, the summarising article [32], pp. 26 f.

[26] Cf., for example, [17] or [20], p. 58.

- A leverage ratio is by no means "model free": highly complex valuation models or even simulation approaches are sometimes needed to measure derivatives on a marked-to-market basis, for example. In a broader context, this is more or less true for all balance-sheet valuations. So even a leverage ratio cannot claim to be the simple, robust rule that proponents of a heuristic approach are looking for.[27]

- It makes it impossible to compare capital adequacy across banks. The adequacy of a bank's capital resources cannot be assessed without measuring the associated risks.

For these and other reasons not mentioned here, the international banking community continues to reject the leverage ratio as a sole indicator and as a binding limit. At most, it may make sense to monitor changes in a bank's leverage ratio, but not its absolute level; this is the approach of the German Banking Act at present.[28] Supervisors have widely differing views on the leverage ratio. Even Haldane/Madouros (Bank of England) by no means call in their famous *"The dog and the frisbee"* speech for a leverage ratio on its own or a minimum leverage ratio set at such a high level that risk-based requirements are overridden and therefore indirectly rendered obsolete (leverage ratio as a frontstop instead of the Basel backstop). Owing to the massive perverse incentives which they too have noted, they talk instead of placing leverage ratios on an equal footing with capital ratios.[29]

3.3 Regulatory Standardised Approaches

Standardised approaches, i.e. approaches which spell out in detail how to calculate capital requirements on the basis of prudential algorithms ("supervisory models"), will always be needed for smaller banks which cannot or do not wish to opt for internal models. But larger banks need standardised approaches too—as a fallback solution if their internal models are or become unsuitable for all or for certain portfolios. Having said that, a standardised approach alone is by no means sufficient for larger banks; the reasons are as follows[30]:

- It is invariably true of a standardised approach that "one size does not fit all banks". Since a standardised approach is not tailored to an individual bank's portfolio structure, it cannot measure certain risks (such as certain basis risks) or can only

---

[27] The discussion about a suitable definition of the leverage ratio also shows that improved definitions invariably lead to significantly greater complexity, cf. [8].

[28] Cf. Section 24 (1) (16) and (1a) (5) of the German Banking Act [27].

[29] Cf. [24], p. 19: "The case against leverage ratios is that they may encourage banks to increase their risk per unit of assets, reducing their usefulness as an indicator of bank failure—a classic Goodhart's Law. Indeed, that was precisely the rationale for seeking risk-sensitivity in the Basel framework in the first place. A formulation which would avoid this regulatory arbitrage, while preserving robustness, would be to place leverage and capital ratios on a more equal footing." A leverage ratio of at least 7 % would be necessary for this purpose, in the authors' view.

[30] Cf. [19], p. 37.

do so very inaccurately. It is normally much less risk-sensitive than an internal model.

- A related problem is that a standardised approach usually works only with comparatively simple portfolios. This results in risk being overstated or understated.
- It normally fails to capture diversification or hedging effects satisfactorily.
- Standardised approaches can thus be more dangerous than internal models because it is often easy to "game the system". Trading revenue, for instance, can be generated seemingly without risk, enabling trading units to inflate risk-adjusted earnings.[31]
- If internal models are no longer used, supervisors will also have to dispense with banks' risk-management expertise.
- Standardised approaches are simple models. But as all proposals for standardised approaches to date have shown, supervisors are by no means better at constructing models than are the banks themselves.

A further alternative would be scenario-based approaches, which are often relatively similar to models, such as those which may currently be used for calculating capital charges for options under the standardised approach to market risk (scenario matrix approach). This alternative, though definitely worth considering, is not being discussed at present and will therefore be only briefly explored in this article. Scenario approaches may be regarded as a kind of "halfway house" between risk-sensitive standardised approaches and internal models. If they are prescribed as a regulatory standardised approach, they may also demonstrate the weaknesses of standardised approaches described above. The key criteria for evaluating such approaches are the scenario generation technique and the process/algorithm used for calculating valuation adjustments on the basis of the scenarios. An especially critical question is to what extent the (tail) loss risk of the instruments and portfolios concerned can be captured. At one end of the spectrum are approaches that merely differentiate between a few scenarios (e.g. base case and adverse case) and make no attempt to estimated a loss distribution. At the other extreme are internal models which simulate such a large number of scenarios that it is possible to estimate a loss distribution on the basis of which a parameter such as VaR or expected shortfall can be calculated. Another important question is whether or not the scenario generation takes account of stressed environmental conditions.

To sum up, standardised approaches usually have considerable failings when it comes to measuring risk, especially the risk associated with large-scale, complex trading activities. On their own, they are not an adequate basis on which to determine appropriate capital requirements.[32] So it may be concluded at this point

---

[31] One example: when supervisors set risk factors in the standardised approach model, basis risk is often ignored because different risk factors are (and must be) mapped to the same regulatory risk factor. This is part of the model simplification process. It is often easy to design a trade to exploit the "difference".

[32] The outlined shortcomings of standardised approaches also mean they have only limited suitability as a floor for model-based capital requirements. Contrary to what is sometimes claimed, model risk would therefore not be reduced by a floor.

that, together or separately, a leverage ratio and standardised approaches are inappropriate and insufficient from a supervisory perspective. Internal models must remain the first choice. Nevertheless, confidence in internal models needs to be significantly strengthened.

4 Ways of Restoring Confidence

4.1 Overview

The first, important step should be to standardise supervisory approval processes to eliminate this major source of variation. A single set of approval and review standards should be developed for application worldwide. A globally consistent procedure needs to be enforced for granting and withdrawing permission to use models. With activities of this kind, supervisors themselves could make a significant contribution to restoring confidence.[33]

A number of further proposals are also under discussion at present. Together, they have the potential to go a long way towards winning back trust:

a. Reducing the variation in model results through standardisation (Sect. 4.2).
b. Enhancing transparency (Sect. 4.3).
c. Highlighting the positive developments as a result of the trading book review (Sect. 4.4).
d. Strengthening the use test concept (Sect. 4.5).
e. A comprehensive approach to model validation (Sect. 4.6).
f. Quantification and capitalisation of model risk (Sect. 4.7).
g. Voluntary commitment by banks to a code of "model ethics" (Sect. 4.8).
h. Other approaches (Sect. 4.9).

4.2 Reducing the Variation in Model Results Through Standardisation

First of all, however, it is important to be aware of the dangers of excessive standardisation[34]:

---

[33] For example: the range of multipliers ("3 + x" multiplier), which convert model results into capital requirements, and the reasons for their application differ widely from one jurisdiction to another.

[34] The Basel Committee is already trying to find a balance between the objectives of "risk sensitivity", "complexity" and "comparability". Standardisation has the potential to reduce the complexity of internal models and increase their comparability. Against that, increasing the complexity of standardised approaches often improves comparability. See [5, 22] on the balancing debate.

- Standardised models can pose a threat to financial stability because they encourage all banks to react in the same way (herd behaviour). Model diversity is a desirable phenomenon from a prudential point of view since it generates less procyclicality.
- Standardised models would frequently be unsuitable for internal use at larger banks, which would consequently need to develop alternative models for internal risk management purposes. As a result, the regulatory model would be maintained purely for prudential purposes (in violation of the use test; see below). This would encourage strategies aimed at reducing capital requirements since the results of this model would not have to, and could not, be used internally.
- It is therefore in the nature of models that a certain amount of variation will inevitably exist.

Nonetheless, it is most certainly possible to standardise models in a way which will reduce their complexity and improve the comparability of their results but will not compromise their suitability for internal use. Here are a few suggestions[35]:

- Develop a market standard for IRC models to avoid variation as a result of differences in the choice of model (proposed standard established by supervisors: see Trading Book Review).
- Reduce the amount of flexibility in how historical data are used. For the standard VaR, one year should be not just the minimum but both the minimum and maximum period. This may well affect different banks in different ways, sometimes increasing capital requirements and sometimes reducing them.
- Standardise the stress period for stressed VaR. The period should be set by supervisors instead of being selected by banks. True, this means the stress period would no longer be optimally suited to the individual portfolio in question. But as the study by the Basel Committee's SIB-TB has shown, similar periods may, as a result of the financial crisis, be considered relevant at the highest portfolio level—namely the second half of 2008 (including Lehman insolvency) and the first half of 2009.[36]

4.3 Enhancing Transparency

Much could also be done to improve transparency. Banks could disclose their modelling methodologies in greater detail, and explain—for example—why changes made to their models have resulted in reduced capital charges. Transparency of this kind will significantly benefit informed experts and analysts. These experts will then be faced with the difficult challenge of preparing their analyses in such a way as to be accessible to the general public. The public at large cannot be expected to be the primary addressees of a bank's disclosures. Someone without specialist knowledge is unlikely to be able to understand a risk report, for instance. Nor is it the task of banks

---

[35] Cf. [23].
[36] Cf. [6], p. 50.

to write their reports in a manner that makes such specialist knowledge unnecessary. This is, however, by no means an argument against improving transparency.

The work of the Enhanced Disclosure Task Force (EDTF) is also a welcome contribution[37] and some banks have already implemented its recommendations in their trading units voluntarily. The slide from the Deutsche Bank's presentation for analysts on 31 January 2013 is just one illustration.[38] This explains, in particular, the changes in market-risk-related RWAs (mRWA flow), i.e. it is made clear what brought about the reduction in capital requirements in the trading area. The reasons include reduced multipliers (for converting model results into capital requirements) on the back of significantly better review results, approval of models (IRB approach, IMM) for some additional products and the consideration of additional netting agreements and collateral in calculations of capital requirements.

Another possible means of improving transparency would be to disclose the history of individual positions with a certain time lag. Serious discussion is nevertheless called for to determine at what point the additional cost of transparency incurred by banks would exceed the additional benefit for stakeholders. From an economic perspective, this may be regarded as a transparency ceiling.

4.4 Highlighting the Positive Developments as a Result of the Trading Book Review

The Basel Committee is currently working on a fundamental review of how capital requirements should be calculated for trading book exposures.[39] It has taken criticism of the existing regime on board and proposes to reduce the leeway granted to banks in the design of their internal models. Without going into the Committee's extensive analysis in detail, here are some key elements of relevance to the questions examined in this article:

- Expected shortfall is to be introduced as a new risk metric calibrated to a period of market stress. The intention is to switch to a coherent measure of risk which can take better account of tail risk.[40] The reference to a stress period is intended to address the issue of "fair-weather models" (the problem facing the turkey in Taleb's "The Black Swan").
- A so-called desk approach is to be introduced for granting and withdrawing approval for models. In the future, model approval is to be decided on a case-by-case basis at trading desk level. This will enable portfolios which are illiquid and/or cannot easily be modelled to be excluded from the model's scope.

---

[37] Cf. [13]. Recommendations for market risk (nos 22–25), cf. pp. 12, 51–55.

[38] Cf. [12], p. 23.

[39] Cf. [2, 3].

[40] Cf. [1], p. 203.

- Model validation will take place at desk level and become even more stringent through backtesting and a new P&L attribution process. This will significantly improve the validation process. At the same time, it will have the effect of raising the barriers to obtaining supervisory approval of internal models.
- All banks using models will also have to calculate requirements using the standardised approach. Supervisors take the view that the standardised approach can serve as a floor, or even a benchmark, for internal models (the level of the floor has not yet been announced). This may provide a further safety mechanism to avoid underestimating risk, even if the standardised approach does not always produce sound results (see above).

4.5 Strengthening the Use Test Concept

Up to now, approval of internal models has been dependent, among other things, on supervisors being convinced that the model is really used for internal risk management purposes. Banks consequently have to demonstrate that the model they have submitted for supervisory approval is their main internal risk management tool. Basically, they have to prove that the internal model used to manage risk is largely identical to the model used to calculate capital charges (use test). The rationale behind this sensible supervisory requirement is that the quality of these risk measurement systems can best be ensured over time if the internal use of the model results is an absolute prerequisite of supervisory approval. As a result of the use test, the bank's own interests are linked to the quality of the model. The design of the model should on no account be driven purely by prudential requirements. Moreover, the reply to the question of how model results are used for internal risk management purposes shows what shape the bank's "risk culture" is in.

The use test concept has been undermined, however, by a development towards more prudentially driven models which began under Basel 2.5 and is even more pronounced under Basel 3.5. This trend should be reversed. At a minimum, the core of the model should be usable internally—that is to say be consistent with the bank's strategies for measuring risk. Conservative adjustments can then be made outside the core.

4.6 A Comprehensive Approach to Model Validation

It should be borne in mind that conventional backtesting methods cannot be performed on IRC models. Instead, the EBA has issued special guidelines based on indirect methods such as stress tests, sensitivity and scenario analyses.[41] A distinction therefore needs to be made between "normal" market risk models

---

[41] Cf. [14], pp. 15 f.

and IRC models. Though validation standards already exist for IRC models, they can by no means be described as comprehensive.

For normal market risk models, a comprehensive approach going beyond purely quantitative backtesting and the P&L attribution process could be supported by banks themselves. Proposals to this effect are already on the table at the Federal Financial Supervisory Authority (BaFin).[42] It would be worth examining whether the minimum requirements for the IRB approach could make an additional contribution. These minimum requirements already pursue a comprehensive quantitative and qualitative approach to validation, though it may not be possible to apply a number of problems needing to be resolved to the area of market risk.[43]

4.7 Quantification and Capitalisation of Model Risk

A further approach might be to quantify and capitalise model risk either in the form of a capital surcharge on model results under pillar 1 or as an additional risk category under pillar 2.

It would be worthwhile discussing the idea of using the diverging result interval of the hypothetical portfolio exercise (see Sect. 2.2) as a quantitative basis for individual capital surcharges. This may be regarded as prudential benchmarking.[44] The portfolios tested in this exercise do not, however, correspond to banks' real individual portfolios, which makes them a questionable basis for individual capital surcharges. As explained above in Sect. 2.2, moreover, it cannot be concluded that the differences are largely due to model weaknesses. The question of how to derive the differences actually due to model risk from the observed "gross" differences is yet to be clarified and will probably be fraught with difficulties. What is more, model risk is not reflected solely in the differences in model results (see below on the nature of model risk, which also covers the inappropriate use of model results, for example, which can result in flawed management decisions).

This raises the question as to whether it may be better to address model risk under pillar 2. If model risk is assumed to arise, first, when statistical models are not used properly and, second, from an inevitable uncertainty surrounding key features of models, then it is likely to be encountered above all in the areas of

- design (model assumptions concerning the distribution of market risk parameters or portfolio losses, for example),
- implementation (e.g. the approximation assumptions necessary for IT purposes),
- internal processes (e.g. complete and accurate coverage of positions, capture of market data, valuation models at instrument level [see below]) and IT systems used by banks to estimate risk, and

---

[42] Cf. [9], pp. 38–49.

[43] Cf. Articles 174, 185 CRR [29].

[44] The EBA is currently preparing a regulatory technical standard to this effect under Article 78 of CRD IV.

- model use.[45]

The authors take the view that solving the question of how to quantify model risk for the purpose of calculating capital charges is a process very much in its infancy and that it is consequently too soon for regulatory action in this field. As in other areas, risk-sensitive capital requirements should be sought; one-size-fits-all approaches, like that called for by the Liikanen Group, should not be pursued because they usually end up setting perverse incentives.

This point notwithstanding, there are already rigid capital requirements for trading activities under pillar 1 which address model risk, namely in the area of prudent valuation. These require valuation adjustments to be calculated on accounting measurements of fair value instruments (additional valuation adjustments, AVAs) and deducted from CET1 capital. This creates a capital buffer to cover model risk associated with valuation models at instrument level (see above).[46] Valuation risk arising from the existence of competing valuation models and from model calibration is addressed by the EBA standard. Deductions for market price uncertainty (Article 8 of the EBA RTS) can also be interpreted as charges for model risk, even if the EBA does not itself use the term.

4.8 Voluntary Commitment by Banks to a Code of "Model Ethics"

A commitment could be made to refrain from aggressive or inappropriate modelling with the sole aim of minimising capital requirements. Banks voluntarily exclude portfolios, such as certain (though by no means all) securitisation portfolios, from the scope of their model if questionable results tend to be generated. This may be regarded as a subitem of the modelling validation issue. The desk approach under Basel 3.5 will help to put this new culture into practice. Since capital requirements will have to be calculated using the standardised approach as well as the IMA, any aggressive modelling should be exposed. At a minimum, banks will have to demonstrate that the standardised approach overstates risk in the portfolio in question. If this cannot be demonstrated, a case of excessively aggressive modelling may be assumed.

4.9 Other Approaches

Other approaches to restoring confidence also deserve a brief mention:

- further incentives to use models appropriately
- opening up of access to trade repository data
- review of models by auditors

---

[45] Cf. [28], pp. 20–23.
[46] Cf. [15], p. 20, Art. 11.

- more stringent new product introduction (NPI) processes.

In addition to the code of "moral ethics" discussed in Sect. 4.8, the following additional incentive to use models appropriately could be considered. Establishing a link between traders' bonuses and model backtesting results could serve to improve the alignment of interests. This idea is also closely connected with the issue of strengthening the use test concept (see Sect. 4.5).

Trade repositories already collect key data, including calculated market values, relating to all derivative contracts, irrespective of whether they are centrally cleared or not. As things stand, banks have no way of accessing the data of other banks. If access were made possible at an anonymised level, for example, banks would be able to carry out internal benchmarking, which could reduce valuation uncertainty and thus model risk (see also Sect. 4.7).

External auditors already review banks' internal models (both instrument and stochastic) when auditing the annual accounts. Ways could be explored of further improving or extending this process, e.g. to include a review of use test compliance.

In the insurance industry, the chief actuary is personally responsible for the correct pricing of new products. This practice could be adopted in the NPI process used in the banking industry. The CRO would then be responsible for pricing products fairly, including products aimed at retail clients. The NPI process could also be made stricter by requiring external reviewers to approve major new products. Finally, the suitability of proxy approaches, which are extremely important in the NPI process, could be examined more stringently and in greater depth.

5 Conclusion

The key conclusions of this article can be summarised as follows:

- A risk-sensitive and model-based approach to calculating capital requirements for banks should be retained.
- Not only should model-based approaches be formally retained, but there should also continue to be a capital incentive to use these approaches (i.e. no overriding leverage ratio, no floor set at too high a level).
- Non-risk-sensitive approaches to calculating capital requirements should, at most, be used in a complementary capacity, serving merely as indicators and not as binding limits. Otherwise, dangerous perverse incentives will arise.
- There are also dangers associated with risk-sensitive standardised approaches because these typically overestimate or underestimate the actual risk.
- Variation in the area of models is something that needs to be lived with to a certain extent. Some standardisation is nevertheless possible, as are other ways of restoring confidence. But it should not compromise the internal usability of models.

Open Access This chapter is distributed under the terms of the Creative Commons Attribution Noncommercial License, which permits any noncommercial use, distribution, and reproduction in any medium, provided the original author(s) and source are credited.

References

1. Artzner, P., Delbaen, F., Eber, J.M., Heath, D.: Coherent measures of risk. Math. Financ. **9**(3), 203–228 (1999)
2. Basel Committee on Banking Supervision: Consultative document—Fundamental review of the trading book (2012)
3. Basel Committee on Banking Supervision: Consultative document—Fundamental review of the trading book: A revised market risk framework (2013)
4. Basel Committee on Banking Supervision: Consultative document—Revised Basel III leverage ratio framework and disclosure requirements (2013)
5. Basel Committee on Banking Supervision: Discussion paper—The regulatory framework: balancing risk sensitivity, simplicity and comparability (2013)
6. Basel Committee on Banking Supervision: Regulatory consistency assessment programme (RCAP)—Analysis of risk-weighted assets for market risk (2013)
7. Basel Committee on Banking Supervision: Regulatory consistency assessment programme (RCAP)—Second report on risk-weighted assets for market risk in the trading book (2013)
8. Basel Committee on Banking Supervision: Basel III leverage ratio framework and disclosure requirements (2014)
9. Bongers, O.: Mindestanforderungen an die Validierung von Risikomodellen. In: Martin, R.W., Quell, P., Wehn, C.: Modellrisiko und Validierung von Risikomodellen, pp. 33–64. Cologne (2013)
10. Bundesverband deutscher Banken (Association of German Banks): Discussion paper: Finanzmarktturbulenzen—Gibt es Weiterentwicklungsmöglichkeiten von einzelnen Methoden im Risikomanagement? (2008)
11. Deutsche Bank AG: 2007 to 2013 Annual Reports
12. Deutsche Bank AG: Investor Relations, presentation for Deutsche Bank analysts' conference call on 31 January 2013, https://www.deutsche-bank.de/ir/de/images/Jain_Krause_4Q2012_Analyst_call_31_Jan_2013_final.pdf
13. Enhanced Disclosure Task Force: Enhancing the Risk Disclosures of Banks—Report of the Enhanced Disclosure Task Force (2013)
14. European Banking Authority (EBA): EBA Guidelines on the Incremental Default and Migration Risk Charge (IRC), EBA/GL/2012/3 (2012)
15. European Banking Authority (EBA): EBA FINAL draft Regulatory Technical Standards on prudent valuation under Article 105(14) of Regulation (EU) No 575/2013 (Capital Requirements Regulation—CRR) (2014)
16. Federal Financial Supervisory Authority (Bundesanstalt für Finanzdienstleistungsaufsicht—BaFin): 2007 Annual Report (2008)
17. Frenkel, M., Rudolf, M.: Die Auswirkungen der Einführung einer Leverage Ratio als zusätzliche aufsichtsrechtliche Beschränkung der Geschäftstätigkeit von Banken, expert opinion for the Association of German Banks (2010)
18. Gaumert, U.: Finanzmarktkrise—Höhere Kapitalanforderungen im Handelsbuch internationaler Großbanken? In: Nagel, R., Serfling, K. (eds.) Banken, Performance und Finanzmärkte, festschrift for Karl Scheidl's 80th birthday, pp. 117–150 (2009)
19. Gaumert, U.: Plädoyer für eine modellbasierte Kapitalunterlegung. In: Die Bank, 5/2013, pp. 35–39 (2013)
20. Gaumert, U., Götz, S., Ortgies, J.: Basel III—eine kritische Würdigung. In: Die Bank, 5/2011, pp. 54–60 (2011)

21. Gaumert, U., Schulte-Mattler, H.: Höhere Kapitalanforderungen im Handelsbuch. In: Die Bank, 12/2009, pp. 58–64 (2009)
22. German Banking Industry Committee: Comments on the BCBS Discussion Paper "The regulatory framework: balancing risk sensitivity, simplicity and comparability" (2013)
23. German Banking Industry Committee: Position paper "Standardisierungsmöglichkeiten bei internen Marktrisikomodellen" (2013)
24. Haldane, A., Madouros, V.: The Dog and the Frisbee. Bank Of England, London (2012)
25. Hartmann-Wendels, T.: Umsetzung von Basel III in europäisches Recht. In: Die Bank, 7/2012, pp. 38–44 (2012)
26. Hellwig, M., Admati, A.: The Bankers' New Clothes. Princeton University Press, Princeton (2013)
27. Kreditwesengesetz: Gesetz über das Kreditwesen—KWG, non-official reading version of Deutschen Bundesbank, Frankfurt am Main, as at 2 January (2014)
28. Quell, P.: Grundsätzliche Aspekte des Modellrisikos. In: Martin, R.W., Quell, P., Wehn, C.: Modellrisiko und Validierung von Risikomodellen, pp. 15–32. Cologne (2013)
29. Regulation (EU) No 575/2013 of the European Parliament and of the Council of 26 June 2013 on prudential requirements for credit institutions and investment firms and amending Regulation (EU) No 648/2012 (CRR—Capital Requirements Regulation)
30. Senior Supervisors Group: Observations on Risk Management Practices during Recent Market Turbulence (2008)
31. Wissenschaftlicher Beirat beim BMWi (Academic advisory board at the Federal Ministry for Economic Affairs and Energy): Reform von Bankenregulierung und Bankenaufsicht nach der Finanzkrise, Berlin, report 03/2010 (2010)
32. Zimmermann, G., Weber, M.: Die Leverage Ratio—Beginn eines Paradigmenwechsels in der Bankenregulierung? In: Risiko Manager 25/26, pp. 26–28 (2012)

Model Risk in Incomplete Markets with Jumps

Nils Detering and Natalie Packham

Abstract We are concerned with determining the model risk of contingent claims when markets are incomplete. Contrary to existing measures of model risk, typically based on price discrepancies between models, we develop value-at-risk and expected shortfall measures based on realized P&L from model risk, resp. model risk and some residual market risk. This is motivated, e.g., by financial regulators' plans to introduce extra capital charges for model risk. In an incomplete market setting, we also investigate the question of hedge quality when using hedging strategies from a (deliberately) misspecified model, for example, because the misspecified model is a simplified model where hedges are easily determined. An application to energy markets demonstrates the degree of model error.

1 Introduction

We are concerned with determining model risk of contingent claims when market models are incomplete. Contrary to existing measures of model risk, based on price discrepancies between models, e.g., [8, 26], we develop measures based on the realized P&L from model risk. This is motivated by financial regulators' plans to introduce extra capital charges for model risk, e.g., [5, 13, 17]. In a complete and frictionless market model, the "residual" P&L observed on a perfectly hedged position is due to pricing and hedging in a misspecified model. The distribution of this P&L can therefore be taken as an input for specifying measures of model risk,

This work was financially supported by the Frankfurt Institute for Risk Management and Regulation (FIRM) and by the Europlace Institute of Finance.

N. Detering (✉)
Department of Mathematics, University of Munich, Theresienstraße 39,
80333 Munich, Germany
e-mail: n.detering@math.lmu.de

N. Packham
Frankfurt School of Finance & Management, Sonnemannstr. 9–11,
60314 Frankfurt am Main, Germany
e-mail: n.packham@fs.de

© The Author(s) 2015
K. Glau et al. (eds.), *Innovations in Quantitative Risk Management*,
Springer Proceedings in Mathematics & Statistics 99,
DOI 10.1007/978-3-319-09114-3_3

such as expected loss, value-at-risk, or expected shortfall, [10]. In an incomplete market, model risk cannot be entirely isolated from market risk by hedging, and further, it is not a priori clear, which hedging strategies are most effective under model uncertainty. The purpose of this paper is to investigate these questions.

The analysis in [10] is primarily focussed on complete and frictionless market models, as this allows for a convenient separation into P&L from market risk and P&L from model risk: Since market risk is hedgeable, any remaining P&L is due to pricing and hedging in a misspecified model. In the setting of incomplete markets, one would rather distinguish between *hedgeable* and *unhedgeable (or residual)* P&L, expressing that the unhedgeable P&L refers to model uncertainty and some unhedged market risk. However, from a practical perspective, as an institution needs to take care of both market risk and model risk—either through hedging or through capital requirements—the distinction is of minor importance.

In addition, the determination and choice of effective hedging strategies in incomplete markets is not as straightforward as the replicating argument in a complete market, but is of high practical relevance. The techniques developed in this paper are suitable to comparing the effectiveness of hedging strategies in incomplete markets under model uncertainty.

Model risk is associated with uncertainty about the model or probability measure that governs the probabilistic behavior of unknown outcomes. In this context, uncertainty refers to *uncertainty in the Knightian sense*, e.g., [16, 23], in which case the *model uncertainty* or *model ambiguity* is expressed by a set of probability measures, each of which defines a valid pricing and hedging model.

A set of axioms for measures of model risk, in the spirit of coherent and convex risk measures [1, 18], was put forward by [8]. A popular measure fulfilling these axioms is a contingent claim's price range across the set of models expressing the model uncertainty. This measure is generalized by [2] to account for a distribution on the model set. It thus allows to incorporate the likelihood of the models into the price range and as such to derive value-at-risk and expected shortfall type measures. However, these measures do not account for the potential losses from model risk realized when hedging in a misspecified model. In a complete market setup, [10] develop value-at-risk and expected shortfall measures on the distribution of losses from model risk, and show that these measures fulfill the axioms for model risk (with the usual exception of value-at-risk not being subadditive).

As a generalization of [10], we develop measures for unhedged risk in incomplete markets, comprising both market and model risk. This applies, for example, when asset price processes are subject to jumps under the pricing measure, where, if at all, perfect replication of contingent claims is possible only under conditions not met in practice (such as infinitely many hedging instruments). Furthermore, in an incomplete market setting, we investigate the question of hedge quality when using hedging strategies from a (deliberately) misspecified model, for example, because the misspecified model is a simplified model where hedges are easily determined. A typical case could be to use a simplified complete market model to determine a replication strategy, when it is known that the actual market is incomplete.

Several simulation studies investigate the risk from hedging in a simplified model, e.g., [11, 24, 25]. However, to the best of our knowledge, this is never compared to the residual risk in the alternative model when following a risk-minimizing strategy. Yet, this comparison is important for selecting an appropriate model for pricing and hedging.

In a case study, we study the respective loss distributions and measures when applied to options on energy futures. Empirical returns in the energy spot and future markets behave in a spiky way and thus need to be modeled with jump processes. However, to reduce the computational cost and to attain a parsimonious model, often simplified continuous asset price processes are assumed. Based on the measures of model risk, we assess the quality and robustness of hedging in a continuous asset price model when the underlying price process has jumps relative to determining hedges in the jump model itself. As asset price models, we employ continuous and pure-jump versions of the Schwartz model [27], calibrated to the spot market at the Nordic energy exchange Nord Pool.

The paper is structured as follows: In Sect. 2, we construct the loss variable and loss distribution relevant for model risk. Section 3 defines measures on the distribution of losses from model risk and relates them to the axioms for measures of model uncertainty introduced by [8]. In Sect. 4, we introduce a way of measuring the relative losses from hedging in a misspecified model as opposed to hedging in the appropriate model. Finally, Sect. 5 contains a case study from the energy market to illustrate the relative loss measure and draw conclusions about the quality of hedging strategies determined in a complete model with continuous asset price processes, when the underlying market is in fact subject to jumps.

2 Losses from Hedged Positions

In this section, we formalize the market setup and the loss process expressing the residual losses from a hedged position. In the case of a complete and frictionless market, these losses correspond to model risk, whereas in the case of an incomplete market, these losses comprise in addition the market risk that is not hedged away.

2.1 Market and Model Setup

We begin with a standard market setup under model certainty, as in e.g., [22]. On a probability space $(\Omega, \mathcal{F}, \overline{\mathbb{Q}})$ endowed with a filtration $(\mathcal{F}_t)_{t \geq 0}$ satisfying the "usual hypotheses" are defined adapted asset price processes $(S_t^j)_{t \geq 0}$, $j = 0, \ldots, d$. The asset with price process S^0 represents the money market account, whereas $S^1, \ldots, S^d$ are risky assets. All prices are discounted, that is, expressed in units of the money market account, and $\overline{\mathbb{Q}}$-martingales, with $\overline{\mathbb{Q}}$ a martingale measure equivalent to the objective probability measure.

Throughout we shall assume that S is a Markov process. This applies to many models commonly used in practice, such as the Black–Scholes model, exponential Lévy models, exponential additive models, and stochastic volatility models, such as the Heston model. We shall see below that the Markov assumption simplifies the analysis considerably.

Fixing a time horizon T, we consider European-type claims with $\mathcal{F}_T$-measurable integrable payoff. Other claims, in particular, path-dependent options, such as Barrier options, can be integrated into the analysis; we refer to [10] for the more general case.

In addition to the risky assets $S = (S^1, \ldots, S^d)$, there may be tradeable options maturing at T written on S, with observable market prices at time 0, so-called *benchmark instruments*. Their $\mathcal{F}_T$-measurable payoffs are denoted by $(H_i)_{i \in I}$, and their observed market prices by $C_i^*, i \in I$, or by $[C_i^{\text{bid}}, C_i^{\text{ask}}], i \in I$, if no unique price is available. These benchmark instruments can be used for static hedging, potentially reducing a claim's model risk.

A trading strategy is a predictable process $\Phi = (\phi^0, \ldots, \phi^d, u_1, \ldots, u_I)$, where $\phi^j = (\phi_t^j)_{t \geq 0}$ denotes the holdings in asset j and $u_i \in \mathbb{R}$ denotes the static holding of benchmark instrument i. The time-t value of the portfolio is $V_t(\Phi) = \sum_{j=0}^d \phi_t^j S_t^j + \sum_{i=1}^I u_i H_t^i$, with $H_t^i, i = 1, \ldots, I$, the time-t prices of the benchmark instruments. To rule out arbitrage opportunities, we require that Φ is admissible. Further, Φ is assumed to be self-financing, that is, $\mathrm{d}V_t(\Phi) = \sum_{j=1}^d \phi_t^j \, \mathrm{d}S_t^j + \sum_{i=1}^I u_i \, \mathrm{d}H_t^i, t \geq 0$.

A contingent claim with $\mathcal{F}_T$-measurable payoff X is hedgeable if there exists a replicating strategy, i.e., a self-financing trading strategy Φ such that $V_T(\Phi) = X$. Hedging eliminates any P&L arising from market risk, and, because of the absence of arbitrage opportunities, the claim's price process and the price of the hedging strategy agree for all $0 \leq t \leq T$. In an incomplete market, in the absence of a replicating strategy, losses from market risk may be eliminated or reduced by super-replicating strategies, e.g., [14], or by risk-minimizing strategies, e.g., [19, 20], but some P&L due to market risk remains.

Aside from market risk, a stakeholder (trader, hedger, shareholder, regulator) may be concerned about model risk when pricing and hedging a contingent claim. Model risk refers to potential losses from mispricing and mishedging, because model $\overline{\mathbb{Q}}$ is possibly misspecified. This uncertainty regarding model $\overline{\mathbb{Q}}$ is captured by a set $\mathcal{Q}$ of martingale measures for the asset price processes, e.g., [8, 9], which may incorporate uncertainty about both the model type and model parameters.

Let

$$\mathcal{C} = \left\{ X \in \sigma(S_T) \mid \sup_{\mathbb{Q} \in \mathcal{Q}} \mathbb{E}\left[X^2\right] < \infty \right\},$$

be the set of contingent claims under consideration, where we require square-integrability, because for claims with finite second moments quadratic minimizing hedging strategies exist, which will be employed later. The set of trading strategies considered is

$$
S = \Bigg\{ \Phi \, | \, \Phi \text{ admissible, predictable, self-financing, } \Phi_t \in \sigma(S_t), \forall t \geq 0
$$

$$
\text{and } \overline{\mathbb{E}} \left[\int_0^T (\phi^j)^2 \, d[S^j, S^j] \right] < \infty, \, j = 0, \ldots, d \Bigg\}.
$$

The condition $\Phi_t \in \sigma(S_t)$ implies that the hedging strategy is a Markov process.

Working on a set of measures requires further conditions, in particular, as the measures in Q need not be absolutely continuous with respect to $\overline{Q}$. More specifically, the asset price processes must be consistent under all measures and specifying trading strategies requires the notion of a stochastic integral with respect to the set of measures.

In case the models in Q are diffusion processes, [28] develop the necessary tools from stochastic analysis, such as existence of a stochastic integral, martingale representation, etc. Although this restricts the joint occurrence of certain probability measures, it does not exclude any particular measure. For our purposes, this limitation does not play a role, as we are primarily interested in choosing a rich set of possible models to cover the model uncertainty. For details, we refer to [10].

In the general case, we pose the following condition on the set of measures Q, which ensures that all objects are well defined when working with uncountably many measures.

Assumption 1 There exists a universal version of the stochastic integral $\int_0^t \phi \, dS$, $\phi \in S$. In addition, for all $\mathbb{Q} \in Q$, the integral coincides $\mathbb{Q}$–a.s. with the usual probabilistic construction and $\int_0^t \phi \, dS$ is $\mathcal{F}_t$-measurable.

2.2 Loss Process

Consider a short position in a claim $X \in \mathcal{C}$ and a trading strategy $\Phi \in S$. The time-T loss of X that we consider is given by

$$
L_T(X, \Phi) := -(V_T(\phi) - Y), \tag{1}
$$

where $V_T(\phi) = V_T((\phi, 0, \ldots, 0))$ and $Y = X - \sum_{i=1}^I u_i H_i$. If $\overline{\mathbb{Q}}$ calibrates to the market prices of the benchmark instruments, i.e., $\overline{\mathbb{E}}[H_i] = C_i^*, i = 1, \ldots, I$, then $L_T(X, \Phi) = -(V_T(\Phi) - X)$, which corresponds to the overall realized loss from the position. However, if $\overline{\mathbb{Q}}$ does not calibrate perfectly to the benchmark instruments, then there is additional instantaneous P&L at time 0 from trading the benchmark instruments. This is not included in Eq. (1), and will be ignored in what follows, as this is booked as (sunk) trading cost and as is does not give rise to further risks.

The goal will be to extend this variable to a loss process $L_t(X, \Phi)$, $t < T$, with Φ a hedging, resp. replicating strategy under $\overline{\mathbb{Q}}$. As both the time-t price, $\overline{\mathbb{E}}[Y|\mathcal{F}_t]$ and the strategy ϕ are defined only $\overline{\mathbb{Q}}$-a.s., one must be explicit in specifying the version to be used when dealing with models that are not absolutely continuous with respect to $\overline{\mathbb{Q}}$. In our setup, we have $\overline{\mathbb{E}}[Y|\mathcal{F}_t] = \overline{\mathbb{E}}[Y|S_t] = f(S_t)$ for some Borel-measurable function f, and likewise for the trading strategy. Since $\mathcal{Q}$ expresses the model uncertainty when employing $\overline{\mathbb{Q}}$ for pricing and hedging, it must not be involved in the choice of the respective versions of the pricing and hedging strategies.

Assumption 2 The versions of $\overline{\mathbb{E}}[Y|S_t]$, $t \leq T$, and ϕ are chosen irrespective of the measures contained in $\mathcal{Q}$.

We further impose linearity conditions on the versions of $\overline{\mathbb{E}}[Y|\mathcal{F}_t]$ and ϕ, which are in general only fulfilled $\overline{\mathbb{Q}}$-a.s. but for all practically relevant models and claims hold for all $\omega \in \Omega$. This will be important for the axiomatic setup in Sect. 3.2.

Assumption 3 Let $X_1, X_2 \in \mathcal{C}$, $\Phi_1 = (\phi_1, u_1^1, \ldots, u_I^1)$, $\Phi_2 = (\phi_2, u_1^2, \ldots, u_I^2) \in \mathcal{S}$ and define $Y_j := X_j - \sum_{i=1}^I u_i^j H_i$, $j = 1, 2$. For all $t \leq T$, it holds that

$$\overline{\mathbb{E}}[aY_1 + bY_2|\mathcal{F}_t](\omega) = a\overline{\mathbb{E}}[Y_1|\mathcal{F}_t](\omega) + b\overline{\mathbb{E}}[Y_2|\mathcal{F}_t](\omega), \quad a, b \in \mathbb{R}, \quad \omega \in \Omega$$

and

$$V_t(a\phi_1(\omega) + b\phi_2(\omega)) = aV_t(\phi_1(\omega)) + bV_t(\phi_2(\omega)), \quad a, b \in \mathbb{R}, \quad \omega \in \Omega.$$

$$\overline{\mathbb{E}}[Y_1|S_T](\omega) = Y_1(\omega), \omega \in \Omega.$$

Assumptions 2 and 3 will be fulfilled in typical cases relevant in practice. Suppose for example that S is a Black–Scholes model under $\overline{\mathbb{Q}}$. Then prices and the replicating strategy of European payoffs can be determined via the Black–Scholes PDE, and these are suitable versions fulfilling the assumptions.

Definition 1 Let $X \in \mathcal{C}$ and $\Phi = (\phi, u_1, \ldots, u_I) \in \mathcal{S}$. The loss process associated with a short position in X and the trading strategy Φ is given by

$$L_t := L_t(X, \Phi) = -(V_t(\phi) - \overline{\mathbb{E}}[Y|S_t])$$

$$= -\left(V_0 + \sum_{j=1}^d \int_0^t \phi^j \, dS^j - \overline{\mathbb{E}}[Y|S_t]\right), \quad 0 \leq t \leq T, \qquad (2)$$

with $Y = X - \sum_{i=1}^I u_i H_i$ and $V_0 = \overline{\mathbb{E}}[Y]$.

If Φ is a replicating strategy under $\overline{\mathbb{Q}}$, then $L_t = 0$ $\overline{\mathbb{Q}}$-a.s., but possibly for some $\mathbb{Q} \in \mathcal{Q}$, $\mathbb{Q}(L_t = 0) < 1$, which expresses that Φ fails to replicate X under $\mathbb{Q}$. A model-free hedging strategy is defined as follows:

Definition 2 The trading strategy $\Phi = ((\phi_t)_{0 \leq s \leq T}, u_1, \ldots, u_I)$ is a *model-free* or *model-independent* replicating strategy for claim X with respect to $\mathcal{Q}$, if $L_t = 0$, $t \geq 0$, $\mathbb{Q}$-a.s., for all $\mathbb{Q} \in \mathcal{Q}$.

Note that our definition of the hedge error based on a continuous time integral separates model risk from a discretization error. When actually calculating the hedge error, it is necessary to use a time grid small enough such that the discretization error is negligible.

The following proposition shows that the overall expected loss at time T from replicating in $\overline{\mathbb{Q}}$ when the market evolves according to $\mathbb{Q}_M$ instead of $\overline{\mathbb{Q}}$ depends only on the price difference.

Proposition 1 1. *The total expected loss from replicating under $\overline{\mathbb{Q}}$ claim X, that is* $\mathbb{E}[L_T]$ *plus the initial transaction cost* $\mathbb{E}[\sum_{i=1}^{I} u_i (H_i^0 - C_i^\star)]$, *when the market evolves according to $\mathbb{Q}_M$ is just the price difference in the two models,* $-(\overline{\mathbb{E}}[X] - \mathbb{E}^{\mathbb{Q}_M}[X])$.
2. *The price range measure, defined by* $\sup_{\mathbb{Q} \in \mathcal{Q}} \mathbb{E}^{\mathbb{Q}}[X] - \inf_{\mathbb{Q} \in \mathcal{Q}} \mathbb{E}^{\mathbb{Q}}[X]$, *can be expressed as* $\sup_{\mathbb{Q}, \overline{\mathbb{Q}}} \mathbb{E}^{\mathbb{Q}}[L_T^{\overline{\mathbb{Q}}}]$, *where* $L_T^{\overline{\mathbb{Q}}}$ *denotes the loss variable from hedging under* $\overline{\mathbb{Q}}$.

Proof See [10].

If a claim cannot be replicated, then—given the static hedging component $\sum_{i=1}^{I} u_i H_i$—a hedging strategy can be defined as a solution $(\hat{V}_0, \hat{\Phi}) \in \mathbb{R} \times \mathcal{S}$ of the optimization problem

$$
\inf_{(V_0 \in \mathbb{R}, \Phi \in \mathcal{S})} \overline{\mathbb{E}}[U(L_T(X, \Phi))] = \inf_{(V_0 \in \mathbb{R}, \Phi \in \mathcal{S})} \overline{\mathbb{E}}\left[U\left(-\left(V_0 + \sum_{j=1}^{d} \int_0^T \phi \, dS^j - Y\right)\right)\right],
$$
(3)

where $U : \mathbb{R} \to \mathbb{R}_+$ weighs the magnitude of the hedge error. The most common choice is $U(x) = x^2$, which minimizes the quadratic hedge error. This so-called *quadratic hedging* has the advantage that the resulting pricing and hedging rules become linear and it is also the analytically most tractable rule. Under this choice of $U(x)$, if S is a martingale, then a solution exists and $\hat{V}_0 = \overline{\mathbb{E}}[Y]$, [20].

Of course, in an incomplete market, $L_T(X, \Phi)$ entails not only losses due to model misspecification, but some losses due to market risk as well, since $\overline{\mathbb{Q}}(L_T(X, \Phi) = 0) < 1$, that is, P&L is incurred even when there is no model uncertainty.

For the explicit determination of $L_t(X, \Phi)$ in some examples, we refer to [10]. It is worth noting that in a complete market setup, the loss process corresponds to the *tracking error* of [15].

2.3 Loss Distribution

The next step is to associate a distribution with the loss variable L_t, $t \leq T$, based on which risk measures such as value-at-risk and expected shortfall can be defined.

This is achieved by considering an extended probability space $(\Omega, \mathcal{F}, \mathbb{P})$, where $\mathcal{F}$ now incorporates in addition the model uncertainty and $\mathbb{P}$ contains information about the degree of uncertainty associated with each model. To make this precise, let $\mathcal{G} \subset \mathcal{F}$ be a σ-algebra such that conditioning on $\mathcal{G}$ eliminates the uncertainty about the pricing measure $\mathbb{Q} \in \mathcal{Q}$. In this setting, the measures in $\mathcal{Q}$ constitute a *regular conditional probability* with respect to $\mathcal{G}$. For existence and construction of this probability space, we refer to [10].

In this setup, the models can be indexed by a random variable $\theta \in \Theta \subseteq \mathbb{R}$, with $\sigma(\theta) = \mathcal{G}$, so that $\mathbb{Q}_\theta = \mathbb{P}(\cdot|\sigma(\theta))$ and

$$\mathbb{P}(B) = \mathbb{E}[\mathbb{P}(B|\sigma(\theta))] = \int_\Omega \mathbb{P}(B|\sigma(\theta))\,d\mathbb{P} = \int_\Theta \mathbb{P}(B|\theta = a)\,\mu(da), \quad B \in \mathcal{F},$$

where μ is the distribution of θ. In particular, losses from hedging in a misspecified model under model uncertainty have distribution function

$$\mathbb{P}(L_t \leq x) = \int_\Theta \mathbb{Q}_a(L_t \leq x)\,\mu(da), \quad 0 \leq t \leq T.$$

The following proposition is proved in [10].

Proposition 2 *A strategy Φ is a model-free hedging strategy for claim X $\mathbb{P}$–a.s. if and only if $\mathbb{P}(L_t = 0) = 1$.*

Hence, model uncertainty is expressed by the unconditional distribution $\mathbb{P}$, whereas model certainty is expressed via the conditional distribution $\mathbb{P}(\cdot|\sigma(\theta))$.

A concrete approach to determining the distribution θ is presented in [10]. Here, probability weights are assigned to the models in $\mathcal{Q}$ via the Akaike Information Criterion (AIC), e.g., [6, 7], which trades off calibration quality against model complexity.

3 Measures of Model Risk

The loss distribution aggregated across the measures in $\mathcal{Q}$ from Sect. 2.3 is the key input to define measures of model risk. For the time being, we continue to work in a setting where a particular model $\overline{\mathbb{Q}}$ is used for pricing and hedging, as this is appropriately quantifies the model risk from a bank's internal perspective.

If a claim cannot be replicated, and the trading strategy Φ is merely a hedging strategy in some risk-minimizing sense, then the loss variable $L_t(X, \Phi)$ from Definition 1 features not only model risk, but also the unhedged market risk. To disentangle model risk from the market risk, one could first determine the market risk from the unhedged part of the claim under $\overline{\mathbb{Q}}$ and set this into relation to the overall residual risk. This requires taking into account potential diversification effects, since

risks are not additive. We shall continue to work under the setup of measuring residual risk, and use the terminology "model risk," although some market risk is also
present.

Market incompleteness can also be seen to be a form of model risk, as—in addition
to the uncertainty on the objective measure—it causes uncertainty on the equivalent
martingale measure. However, hedging strategies would typically be chosen that are
risk minimizing not under the martingale measure, but risk minimizing under the
objective measure. In the case of continuous asset prices, this implies that hedging
is done under the minimal-martingale measure, which is uniquely determined. In
practice, it is more common to choose an equivalent measure that calibrates sufficiently well, and in this case one could argue that incompleteness also increases
model uncertainty. In our setup, this would be reflected by a larger set $\mathcal{Q}$.

3.1 Value-at-Risk and Expected Shortfall

The usual value-at-risk and expected shortfall measures are defined as follows:

Definition 3 Let $L_t(X, \Phi)$ be the time-t loss from the strategy Φ that hedges claim
X under $\overline{\mathbb{Q}}$. Given a confidence level $\alpha \in (0, 1)$,

1. *Value-at-risk (VaR)* is given by

$$\mathrm{VaR}_\alpha(L_t(X, \Phi)) = \inf\{l \in \mathbb{R} : \mathbb{P}(L_t(X, \Phi) > l) \le 1 - \alpha\},$$

   that is, VaR_α is just the α-quantile of the loss distribution;
2. *Expected shortfall (ES)* is given by

$$\mathrm{ES}_\alpha(L_t(X, \Phi)) = \frac{1}{(1 - \alpha)} \int_\alpha^1 \mathrm{VaR}_u(L_t(X, \Phi))\, du.$$

In the presence of benchmark instruments, the hedging strategy in model $\overline{\mathbb{Q}}$ may
not be unique. If the claim X can be replicated, then $\Pi = \{\Phi \in \mathcal{S} : \overline{\mathbb{Q}}(L_t(X, \Phi) =$
$0) = 1, t \le T\}$ is the set of replicating strategies for claim X in model $\overline{\mathbb{Q}}$. Otherwise,
we focus on quadratic hedging and define $\Pi = \{\Phi = (\phi^0, \dots, \phi^j, u_1, \dots, u_I) \in$
$\mathcal{S}, (u_1, \dots, u_I) \in \mathbb{R}^I : \Phi = \hat{\Phi} \text{ under } \overline{\mathbb{Q}}\}$, where $\hat{\Phi}$ refers to the quadratic risk-
minimizing strategy attaining the infimum in (3) with $U(x) = x^2$. Because in an
incomplete market, the loss from hedging entails some market risk aside from model
risk, the benchmark instruments play a more important role than in complete market,
as they are not necessarily redundant, but may reduce the hedge error under $\overline{\mathbb{Q}}$.

To abstract from the particular hedging strategy chosen, we define measures that
quantify the minimal degree of model dependence, indicating that when pricing and
hedging under measure $\overline{\mathbb{Q}}$, the model dependence cannot be further reduced. This is

reasonable in the sense that it is not of interest whether a position is indeed hedged or not. Rather the hedging argument serves only to eliminate (or reduce, in case the claim cannot be replicated) P&L from market risk. Choosing the minimal degree allows to appropriately capture claims that can be replicated in a model-free way.

Definition 4 Concrete measures capturing the model uncertainty when pricing and hedging claim X according to model $\overline{\mathbb{Q}}$ are given by

1. $\mu_{\text{SQE},t}^{\overline{\mathbb{Q}}}(X) = \inf_{\Phi \in \Pi} \mathbb{E}[L_t(X, \Phi)^2]$,
2. $\mu_{\text{VaR},\alpha,t}^{\overline{\mathbb{Q}}}(X) = \inf_{\Phi \in \Pi} \text{VaR}_\alpha(|L_t(X, \Phi)|)$,
3. $\mu_{\text{ES},\alpha,t}^{\overline{\mathbb{Q}}}(X) = \inf_{\Phi \in \Pi} \text{ES}_\alpha(|L_t(X, \Phi)|)$.
4. $\rho_{\text{VaR},\alpha,t}^{\overline{\mathbb{Q}}}(X) = \inf_{\Phi \in \Pi} \max(\text{VaR}_\alpha(L_t(X, \Phi)), 0)$,
5. $\rho_{\text{ES},\alpha,t}^{\overline{\mathbb{Q}}}(X) = \inf_{\Phi \in \Pi} \max(\text{ES}_\alpha(L_t(X, \Phi)), 0)$.

The measures $\mu_{\text{VaR},\alpha,t}^{\overline{\mathbb{Q}}}$ and $\mu_{\text{ES},\alpha,t}^{\overline{\mathbb{Q}}}$ capture model uncertainty in an absolute sense, and are thus measures of the magnitude or degree of model uncertainty. The measures $\rho_{\text{VaR},\alpha,t}^{\overline{\mathbb{Q}}}$ and $\rho_{\text{ES},\alpha,t}^{\overline{\mathbb{Q}}}$ consider losses only. As such, they are suitable for defining a capital charge against losses from model risk.

Contrary to the case of bank internal risk measurement, a regulator may wish to measure model risk independently of a particular pricing or hedging measure, taking a more prudent approach. To abstract from the pricing measure, one would first define the set $\mathcal{Q}_H \subseteq \mathcal{Q}$ of potential pricing and hedging measures (e.g., measures that calibrate sufficiently well) and then define the risk measure in a worst-case sense as follows:

Definition 5 Let $\mu_t^{\mathbb{Q}_H}(X)$ be a measure of model uncertainty when pricing and hedging X according to model $\mathbb{Q}_H \in \mathcal{Q}_H$. The model uncertainty of claim X is given by

$$\mu_t(X) = \sup_{\mathbb{Q}_H \in \mathcal{Q}_H} \mu_t^{\mathbb{Q}_H}(X). \tag{4}$$

Capital charges can then be determined from either $\mu_{\text{VaR},\alpha,t}^{\overline{\mathbb{Q}}}(X)$, resp. $\mu_{\text{ES},\alpha,t}^{\overline{\mathbb{Q}}}(X)$, or from $\mu_{\text{VaR},\alpha,t}(X)$, resp. $\mu_{\text{ES},\alpha,t}(X)$.

3.2 Axioms for Measures of Model Risk

Cont [8] introduces a set of axioms for measures of model risk. A measure satisfying these axioms is called a *convex measure of model risk*. The axioms follow the general notion of convex risk measures, [18, 21], but are adapted to the special case of model risk. In particular, these axioms take into account the possibility of static hedging with liquidly traded option and of hedging in a model-free way. More specifically, the axioms postulate that an option that can be statically hedged with liquidly traded options is assigned a model risk bounded by the cost of replication, which can be expressed in terms of the bid-ask spread. Consequently, partial static hedging for a claim reduces model risk. Further, the possibility of model-free hedging with the underlying asset reduces model risk to zero. Finally, to express that model risk can be reduced through diversification, convexity is required.

Here we only state the following result, which ensures that our measures fulfill the axioms proposed in Cont [8]. The proof is given in [10] for complete markets and can be easily generalized to an incomplete market.

Proposition 3 *The measures* $\mu_{\mathrm{SQE},t}^{\overline{\mathbb{Q}}}(X)$, $\mu_{\mathrm{ES},\alpha,t}^{\overline{\mathbb{Q}}}(X)$ *and* $\rho_{\mathrm{ES},\alpha,t}^{\overline{\mathbb{Q}}}(X)$ *satisfy the axioms of model uncertainty. The measures* $\mu_{\mathrm{VaR},\alpha,t}^{\overline{\mathbb{Q}}}(X)$ *and* $\rho_{\mathrm{VaR},\alpha,t}^{\overline{\mathbb{Q}}}(X)$ *satisfy Axioms 1, 2, and 4.*

4 Hedge Differences

Instead of considering the P&L arising from model misspecification as in Sect. 2.2, one might be interested in a direct comparison of hedging strategies implied by different models. For example, one might wish to assess the quality of hedging strategies determined from a deliberately misspecified, but simpler model, in a more appropriate, but more involved model.

We first explain the idea with respect to one alternative model $\mathbb{Q}_M \in \mathcal{Q}$ and outline then how measures with respect to the entire model set can be built. As before, $\overline{\mathbb{Q}}$ is the model for pricing and hedging and, fixing a claim $X \in \mathcal{C}$, Π is the set of quadratic risk-minimizing (QRM) hedging strategies for X under $\overline{\mathbb{Q}}$ (containing various hedging strategies, depending on how static hedges with the benchmark instruments are chosen).

We seek an answer to the following question: If the market turns out to follow $\mathbb{Q}_M$, what is the loss incurred by hedging in $\overline{\mathbb{Q}}$ instead of hedging in $\mathbb{Q}_M$? Let $\Phi = (\phi, u_1, \ldots, u_I) \in \Pi$ be the QRM strategy for $Y = X - \sum_{i=1}^{I} u_i H_i$, and let Φ_M be the respective QRM strategy for Y derived under $\mathbb{Q}_M$. The *relative* difference of the hedge portfolio compared to the hedge portfolio when using the strategy of $\mathbb{Q}_M$ is given by

$$L_t^{\Delta}(X, \Phi, \Phi_M) = \mathbb{E}^{\mathbb{Q}_M}[Y] - \bar{\mathbb{E}}[Y] + \sum_{j=1}^{d} \int_0^t (\phi_M^j - \phi^j)\, \mathrm{d}S^j. \tag{5}$$

This variable differs from $L_t(X, \Phi)$, cf. Eq. (2), in that it expresses the difference between the hedging strategies Φ and Φ_M, whereas $L_t(X, \Phi)$ describes the difference between the hedging strategy Φ and the claim X.[1]

The next proposition provides some insight on the different nature of the two variables.

Proposition 4 *The following properties hold for the processes $L^\Delta(X, \Phi, \Phi_M)$ and $L(X, \Phi)$:*

1. $L^\Delta(X, \Phi, \Phi_M)$ *is a* $\mathbb{Q}_M$*-martingale with* $L^\Delta_0(X, \Phi, \Phi_M) = \mathbb{E}^{\mathbb{Q}_M}[Y] - \bar{\mathbb{E}}[Y]$
2. $\mathbb{E}^{\mathbb{Q}_M}[L_T(X, \Phi)] = \mathbb{E}^{\mathbb{Q}_M}[Y] - \bar{\mathbb{E}}[Y]$
3. $L^\Delta_T(X, \Phi, \Phi_M) = L_T(X, \Phi)$ $\mathbb{Q}_M$*-a.s. if Y can be replicated under* $\mathbb{Q}_M$
4. $L^\Delta_t(X, \Phi, \Phi_M) - L_t(X, \Phi) = \mathbb{E}^{\mathbb{Q}_M}[Y|\mathcal{F}_t] - \bar{\mathbb{E}}[Y|\mathcal{F}_t]$ $\mathbb{Q}_M$*-a.s. if Y can be replicated under* $\mathbb{Q}_M$.

Proof 1. This follows directly from the definition of $L^\Delta(X, \Phi, \Phi_M)$ and the fact that Φ^M and Φ are in $\mathcal{S}$.
2. See Proposition 1.
3. If Y can be replicated, then $Y = \mathbb{E}^{\mathbb{Q}_M}[Y] + \sum_{j=1}^d \int_0^T \phi^j_M \, dS^j$ $\mathbb{Q}_M$–a.s., and consequently $L^\Delta_T(X, \Phi, \Phi_M) = Y - (\bar{\mathbb{E}}[Y] + \sum_{j=1}^d \int_0^T \phi^j \, dS^j)$ $\mathbb{Q}_M$–a.s.. The claim follows by observing that $L_t(X, \Phi) = -(\bar{\mathbb{E}}[Y] + \sum_{j=1}^d \int_0^T \phi^j \, dS^j - Y)$.
4. Using that $L^\Delta_t(X, \Phi, \Phi_M) = \mathbb{E}^{\mathbb{Q}_M}[Y|\mathcal{F}_t] - (\bar{\mathbb{E}}[Y] + \sum_{j=1}^d \int_0^t \phi^j \, dS^j$ $\mathbb{Q}_M$–a.s., since Y can be replicated under $\mathbb{Q}_M$, the claim follows with the definition of $L_t(X, \Phi)$.

Observe that the variable $L_t(X, \Phi)$ is neither a sub-martingale nor a super-martingale as shown in the example in [10, Sect. 3.5.].

As an example, Fig. 1 shows the distributions of $L_t(X, \Phi)$ and $L^\Delta_t(X, \Phi, \Phi_M)$ for an at-the-money call option $X = (S_T - K)^+$ with $S_0 = K = 1$, with expiry T in 3 months, at time $t = T/2$, dynamically hedged with the underlying asset, i.e., $\Phi = (\phi)$, resp. $\Phi_M = (\phi_M)$. Under the misspecified model $\bar{\mathbb{Q}}$, the asset price process corresponds to a geometric Brownian motion with 20% volatility, whereas under $\mathbb{Q}_M$ the asset price process follows a geometric Brownian motion with 25% volatility. The correlation of the two loss variables is 67.97%. At maturity T, both variables agree.

Generalizing the relative hedge difference to a set of models is not straightforward, as the loss variable $L^\Delta_t(X, \Phi, \Phi_M)$ depends explicitly on $\mathbb{Q}_M$ and, as such, a version of the variable that is valid under all models cannot be constructed. [12] shows how a loss *distribution* under model uncertainty can be constructed, which can then be used to define the usual risk measures such as value-at-risk and expected shortfall.

---

[1] There is no need to pose specific conditions on the version of the hedging strategy Φ_M chosen, since in the following only properties of $L^\Delta_t(X, \Phi, \Phi_M)$ under $\mathbb{Q}_M$ are analyzed.

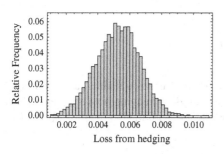 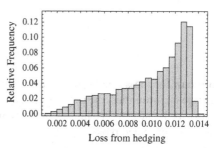

Fig. 1 Loss at $t = 1/2T$ from dynamically hedging an at-the-money call option with a maturity T of 3 months based on 10,000 simulations and 1,000 time steps. *Left* Distribution of $L_t(X, \Phi)$, $\mathbb{E}[L_t(X, \Phi)] = 0.0053$. *Right* Distribution of $L_t^\Delta(X, \Phi, \Phi_M)$, $\mathbb{E}[L_t^\Delta(X, \Phi, \Phi_M)] = 0.0099$, which equals the initial price difference

5 Application to Energy Markets

As a real-worked example, we study the loss variables and risks from hedging options on futures in energy markets. The spot and future prices in energy markets are extremely volatile and show large spikes, and a realistic model for the price dynamics should therefore involve jumps. However, continuous models based on Brownian motions are not only computationally more tractable, but prevalent in practice. Our analysis sheds light on the risks of hedging in a simplified continuous model instead of a model involving jumps.

Assume given a probability space $(\Omega, (\mathcal{F}_t)_{0 \leq t \leq T})$ with a measure $\tilde{\mathbb{P}}$ on which a two-dimensional Lévy process $(L_t) = (L_{1,t}, L_{2,t})_{t>0}$ with independent components is defined. A popular two-factor model for the energy spot price is developed by Schwartz and Smith [27]. The spot is driven by a short-term mean reverting factor to account for short-term energy supply and energy demand and a long-term factor for changes in the equilibrium price level. In its extended form, [4, Sect. 5], the logarithm of the spot price is

$$\log S_t = \Lambda_t + X_t + Y_t \tag{6}$$

with $(\Lambda_t)_{t>0}$ a deterministic seasonality function, $(X_t)_{t>0}$ a Lévy driven Ornstein Uhlenbeck process with dynamics $dX_t = -\lambda X_t dt + dL_{1,t}$ and $(Y_t)_{t>0}$ defined by $dY_t = dL_{2,t}$. We further assume that the cumulant function $\Psi(z) := \log(\mathbb{E}[e^{\langle z, L_1 \rangle}])$ is well defined for $z = (z_1, z_2) \in \mathbb{R}^2$, $|z| \leq C$, for $C \in \mathbb{R}$. Due to the independence of L_1 and L_2, the cumulant transforms of both processes add up and we have $\Psi(z) = \Psi_1(z_1) + \Psi_2(z_2)$ where Ψ_1 and Ψ_2 is the cumulant for L_1 and L_2, respectively.

We consider the pricing and hedging of options on the future contract. In contrast to, for example, equity markets, the future contract in energy delivers over a period of time $[T_1, T_2]$ instead of a fixed time point by defining a payout

$$\frac{1}{T_2 - T_1} \int_{T_1}^{T_2} S_r \, dr \tag{7}$$

in return for the agreed future price. While the spot is not tradable due to lack of storage opportunities, the future is tradable and used for hedging both options on the future itself and options directly on the spot price. Assuming that the future price F_t equals its expected payout

$$F_t = \mathbb{E}^{\mathbb{Q}} \left[\frac{1}{T_2 - T_1} \int_{T_1}^{T_2} S_r \, dr \Big| \mathcal{F}_t \right] \tag{8}$$

under a pricing measure $\mathbb{Q} \sim \tilde{\mathbb{P}}$, the value F_t is derived in analytic form in [4]. Under the assumption that L_1 and L_2 are normal inverse Gaussian (NIG) distributed Lévy processes an approximate process $(\widehat{F}_t^L)_{t<T_2}$ is determined in [4] by matching first and second moments such that $(\widehat{F}_t^L)_{t<T_2}$ is of exponential additive type. We assume in this application that $\mathbb{Q} = \tilde{\mathbb{P}}$. The value of $\widehat{F}_t^L$ is then

$$\widehat{F}_t^L = F_0 \exp\left(-\int_0^t \Psi_1(\Sigma_1^L(s)) + \Psi_2(\Sigma_2^L(s)) \, ds + \int_0^t \Sigma_1^L(s) \, dL_{1,s} + \int_0^t \Sigma_2^L(s) \, dL_{2,s} \right) \tag{9}$$

with time-dependent, deterministic functions $\Sigma_1^L(t)$ and $\Sigma_2^L(t)$. The process $\widehat{F}^L$ depends on the interval $[T_1, T_2]$, but in order to avoid overloading the notation and since we shall only consider a single delivery period in our example, we simply write $\widehat{F}^L$, Σ_1^L and Σ_2^L. The market under this model is incomplete and claims can in general only be hedged with risk-minimizing strategies. Integral representations for prices and quadratic risk-minimizing hedge positions of call and put payoffs can be derived, and we refer the reader to [4, Prop.3.9.] for further details and the explicit formulas.

As a pricing and hedging model, we consider a simplified version of (6), which is driven by two (nonstandard) independent $\overline{\mathbb{Q}}$-Brownian motions $(B_{1,t})_{t>0}$ and $(B_{2,t})_{t>0}$ defined on $(\Omega, (\mathcal{F}_t)_{0\leq t\leq T})$ and we derive, again by moment matching, an analog approximate future price process $\widehat{F}^B$ of the form

$$\widehat{F}_t^B = F_0 \exp\left(-\int_0^t \Psi_1^B((\Sigma_1^B(s)) + \Psi_2^B((\Sigma_2^B(s)) \, ds + \int_0^t \Sigma_1^B(s) \, dB_{1,s} + \int_0^t \Sigma_2^B(s) \, dB_{2,s} \right) \tag{10}$$

with time-dependent, deterministic functions $\Sigma_1^B(t)$ and $\Sigma_2^B(t)$ and with $\Psi_1^B(z)$ and $\Psi_2^B(z)$ being the cumulant transforms of $B_{1,1}$ and $B_{2,1}$.

Although the model has two sources of randomness, it is a complete model under the filtration generated by the future price itself as the next proposition shows, which means that all practically relevant claims can be replicated.

Proposition 5 *Let* $(\mathcal{G}_t)_{t<T}$ *be the filtration generated by* $\widehat{F}^B$ *up to time* t, *i.e.,* $\mathcal{G}_t := \sigma\{\widehat{F}_s^B, s \leq t\}$. *Then the market consisting of* $\widehat{F}^B$ *and a constant riskless bank account is a complete financial market with respect to* $(\mathcal{G}_t)_{t<T}$.

Proof See [12].

We estimate the parameters for both models based on future and spot data from Nord Pool energy exchange. We use average daily system peak load electricity spot prices for the period from January 2011 until May 2013 (prices as shown on Bloomberg page "ENOSOSPK") and weekday prices for front month and second month future contracts. For details on the estimation procedure, we refer to [4, Sect. 5.2.]. In Table 1, we collect the parameter estimates for the two factors of both models, the simplified model with two nonstandard Brownian motions and the model with two independent NIG-Lévy processes. The estimates for the Brownian factor are only the drift term μ and the volatility term σ. The NIG distribution is a four-parameter distribution with scale parameter δ, tail heaviness α, skew parameter β, and the location parameter ν, see [3].

Figure 2 shows the empirical return distributions of both factors together with the density function of the estimated distribution. It is obvious that the NIG distribution provides a significantly better fit to the empirical returns than the normal distribution.

The claim to be hedged is an option on a future with a one-week delivery period trading one month prior to expiry, so that $T_1 = 23$ and $T_2 = 30$. Based on the parameter estimates, we determine scaling terms $\Sigma_1^L(t)$ and $\Sigma_2^L(t)$ for the dynamics of $\widehat{F}_t^L$ and scaling terms $\Sigma_1^B(t)$ and $\Sigma_2^B(t)$ for the dynamics of $\widehat{F}^B$, respectively. Assuming that the measures $\mathbb{Q}$ and $\overline{\mathbb{Q}}$ are orthogonal, we define an aggregating process $\widehat{F}$ such that $\widehat{F} = \widehat{F}^B$ $\overline{\mathbb{Q}}$-a.s. and $\widehat{F} = \widehat{F}_t^L$ $\mathbb{Q}$-a.s.. Pricing and hedging is performed under $\overline{\mathbb{Q}}$, and there is only one alternative measure, denoted by $\mathbb{Q}$. Our model set is thus $\mathcal{Q} = \{\overline{\mathbb{Q}}, \mathbb{Q}\}$. Applying the Akaike Information Criterion (AIC), we assign a probability distribution to the model set $\mathcal{Q}$. It turns out that model $\mathbb{Q}$ gets assigned a probability of basically 1 due to its much better fit of the returns and we simulate according to this model.

We consider an at-the-money call option $X := (\widehat{F}_{T_2} - F_0)^+$ and calculate the hedge positions implied by $\overline{\mathbb{Q}}$. For the simulation of the process under $\mathbb{Q}$, we use 600 time steps in order to reduce the discretization error. We investigate the distribution of $L_T^{\Delta}(X, \Phi, \Phi_{\mathbb{Q}})$ and $L_T(X, \Phi)$, with Φ and $\Phi_{\mathbb{Q}}$ dynamic hedging strategy as there are no benchmark instruments. As implied by Proposition 5, the hedging strategy is actually a perfect hedge under the model $\overline{\mathbb{Q}}$.

Figure 3 shows on the left-hand side the distributions under $\mathbb{Q}$ of $L_T(X, \Phi)$ and $L_T^{\Delta}(X, \Phi, \Phi_{\mathbb{Q}})$. To compare, Fig. 3 shows the distribution under $\mathbb{Q}$ of the hedge error $L_T(X, \Phi^{\mathbb{Q}})$ when hedging under $\mathbb{Q}$ (top right). Here, the hedge error is introduced by market incompleteness.

Table 1 Estimated parameters for the NIG distributions of $L_{1,t}$ and $L_{2,t}$ and parameters for the normal distributions of $B_{1,t}$ and $B_{2,t}$

	$\hat{\alpha}$	$\hat{\beta}$	$\hat{\nu}$	$\hat{\delta}$
$L_{2,t}$	1.9240	−0.8860	0.0176	0.0622
$L_{1,t}$	33.3008	−1.0988	−0.0009	0.0071
	$\hat{\sigma}$	$\hat{\mu}$		
$B_{2,t}$	0.2328	−0.0004		
$B_{1,t}$	0.0133	0.0002		

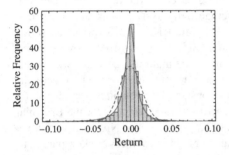 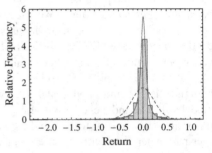

Fig. 2 Empirical distributions of long-term factor (*left*) and short-term factor (*right*) together with fitted NIG distribution (*solid line*) and normal distribution (*dashed line*)

It turns out that the loss due to the misspecified model $\overline{\mathbb{Q}}$ is minor compared to the loss due to the incompleteness. The loss due to model misspecification as measured by $L_T^\Delta(X, \Phi, \Phi_{\mathbb{Q}})$ has a mean-squared value of $\mu_{SQE,t}^{\overline{\mathbb{Q}},\Delta}(X) = \mathbb{E}^{\mathbb{Q}}[(L_T^\Delta(X, \Phi, \Phi_{\mathbb{Q}}))^2] = 9.50$. The mean-squared hedge error from hedging under the misspecified model is greater with $\mu_{SQE,t}^{\overline{\mathbb{Q}}}(X) = \mathbb{E}^{\mathbb{Q}}[(L_T(X, \Phi))^2] = 34.61$. Although the magnitude appears high, it is relativized by the fact that even under correct model specification the mean-squared hedge error $\mathbb{E}^{\mathbb{Q}}[(L_T(X, \Phi_{\mathbb{Q}}))^2]$ is 25.54. The initial prices under the two models are $\mathbb{E}^{\overline{\mathbb{Q}}}[X] = 10.954$ and $\mathbb{E}^{\mathbb{Q}}[X] = 8.068$, respectively. If we consider the variance of the loss variables, which corrects for the mean, it turns out that the impact from the misspecified hedge is rather low. For the variable $L_T^\Delta(X, \Phi, \Phi_{\mathbb{Q}})$, we get $\text{Var}(L_T^\Delta(X, \Phi, \Phi_{\mathbb{Q}})) = 1.07$. We find that $\text{Var}(L_T(X, \Phi))$ and $\mathbb{E}^{\mathbb{Q}}[(L_T(X, \Phi_{\mathbb{Q}}))^2] = \text{Var}(L_T(X, \Phi_{\mathbb{Q}}))$ are similar with 25.71 and 25.56, respectively. The lower right of Fig. 3 shows a scatter plot of $(L_T(X, \Phi)$ and $L_T(X, \Phi_{\mathbb{Q}})$. The two variables show a correlation of 97.91 %, implying a strong linear dependence between the hedge error under model $\mathbb{Q}$ (market risk) and the hedge error due to using the misspecified model $\overline{\mathbb{Q}}$.

The fact that the impact due to hedging in the wrong model is relatively low in this case study should not be misinterpreted. It confirms a stylized fact that is well known for diffusion processes (see [15]), namely that, hedging is robust, as long as the overall variance of the underlying is described sufficiently well by the model.

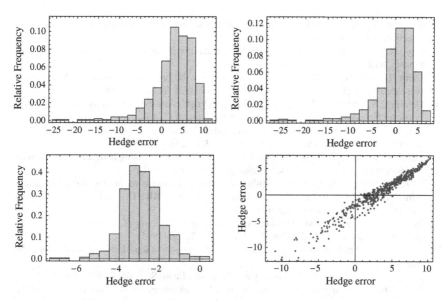

Fig. 3 *Upper left* $\mathbb{Q}(L_T(X, \Phi) < \cdot)$. *Upper right* $\mathbb{Q}(L_T(X, \Phi_{\mathbb{Q}}) < \cdot)$. *Lower left* $\mathbb{Q}(L_T^{\Delta}(X, \Phi, \Phi_{\mathbb{Q}}) < \cdot)$. *Lower right* Scatter plot of $L_T(X, \Phi)$ and $L_T(X, \Phi_{\mathbb{Q}})$

The overall volatility in our setup is the same for both models due to the moment matching procedure and uncertainty in this volatility is likely to result in greater model risk. The study makes also clear that the hedging error due to incompleteness cannot be neglected.

Open Access This chapter is distributed under the terms of the Creative Commons Attribution Noncommercial License, which permits any noncommercial use, distribution, and reproduction in any medium, provided the original author(s) and source are credited.

References

1. Artzner, P., Delbaen, F., Eber, J., Heath, D.: Coherent measures of risk. Math. Financ. **9**(3), 203–228 (1999)
2. Bannör, K., Scherer, M.: Capturing parameter uncertainty with convex risk measures. Eur. Actuar. J. **3**, 97–132 (2013)
3. Barndorff-Nielsen, O.E.: Processes of normal inverse Gaussian type. Financ. Stochast. **2**, 41–68 (1998)
4. Benth, F.E., Detering, N.: Pricing and hedging Asian-style options in energy. Financ. Stochast. (2013)
5. BIS: Revisions to the Basel II market risk framework. Basel committee on banking supervision, Bank for International Settlements, February (2011)
6. Burnham, K., Anderson, D.: Model Selection and Multimodel Inference: A Practical Information-Theoretic Approach, 2nd edn. Springer, New York (2002)

7. Burnham, K., Anderson, D.: Multimodel inference—understanding AIC and BIC in model selection. Sociol. Methods Res. **33**(2), 261–304 (2004)
8. Cont, R.: Model uncertainty and its impact on the pricing of derivative instruments. Math. Financ. **16**(3), 519–547 (2006)
9. Denis, L., Martini, C.: A theoretical framework for the pricing of contingent claims in the presence of model uncertainty. Ann. Appl. Probab. **16**(2), 827–852 (2006)
10. Detering, N., Packham, N.: Measuring the model risk of contingent claims. Working Paper, Frankfurt School of Finance & Management (submitted) (2013)
11. Detering, N., Weber, A., Wystup, U.: Return distributions of equity-linked retirement plans under jump and interest rate risk. Eur. Actuar. J. **3**(1), 203–228 (2013)
12. Detering, N.: Measuring the model risk of quadratic risk minimizing hedging strategies with an application to energy markets. Working paper, February (2014)
13. EBA: Discussion paper on draft regulatory technical standards on prudent valuation, under Article 100 of the draft Capital Requirements Regulation (CRR). Discussion Paper, European Banking Authority, November (2012)
14. El Karoui, N., Quenez, M.: Dynamic programming and pricing of contingent claims in an incomplete market. SIAM J. Control Optim. **33**(1), 29–66 (1995)
15. El Karoui, N., Jeanblanc-Picqué, M., Shreve, S.: Robustness of the Black and Scholes formula. Math. Financ. **8**(2), 93–126 (1998)
16. Epstein, L.: A definition of uncertainty aversion. Rev. Econ. Stud. **66**(3), 579–608 (1999)
17. Federal Reserve: Supervisory guidance on model risk management. Board of Governors of the Federal Reserve System, Office of the Comptroller of the Currency, SR Letter 11-7 Attachment, April (2011)
18. Föllmer, H., Schied, A.: Convex measures of risk and trading constraints. Financ. Stochast. **6**(4), 429–447 (2002)
19. Föllmer, H., Schweizer, M.: Hedging of contingent claims under incomplete information. In: Davis, M., Elliott, R. (eds.) Applied Stochastic Analysis, Stochastics Monographs, vol. 5, pp. 389–414. Gordon and Breach, London (1991)
20. Föllmer, H., Sondermann, D.: Hedging of non-redundant contingent claims. In: Hildenbrand, W., MasCollel, A. (eds.) Contributions to Mathematicla Economics, pp. 205–223. North-Holland, Amsterdam (1986)
21. Frittelli, M., Gianin, E.R.: Putting order in risk measures. J. Bank. Financ. **26**(7), 1473–1486 (2002)
22. Jeanblanc, M., Yor, M., Chesney, M.: Mathematical Methods for Financial Markets. Springer, Berlin (2009)
23. Knight, F.H.: Risk, Uncertainty and Profit. Houghton Mifflin, Boston (1921)
24. Melino, A., Turnbull, S.M.: Misspecification and the pricing and hedging of long-term foreign currency options. J. Int. Money Financ. **14**(3), 373–393 (1995)
25. Nalholm, M., Poulsen, R.: Static hedging and model risk for barrier options. J. Futures Mark. **26**(5), 449–463 (2006)
26. Schoutens, W., Simons, E., Tistaert, J.: A perfect calibration! now what? Wilmott Mag. **2**, 66–78 (2004)
27. Schwartz, E.S., Smith, J.E.: Short-term variations and long-term dynamics in commodity prices. Manag. Sci. **46**(7), 893–911 (2000)
28. Soner, M., Touzi, N., Zhang, J.: Quasi-sure stochastic analysis through aggregation. Electron. J. Probab. **16**, 1844–1879 (2011). Article number 67

Part II
Financial Engineering

Bid-Ask Spread for Exotic Options under Conic Finance

Florence Guillaume and Wim Schoutens

Abstract This paper puts the concepts of model and calibration risks into the perspective of bid and ask pricing and marketed cash-flows which originate from the conic finance theory. Different asset pricing models calibrated to liquidly traded derivatives by making use of various plausible calibration methodologies lead to different risk-neutral measures which can be seen as the test measures used to assess the (un)acceptability of risks.

Keywords Calibration risk · Model risk · Exotic bid-ask spread · Conic finance · Metric-free calibration risk measure

1 Introduction

The publication of the pioneering work of Black and Scholes in 1973 sparked off an unprecedented boom in the derivative market, paving the way for the use of financial models for pricing financial instruments and hedging financial positions. Since the late 1970s, incited by the emergence of a liquid market for plain-vanilla options, a multitude of option pricing models has seen the day, in an attempt to mimic the stylized facts of empirical returns and implied volatility surfaces. The need for such advanced pricing models, ranging from stochastic volatility models to models with jumps and many more, has even been intensified after Black Monday, which evidenced the inability of the classical Black–Scholes model to explain the intrinsic smiling nature of implied volatility. The following wide panoply of models has inescapably given rise to what is commonly referred to as model uncertainty or, by malapropism, model risk. The ambiguity in question is the *Knightian uncertainty* as defined by Knight [17], i.e., the uncertainty about the true process generating the data,

F. Guillaume (✉)
University of Antwerp, Middelheimlaan 1, 2020 Antwerpen, Belgium
e-mail: florence.guillaume@uantwerpen.be

W. Schoutens
K.U.Leuven, Celestijnenlaan 200 B, 3001 Leuven, Belgium
e-mail: Wim@Schoutens.be

© The Author(s) 2015

K. Glau et al. (eds.), *Innovations in Quantitative Risk Management*,
Springer Proceedings in Mathematics & Statistics 99,
DOI 10.1007/978-3-319-09114-3_4

as opposed to the notion of risk dealing with the uncertainty on the future scenario of a given stochastic process. This relatively new kind of "risk" has significantly increased this last decade due to the rapid growth of the derivative market and has led in some instances to colossal losses caused by the misvaluation of derivative instruments. Recently, the financial community has shown an accrued interest in the assessment of model and parameter uncertainty (see, for instance, Morini [19]). In particular, the Basel Committee on Banking Supervision [2] has issued a directive to compel financial institutions to take into account the uncertainty of the model valuation in the mark-to-model valuation of exotic products. Cont [6] set up the theoretical basis of a quantitative framework built upon coherent or convex risk measures and aimed at assessing model uncertainty by a worst-case approach.[1] Addressing the question from a more practical angle, Schoutens et al. [22] illustrated on real market data how models fitting the option surface equally well can lead to significantly different results once used to price exotic instruments or to hedge a financial position.

Another source of risk for the price of exotics originates from the choice of the procedure used to calibrate a specific model on the market reality. Indeed, although the standard approach consists of solving the so-called inverse problem, i.e., quoting Cont [7], of finding the parameters for which *the value of benchmark instruments, computed in the model, corresponds to their market prices*, alternative procedures have seen the day. The ability of the model to replicate the current market situation could rather be specified in terms of the distribution goodness of fit or in terms of moments of the asset log-returns as proposed by Eriksson et al. [9] and Guillaume and Schoutens [12]. In practice, even solving the inverse problem requires making a choice among several equally suitable alternatives. Indeed, matching perfectly the whole set of liquidly traded instruments is typically not plausible such that one looks for an "optimal" match, i.e., for the parameter set which replicates as well as possible the market price of a set of benchmark instruments. Put another way, we minimize the *distance* between the model and the market prices of those standard instruments. Hence, the calibration exercise first requires not only the definition of the concept of a distance and its metric but also the specification of the benchmark instruments. Benchmark instruments usually refer to liquidly traded instruments. In equity markets, it is a common practice to select liquid European vanilla options. But even with such a precise specification, several equally plausible selections can arise. We could for instance select out-of-the-money options with a positive bid price, following the methodology used by the Chicago Board Options Exchange (CBOE [4]) to compute the VIX volatility index, or select out-of-the-money options with a positive trading volume, or ... Besides, practitioners sometimes resort to time series or market quotes to fix some of the parameters beforehand, allowing for a greater stability of the calibrated parameters over time. In particular, the recent emergence of a liquid market for volatility derivatives has made this methodology possible to calibrate stochastic volatility models. Such an alternative has been investigated in Guillaume and Schoutens [11] under the Heston stochastic volatility model, where

---

[1] Another framework for risk management under Knightian uncertainty is based on the concept of g-expectations (see, for instance, Peng [20] and references therein).

the spot variance and the long-run variance are inferred from the spot value of the VIX volatility index and from the VIX option price surface, respectively. Another example is Brockhaus and Long [3] (see also Guillaume and Schoutens [13]) who propose to choose the spot variance, the long-run variance, and the mean reverting rate of the Heston stochastic volatility model in order to replicate as well as possible the term structure of model-free variance swap prices, i.e., of the return expected future total variance. Regarding the specification of the distance metric, several alternatives can be found in the literature. The discrepancy could be defined as relative, absolute, or in the least-square sense differences and expressed in terms of price or implied volatility. Detlefsen and Härdle [8] introduced the concept of calibration risk (or should we say *calibration uncertainty*) arising from the different (plausible) specifications of the objective function we want to minimize. Later, Guillaume and Schoutens [10] and Guillaume and Schoutens [11] extended the concept of calibration risk to include not only the choice of the functional but also the calibration methodology and illustrated it under the Heston stochastic volatility model.

In order to measure the impact of model or parameter ambiguity on the price of structured products, several alternatives have been proposed in the financial litera- ture. Cont [6] proposed the so-called worst-case approach where the impact of model uncertainty on the value of a claim is measured by the difference between the supre- mum and infimum of the expected claim price over all pricing models consistent with the market quote of a set of benchmark instruments (see also Hamida and Cont [16]). Gupta and Reisinger [14] adopted a Bayesian approach allowing for a distribution of exotic prices resulting directly from the posterior distribution of the parameter set obtained by updating a plausible prior distribution using a set of liquidly traded instru- ments (see also Gupta et al. [15]). Another methodology allowing for a distribution of exotic prices, but based on risk-capturing functionals has recently been proposed by Bannör and Scherer [1]. This method differs from the Bayesian approach since the distribution of the parameter set is constructed explicitly by allocating a higher proba- bility to parameter sets leading to a lower discrepancy between the model and market prices of a set of benchmark instruments. Whereas the Bayesian approach requires a parametric family of models and is consequently appropriate to assess parameter uncertainty, the two alternative proxies (i.e., the worst-case and the risk-capturing functionals approaches) can be considered to quantify the ambiguity resulting from a broader set of models with different intrinsic characteristics. These three approaches share the characteristic that the plausibility of any pricing measure $\mathcal{Q}$ is assessed by considering the average distance between the model and market prices, either by allocating a probability weight to each measure $\mathcal{Q}$ which is proportional to this distance or by selecting the measures $\mathcal{Q}$ for which the distance falls within the aver- age bid-ask spread. Hence, the resulting measure of uncertainty implicitly depends on the metric chosen to express this average distance. We will adopt a somewhat different methodology, although similar to the ones above-mentioned. We start from a set of plausible calibration procedures and we consider the resulting risk-neutral probability measures (i.e., the optimal parameter sets) as the test measures used to assess the (un)acceptability of any zero cost cash-flow X. In other words, these pric- ing measures can be seen as the ones defining the cone of acceptable cash-flows;

where X is acceptable or marketed, denoted by $X \in \mathscr{A}$, if its expectation under any of the test measures $\mathscr{Q}$ is nonnegative:

$$X \in \mathscr{A} \Leftrightarrow E_{\mathscr{Q}}[X] \geq 0 \; \forall \mathscr{Q} \in \mathscr{M}.$$

This allows us to define the cone of marketed cash-flows in a market-consistent way rather than parametrically in terms of some family of concave distortion functions as proposed by Cherny and Madan [5]. We can even play with the minimum proportion p of model prices included within their bid-ask spread in order to change the amplitude of the cone of acceptability by requiring that at least $\lceil pM \rceil$ model prices are within their market spread for $\mathscr{Q}$ to be included in the set of test measures $\mathscr{M}$:

$$\mathscr{Q} \in \mathscr{M} \Leftrightarrow \# \left\{ \widehat{P}_i^{\mathscr{Q}} \in [b_i, a_i], i = 1, \ldots, M \right\} \geq \lceil pM \rceil,$$

where $\widehat{P}_i^{\mathscr{Q}}$, a_i, b_i, $i = 1, \ldots, M$ denote the model price under the pricing measure $\mathscr{Q}$, the quoted ask price, and the quoted bid price of the M benchmark instruments, respectively. The higher the proportion, the smaller the set of test measures $\mathscr{M}$ and hence, the wider the cone of acceptability. We opt for a threshold expressed as a percentage rather than as an average distance since we want our specification to be free of any distance metric. Indeed, the set $\mathscr{M}$ will be built by considering different objective functions (expressed as price or implied volatility differences, as absolute, relative, or in the least-square sense differences, ...) such that we do not want to favor any of these metrics, to the detriment of the others. The impact of model or parameter uncertainty on the price of exotic (i.e., illiquid) instruments is then assessed by adopting a worst-case approach as in Cont [6]:

$$s(p) = \max_{\mathscr{Q} \in \mathscr{M}} \left\{ EP^{\mathscr{Q}} \right\} - \min_{\mathscr{Q} \in \mathscr{M}} \left\{ EP^{\mathscr{Q}} \right\}, \tag{1}$$

provided that $\mathscr{M} \neq \emptyset$; where $EP^{\mathscr{Q}}$ denotes the exotic price under the pricing measure $\mathscr{Q}$. The model uncertainty can thus be quantified by the bid-ask spread of illiquid products. Indeed, the cash-flow of selling a claim with payoff X at time T at its ask price is acceptable for the market if $E_{\mathscr{Q}}[a - \exp(-rT)X] \geq 0, \forall \mathscr{Q} \in \mathscr{M}$, i.e., if $a \geq \exp(-rT) \max_{\mathscr{Q} \in \mathscr{M}} \{ E_{\mathscr{Q}}[X] \}$. For the sake of competitiveness, the ask price is set at the minimum value, i.e.,

$$a = \exp(-rT) \max_{\mathscr{Q} \in \mathscr{M}} \{ E_{\mathscr{Q}}[X] \}.$$

Similarly, the cash-flow of buying a claim with payoff X at time T at its bid price is acceptable for the market if $E_{\mathscr{Q}}[-b + \exp(-rT)X] \geq 0, \forall \mathscr{Q} \in \mathscr{M}$, i.e., taking the maximum possible value for competitiveness reasons

$$b = \exp(-rT) \min_{\mathscr{Q} \in \mathscr{M}} \{ E_{\mathscr{Q}}[X] \}.$$

The impact of model uncertainty can be expressed as a function of the severity of the percentage threshold p. We note that decreasing the threshold ultimately boils down to considering a thinner set of benchmark instruments since the model price has to fall within the market bid-ask spread for a smaller number of calibration instruments in order for a pricing measure to be selected. In particular, such a relaxation typically results in the "elimination" of the most illiquid calibration instruments, i.e., deep out-of-the-money options in the case of equity markets (see Fig. 2).

For the numerical study, we consider the Variance Gamma (VG) model of Madan et al. [18] only, although the methodology can be equivalently used to assess calibration or/and model uncertainty. The calibration instrument set consists of liquid out-of-the-money options: moving away from the forward price, we select put and call options with a positive bid price and with a strike lower and higher than the forward price, respectively, and this until we encounter two successive options with zero bid. Denoting by $P_i = \frac{a_i+b_i}{2}$ the mid-price of option i and by σ_i its implied volatility, the set of measures $\mathscr{M}$ results from the following specifications for the objective function we minimize (i.e., for the distance and its metric):

1. Root-mean square error (RMSE)

   a. price specification

$$\text{RMSE} = \sqrt{\sum_{i=1}^{M} \omega_i \left(P_i - \widehat{P_i}\right)^2}$$

   b. implied volatility specification

$$\text{RMSE}^\sigma = \sqrt{\sum_{i=1}^{M} \omega_i \left(\sigma_i - \widehat{\sigma_i}\right)^2}$$

2. Average relative percentage error (ARPE)

   a. price specification

$$\text{ARPE} = \sum_{i=1}^{M} \omega_i \frac{\left|P_i - \widehat{P_i}\right|}{P_i}$$

   b. implied volatility specification

$$\text{ARPE}^\sigma = \sum_{i=1}^{M} \omega_i \frac{\left|\sigma_i - \widehat{\sigma_i}\right|}{\sigma_i}$$

3. Average absolute error (APE)

   a. price specification

$$\text{APE} = \frac{1}{\bar{P}} \sum_{i=1}^{M} \omega_i \left| P_i - \widehat{P}_i \right|$$

b. implied volatility specification

$$\text{APE}^\sigma = \frac{1}{\bar{\sigma}} \sum_{i=1}^{M} \omega_i \left| \sigma_i - \widehat{\sigma}_i \right|,$$

where $\bar{P}$ and $\bar{\sigma}$ denote the average option price and the average implied volatility, respectively.

Each of these six objective functions can again be subdivided into an unweighted functional for which the weight $\omega_i = \omega = \frac{1}{M}$ $\forall i$ and a weighted functional for which the weight ω_i is proportional to the trading volume of option i. We furthermore consider the possibility of adding an extra penalty term to the objective function in order to force the model prices to lie within their market bid-ask spread. Besides these standard specifications (in terms of the price or the implied volatility of the calibration instruments), we consider the so-called moment matching market implied calibration proposed by Guillaume and Schoutens [12] and which consists in matching the moments of the asset log-return which are inferred from the implied volatility surface. As the VG model is fully characterized by three parameters, we consider three standardized moments, namely the variance, the skewness, and the kurtosis. Since as shown by Guillaume and Schoutens [12], the variance can always be perfectly matched, we either allocate the same weight to the matching of the skewness and the kurtosis or we match uppermost the lower moment, i.e., the skewness. This leads to a total of 26 plausible calibration procedures, each of them leading to a test measure $\mathcal{Q} \in \mathcal{M}$ provided that the proportion of model prices falling within their market bid-ask spread is at least equal to the threshold p.

2 Exotic Bid-Ask Spread

For the numerical study, we consider daily S&P 500 option surfaces for a timespan ranging from October 2008 to October 2009, including ,therefore, the recent credit crunch[2]. We calibrate the VG model daily on the quoted (liquid) maturity which is the closest to the reference maturity of three months. Note that we only consider maturities for which the total trading volume of out-of-the-money options exceeds 1,000 contracts which allows to avoid the extreme situation of an undetermined calibration problem where the number of parameters to calibrate is higher than the number of benchmark instruments. This also ensures that the number of option prices is large enough (and so the strike range wide and refined enough) to guarantee a sufficient precision for the derived market implied moments. For each of the trading days

---

[2] The data are taken from the KU Leuven data collection which is a private collection of historical daily spot and option prices of major US equity stocks and indices.

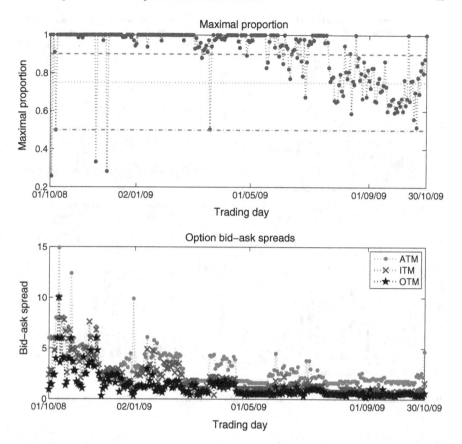

Fig. 1 Maximum proportion π of option prices replicated within their bid-ask spread (*upper*) and option bid-ask spreads (*below*)

included in the sample period, we successively perform the 26 calibration methodologies, which leads to 26 optimal parameter sets. We then select those for which the proportion of model prices falling within their market bid-ask spread is at least p. The higher the threshold p, the fewer the test measures $\mathcal{Q} \in \mathcal{M}$ and hence, the thinner the exotic bid-ask spreads. Figure 1 shows the highest proportion π of option prices replicated within their bid-ask spread for the 26 above-mentioned calibration procedures:

$$\pi = \frac{1}{M} \max_{\mathcal{Q}} \# \left\{ \widehat{P}_i^{\mathcal{Q}} \in [b_i, a_i], i = 1, \ldots, M \right\}.$$

If $\pi < p$, then $\mathcal{M}$ is an empty set and there does not exist exotic spread for that particular threshold p as defined by (1). Hence, when selecting the proportion threshold p, we should keep in mind the trade-off between the in-spread precision and the number

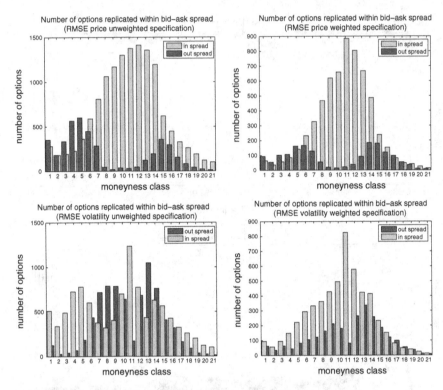

Fig. 2 Number of options for which the model price falls within the quoted bid-ask spread

of test measures. Indeed, the higher the proportion, the higher the precision but the fewer the measures selected as test measure, which can in turn lead to an underestimation of the calibration uncertainty measured as the exotic bid-ask spreads. From Fig. 1, we observe that π is significantly higher during the heart of the recent credit crunch, i.e., from the beginning of the sample period until mid 2009. This can easily be explained by the typically wider bid-ask spreads observed during market distress periods. Indeed, as shown on the lower panel of Fig. 1, the quoted spread for at-the-money, in-the-money ($K = 0.75\ S_0$), and out-of-the-money ($K = 1.25\ S_0$) options has significantly shrunk after the troubled period of October 2008–July 2009.

Figure 2 shows the number of vanilla options whose model price falls within the quoted bid-ask spread as a function of the option moneyness for four of the calibration procedures under investigation, namely the weighted and unweighted RMSE price and implied volatility specifications without penalty term. To assess the impact of moneyness on the model ability to replicate option prices within their bid-ask spread, we split the strike range into 21 classes: $\frac{K}{S_0} < 0.5$, $0.5 \leq \frac{K}{S_0} < 0.55$, $0.55 \leq \frac{K}{S_0} < 0.6, \ldots, 1.45 \leq \frac{K}{S_0} < 1.5$, and $\frac{K}{S_0} > 1.5$. We clearly see that, at least for the price specifications, option prices falling outside their quoted bid-ask spread are mainly observed for deep out-of-the-money calls and puts. This trend is even more marked and present in the implied volatility specifications when we add a penalty term in the objective function to constraint the model price within the market spread. Hence, increasing the proportion threshold p mainly boils down to limit the set of calibration instruments to close to the money vanilla options.

In order to illustrate the impact of parameter uncertainty on the bid-ask spread of exotics, we consider the following path dependent options (with a maturity of $T = 3$ months):

1. **Asian option**

   The payoff of Asian options depends on the arithmetic average of the stock price from the emission to the maturity date of the option. The fair price of the Asian call and put options with maturity T is given by

   $$AC = \exp(-rT)E_{\mathscr{Q}}[(\underset{0 \le t \le T}{\text{mean }} S_t - K)^+] \quad AP = \exp(-rT)E_{\mathscr{Q}}[(K - \underset{0 \le t \le T}{\text{mean }} S_t)^+].$$

2. **Lookback call option**

   The payoff of lookback call and put options corresponds to the call and put vanilla payoff where the strike is taken equal to the lowest and highest levels the stock has reached during the option lifetime, respectively. The fair price of the lookback call and put with maturity T is given by

   $$LC = \exp(-rT)E_{\mathscr{Q}}\left[(S_T - m_T^S)^+\right] \quad LP = \exp(-rT)E_{\mathscr{Q}}\left[(M_T^S - S_T)^+\right],$$

   respectively, where m_t^X and M_X^t denote the minimum and maximum processes of the process $X = \{X_t, 0 \le t \le T\}$, respectively:

   $$m_t^X = \inf\{X_s, 0 \le s \le t\} \quad M_t^X = \sup\{X_s, 0 \le s \le t\}.$$

3. **Barrier call option**

   The payoff of a one-touch barrier option depends on whether the underlying stock price reaches the barrier H during the lifetime of the option. We illustrate the findings by looking at the up-and-in call and the down-and-in put price:

   $$UIBC = \exp(-rT)E_{\mathscr{Q}}\left[(S_T - K)^+ \mathbf{1}\left(M_T^S \ge H\right)\right]$$

   $$DIBP = \exp(-rT)E_{\mathscr{Q}}\left[(K - S_T)^+ \mathbf{1}\left(m_T^S \le H\right)\right].$$

4. **Cliquet option**

   The payoff of a cliquet option depends on the sum of the stock returns over a series of consecutive time periods; each local performance being first floored and/or capped. Moreover, the final sum is usually further floored and/or capped to guarantee a minimum and/or maximum overall payoff such that cliquet options protect investors against downside risk while allowing them for significant upside potential. The Cliquet we consider has a fair price given by

$$\text{Cliquet} = \exp(-rT)E_{\mathscr{Q}}\left[\max\left(0, \sum_{i=1}^{N}\min\left(\text{cap}, \max\left(\text{floor}, \frac{S_{t_i} - S_{t_{i-1}}}{S_{t_{i-1}}}\right)\right)\right)\right].$$

For sake of comparison, we also price a 3 months at-the-money call option. Note that this option does not generally belong to the set of benchmark instruments since, most of the time, we can not observe a market quote for the option with the exact same maturity and moneyness.

The path dependent nature of exotic options requires the use of the Monte Carlo procedure to simulate sample paths of the underlying index. The stock price process

$$S_t = \frac{S_0 \exp((r - q)t + X_t)}{E_{\mathscr{Q}}[\exp(X_t)]}, \quad X \sim VG(\sigma, \nu, \theta)$$

is discretized by using a first order Euler scheme (for more details on the simulation, see Schoutens [21]). The (standard) Monte Carlo simulation is performed by considering one million scenarios and 252 trading days a year.

The bid and ask prices and the relative bid-ask spread (dollar bid-ask spread expressed as a proportion of the mid-price) of different exotic options are shown on Figs. 3 and 4, respectively, and this for a proportion threshold p equal to 0.5, 0.75, and 0.9. For sake of comparison, Fig. 5 shows the same results but for the 3 months at-the-money call option. The figures clearly indicate that the impact of parameter uncertainty is much more marked for path-dependent derivatives than for (non-quoted) vanilla options. Indeed, the relative bid-ask spread is of a magnitude order at least 10 times higher for the Asian call, lookback call, barrier call, and cliquet than for the vanilla call option. Besides, we observe that a far above average call relative spread does not necessarily imply a far above average percentage spread for path dependent options. In order to assess the consistency of our findings, we have reproduced the Monte Carlo simulation 400 times for one fixed quoting day (namely October, 1, 2008) with different sets of sample paths and computed the option relative spreads for each simulation. Figure 6 shows the resultant histogram for each relative spread and clearly brings out the consistency of the results: the relative spread is far more significant for the exotic options than for the vanilla options whatever the set of sample paths considered. The consistency of the Monte Carlo study is besides guaranteed by the fact that we used the same set of sample paths to price each option. Table 1 which shows the average price, standard deviation, and relative spread (across the 400 Monte Carlo simulations) for the price weighted RMSE functional confirms that the exotic bid-ask spreads are due to the nature of the exotic options rather than to the intrinsic uncertainty of Monte Carlo simulations. Indeed, the Monte Carlo relative spread given in Table 1 is significantly smaller than the option spread depicted on Fig. 6, and this for each exotic option. Table 2 shows the average of the relative spread over the whole period under investigation, and this for the different options under consideration. We clearly observe that the threshold p

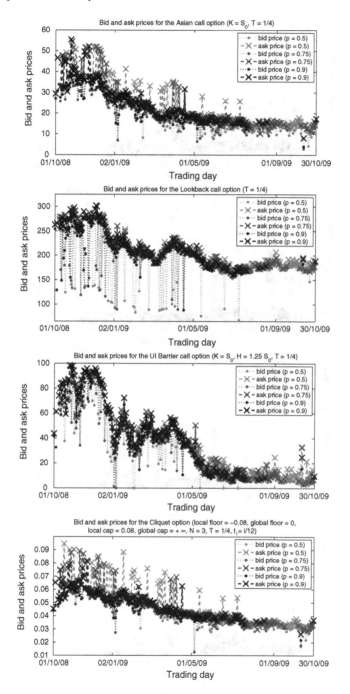

Fig. 3 Evolution of exotic bid and ask prices through time

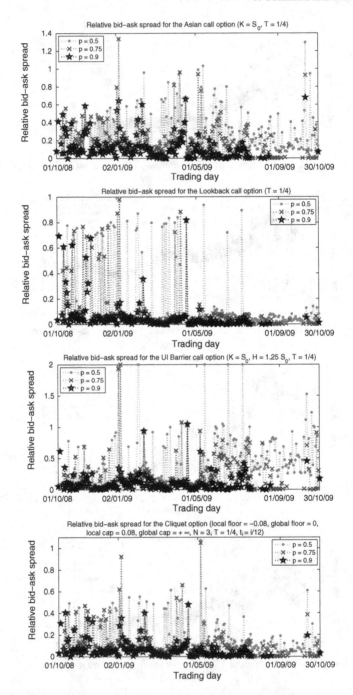

Fig. 4 Evolution of exotic relative bid–ask spreads (in absolute value) through time

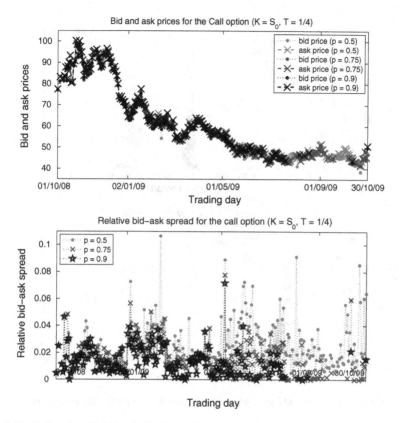

Fig. 5 Evolution of vanilla bid and ask prices and relative bid-ask spread (in absolute value) through time

impacts more severely the spread of the path-dependent options. Indeed, decreasing p leads to a sharper increase of the relative bid-ask spread for the exotic options than for the European call and put options. Besides, the calibration risk is predominant for the up-and-in barrier call option and, to a smaller extent, for the Asian options. Table 3 shows the 95 % quantile of relative bid-ask spreads. We clearly see that in terms of extreme events, the more risky options are the up-and-in barrier call option and the lookback options. By way of conclusion, our findings clearly illustrate the impact of the calibration methodology on the price of exotic options, suggesting that risk managers should take into account calibration uncertainty when assessing the safety margin.

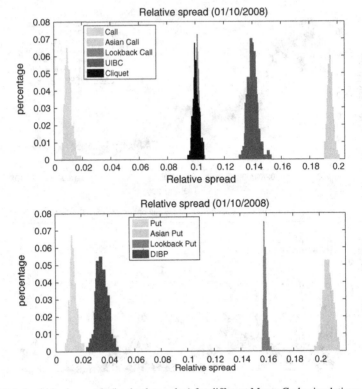

Fig. 6 Relative bid-ask spreads (in absolute value) for different Monte Carlo simulations

Table 1 Monte Carlo precision

	Call	Put	Asian call	Asian put	Lookback call	Lookback put	UIBC	DIBP	Cliquet
Mean	77.201	73.647	29.270	27.487	254.70	283.32	43.690	57.681	0.0449
Std	0.1000	0.1202	0.0345	0.0552	0.1505	0.1856	0.0922	0.1225	5E–05
Rel. spread[a]	0.0078	0.0092	0.0073	0.0118	0.0034	0.0035	0.0131	0.0120	0.0066

[a] The Monte Carlo relative spread is defined as the maximum minus the minimum price divided by the average price across the 400 Monte Carlo simulations

Table 2 Average relative bid-ask spreads (in %)

p	Call	Put	Asian call	Asian put	Lookback call	Lookback put	UIBC	DIBP	Cliquet
0.5	2.59	2.53	29.82	27.41	17.89	24.97	43.80	7.20	17.26
0.75	1.66	1.72	19.81	18.68	11.09	16.78	22.11	3.75	11.50
0.9	1.37	1.46	12.18	11.77	5.97	9.75	10.31	2.44	6.55

Table 3 95 % quantile of relative bid-ask spreads (in %)

p	Call	Put	Asian call	Asian put	Lookback call	Lookback put	UIBC	DIBP	Cliquet
0.5	5.75	5.24	77.95	67.56	79.19	90.87	102.32	28.41	49.75
0.75	3.89	3.67	51.99	51.98	68.04	81.13	72.11	12.88	40.64
0.9	3.15	3.26	40.46	40.44	27.28	43.58	33.03	5.06	24.36

3 Conclusion

This paper sets the theoretical foundation of a new framework aimed at assessing the impact of calibration uncertainty. The main advantage of the proposed methodology resides in its metric-free nature since the selection of test measures does not depend on any specified distance. Besides, the paper links the concept of uncertainty and the recently developed conic finance theory by defining the test measures used to construct the cone of acceptable cash-flows as the pricing measures resulting from any plausible calibration methodology such that model and parameter uncertainties are naturally measured as bid-ask spreads. The numerical study has highlighted the significant impact of parameter uncertainty for a wide range of path-dependent options under the popular VG model.

Open Access This chapter is distributed under the terms of the Creative Commons Attribution Noncommercial License, which permits any noncommercial use, distribution, and reproduction in any medium, provided the original author(s) and source are credited.

References

1. Bannör, K.F., Scherer, M.: Capturing parameter risk with convex risk measures. Eur. Actuar. J. **3**, 97–132 (2013)
2. Basel Committee on Banking Supervision: Revisions to the Basel II market risk framework. Technical report, Bank for International Settlements (2009)
3. Brockhaus, O., Long, D.: Volatility swaps made simple. Risk **13**(1), 92–95 (2000)
4. CBOE. VIX: CBOE volatility index. Technical report, Chicago (2003)
5. Cherny, A., Madan, D.B.: Markets as a counterparty: an introduction to conic finance. Int. J. Theor. Appl. Financ. **13**, 1149–1177 (2010)
6. Cont, R.: Model uncertainty and its impact on the pricing of derivative instruments. Math. Financ. **16**, 519–547 (2006)
7. Cont, R.: Model calibration. Encyclopedia of Quantitative Finance, pp. 1210–1219. Wiley, Chichester (2010)
8. Detlefsen, K., Härdle, W.K.: Calibration risk for exotic options. J. Deriv. **14**, 47–63 (2007)
9. Eriksson, A., Ghysels, E., Wang, F.: The normal inverse gaussian distribution and the pricing of derivatives. J. Deriv. **16**, 23–37 (2009)
10. Guillaume, F., Schoutens, W.: Use a reduce Heston or reduce the use of Heston? Wilmott J. **2**, 171–192 (2010)
11. Guillaume, F., Schoutens, W.: Calibration risk: illustrating the impact of calibration risk under the heston model. Rev. Deriv. Res. **15**, 57–79 (2012)

12. Guillaume, F., Schoutens, W.: A moment matching market implied calibration. Quant. Financ. **13**, 1359–1373 (2013)
13. Guillaume, F., Schoutens, W.: Heston model: the variance swap calibration. J. Optim. Theory Appl. (2013) (to appear)
14. Gupta, A., Reisinger, C.: Robust calibration of financial models using bayesian estimators. J. Comput. Financ. (2013) (to appear)
15. Gupta, A., Reisinger, C., Whitley, A.: Model uncertainty and its impact on derivative pricing. In: Böcker, K. (ed.) Rethinking Risk Measurement and Reporting, pp. 625–663. Nick Carver (2010)
16. Hamida, S.B., Cont, R.: Recovering volatility from option prices by evolutionary optimization. J. Comput. Financ. **8**, 43–76 (2005)
17. Knight, F.: Risk, Uncertainty and Profit. Houghton Mifflin Co., Boston (1920)
18. Madan, D.B., Carr, P., Chang, E.: The variance gamma process and option pricing. Eur. Financ. Rev. **2**, 79–105 (1998)
19. Morini, M.: Understanding and Managing Model Risk. A Practical Guide for Quants, Traders and Validators. Wiley, New York (2011)
20. Peng, S.: Nonlinear expectation theory and stochastic calculus under knightian uncertainty. In: Bensoussan, A., Peng, S., Sung, J. (eds.) Real Options, Ambiguity, Risk and Insurance, pp. 144–184. IOS Press BV, Amsterdam (2013)
21. Schoutens, W.: Lévy Processes in Finance: Pricing Financial Derivatives. Wiley, New York (2003)
22. Schoutens, W., Simons, E., Tistaert, J.: A perfect calibration! now what? Wilmott Mag. **3**, 66–78 (2004)

Derivative Pricing under the Possibility of Long Memory in the supOU Stochastic Volatility Model

Robert Stelzer and Jovana Zavišin

Abstract We consider the supOU stochastic volatility model which is able to exhibit long-range dependence. For this model, we give conditions for the discounted stock price to be a martingale, calculate the characteristic function, give a strip where it is analytic, and discuss the use of Fourier pricing techniques. Finally, we present a concrete specification with polynomially decaying autocorrelations and calibrate it to observed market prices of plain vanilla options.

Keywords Calibration · Fourier pricing · Lévy basis · Long memory · Superposition of Ornstein–Uhlenbeck-type processes · Stochastic volatility

AMS Subject Classification 2010 Primary: 91G20, 60G51 · Secondary: 91B25

1 Introduction

The Ornstein–Uhlenbeck (OU)-type stochastic volatility (SV) model introduced in [3] is one of the most popular stochastic volatility models for prices of financial assets driven by a Lévy process (see, e.g., [11, 25]). It covers many of the stylized facts typically encountered in financial data (cf. [10, 14]). Over the years many variants have been introduced, for instance a variant with two sided jumps in [1] or a multivariate extension in [21].

The views expressed herein are those of the authors and do not necessarily reflect the views of the institutions mentioned below.

R. Stelzer (✉)
Institute of Mathematical Finance, Ulm University, Helmholtzstr. 18,
89081 Ulm, Germany
e-mail: robert.stelzer@uni-ulm.de

J. Zavišin
risklab GmbH/Allianz Global Investors, Seidlstr. 24-24a,
80335 Munich, Germany
e-mail: jovana.zavisin@risklab.com

© The Author(s) 2015

K. Glau et al. (eds.), *Innovations in Quantitative Risk Management*,
Springer Proceedings in Mathematics & Statistics 99,
DOI 10.1007/978-3-319-09114-3_5

In this paper, we consider a variant of the model which additionally can cover the stylized fact of long-range dependence (or slower than exponentially decaying autocorrelations), the supOU stochastic volatility model. In this model, we specify the volatility as a superposition of Ornstein–Uhlenbeck (thus "supOU") processes, which have been introduced in [2]. Various features of this volatility model (in a multidimensional setting) have been considered in [4, 5, 18, 26].

Typically long-range dependence is obtained by using fractional Brownian motion or fractional Lévy processes as the driving noises, see, e.g., [6, 7] for a critical discussion of such models for financial markets. In such models one cannot have jumps, as fractional Lévy processes (cf. [16]) have continuous paths, and one is bound to have long memory. In our supOU model, one has a natural extension of the OU-type model that exhibits jumps and, depending on the parameters, can exhibit short or long memory. However, our model shares one disadvantage with fractional process based models, viz. that it is no longer Markovian. In this context, one should bear in mind that most Markov processes one employs to model volatilities are geometrically ergodic and thus cannot exhibit long memory, although there exists also Markov process with polynomial mixing coefficients and even long memory (see, e.g., [27]).

The focus of the present paper is on derivative pricing in and calibration of the univariate supOU SV model similar to the papers [19, 20] in the (multivariate) OU-type SV model. To this end, we first briefly review the model in Sect. 2. In Sect. 3, we give conditions on the parameters such that the discounted stock price process is a martingale which implies that under these conditions the model can be used to describe the risk neutral dynamics of a financial asset. Thereafter, we start Sect. 4 with a review of Fourier pricing. Then, we give the characteristic function of the log asset price in the supOU SV model and show conditions for the moment generating function to be sufficiently regular so that Fourier pricing is applicable. Finally, we present a concrete specification, the Γ-supOU SV model, in Sect. 5 and discuss its calibration to market data which we illustrate with a small example using options on the DAX. Finally, we discuss a subtle issue regarding how to employ the calibrated model to calculate prices of European options with a general maturity.

2 A Review of the supOU Stochastic Volatility Model

We briefly review the definition and the most important known facts of the supOU stochastic volatility model introduced in [5]. More background on supOU processes can be found in [2, 4, 13, 26].

In the following, $\mathbb{R}_-$ denotes the set of negative real numbers and $\mathscr{B}_b(\mathbb{R}_- \times \mathbb{R})$ denotes the bounded Borel sets of $\mathbb{R}_- \times \mathbb{R}$.

Definition 2.1 A family $\Lambda = \{\Lambda(B) : B \in \mathscr{B}_b(\mathbb{R}_- \times \mathbb{R})\}$ of real-valued random variables is called a real-valued Lévy basis (infinitely divisible independently scattered random measure) on $\mathbb{R}_- \times \mathbb{R}$ if:

- the distribution of $\Lambda(B)$ is infinitely divisible for all $B \in \mathscr{B}_b(\mathbb{R}_- \times \mathbb{R})$,
- for any $n \in \mathbb{N}$ and pairwise disjoint sets $B_1, \ldots, B_n \in \mathscr{B}_b(\mathbb{R}_- \times \mathbb{R})$ the random variables $\Lambda(B_1), \ldots, \Lambda(B_n)$ are independent,
- for any sequence of pairwise disjoint sets $B_n \in \mathscr{B}_b(\mathbb{R}_- \times \mathbb{R})$ with $n \in \mathbb{N}$ satisfying $\cup_{n \in \mathbb{N}} B_n \in \mathscr{B}_b(\mathbb{R}_- \times \mathbb{R})$ the series $\sum_{n=1}^\infty \Lambda(B_n)$ converges a.s. and $\Lambda(\cup_{n \in \mathbb{N}} B_n) = \sum_{n=1}^\infty \Lambda(B_n)$.

We consider only Lévy bases with characteristic functions of the form

$$\mathbb{E}(\exp(iu\Lambda(B))) = \exp(\varphi(u)\Pi(B))$$

for all $u \in \mathbb{R}$ and all $B \in \mathscr{B}_b(\mathbb{R}_- \times \mathbb{R})$, where $\Pi = \pi \times \lambda$ is the product of a probability measure π on $\mathbb{R}_-$ and the Lebesgue measure λ on $\mathbb{R}$ and

$$\varphi(u) = iu\gamma_0 + \int_{\mathbb{R}_+} \left(e^{iux} - 1 \right) v(dx)$$

is the cumulant transform of an infinitely divisible distribution on $\mathbb{R}_+$ with Lévy-Khintchine triplet $(\gamma_0, 0, v)$, which is also the characteristic triplet of the underlying Lévy process $L_t = \Lambda(\mathbb{R}_- \times (0, t])$ and $L_{-t} = \Lambda(\mathbb{R}_- \times (-t, 0))$ for $t \in \mathbb{R}_+$ (see, e.g., [24] for the relevant background on infinitely divisible distributions and Lévy processes). We call the triplet (γ_0, v, π) the *generating triplet*. Note that this means that $\gamma_0 \geq 0$, $v(\mathbb{R}\backslash\mathbb{R}_+) = 0$, and $\int_{|x| \leq 1} |x|v(dx) < \infty$.

If L is a pure jump Lévy process with triplet $(0, 0, v)$ and jump measure $N(ds, dx)$, then turning the Poisson point process of jumps in $\mathbb{R} \times \mathbb{R}_+ \backslash \{0\}$ to one in $\mathbb{R} \times \mathbb{R}_+ \backslash \{0\} \times \mathbb{R}_-$ by marking all jumps with independent marks distributed according to π produces the jump measure of a Lévy basis with triplet (γ_0, v, π).

In the supOU process defined now, this can be understood as assigning every jump of a Lévy process an individual exponential decay rate. We restrict our attention to positive supOU processes as this is natural when using them to model a variance changing over time.

Theorem 2.2 ([2, 4, 13]) *Let Λ be an $\mathbb{R}_+$-valued Lévy basis on $\mathbb{R}_- \times \mathbb{R}$ with generating triplet (γ_0, v, π). Assume*

$$\int_{|x|>1} \ln(|x|)v(dx) < \infty, \quad and \quad -\int_{\mathbb{R}_-} \frac{1}{A}\pi(dA) < \infty.$$

Then the process $\Sigma = (\Sigma_t)_{t \in \mathbb{R}}$ given by

$$\Sigma_t = \int_{\mathbb{R}_-} \int_{-\infty}^t e^{A(t-s)} \Lambda(dA, ds)$$

is well defined as a Lebesgue integral for all $t \in \mathbb{R}$ and it is stationary.

Moreover, $\Sigma_t \geq 0$ for all $t \in \mathbb{R}$ and the distribution of Σ_t is infinitely divisible with characteristic function given by $\mathbb{E}\left(e^{iu\Sigma_t}\right) = e^{iu\gamma_{\Sigma,0} + \int_{\mathbb{R}_+}\left(e^{iux}-1\right)v_\Sigma(dx)}$, *for all $u \in \mathbb{R}$ where*

$$\gamma_{\Sigma,0} = \int_{\mathbb{R}_-}\int_0^\infty e^{As}\gamma_0 ds\pi(dA), \, v_\Sigma(B) = \int_{\mathbb{R}_-}\int_0^\infty\int_{\mathbb{R}_+}\mathbf{1}_B\left(e^{As}x\right)v(dx)ds\pi(dA),$$

for all $B \in \mathscr{B}(\mathbb{R})$.

As shown in [4, Theorem 3.12] the supOU process is adapted to the filtration generated by Λ and has locally bounded paths. Provided π has a finite first moment, one can take a supOU process to have càdlàg paths.

Definition 2.3 Let W be a standard Brownian motion, $a = (a_t)_{t\in\mathbb{R}_+}$ a predictable real-valued process, Λ an $\mathbb{R}_+$-valued Lévy basis on $\mathbb{R}_- \times \mathbb{R}$ independent of W with generating triplet (γ_0, v, π) and let L be its underlying Lévy process. Let Σ be a non-negative càdlàg supOU process and $\rho \in \mathbb{R}$. Assume that the logarithmic price process $X = (X_t)_{t\in\mathbb{R}_+}$ is given by

$$X_t = X_0 + \int_0^t a_s ds + \int_0^t \Sigma_s^{\frac{1}{2}} dW_s + \rho(L_t - \gamma_0 t),$$

where X_0 is independent of Λ. Then we say that X follows a univariate supOU stochastic volatility model and refer to it by $SVsupOU(a, \rho, \gamma_0, v, \pi)$.

In the following, we always use as filtration the one generated by W and Λ.

In Definition 2.3 X is supposed to be the log price of some financial asset and ρ is the typically negative correlation between jumps in the volatility and log asset prices modeling the leverage effect. To ensure that the absolutely continuous drift is completely given by a_t, we subtract the drift γ_0 from the Lévy process noting that this can be done without loss of generality.

In [5], it has been shown that the model is able to exhibit long-range dependence in the squared log returns. The typical example leading to a polynomial decay of the autocovariance function of the squared returns and to long-range dependence for certain choices of the parameter is to take π as a Gamma distribution mirrored at the origin. [13, 26] discuss in general which properties of π result in long-range dependence.

3 Martingale Conditions

Now we assume given a market with a deterministic numeraire (or bond) with price process e^{rt} for some $r \geq 0$ and a risky asset with price process S_t.

We want to model the market by a supOU stochastic volatility model under the risk neutral dynamics. Thus, we need to understand when $\hat{S}_t = e^{-rt}e^{X_t}$ is a martingale for the filtration $\mathbb{G} = (\mathscr{G}_t)_{t\in\mathbb{R}_+}$ generated by the Wiener process and the Lévy basis, i.e., $\mathscr{G}_t = \sigma(\{\Lambda(A), W_s : s \in [0, t] \text{ and } A \in \mathscr{B}_b(\mathbb{R}_- \times (-\infty, t])\})$ for $t \in \mathbb{R}_+$. Implicitly, we understand that the filtration is modified such that the usual hypotheses (see, e.g., [22]) are satisfied.

Theorem 3.1 (Martingale condition) *Consider a market as described above. Suppose that*

$$\int_{x>1} \left(e^{\rho x} - 1\right) v(\mathrm{d}x) < \infty. \tag{1}$$

If the process $a = (a_t)_{t\in\mathbb{R}_+}$ *satisfies*

$$a_t = r - \frac{1}{2}\Sigma_t - \int_{\mathbb{R}_+} \left(e^{\rho x} - 1\right) v(\mathrm{d}x), \tag{2}$$

then the discounted price process $\hat{S}$ *is a martingale.*

Proof The arguments are straightforward adaptations of the ones in [19, Proposition 2.10] or [20, Sect. 3]. ∎

4 Fourier Pricing in the supOU Stochastic Volatility Model

Our aim now is to use the Fourier pricing approach in the supOU stochastic volatility model for calculating prices of European derivatives.

4.1 A Review on Fourier Pricing

We start with a brief review on the well-known Fourier pricing techniques introduced in [9, 23].

Let the price process of a financial asset be modeled as an exponential semimartingale $S = (S_t)_{0\le t\le T}$, i.e., $S_t = S_0 e^{X_t}$, $0 \le t \le T$ where $X = (X_t)_{0\le t\le T}$ is a semimartingale.

Let r be the risk-free interest rate and let us assume that we are directly working under an equivalent martingale measure, i.e., the discounted price process $\hat{S} = (\hat{S}_t)_{0\le t\le T}$ given by $\hat{S}_t = S_0 e^{X_t - rt}$ is a martingale.

We call the process X the underlying process and without loss of generality we can assume that $X_0 = 0$. We denote by s minus the logarithm of the initial value of S, i.e., $s = -\log(S_0)$.

Let $\hat{f}$ denote the Fourier transform of the function f, i.e., $\hat{f}(u) = \int_{\mathbb{R}} e^{iux} f(x)\mathrm{d}x$.

Let now $f : \mathbb{R} \to \mathbb{R}_+$ be a measurable function that we refer to as the payoff function. Then, the arbitrage-free price of the derivative with payoff $f(X_T - s)$ and maturity T at time zero is the conditional expected discounted payoff under the chosen equivalent martingale measure, i.e., $V_f(X_T; s) = e^{-rt}\mathbb{E}\left(f(X_T - s)|\mathscr{G}_0\right)$.

The following theorem gives the valuation formula for the price of the derivative paying $f(X_T - s)$ at time T.

Theorem 4.1 ([12] Theorem 2.2, Remark 2.3) *Let $f : \mathbb{R} \to \mathbb{R}_+$ be a payoff function and let $g_R(x) = e^{-Rx} f(x)$ for some $R \in \mathbb{R}$ denote the dampened payoff function. Define $\Phi_{X_T|\mathscr{G}_0}(u) := \mathbb{E}\left(e^{uX_T}|\mathscr{G}_0\right), u \in \mathbb{C}$. If*

$$(i)\; g_R \in L^1(\mathbb{R}) \cap L^\infty(\mathbb{R}),\; (ii)\; \Phi_{X_T|\mathscr{G}_0}(R) < \infty,\; (iii)\; \Phi_{X_T|\mathscr{G}_0}(R + i\cdot) \in L^1(\mathbb{R}),$$

then $V_f(X_T; s) = \frac{e^{-rt-Rs}}{2\pi} \int_{\mathbb{R}} e^{-ius} \Phi_{X_T|\mathscr{G}_0}(R + iu)\hat{f}(iR - u)du.$

It is well known that for a European Call option with maturity T and strike $K > 0$ condition (i) is satisfied for $R > 1$ and that for the payoff function $f(x) = \max(e^x - K, 0) =: (e^x - K)^+$ the Fourier transform is $\hat{f}(u) = \frac{K^{1+iu}}{iu(1+iu)}$ for $u \in \mathbb{C}$ with $\mathrm{Im}(u) \in (1, \infty)$.

In the following, we calculate the characteristic/moment generating function for the supOU SV model and show conditions when the above Fourier pricing techniques are applicable.

4.2 The Characteristic Function

Consider the general supOU SV model with drift of the form $a_t = \mu + \gamma_0 + \beta \Sigma_t$. Note that then the discounted stock price is a martingale if and only if $\beta = -1/2$ and $\mu + \gamma_0 = r - \int_{\mathbb{R}_+} (e^{\rho x} - 1) \nu(dx)$.

Standard calculations as in [19, Theorem 2.5] or [20] give the following result which is the univariate special case of a formula reported in [4, Sect. 5.2].

Theorem 4.2 *Let $X_0 \in \mathbb{R}$ and let the log-price process X follow a supOU SV model of the above form. Then, for every $t \in \mathbb{R}_+$ and for all $u \in \mathbb{R}$ the characteristic function of X_t given $\mathscr{G}_0$ is given by*

$$\Phi_{X_t|\mathscr{G}_0}(iu) = \mathbb{E}\left(e^{iuX_t}|\mathscr{G}_0\right) \tag{3}$$

$$= \exp\left\{i\left(u(X_0 + \mu t) + \left(u\beta + \frac{i}{2}u^2\right)\int_{\mathbb{R}_-}\int_{-\infty}^{0} \frac{1}{A}\left(e^{A(t-s)} - e^{-As}\right)\Lambda(dA, ds)\right)\right.$$

$$\left. + \int_{\mathbb{R}_-}\int_{0}^{t} \varphi\left(\frac{e^{A(t-s)}}{A}\left(u\beta + \frac{i}{2}u^2\right) - \left(\frac{1}{A}\left(u\beta + \frac{i}{2}u^2\right) - \rho u\right)\right)ds\pi(dA)\right\}.$$

Note that in contrast to the case of the OU-type stochastic volatility model, where (X, Σ) is a strong Markov process, in the supOU stochastic volatility model Σ is not Markovian. Thus, conditioning on X_0 and Σ_0 is not equivalent to conditioning upon $\mathscr{G}_0$. Therefore, $\Phi_{X_t | \mathscr{G}_0}(iu)$ is not simply a function of X_0, Σ_0. Instead, the whole past of the Lévy basis enters via the $\mathscr{G}_0$-measurable

$$
z_t := \int_{\mathbb{R}_-} \int_{-\infty}^{0} \frac{1}{A} \left(e^{A(t-s)} - e^{-As} \right) \Lambda(dA, ds),
$$

which has a similar role as the initial volatility Σ_0 in the OU-type stochastic volatility model. Like Σ_0 in the OU-type models, z_t can be treated as an additional parameter to be determined when calibrating the model to market option prices. We can immediately see that thus the number of parameters to be estimated increases with each additional maturity. As it will become clear later, the following observation is important.

Lemma 4.3 $z_{t_1} \leq z_{t_2}$, for all $t_1, t_2 \in \mathbb{R}_+$ such that $t_1 \leq t_2$.

Proof For $t \in \mathbb{R}_+$ and $s \leq t$ we have $\frac{1}{A} \left(e^{A(t-s)} - e^{-As} \right) = \frac{e^{-As}}{A} \left(e^{At} - 1 \right)$ and for $t_1 \leq t_2$ one sees $e^{At_2} - 1 \leq e^{At_1} - 1 \leq 0$ since $A < 0$. This implies that for $s \leq t_1 \leq t_2$ $\frac{e^{-As}}{A} \left(e^{At_1} - 1 \right) \leq \frac{e^{-As}}{A} \left(e^{At_2} - 1 \right)$ and thus $z_{t_1} \leq z_{t_2}$.

4.3 Regularity of the Moment Generating Function

In order to apply Fourier pricing, we now show where the moment generating function $\Phi_{X_T | \mathscr{G}_0}$ is analytic.

Let $\theta_L(u) = \gamma_0 u + \int_{\mathbb{R}_+} (e^{ux} - 1) \nu(dx)$ be the cumulant transform of the Lévy basis (or rather its underlying subordinator). If $\int_{x \geq 1} e^{rx} \nu(dx) < \infty$ for all $r \in \mathbb{R}$ such that $r < \varepsilon$ for some $\varepsilon > 0$, then the function θ_L is analytic in the open set $S_L := \{z \in \mathbb{C} : \operatorname{Re}(z) < \varepsilon\}$, as can be seen, e.g., from the arguments at the start of the proof of [19, Lemma 2.7].

Theorem 4.4 *Let the measure ν satisfy*

$$
\int_{x \geq 1} e^{rx} \nu(dx) < \infty \quad \text{for all } r \in \mathbb{R} \text{ such that } r < \varepsilon \tag{4}
$$

for some $\varepsilon > 0$. Then the function $\Theta(u) = \int_{\mathbb{R}_-} \int_0^t \theta_L(u f_u(A, s)) ds \pi(dA)$ is analytic on the open strip

$$S := \{u \in \mathbb{C}, \ |\operatorname{Re}(u)| < \delta\} \ \text{with} \ \delta := -|\beta| - \frac{|\rho|}{t} + \sqrt{\Delta}, \tag{5}$$

where $\Delta := \left(|\beta| + \frac{|\rho|}{t}\right)^2 + \frac{2\varepsilon}{t}$.

The rough idea of the proof is similar to [19, Theorem 2.8], but the fact that we now integrate over the mean reversion parameter adds significant difficulty, as now bounds independent of the mean reversion parameter need to be obtained and a very general holomorphicity result for integrals has to be employed.

Proof Define

$$f_u(A, s) = \mathbf{1}_{[0,t]}(s) \left(\frac{e^{A(t-s)}}{A} \left(\beta + \frac{u}{2}\right) - \left(\frac{1}{A}\left(\beta + \frac{u}{2}\right) - \rho\right) \right). \tag{6}$$

We first determine $\delta > 0$ such that for all $u \in \mathbb{R}$ with $|u| < \delta$ it holds that $|u f_u(A, s)| < \varepsilon$. We have

$$|u f_u(A, s)| \le \left| \frac{e^{A(t-s)} - 1}{A} \right| \left(|\beta||u| + \frac{u^2}{2} \right) + |\rho||u| \tag{7}$$

by the triangle inequality. In order to find the upper bound for the latter term, we first note that elementary analysis shows

$$\left| \frac{e^{A(t-s)} - 1}{A} \right| \le t \tag{8}$$

for all $A < 0$ and $s \in [0, t]$. Thus, we have to find $\delta > 0$ such that $|u f_u(A, s)| \le t \left(|\beta||u| + \frac{u^2}{2} \right) + |\rho||u| < \varepsilon,$ for all $u \in \mathbb{R}$ with $|u| < \delta$, i.e., to find the solutions of the quadratic equation

$$\frac{t}{2} u^2 + (t|\beta| + |\rho|) |u| - \varepsilon = 0. \tag{9}$$

Since for $u = 0$ the sign of (9) is negative, i.e., (9) is equal to $-\varepsilon$, we know that there exist one positive and one negative solution. The positive one is δ as given in (5).

Now let $u \in S$, i.e., $u = v + iw$ with $v, w \in \mathbb{R}$, $|v| < \delta$. Observe that $\operatorname{Re}(u f_u(A, s)) = v f_v(A, s) - \frac{w^2}{2} \left(\frac{e^{A(t-s)} - 1}{A} \right)$ and $\frac{e^{A(t-s)} - 1}{A} \ge 0$ for all $s \in [0, t]$ and $A < 0$. Hence, $\operatorname{Re}(u f_u(A, s)) \le v f_v(A, s)$. This implies that

$$\int_{x \ge 1} e^{\operatorname{Re}(u f_u(A,s))x} \nu(dx) \le \int_{x \ge 1} e^{v f_v(A,s)x} \nu(dx) < \infty$$

due to $|vf_v(A,s)| < \varepsilon$ for $|v| < \delta$ and condition (4). Hence for $u \in S$ the function $\theta_L(uf_u(A,s)) = \gamma_0 uf_u(A,s) + \int_{\mathbb{R}_+} \left(e^{uf_u(A,s)x} - 1\right) v(dx)$ is well defined. $uf_u(A,s)$ is a polynomial of u and thus it is an analytic function in $\mathbb{C}$, for all $s \in [0,t]$ and $A < 0$. The function θ_L is analytic in the set $S_L = \{z \in \mathbb{C} : |\mathrm{Re}(z)| < \varepsilon\}$.

Thus, the function $\theta_L(uf_u(A,s))$ is analytic in S, for all $s \in [0,t]$ and $A < 0$. By the holomorphicity theorem for parameter dependent integrals (see, e.g., [15]), we can conclude that $\int_0^t \theta_L(uf_u(A,s))ds$ is analytic in S, for all $A < 0$.

Defining $\varphi(u,A) := \int_0^t \theta_L(uf_u(A,s))ds$ we now apply [17] to prove that $\Theta(u) = \int_{\mathbb{R}_-} \int_0^t \theta_L(uf_u(A,s))ds\pi(dA) = \int_{\mathbb{R}_-} \varphi(u,A)\pi(dA)$ is analytic in S. Its conditions A_1 and A_2 are obviously satisfied. It remains to prove that condition A_3 holds, i.e., that $\int_{\mathbb{R}_-} |\varphi(u,A)|\pi(dA)$ is locally bounded. First, observe that

$$|\theta_L(uf_u(A,s))| \leq |\gamma_0 uf_u(A,s)| + \int\limits_{x \leq 1} \left|e^{uf_u(A,s)x} - 1\right| v(dx)$$
$$+ \int\limits_{x > 1} \left|e^{uf_u(A,s)x} - 1\right| v(dx). \tag{10}$$

Using (8), we can bound the first summand in (10) by:

$$|\gamma_0 uf_u(A,s)| \leq |\gamma_0| \left(t\left(|\beta||u| + \frac{|u|^2}{2}\right) + |\rho||u|\right) =: B_1(u).$$

For the second summand, using Taylor's theorem we have that $\left|e^{uf_u(A,s)x} - 1\right| \leq |uf_u(A,s)||x| + O(|uf_u(A,s)|^2|x|^2)$. Since $|uf_u(A,s)| \leq t\left(|\beta||u| + \frac{|u|^2}{2}\right) + |\rho||u|$, for the remainder term of Taylor's formula we have

$$O(|uf_u(A,s)|^2|x|^2) \leq O\left(\left|t\left(|\beta||u| + \frac{|u|^2}{2}\right) + |\rho||u|\right|^2 |x|^2\right),$$

where the latter term converges to zero as $x \to 0$. If we define

$$K(u) := t\left(|\beta||u| + \frac{|u|^2}{2}\right) + |\rho||u|$$

we obtain that

$$\int\limits_{x \leq 1} \left|e^{uf_u(A,s)x} - 1\right| v(dx) \leq K(u) \int\limits_{x \leq 1} xv(dx) + \int\limits_{x \leq 1} O\left(K(u)^2|x|^2\right) v(dx) =: B_2(u),$$

which is finite due to the properties of the measure v.

Let $S_n := \{\mathbb{C} \ni u = v + iw : \ |v| \le \delta - 1/n\} \subseteq S$. Since the function $v f_v(A, s)$ is continuous on the compact set $V_n = \{v \in \mathbb{R} : |v| \le \delta - 1/n\}$, it attains its minimum and maximum on that set, i.e., there exists $v^* \in V_n$ such that $v f_v(A, s) \le v^* f_{v^*}(A, s) \le |v^* f_{v^*}(A, s)| =: K_n(u)$ for all $v \in V_n$. Note that $v^* \in V_n$ implies that $K_n(u) < \varepsilon$. Since $\mathrm{Re}(u f_u(A, s)) \le v f_v(A, s)$ and $\left| e^{u f_u(A,s)x} \right| = e^{\mathrm{Re}(u f_u(A,s))x} \le e^{K_n(u)x}$, it follows that

$$\int_{x>1} \left| e^{u f_u(A,s)x} - 1 \right| v(dx) \le \int_{x>1} e^{K_n(u)x} v(dx) + \int_{x>1} v(dx) =: B_{3,n}(u),$$

which is finite due to (4) and the properties of the measure v.

Since $B_1(u)$, $B_2(u)$, and $B_{3,n}(u)$ do not depend neither on s nor on A, we have $|\varphi(u, A)| \le t(B_1(u) + B_2(u) + B_{3,n}(u))$ and

$$\int_{\mathbb{R}_-} t(B_1(u) + B_2(u) + B_{3,n}(u))\pi(dA) = t(B_1(u) + B_2(u) + B_{3,n}(u)) < \infty,$$

so the function $t(B_1(u) + B_2(u) + B_{3,n}(u))$ is integrable with respect to π. Since $\varphi(u, A)$ is analytic and thus a continuous function on S_n, for all $A < 0$, it also holds that $|\varphi(u, A)|$ is continuous on S_n, for all $A < 0$. By the dominated convergence theorem, it follows that $\int_{\mathbb{R}_-} |\varphi(u, A)| \pi(dA)$ is continuous and thus a locally bounded function on S_n. Since $n \in \mathbb{N}$ was arbitrary, it follows that the function is continuous and locally bounded on S, which completes the proof.

Now, we can easily give conditions ensuring that (ii) in Theorem 4.1 is satisfied.

Corollary 4.5 *Let $\int_{x \ge 1} e^{rx} v(dx) < \infty$ for all $r \in \mathbb{R}$ such that $r < \varepsilon$ for some $\varepsilon > 0$. Then the moment generating function $\Phi_{X_T | \mathscr{G}_0}$ is analytic on the open strip $S := \{u \in \mathbb{C} : |\mathrm{Re}(u)| < \delta\}$ with $\delta := -|\beta| - \frac{|\varrho|}{T} + \sqrt{\Delta}$ where $\Delta := \left(|\beta| + \frac{|\varrho|}{T} \right)^2 + \frac{2\varepsilon}{T}$. Furthermore,*

$$\Phi_{X_T | \mathscr{G}_0}(u) = \tag{11}$$

$$\exp \left\{ u(X_0 + \mu T) + \left(u\beta + \frac{1}{2}u^2 \right) \int_{\mathbb{R}_-} \int_{-\infty}^{0} \frac{1}{A} \left(e^{A(T-s)} - e^{-As} \right) \Lambda(dA, ds) + \Theta(u) \right\}$$

for all $u \in S$.

Proof Follows from Theorems 4.2 and 4.4 noting that an analytic function is uniquely identified by its values on a line and [19, Lemma A.1]. □

Very similar to [19, Theorem 6.11], we can now prove that also condition (iii) in Theorem 4.1 is satisfied for the supOU SV model.

Theorem 4.6 *If $u \in \mathbb{C}$, $u = v + iw$ and $u \in S$ as defined in Theorem 4.4, then the map*

$$w \mapsto \Phi_{X_T | \mathscr{G}_0}(v + iw)$$

is absolutely integrable.

5 Examples

5.1 Concrete Specifications

If we want to price a derivative by Fourier inversion, then this means in the supOU SV model that we have to calculate the inverse Fourier transform by numerical integration and inside this the double integral in $\Theta(u) = \int_{\mathbb{R}_-} \int_0^t \theta_L(u f_u(A, s)) ds \pi(dA)$. If we want to calibrate our model to market data, the optimizer will repeat this procedure very often and so it is important to consider specifications where at least some of the integrals can be calculated analytically.

Actually, it is not hard to see that one can use the standard specifications for v of the OU-type stochastic volatility model (see [3, 11, 20, 25]) which are named after the resulting stationary distribution of the OU-type processes.

As in the case of a Γ-OU process we can choose the underlying Lévy process to be a compound Poisson process with the characteristic triplet $(\gamma_0, 0, abe^{-bx}\mathbf{1}_{\{x>0\}}dx)$ with $a, b > 0$ where abusing notation we specified the Lévy measure by its density. Furthermore, we assume that A follows a "negative" Γ-distribution, i.e., that π is the distribution of BR, where $B \in \mathbb{R}_-$ and $R \sim \Gamma(\alpha, 1)$ with $\alpha > 1$ which is the specification typically used to obtain long memory/a polynomial decay of the autocorrelation function. We refer to this specification as the Γ -*supOU SV model*.

Using (6) we have

$$\Theta(u) = u \int_{\mathbb{R}_-} \int_0^t \gamma_0 f_u(A, s) ds \pi(dA) + \int_{\mathbb{R}_-} \int_0^t \int_{\mathbb{R}_+} \left(e^{u f_u(A,s)x} - 1 \right) v(dx) ds \pi(dA).$$

For the first summand in $\Theta(u)$ we see

$$u \int_{\mathbb{R}_-} \int_0^t \gamma_0 f_u(A, s) ds \pi(dA) = \gamma_0 \left(\underbrace{\int_{\mathbb{R}_-} \int_0^t \frac{e^{A(t-s)}}{A} \left(u\beta + \frac{u^2}{2} \right) ds \pi(dA)}_{I_1} \right.$$

$$\left. \underbrace{- \int_{\mathbb{R}_-} \int_0^t \frac{1}{A} \left(u\beta + \frac{u^2}{2} \right) ds \pi(dA)}_{I_2} + \underbrace{\int_{\mathbb{R}_-} \int_0^t \rho u \, ds \pi(dA)}_{I_3} \right).$$

For the three parts, we can now show:

$$I_1 = \left(u\beta + \frac{u^2}{2}\right)\frac{(1 - Bt)^{2-\alpha} - 1}{B^2(\alpha - 1)(\alpha - 2)} \text{ if } \alpha \neq 2,$$

$$I_1 = -\frac{\left(u\beta + \frac{u^2}{2}\right)}{B^2}\ln(1 - Bt) \text{ if } \alpha = 2,$$

$$I_2 = \frac{t\left(u\beta + \frac{u^2}{2}\right)}{B(\alpha - 1)}, \qquad I_3 = \rho u \int_0^t \int_{\mathbb{R}_-} ds\pi(dA) = \rho u t.$$

Furthermore setting $C(A) := \frac{1}{A}\left(u\beta + \frac{u^2}{2}\right) - \rho u$ one obtains for the second summand in Θ

$$\int_{\mathbb{R}_-}\int_0^t\int_{\mathbb{R}_+}\left(e^{uf_u(A,s)x} - 1\right)abe^{-bx}dxds\pi(dA)$$

$$= a\int_{\mathbb{R}_-}\frac{1}{A(b + C(A))}\left(b\ln\left(\frac{b - \rho u}{b - \frac{e^{At}}{A}\left(u\beta + \frac{u^2}{2}\right) + C(A)}\right) - AC(A)t\right)\pi(dA).$$

Unfortunately, we have been unable to obtain a more explicit formula for this integral, and so it has to be calculated numerically. In our example later on we have used the standard Matlab command "integral" for this. Note that the well-behavedness of this numerical integration depends on the choice of π. For our choice, π being a negative Gamma distribution implies roughly (i.e., up to a power) an exponentially fast decaying integrand for $A \to \infty$, whereas the behavior at zero appears to be hard to determine.

We can also choose the underlying Lévy process as in an IG-OU model with parameters δ and γ, while keeping the choice of the measure π the same. In this case, we have $\nu(dx) = \frac{1}{2\sqrt{2\pi}}\delta\left(x^{-1} + \gamma^2\right)x^{-\frac{1}{2}}\exp\left(-\frac{1}{2}\gamma^2 x\right)\mathbf{1}_{\{x>0\}}dx$ and the only difference compared to the previous case is in the calculation of the triple integral which also can be partially calculated analytically so that only a one-dimensional numerical integration is necessary.

5.2 Calibration and an Illustrative Example

In this chapter, we calibrate the Γ−supOU SV model to market prices of European plain vanilla call options written on the DAX.

Let $t_1, t_2, \ldots, t_M$ be the set of different times to maturity (in increasing order) for which we have market option prices. The parameters to be determined by calibration

Table 1 Calibrated supOU SV model parameters for DAX data of August 19, 2013

	ρ	a	b	B	α	γ_0	
	-10.8797	0.2225	29.4025	-0.0004	4.3632	0.0000	
z_{t_1}	z_{t_2}	z_{t_3}	z_{t_4}	z_{t_5}	z_{t_6}	z_{t_7}	z_{t_8}
0.0012	0.0026	0.0038	0.0054	0.0093	0.0136	0.0225	0.0328

are $(\rho, a, b, B, \alpha, \gamma_0, z_{t_1}, \ldots, z_{t_M})$, where ρ describes the leverage, a and b are parameters of the measure v, B, and α are parameters of the measure π and γ_0 is the drift parameter. Finally, $z_{t_1}, \ldots, z_{t_M}$ are $z_{t_i} = \int_{\mathbb{R}_-} \int_{-\infty}^0 \frac{1}{A} \left(e^{A(t_i - s)} - e^{-As} \right) \Lambda(dA, ds)$, $i = 1, \ldots, M$.

We calibrate by minimizing the root mean squared error between the Black–Scholes implied volatilities corresponding to market and model prices, i.e., RMSE $= \sqrt{\sum_{i=1}^M \sum_{j=1}^{N_i} \left(\text{blsimpv} \left(C_{ij}^M \right) - \text{blsimpv} \left(C_{ij} \right) \right)^2 / \sum_{i=1}^M N_i}$, where M is the number of different times to maturity, N_i is the number of options for each maturity, $\left\{ C_{ij}^M \right\}$ is the set of market prices and $\{C_{ij}\}$ is the set of model prices, $i = 1, \ldots, N_M$, $j = 1, \ldots, M$. Of course, minimizing the difference between Black–Scholes implied volatilities is just one possible choice for the objective function. We note that this data example is only supposed to be an illustrative proof of concept and that using other objective functions including in particular weights for the different options should improve the results.

We use closing prices of 200 DAX options on August 19, 2013. The level of DAX on that day was 8366.29. The data source was Bloomberg Finance L.P. and all the options were listed on EUREX.

For the instantaneous risk-free interest rate, we used the 3-month LIBOR rate, which was 0.15173 %. The maturities of the options were 31, 59, 87, 122, 213, 304, 486, and 668 days. The calibration procedure was performed in MATLAB. To avoid being stuck in local minima the calibration was run several times with different initial values and the overall minimum RMSE was taken.

The implied parameters from the calibration procedure are given in Table 1. The fit is good: The RMSE is 0.0046. We plot market against model Black–Scholes implied volatilities in Fig. 1. Although the RMSE is very low and in plots of market against fitted model prices (not shown here) one sees basically no differences, Fig. 1 shows that our model fits the implied volatilities for medium and long maturities very well, but the quality of the fit for shorter maturities is lower.

The vector of the parameters $\{z_{t_i}\}_{i=1,\ldots,M}$ is indeed increasing with maturity (cf. Lemma 4.3), although we actually refrained from including this restriction into our optimization problem. The autocorrelation function of the Γ-supOU model exhibits long memory for $\alpha \in (1, 2)$ (cf. [26, Sect. 2.2]). Since the calibration returns $\alpha = 4.3632$, our market data are in line with a rather slow polynomial decay of the autocorrelation function, which is in contrast to the exponential decay of the autocorrelation function in the OU-type SV model, but the calibrated model does not

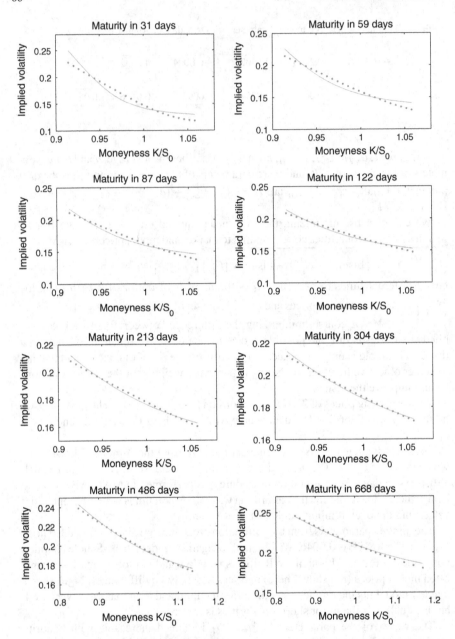

Fig. 1 Calibration of the supOU model to call options on DAX: The Black–Scholes implied volatilities. The implied volatilities from market prices are depicted by a *dot*, the implied volatilities from model prices by a *solid line*

exhibit long memory. One should be very careful not to overinterpret these findings, as no confidence intervals/hypothesis tests are available in connection with such a standard calibration.

The leverage parameter ρ is negative, which implies a negative correlation between jumps in the volatility and returns. Hence, the typical leverage effect is present. The drift parameter of the underlying Lévy basis γ_0 is estimated to be practically zero. So our calibration suggests that a driftless pure jump Lévy basis may be quite adequate to use.

Let us briefly turn to a comparison with the OU-type stochastic volatility model (cf. [19] or [20]) noting that a detailed comparison with various other models is certainly called for, but beyond the scope of the present paper. For some $\beta < 0$ looking at a sequence of Γ-supOU models with $\alpha_n = n$, $B_n = \beta/n$ and all others parameters fixed, shows that the mean reversion probability measures π_n converge weakly to the delta distribution at β. So the OU model is in some sense a limiting case of the supOU model. However, the limiting model is very different from all approximating models, as it is Markovian, has the same decay rate for all jumps, whereas the approximating supOU models have all negative real numbers as possible decay rates for individual jumps. This implies that in connection with real data the behavior of the OU and the supOU model can well be rather different. Calibrating a Γ-OU model to our DAX data set (so the only parameter now different is π, which is a Dirac measure) returns actually a globally better fit (the RMSE is 0.0037). Looking at the plots of market against model implied volatilities they all look quite similar (Fig. 2 shows only the last four largest maturities) to the ones in Fig. 1, although the fit for the early maturities is definitely better when looking closely. Yet, there is one big exception, the last maturity, where the supOU model fits much better. Whereas the rate of the underlying compound Poisson process is $a = 0.2225$ in the supOU model, it is 1.2671 in the OU model. The mean of the decay rates is -0.0017 in the supOU model and the decay rate of the OU case is -1.3906. Noting that the standard deviation of the decay rates is 0.0008 in the supOU model, the two calibrated models are indeed in many respects rather different.

Remark 5.1 (How to price options with general maturities?) After having calibrated a model to observed liquid market prices one often wants to use it to price other (exotic) derivatives. Looking at a European derivative with payoff $f(S_T)$ for some measurable function f and maturity $T > 0$, one soon realizes that we can only obtain its price directly if $T \in \{t_1, t_2, \ldots, t_M\}$, as only then we know z_T, thus the characteristic function $\Phi_{X_T|\mathscr{G}_0}$ and therefore the distribution of the price process at time T conditional on our current information $\mathscr{G}_0$. This is not desirable and the problem is that we assume that we know $\mathscr{G}_0$ in theory, but we have only limited information in the market prices which we can use to get only parts of the information in $\mathscr{G}_0$.

It seems that to get z_t for all $t \in \mathbb{R}_+$ one needs to really know the whole past of Λ, i.e., all jumps before time 0 and the associated times and decay rates. This is clearly not feasible. A detailed analysis on the dependence of z_t on t is beyond the scope of this paper. But we briefly want to comment on possible ad hoc solutions

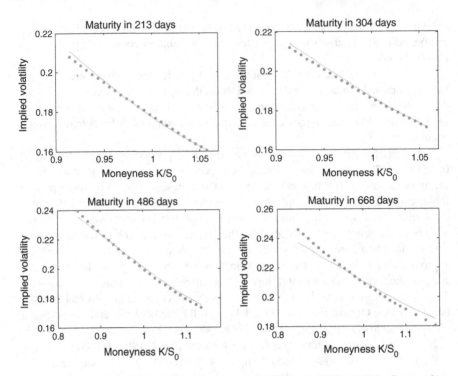

Fig. 2 Calibration of the OU model to call options on DAX: The implied volatilities from market prices are depicted by a *dot*, the implied volatilities from model prices by a *solid line*. Last four maturities only

to "estimate" z_T based on $\{z_{t_i}\}_{i=1,\dots,M}$. The first one is to either interpolate or fit a parametric curve $t \mapsto z_t$ to the "observed" $\{z_{t_i}\}_{i=1,\dots,M}$. If one also ensures the decreasingness in t in this procedure, one should get a reasonable approximation, especially when the grid $\{t_i\}_{i=1,\dots,M}$ is fine and one considers maturities in $[t_1, t_M]$.

From the probabilistic point of view, one wants to compute $E(z_T | \{z_{t_i}\}_{i=1,\dots,M})$ for $T \notin \{t_1, t_2, \dots, t_M\}$. Whether and how this conditional expectation can be calculated, is again a question for future investigations. But what one can calculate easily is the best (in the L^2 sense) linear predictor of z_T given $\{z_{t_i}\}_{i=1,\dots,M}$. One simply needs to straightforwardly adapt standard time series techniques (like the innovations algorithm or linear L^2 filtering, see, e.g., [8]) noting that one has

$$\text{cov}(z_t, z_u) = \int_{\mathbb{R}_-} \int_{\mathbb{R}_-} \frac{e^{-2As}}{A^2} (e^{At} - 1)(e^{Au} - 1) \, ds \, \pi(dA) \int_{\mathbb{R}_+} x^2 v(dx) \; \forall t, u \in \mathbb{R}_+.$$

Open Access This chapter is distributed under the terms of the Creative Commons Attribution Noncommercial License, which permits any noncommercial use, distribution, and reproduction in any medium, provided the original author(s) and source are credited.

References

1. Bannör, K.F., Scherer, M.: A BNS-type stochastic volatility model with two-sided jumps with applications to FX options pricing. Wilmott **2013**, 58–69 (2013)
2. Barndorff-Nielsen, O.: Superposition of Ornstein–Uhlenbeck type processes. Theory Probab. Appl. **45**, 175–194 (2001)
3. Barndorff-Nielsen, O., Shephard, N.: Non-Gaussian Ornstein-Uhlenbeck-based models and some of their uses in financial economics (with discussion). J. R. Stat. Soc. B Stat. Methodol. **63**, 167–241 (2001)
4. Barndorff-Nielsen, O., Stelzer, R.: Multivariate supOU processes. Ann. Appl. Probab. **21**(1), 140–182 (2011)
5. Barndorff-Nielsen, O., Stelzer, R.: The multivariate supOU stochastic volatility model. Math. Finance **23**, 275–296 (2013)
6. Bender, C., Sottinen, T., Valkeila, E.: Arbitrage with fractional Brownian motion? Theory Stoch. Process. **13**(1–2), 23–34 (2007)
7. Björk, T., Hult, H.: A note on Wick products and the fractional Black-Scholes model. Finance Stoch. **9**(2), 197–209 (2005)
8. Brockwell, P.J., Davis, R.A.: Time Series: Theory and Methods, vol. 2. Springer, New York (1991)
9. Carr, P., Madan, D.B.: Option valuation using the fast Fourier transform. J. Comput. Finance **2**, 61–73 (1999)
10. Cont, R.: Empirical properties of asset returns: stylized facts and statistical issues. Quant. Finance **1**, 223–236 (2001)
11. Cont, R., Tankov, P.: Financial Modelling with Jump Processes. CRC Financial Mathematical Series. Chapman & Hall, London (2004)
12. Eberlein, E., Glau, K., Papapantoleon, A.: Analysis of Fourier transform valuation formulas and applications. Appl. Math. Finance **17**, 211–240 (2010)
13. V, Fasen, Klüppelberg, C.: Extremes of supOU processes. In: Benth, F.E., Di Nunno, G., Lindstrom, T., Øksendal, B., Zhang, T. (eds.) Stochastic Analysis and Applications: The Abel Symposium 2005. Abel Symposia, vol. 2, pp. 340–359. Springer, Berlin (2007)
14. Guillaume, D.M., Dacorogna, M.M., Davé, R.D., Müller, U.A., Olsen, R.B., Pictet, O.V.: From the bird's eye to the microscope: a survey of new stylized facts of the intra-daily foreign exchange markets. Finance Stoch. **1**, 95–129 (1997)
15. Königsberger, K.: Analysis 2. Springer, Heidelberg (2004)
16. Marquardt, T.: Fractional Lévy processes with an application to long memory moving average processes. Bernoulli **12**, 1099–1126 (2006)
17. Mattner, L.: Complex differentiation under the integral. Nieuw Archief voor Wiskunde **5/2**(2), 32–35 (2001)
18. Moser, M., Stelzer, R.: Tail behavior of multivariate Lévy driven mixed moving average processes and related stochastic volatility models. Adv. Appl. Probab. **43**, 1109–1135 (2011)
19. Muhle-Karbe, J., Pfaffel, O., Stelzer, R.: Option pricing in multivariate stochastic volatility models of OU type. SIAM J. Finance Math. **3**, 66–94 (2011)
20. Nicolato, E., Venardos, E.: Option pricing in stochastic volatility models of the Ornstein-Uhlenbeck type. Math. Finance **13**, 445–466 (2003)
21. Pigorsch, C., Stelzer, R.: A multivariate Ornstein–Uhlenbeck type stochastic volatility model. Working paper (2009) http://www.uni-ulm.de/mawi/finmath/people/stelzer/publications.html
22. Protter, P.: Stochastic Integration and Differential Equations. Stochastic Modelling and Applied Probability, vol. 21, 2nd edn. Springer, New York (2004)
23. Raible, S.: Lévy Processes in Finance: Theory, Numerics and Empirical Facts. Dissertation, Mathematische Fakultät, Albert-Ludwigs-Universität Freiburg i. Br., Freiburg, Germany, (2000)
24. Sato, K.: Lévy Processes and Infinitely Divisible Distributions, volume 68 of Cambridge Studies in Advanced Mathematics. Cambridge University Press, Cambridge (1999)

25. Schoutens, W.: Lévy Processes in Finance—Pricing Financial Derivatives. Wiley, Chicester (2003)
26. Stelzer, R, Tosstorff, T, Wittlinger, M.: Moment based estimation of supOU processes and a related stochastic volatility model. submitted for publication (2013) http://arxiv.org/abs/1305.1470v1
27. Veretennikov, A.Y.: On lower bounds for mixing coefficients of Markov diffusions. In: Kabanov, Y., Lipster, R., Stoyanov, J. (eds.) From Stochastic Calculus to Mathematical Finance—The Shiryaev Festschrift, pp. 33–68. Springer, Berlin (2006)

A Two-Sided BNS Model for Multicurrency FX Markets

Karl Friedrich Bannör, Matthias Scherer and Thorsten Schulz

Abstract We present a multivariate jump-diffusion model incorporating stochastic volatility and two-sided jumps for multicurrency FX markets, which is an extension of the univariate Γ-OU-BNS model introduced by [2]. The model can be considered a multivariate variant of the two-sided Γ-OU-BNS model (cf. [1]). We discuss FX option pricing and provide a calibration exercise, modeling two FX rates with a common currency by a bivariate model and calibrating the dependence parameters to the implied FX volatility surface.

Keywords Barndorff–Nielsen–Shephard model · Stochastic volatility · Multivariate model · Jump-diffusion model · Multicurrency FX markets

1 Introduction

For derivatives valuation, the Black–Scholes model, presented in the seminal paper [4], generated a wave of stochastic models for the description of stock-prices. Since the assumptions of the Black–Scholes model (normally distributed log-returns, independent returns) cannot be observed in neither time series of stock-prices nor option markets (implicitly expressed in terms of the volatility surface), several alternative models have been developed trying to overcome these assumptions. Some

K.F. Bannör
Deloitte & Touche GmbH Wirtschaftsprüfungsgesellschaft, Rosenheimer Platz 4, 81669 München, Germany
e-mail: kbannoer@deloitte.de

M. Scherer · T. Schulz(✉)
Technische Universität München, Parkring 11, 85748 Garching-Hochbrück, Germany
e-mail: scherer@tum.de

T. Schulz
e-mail: t.schulz@tum.de

© The Author(s) 2015
K. Glau et al. (eds.), *Innovations in Quantitative Risk Management*,
Springer Proceedings in Mathematics & Statistics 99,
DOI 10.1007/978-3-319-09114-3_6

models, as, e.g., [9, 23] account for stochastic volatility, while others as, e.g., [12, 16] enrich the original Black–Scholes model with jumps. Both approaches have been combined in the models of, e.g., [3, 6]. Another approach combining stochastic volatility and negative jumps in both volatility and asset-price process, employing Lévy subordinator-driven Ornstein–Uhlenbeck processes, is available with the Barndorff–Nielsen–Shephard (BNS) model class, presented in [2] and extended in several papers (e.g. [18]). A multivariate extension of the BNS model class employing matrix subordinators is designed in [20] and pricing in this model is scrutinized in [17]. In the special case of a Γ-OU-BNS model, a tractable variant of a multivariate BNS model based on subordination of compound Poisson processes was developed by [15]. This model allows for a separate calibration of the single assets (following a univariate Γ-OU-BNS model) and the dependence structure.

Besides for options on stocks, these models have also been used to price derivatives on other underlyings. When modeling foreign exchange (FX) rates instead of stock-prices, one has to cope with the introduction of two different interest rates as well as identifying the actual tradeable assets. The Black–Scholes model was adapted to FX markets by [8]. Many of the models mentioned above have been employed for FX rates modeling as, e.g., [3, 9]. Since the original BNS model assumes only downward jumps in the asset-price process, [1] extend the BNS model class to additionally incorporate positive jumps, which is needed for the realistic modeling of FX rates and calibrates much better to FX option surfaces.

In this paper, we unify the extensions of the BNS model from [1, 15] and introduce a multivariate Γ-OU-BNS model with time-changed compound Poisson drivers incorporating dependent jumps in both directions, both generalizing the univariate two-sided Γ-OU-BNS model and the multivariate "classical" Γ-OU-BNS model. Since the two-sided Γ-OU-BNS model seems to be particularly suitable for the modeling of FX rates, we consider a multivariate two-sided Γ-OU-BNS model a sensible choice for the valuation of multivariate FX derivatives such as best-of-two options. Since the multivariate two-sided model accounts for joint and single jumps in the FX rates, the jump behavior of modeled FX rates resembles reality better than models only employing joint or single jumps, as illustrated in Fig. 1. Furthermore, a multivariate two-sided BNS model for FX rates with a common currency also implies a jump-diffusion model for an FX rate via quotient or product processes. A crucial feature of our multivariate approach is the separability of the univariate models from the dependence structure, i.e. one has two sets of parameters that can be determined in consecutive steps: parameters determining each univariate model and parameters determining the dependence. This feature provides tractability for practical applications like simulation or calibration on the one side, but also simplifies interpretability of the model parameters on the other side.

Instead of modeling the FX spot rates only, one could model FX forward rates to get a model setup suited for pricing cross-currency derivatives depending on FX forward rates, as for example cross-currency swaps. Multicurrency models built upon FX forward rates (see e.g. [7]) on the one hand support flexibility to price such derivatives, on the other hand, however, these models do not provide the crucial

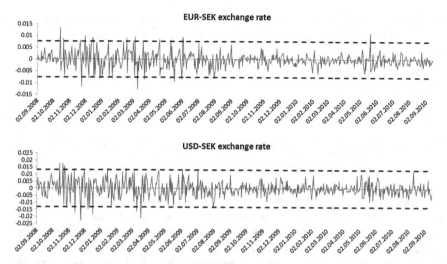

Fig. 1 The logarithmic returns of EUR-SEK and USD-SEK FX rates over time. Assuming that every logarithmic return exceeding three standard deviations (*dashed lines*) from the mean can be interpreted as a jump (obviously, smaller jumps occur as well, but may be indistinguishable from movement originating in the Brownian noise), one can see that joint as well as separate jumps in the EUR-SEK and the USD-SEK logarithmic returns occur. Clearly, this 3-standard deviation criterion is just a rule of thumb, however, [10] investigated the necessity of both common and individual jumps in a statistical thoroughly manner. Hence, a multivariate FX model capturing the stylized facts of both joint and separate jumps can be valuable. The data was provided by Thomson Reuters

property of separating the dependence structure from the univariate models, which makes it extremely difficult to calibrate such a multivariate model in a sound manner.

The remaining paper is organized as follows: In Sect. 2, we recall the two-sided Barndorff–Nielsen–Shephard model constructed in [1] and outline stylized facts of its trajectories. In Sect. 3, we introduce a multivariate version of the two-sided Γ-OU-BNS model, using the time change construction from [15] to incorporate dependence between the jump drivers. Section 4 focuses on the specific obstacles occuring when modeling FX rates in a multivariate two-sided Γ-OU-BNS model, particularly the dependence structure of joint jumps and the implied model for a third FX rate which may be induced. In Sect. 5, we describe a calibration of the model to implied volatility surfaces and show how the model can be used to price multivariate derivatives. We then evaluate the model in a numerical case study. Finally, Sect. 6 concludes.

2 The Two-Sided Barndorff–Nielsen–Shephard Model Class

We briefly motivate the construction and main features of the two-sided BNS model class. The classical BNS model accounts for the *leverage effect*, a feature of stock returns, by incorporating negative jumps in the asset-price process, accompanied by upward jumps in the stochastic variance. While downward jumps might be sufficient in the case of modeling stock-price dynamics, it is not suitable when modeling FX rates, where one-sided jumps contradict economic intuition. Hence, [1] develop an extension of the BNS model which allows for two-sided jumps and is able to capture the symmetric nature of FX rates.

We say that a stochastic process $\{S_t\}_{t \geq 0}$ follows a *two-sided BNS model* (abbreviated *BNS2 model*), if the log-price $X_t := \log S_t$ follows the dynamics of the SDEs

$$dX_t = (\mu + \beta\sigma_t^2)\,dt + \sigma_t\,dW_t + \rho_+\,dZ_t^+ + \rho_-\,dZ_t^-,$$
$$d\sigma_t^2 = -\lambda\sigma_t^2\,dt + dZ_t^+ + dZ_t^-,$$

with independent Lévy subordinators $Z^+ = \{Z_t^+\}_{t \geq 0}$ and $Z^- = \{Z_t^-\}_{t \geq 0}$ and $W = \{W_t\}_{t \geq 0}$ being a Brownian motion independent of Z^+ and Z^-, $\mu \in \mathbb{R}$, $\lambda > 0$, $\rho_+ > 0$, $\rho_- < 0$.[1] If the Lévy drivers Z^+, Z^- are independent copies of each other, we call the model a *reduced two-sided BNS model*. If, additionally, $\rho_+ = -\rho_-$ we have a symmetric situation, upward jumps occurring similarly likely as downward jumps. Furthermore, the average absolute jump sizes in the log-prices coincide. Thus, we call the model a *symmetric BNS model* or *SBNS model*. In a calibration exercise of [1], the SBNS model produced decent calibration results, while limiting the number of parameters to five.

In contrast to the classical BNS model, the BNS2 model has two independent Lévy subordinators Z^+, Z^- incorporating jumps in the asset-price process in opposite directions, but both accounting for upward jumps in the variance process $\sigma^2 = \{\sigma_t^2\}_{t \geq 0}$. Thus, shocks in the asset-price are always accompanied by upward jumping variance, regardless of the jump direction. Furthermore, the variance process is still a Lévy subordinator driven Ornstein–Uhlenbeck process. As discussed in [1], the symmetric nature of the two-sided BNS model makes it particularly suitable for FX rates modeling and calibrates well to option surfaces on FX rates.

An important example is the special case where the Lévy drivers Z^+, Z^- are compound Poisson processes with exponential jump heights. In this case we call the model a two-sided Γ-OU-BNS model. The log-price of a two-sided Γ-OU-BNS model has a closed-form characteristic function (cf. [1]), hence allows for rapid calibration to vanilla prices by means of Fourier-pricing methods as introduced in [5, 21]. A typical trajectory of the two-sided Γ-OU-BNS model can be found in Fig. 2. It can clearly be seen that shocks in the FX rate process, e.g. caused by macroeconomic turbulences or unanticipated interest rate movements, cause a sudden rise in volatility. As time goes by without the arrival of new shocks, volatility is calming down again.

---

[1] Compared to the original formulation of the model in [1] and the original BNS model from [18], we do not change the clock of the subordinators to $t \mapsto \lambda t$. This formulation is equivalent and more handy in the upcoming multivariate construction.

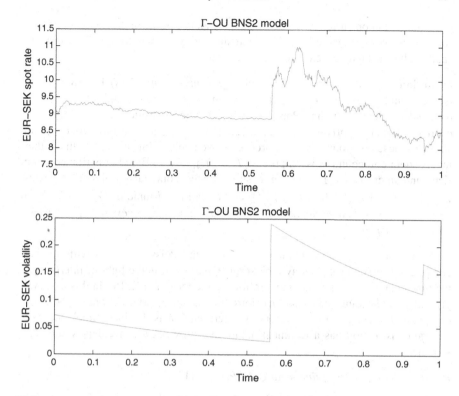

Fig. 2 Sample path of a two-sided BNS model, generated from calibrated parameters. The FX rate process exhibits positive and negative jumps

3 A Tractable Multivariate Extension of the Two-Sided Γ-OU-BNS Model

We now present a multivariate two-sided Γ-OU-BNS model, where the univariate processes still follow the dynamics of a two-sided Γ-OU-BNS model. Here, all univariate FX rate processes live on the same probability space and the probability measure is assumed to be a pricing measure. Besides establishing dependence between the driving Brownian motions, we want to incorporate dependence to the Lévy drivers, thus establishing dependence among the price jumps as well as among the variance processes. Jumps in FX rates are mainly driven by unanticipated macroeconomic events (e.g. interest-rate decisions of some central bank) in one of the monetary areas. If we consider a multivariate model with one common currency, e.g. modeling the EUR-USD and the EUR-CHF exchange rates, it is likely that jumps caused by macroeconomic events in the common currency monetary area have an impact on all exchange rates, e.g. the debt crisis of Eurozone countries should affect both the EUR-USD as well as the EUR-CHF exchange rate. Hence, dependence of the jump processes seems to be a desirable feature of a multivariate model for FX rates with common currency. To establish dependence between the compound Poisson drivers, we employ the time-change methodology presented in [15], yielding an analytically tractable and easy-to-simulate setup.

Definition 1 (*Time-changed CPPs with exponential jump sizes*) Let $c_0, \eta_1, \ldots,$ $\eta_d > 0$ and $c_1, \ldots, c_d \in (0, c_0)$. Furthermore, let $d \in \mathbb{N}$ and $Y^{(1)}, \ldots, Y^{(d)}$ be d independent compound Poisson processes with intensities $c_1/(c_0 - c_1), \ldots,$ $c_d/(c_0 - c_d)$ and $\text{Exp}(c_0\eta_1/(c_0 - c_1)), \ldots, \text{Exp}(c_0\eta_d/(c_0 - c_d))$-distributed jump sizes. To these compound Poisson processes, we apply a time change with another independent compound Poisson process $T = \{T_t\}_{t \geq 0}$ with $\text{Exp}(1)$-distributed jump sizes and intensity c_0. Define the T-subordinated compound Poisson processes $Z^{(1)}, \ldots, Z^{(d)}$ by $\{Z_t^{(j)}\}_{t \geq 0} := \{Y_{T_t}^{(j)}\}_{t \geq 0}$. We call the d-tuple of $(Z^{(1)}, \ldots, Z^{(d)})$ a *time-change-dependent multivariate compound Poisson process with parameters* $(c_0, c_1, \ldots, c_d, \eta_1, \ldots, \eta_d)$.

At first sight, the subordination of a compound Poisson process with another compound Poisson process may look strange, particular in the light of interpreting the time change as "business time", following the idea of [14]. But in this case, we primarily use the joint subordination to introduce dependence via joint jumps between compound Poisson processes without the interpretation as "business time", the time change construction has a technical nature and provides a convenient simulation scheme.

Remark 1 (*Properties of time-changed CPPs, cf.* [15])

(i) Each coordinate of the T-subordinated compound Poisson process $Z^{(j)}$ is again a compound Poisson process with intensities c_j and jump size distribution $\text{Exp}(\eta_j)$ for all $j = 1, \ldots, d$.

(ii) For $c_{\max} := \max_{1 \leq j \leq d} \{c_j\}$, the correlation coefficient of $(Z^{(j)}, Z^{(k)})$, $1 \leq j \leq d$, $1 \leq k \leq d$, $j \neq k$ is given by

$$\text{Corr}\left[Z_t^{(j)}, Z_t^{(k)}\right] = \frac{\sqrt{c_j c_k}}{c_0} = \kappa \frac{\sqrt{c_j c_k}}{c_{\max}},$$

with $\kappa := c_{\max}/c_0 \in (0, 1)$. We call κ the *time-change correlation parameter*. In particular, correlation coefficients ranging from zero to $\sqrt{c_j c_k}/c_{\max}$ are possible, and the correlation does not depend on the point in time t.

(iii) Due to the common time change, the compound Poisson processes $Z^{(1)}, \ldots, Z^{(d)}$ are stochastically dependent. Moreover, it can be shown that the dependence structure of the d-dimensional process $(Z^{(1)}, \ldots, Z^{(d)})$ is driven solely by the time-change correlation parameter κ.

A striking advantage of introducing dependence among the jumps in this manner is that the time-changed processes $Z^{(1)}, \ldots, Z^{(d)}$ remain in the class of compound Poisson processes with exponential jump heights, which ensures that the marginal processes maintain a tractable structure. In particular, the characteristic functions of the univariate log-price processes in a two-sided Γ-OU-BNS model are still at hand. Moreover, the univariate processes $Z^{(1)}, \ldots, Z^{(d)}$ can be simulated as ordinary compound Poisson processes with exponentially distributed jump heights and the Laplace transform is given. Hence, we can now define a multidimensional two-sided Γ-OU-BNS model with dependent jumps.

Definition 2 (*Multivariate two-sided Γ-OU-BNS model*) A d-dimensional stochastic process $\{S_t\}_{t \geq 0}$ with $S_t = (S_t^{(1)}, \ldots, S_t^{(d)})$ follows a *multivariate two-sided Γ-OU-BNS model with time-change-dependent volatility drivers*, if the dynamics of the log-price vector $X_t = (X_t^{(1)}, \ldots, X_t^{(d)}) = (\log S_t^{(1)}, \ldots, \log S_t^{(d)})$ are governed by the following SDEs:

$$dX_t^{(j)} = \left(\mu_j + \beta_j \left(\sigma_t^{(j)}\right)^2\right) dt + \sigma_t^{(j)} dW_t^{(j)} + \rho_+^{(j)} dZ_t^{+(j)} + \rho_-^{(j)} dZ_t^{-(j)},$$

$$d\left(\sigma_t^{(j)}\right)^2 = -\lambda_j \left(\sigma_t^{(j)}\right)^2 dt + dZ_t^{+(j)} + dZ_t^{-(j)},$$

with $(W^{(1)}, \ldots, W^{(d)})$ being correlated Brownian motions with correlation matrix Σ and for all $1 \leq j \leq d$, $\mu_j, \beta_j \in \mathbb{R}$, $\rho_+^{(j)} > 0$, $\rho_-^{(j)} < 0$, $\lambda_j > 0$, and $(Z^{+(1)}, Z^{-(1)}), \ldots, (Z^{+(d)}, Z^{-(d)})$ are pairs of independent compound Poisson processes with exponential jumps. Furthermore, the $2d$-dimensional Lévy process $(Z^{+(1)}, Z^{-(1)}, \ldots, Z^{+(d)}, Z^{-(d)})$ splits up in two time-change-dependent d-tuples of compound Poisson processes (cf. Definition 1).

At first glance, Definition 2 looks cumbersome, but it is necessary to capture all combinations of possible dependence. As a simplifying example, one might think about introducing dependence between $(Z^{+(1)}, \ldots, Z^{+(d)})$ on the one hand and between $(Z^{-(1)}, \ldots, Z^{-(d)})$ on the other hand. In this case, positive jumps of the processes are mutually dependent and negative jumps are mutually dependent, but positive jumps occur independently of negative jumps. A closer examination how to establish the dependence structure between the time-change-dependent compound Poisson processes is made in the following section, since dependence between the jumps has to be introduced in a sound economic manner.

This construction can further be generalized by employing Lévy processes, coupled by Lévy copulas (cf. [11]). For the present investigation, however, we prefer the time-change construction presented in Definition 1, since this construction provides an immediate stochastic representation of the dependence structure. Thus, a straightforward simulation scheme is provided and at least some analytical tractability when doing computational exercises is ensured, which may be more complicated when employing general Lévy copulas.

Remark 2 (*Calibration of the univariate processes*) An immediate corollary from the compound Poisson structure of the univariate jump processes $(Z^{+(j)}, Z^{-(j)})$, $j = 1, \ldots, d$, is that the univariate log-price processes $\{X_t^{(j)}\}_{t \geq 0}$, $j = 1, \ldots, d$, still follow a univariate two-sided Γ-OU-BNS model and the parameters of the univariate processes may be calibrated separately to univariate derivative prices.

The dependence parameters, which are the correlation matrix Σ of the Brownian motions and the time-change correlation parameters $\check{\kappa}$ and $\hat{\kappa}$ that determine the dependence structure of the time-change-dependent multivariate compound Poisson processes, can be calibrated separately afterwards without altering the already fixed marginal distributions. This simplifies the model calibration and is a convenient feature for practical purposes, because it automatically ensures that univariate derivative prices are fitted to the multivariate model.

4 Modeling Two FX Rates with a Bivariate Two-Sided Γ-OU-BNS Model

In this section, we discuss the modeling of FX rates with a bivariate two-sided Γ-OU-BNS model. Particularly, we discuss how to soundly introduce dependence between the Lévy drivers and investigate a possible "built-in" model induced by the model for the two FX rates. We concentrate on the case of two currency pairs, which illustrates the problems of choosing the jump dependence structure best.

To ensure familiarity with the FX markets wording, we recall that an FX rate is the exchange rate between two currencies, expressed as a fraction. The currency in the numerator of the fraction is called (by definition) *domestic currency*, while the currency in the denominator of the fraction is called *foreign currency*.[2] The role each currency plays in an FX rate is defined by market conventions and is often due to historic reasons, so economic interpretations are not necessarily helpful. A more detailed discussion of market conventions of FX rates and derivatives is provided in [22], a standard textbook on FX rates modeling is [13].

4.1 The Dependence Structure of the Lévy Drivers

Analogously to the multivariate classical Γ-OU-BNS model described in the previous section, we use the time-change construction to introduce dependence between the compound Poisson drivers in the bivariate two-sided Γ-OU-BNS model. Since we want to model dependence between the jumps in different FX rates, we have to choose the coupling of the compound Poisson drivers carefully and in a way to capture economic intuition: When modeling two FX rates, we may want to establish an adequate kind of dependence between the different drivers, accounting separately for positive and negative jumps in the respective FX rate. Depending on which currency is foreign or domestic in the two currency pairs of the FX rates, dependence may be

---

[2] The wording "foreign" and "domestic" currency does not necessarily reflect whether the currency is foreign or domestic from the point of view of a market participant. The currency EUR, e.g., is *always* foreign currency by market convention. Sometimes, the foreign currency is called *underlying currency*, while the domestic currency is called *accounting* or *base currency*.

introduced in a different manner to result in sound economic situations. Hence, we can distinguish between the following combinations that may occur for two different FX rates:

1. There are no common currencies, e.g. in the case of EUR-CHF and USD-JPY.
2. In both FX rates the common currency is the foreign (resp. domestic) currency, e.g. EUR-USD and EUR-CHF (EUR-CHF and USD-CHF, respectively).
3. The common currency is the domestic currency in one FX rate and the foreign currency in the other FX rate, e.g. EUR-USD and USD-CHF.

For the sake of simplicity, we restrict ourselves to the second case, which occurs in a detailed numerical study in the following section. The other cases can be treated analoguously.

In case of a common foreign currency, a sudden macroeconomic event strengthening (resp. weakening) the common currency should result in an upward (resp. downward) jump of both FX rates. Hence, it may be a sensible choice to couple the drivers for the positive jumps and to separately couple the drivers for the negative jumps respectively, to ensure the occurrence of joint upward and downward jumps.

4.2 Implicitly Defined Models

When two FX rates are modeled and among the two rates there is a common currency, this bivariate model always implicitly defines a model for the missing currency pair which is not modeled directly, e.g. when modeling EUR-USD and EUR-CHF exchange rates simultaneously, the quotient process automatically implies a model for the USD-CHF exchange rate. Similar to the bivariate Garman–Kohlhagen model, modeling two FX rates directly by a bivariate two-sided BNS model does not necessarily imply a similar model for the quotient or product process from the same family, but the main structure of a jump-diffusion-type model is maintained.

Lemma 1 (Quotient and product process of a two-sided BNS model) *Given two asset-price processes $\{S_t^{(1)}\}_{t \geq 0}$ and $\{S_t^{(2)}\}_{t \geq 0}$ modeled by a multivariate two-sided Γ-OU-BNS models, the product and quotient processes $\{S_t^{(1)} S_t^{(2)}\}_{t \geq 0}$ resp. $\{S_t^{(1)}/S_t^{(2)}\}_{t \geq 0}$ are both of jump-diffusion type.*

Proof Follows directly from $\log(S_t^{(1)} S_t^{(2)}) = X_t^{(1)} + X_t^{(2)}$ and $\log(S_t^{(1)}/S_t^{(2)}) = X_t^{(1)} - X_t^{(2)}$.

Due to symmetry in FX rates, the implied model for the third missing FX rate can be used to calibrate the parameters steering the dependence, namely, the correlation between the Brownian motions as well as the time-change correlation parameters, or equivalently the intensities of the time-change processes. Additionally, the calibration performance of the implied model to plain vanilla options yields a plausibility check whether the bivariate model may be useful for the evaluation of true bivariate options, e.g. best-of-two options or spread options.

5 Application: Calibration to FX Rates and Pricing of Bivariate FX Derivatives

In this section, we describe the calibration process of a bivariate two-sided BNS model to market prices of univariate FX derivatives, which allows us to completely specify the model. Furthermore, we describe how to price bivariate FX options like, e.g., best-of-two options in a bivariate two-sided BNS model.

5.1 Data

As input data for our calibration exercise we use option data on exchange rates concerning the three currencies EUR, USD, and SEK. Since the EUR-USD exchange rate can be regarded as an implied exchange rate, i.e.

$$\frac{\text{USD}}{\text{EUR}} = \frac{\text{SEK}/\text{EUR}}{\text{SEK}/\text{USD}},$$

we model the two exchange rates EUR-SEK and USD-SEK directly with two-sided Γ-OU-BNS models as suggested in [1]. For each currency pair EUR-SEK, USD-SEK, and EUR-USD, we have the implied volatilities of 204 different plain vanilla options (different maturities, different moneyness) available as input data. The option data is as of August 13, 2012, and was provided by Thomson Reuters.

5.2 Model Setup

We consider a market with two traded assets, namely $\{\exp(r_{\text{USD}}t)S_t^{\text{USDSEK}}\}_{t\geq0}$ and $\{\exp(r_{\text{EUR}}t)S_t^{\text{EURSEK}}\}_{t\geq0}$, where S_t^{USDSEK}, S_t^{EURSEK} denote the exchange rates at time t and r_{USD}, r_{EUR}, r_{SEK} denote the risk free interest rates in the corresponding monetary areas. These assets can be seen as the future value of a unit of the respective foreign currency (in this case USD or EUR), valued in the domestic currency (which is SEK). Assume a risk-neutral measure $\mathbb{Q}^{\text{SEK}}$ to be given with numéraire process $\{\exp(r_{\text{SEK}}t)\}_{t\geq0}$, i.e. $\{\exp((r_{\text{USD}} - r_{\text{SEK}})t)S_t^{\text{USDSEK}}\}_{t\geq0}$ and $\{\exp((r_{\text{EUR}} - r_{\text{SEK}})t)S_t^{\text{EURSEK}}\}_{t\geq0}$ are martingales with respect to $\mathbb{Q}^{\text{SEK}}$, governed by the SDEs

$$dX^{\star SEK} = \left(r_{SEK} - r_\star - \frac{(\sigma_t^{\star SEK})^2}{2} - \frac{c_{\star SEK}^+ \rho_{\star SEK}^+}{\eta_{\star SEK}^+ - \rho_{\star SEK}^+} + \frac{c_{\star SEK}^- \rho_{\star SEK}^-}{\eta_{\star SEK}^- + \rho_{\star SEK}^-} \right) dt$$
$$+ \sigma_t^{\star SEK} dW_t^{\star SEK} + \rho_{\star SEK}^+ dZ_t^{+\star SEK} - \rho_{\star SEK}^- dZ_t^{-\star SEK},$$
$$d\sigma_t^{2\star SEK} = -\lambda_{\star SEK}\sigma_t^{2\star SEK} dt + dZ_t^{+\star SEK} + dZ_t^{-\star SEK},$$

for $\lambda_{\star SEK}, \rho_{\star SEK}^+, \rho_{\star SEK}^- > 0$, $\star \in \{EUR, USD\}$, $\{W_t^{EURSEK}, W_t^{USDSEK}\}_{t \geq 0}$ being a two-dimensional Brownian motion with correlation $r \in [-1, 1]$, and $\{Z_t^{+EURSEK}, Z_t^{+USDSEK}\}_{t \geq 0}$ and $\{Z_t^{-EURSEK}, Z_t^{-USDSEK}\}_{t \geq 0}$ being (independent) two-dimensional time-change dependent compound Poisson processes with parameters

$$(\max(c_{EURSEK}^+, c_{USDSEK}^+)/\kappa^+, c_{EURSEK}^+, c_{USDSEK}^+, \eta_{EURSEK}^+, \eta_{USDSEK}^+)$$
$$\text{and} \quad (\max(c_{EURSEK}^-, c_{USDSEK}^-)/\kappa^-, c_{EURSEK}^-, c_{USDSEK}^-, \eta_{EURSEK}^-, \eta_{USDSEK}^-),$$

where κ^+ and κ^- are the time-change correlation parameters (following the framework in Sect. 4.1). Hence, the EUR-SEK, EUR-USD exchange rates follow a bivariate SBNS model. The implied exchange rate process S^{EURUSD} is given by

$$\left\{S_t^{EURUSD}\right\}_{t \geq 0} = \left\{\frac{S_t^{EURSEK}}{S_t^{USDSEK}}\right\}_{t \geq 0}.$$

Due to the change-of-numéraire formula for exchange rates (cf. [19]), the process $\{\exp((r_{EUR} - r_{USD})t)S_t^{EURUSD}\}_{t \geq 0}$ is a martingale with respect to $\mathbb{Q}^{USD}$, where $\mathbb{Q}^{USD}$ is determined by the Radon–Nikodým derivative

$$\left.\frac{d\mathbb{Q}^{USD}}{d\mathbb{Q}^{SEK}}\right|_t = \frac{S_t^{USDSEK} \exp(r_{USD}t)}{S_0^{USDSEK} \exp(r_{SEK}t)}.$$

5.3 Calibration

For calibration purposes, we use the volatility surfaces of the EUR-SEK and USD-SEK exchange rates to fit the univariate parameters. Due to the consistency relationships which have to hold between the exchange rates, we can calibrate the dependence parameters by fitting them to the volatility surface of EUR-USD. Even in presence of other "bivariate options" (e.g. best-of-two options), we argue that European options on the quotient exchange rate currently provide the most liquid and reliable data for a calibration.

The calibration of the presented multivariate model is done in two steps. Due to the fact that the marginal distributions can be separated from the dependence structure within our models, it is possible to keep the parameters governing the dependence

Table 1 Calibrated parameters in the two univariate FX models

$\star$	$S_0^{\star \text{SEK}}$	$\sigma_0^{\star \text{SEK}}$	$c_{\star \text{SEK}}$	$\eta_{\star \text{SEK}}$	$\lambda_{\star \text{SEK}}$	$\rho_{\star \text{SEK}}$	#options	Error (%)
EUR	8.229	0.074	0.71	62.13	3.25	1.66	204	1.08
USD	6.664	0.078	1.15	40.81	2.19	1.22	204	3.17

separated from the parameters governing the marginal distributions. Therefore, in a first step we independently calibrate both univariate models for the EUR-SEK and USD-SEK exchange rates, and in a second step we calibrate the parameters driving the dependence structure. In doing so, the fixed univariate parameters are not affected by the second step. Since there is little market data of multi-currency options, this two step method is very appealing: we can disintegrate one big calibration problem in two smaller ones. The univariate models are calibrated to volatility surfaces of the EUR-SEK and USD-SEK exchange rates via minimizing the relative distance of the model implied option prices to market prices, with equal weight on every option. Option prices in the univariate two-sided BNS models are obtained via Fourier inversion (cf. [5, 21]) by means of the characteristic function of the log-prices.

Table 1 gives an overview of the calibration result of the univariate models. To reduce the number of parameters, we use symmetric two-sided Γ-OU-BNS models as described in [1]. Furthermore, we assume that the time-change correlation parameters κ^+ and κ^- coincide; maintaining the symmetric structure. The relative error in model prices with respect to market prices of the 204 options can be seen as calibration error. The average relative error in the EUR-SEK-model is about one percent, and in the USD-SEK-model it is around three percent. Hence, the univariate models fit the FX market reasonably well. Each univariate calibration requires about 20 s.

The calibration of the parameters governing the dependence is done by means of the third implied exchange rate, namely by the volatility surface of EUR-USD. Model prices of EUR-USD-options with payout function f at time t can be obtained by a Monte-Carlo simulation of the following expected value:

$$
\mathbb{E}_{\mathbb{Q}^{\text{USD}}} \left[f(S_t^{\text{EURUSD}}) \exp\left(-r_{\text{USD}}t\right) \right]
$$
$$
= \mathbb{E}_{\mathbb{Q}^{\text{SEK}}} \left[f\left(\frac{S_t^{\text{EURSEK}}}{S_t^{\text{USDSEK}}}\right) \frac{S_t^{\text{USDSEK}}}{S_0^{\text{USDSEK}}} \exp(-r_{\text{SEK}}t) \right] \tag{1}
$$

Here, we used 100,000 simulations to calibrate the dependence parameters. The execution of the overall optimization procedure takes around four hours. The calibration error of the dependence parameters in terms of average relative error is roughly nine percent, which is still a good result giving consideration to the fact that we try to fit 204 market prices by means of just two parameters in an implicitly specified model. A more complex model, obtained by relaxing the condition that κ^+ and κ^-

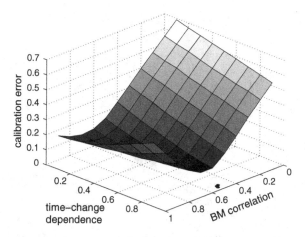

Fig. 3 The best matching correlation between the two Brownian motions is 0.52 and the optimal time-change dependence parameter is $\kappa = 0.96$. This corresponds to a calibration error of around nine percent for the 204 options on this currency pair

coincide, leads to even smaller calibration errors. However, we keep the model as simple as possible to maintain tractability. Figure 3 illustrates the calibration error of this second step depending on different choices of the dependence parameters. Eventually, the whole model is fixed.

Now, we are able to price European multi-currency options, for instance a best-of-two call option with a payoff at time t given by

$$\max\left\{ \max\left\{ \frac{e^{X_T^{\text{USDSEK}}}}{S_0^{\text{USDSEK}}} - K, 0 \right\}, \max\left\{ \frac{e^{X_T^{\text{EURSEK}}}}{S_0^{\text{EURSEK}}} - K, 0 \right\} \right\},$$

i.e. we consider the maximum of two call options with strike $K > 0$ on two exchange rates. This option can be used as an insurance against a weakening SEK, because one gets a payoff if the relative performance of one exchange rate, USD-SEK or EUR-SEK, is greater than $K - 1$. Pricing is done by a Monte-Carlo simulation that estimates the expected value in Eq. (1). We used 100,000 scenarios to price this option, which takes about four minutes. Figure 4 shows option prices of the best-of-two call option dependent on various choices of the dependence parameters.

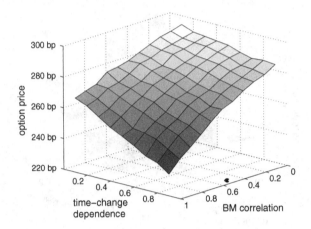

Fig. 4 Prices of a best-of-two call option where $K = 1.1$ and $T = 1$. One observes that both dependence parameters play an important role for the price of this option. For the optimal parameter setting (Brownian motion correlation is 0.52, $\kappa = 0.96$), the fair price of this option is 261 bp

6 Conclusion and Outlook

We introduced a multi-dimensional FX rate model generalizing the univariate two-sided BNS model in a way that each FX rate is still modeled as a two-sided BNS model. Thus, the parameters driving the dependence structure can be separated from the marginal distributions. This simplifies the calibration of the overall model tremendously, such that the multicurrency model can be calibrated to plain vanilla FX option prices. As an outlook for further research, we wonder whether there exists a measure change from the real world measure to the martingale measure we assumed to exist in the first place.

Acknowledgments We thank the KPMG Center of Excellence in Risk Management for making this work possible. Furthermore, we thank the TUM Graduate School for supporting these studies and two anonymous referees for valuable comments.

Open Access This chapter is distributed under the terms of the Creative Commons Attribution Noncommercial License, which permits any noncommercial use, distribution, and reproduction in any medium, provided the original author(s) and source are credited.

References

1. Bannör, K., Scherer, M.: A BNS-type stochastic volatility model with two-sided jumps, with applications to FX options pricing. Wilmott **2013**(65), 58–69 (2013)
2. Barndorff-Nielsen, O.E., Shephard, N.: Non-Gaussian Ornstein-Uhlenbeck-based models and some of their uses in financial economics. J. R. Stat. Soc.: Ser. B (Stat. Methodol.) **63**, 167–241 (2001)

3. Bates, D.: Jumps and stochastic volatility: exchange rate processes implicit in Deutsche Mark options. Rev. Financ. Stud. **9**(1), 69–107 (1996)
4. Black, F., Scholes, M.: The pricing of options and corporate liabilities. J. Polit. Econ. **81**(3), 637–654 (1973)
5. Carr, P., Madan, D.: Option valuation using the fast Fourier transform. J. Comput. Financ. **2**, 61–73 (1999)
6. Duffie, D., Pan, J., Singleton, K.: Transform analysis and asset pricing for affine jump-diffusions. Econometrica **68**, 1343–1376 (2000)
7. Eberlein, E., Koval, N.: A cross-currency Lévy market model. Quant. Financ. **6**(6), 465–480 (2006)
8. Garman, M., Kohlhagen, S.: Foreign currency option values. J. Int. Money Financ. **2**, 231–237 (1983)
9. Heston, S.: A closed-form solution for options with stochastic volatility with applications to bond and currency options. Rev. Financ. Stud. **6**(2), 327–343 (1993)
10. Jacod, J., Todorov, V.: Testing for common arrivals of jumps for discretely observed multidimensional processes. Ann. Stat. **37**(4), 1792–1838 (2009)
11. Kallsen, J., Tankov, P.: Characterization of dependence of multidimensional Lévy processes using Lévy copulas. J. Multivar. Anal. **97**(7), 1551–1572 (2006)
12. Kou, S.G.: A jump-diffusion model for option pricing. Manag. Sci. **48**(8), 1086–1101 (2002)
13. Lipton, A.: Mathematical Methods For Foreign Exchange: A Financial Engineer's Approach. World Scientific Publishing Company, River Edge (2001)
14. Luciano, E., Schoutens, W.: A multivariate jump-driven financial asset model. Quant. Financ. **6**(5), 385–402 (2006)
15. Mai, J.-F., Scherer, M., Schulz, T.: Sequential modeling of dependent jump processes. Wilmott **2014**(70), 54–63 (2014)
16. Merton, R.: Option pricing when underlying stock returns are discontinuous. J. Financ. Econ. **3**(1), 125–144 (1976)
17. Muhle-Karbe, J., Pfaffel, O., Stelzer, R.: Option pricing in multivariate stochastic volatility models of OU type. SIAM J. Financ. Math. **3**, 66–94 (2012)
18. Nicolato, E., Venardos, E.: Option pricing in stochastic volatility models of the Ornstein-Uhlenbeck type. Math. Financ. **13**(4), 445–466 (2003)
19. Pelsser, A.: Mathematical foundation of convexity correction. Quant. Financ. **3**(1), 59–65 (2003)
20. Pigorsch, C., Stelzer, R.: A multivariate generalization of the Ornstein-Uhlenbeck stochastic volatility model, working paper (2008)
21. Raible, S.: Lévy Processes in finance: theory, numerics, and empirical facts, PhD thesis, Albert-Ludwigs-Universität Freiburg i. Br. (2000)
22. Reiswich, D., Wystup, U.: FX volatility smile construction. Wilmott **60**, 58–69 (2012)
23. Stein, E., Stein, J.: Stock price distributions with stochastic volatility: an analytic approach. Rev. Financ. Stud. **4**(4), 727–752 (1991)

Modeling the Price of Natural Gas with Temperature and Oil Price as Exogenous Factors

Jan Müller, Guido Hirsch and Alfred Müller

Abstract The literature on stochastic models for the spot market of gas is dominated by purely stochastic approaches. In contrast to these models, Stoll and Wiebauer [14] propose a fundamental model with temperature as an exogenous factor. A model containing only deterministic, temperature-dependent and purely stochastic components, however, still seems not able to capture economic influences on the price. In order to improve the model of Stoll and Wiebauer [14], we include the oil price as another exogenous factor. There are at least two fundamental reasons why this should improve the model. First, the oil price can be considered as a proxy for the general state of the world economy. Furthermore, pricing formulas in oil price indexed gas import contracts in Central Europe are covered by the oil price component. It is shown that the new model can explain price movements of the last few years much better than previous models. The inclusion of oil price and temperature in the regression of a least squares Monte Carlo method leads to more realistic valuation results for gas storages and swing options.

Keywords Gas spot price · Oil price model · Temperature · Gas storage valuation · Least squares Monte Carlo · Seasonal time series model

1 Introduction

During the last years trading in natural gas has become more important. The traded quantities over-the-counter and on energy exchanges have strongly increased and new products have been developed. For example, swing options increase the flexibility of

J. Müller (✉) · A. Müller
Department Mathematik, Universität Siegen, Walter-Flex-Str. 3, 57072 Siegen, Germany
e-mail: jan.mueller@uni-siegen.de

A. Müller
e-mail: mueller@mathematik.uni-siegen.de

G. Hirsch
EnBW Energie Baden-Württemberg AG, Durlacher Allee 93, 76131 Karlsruhe, Germany
e-mail: gu.hirsch@enbw.com

© The Author(s) 2015
K. Glau et al. (eds.), *Innovations in Quantitative Risk Management*,
Springer Proceedings in Mathematics & Statistics 99,
DOI 10.1007/978-3-319-09114-3_7

109

suppliers and they are used as an instrument for risk management purposes. Important facilities for the security of supply are gas storages.

These are two examples of complex American-style real options that illustrate the need for reliable pricing methods. Both options rely on nontrivial trading strategies where exercise decisions are taken under uncertainty. Therefore, analytic pricing formulas cannot be expected. The identification of an optimal trading strategy under uncertainty is a typical problem of stochastic dynamic programming, but even then numerical solutions are difficult to obtain due to the curse of dimensionality. Therefore, simulation-based approximation algorithms have been successfully applied in this area. Longstaff and Schwartz [9] introduced the least square Monte Carlo method for the valuation of American options. Meinshausen and Hambly [10] extended the idea to Swing options, and Boogert and de Jong [5] applied it to the valuation of gas storages. Their least squares Monte Carlo algorithm requires a stochastic price model for daily spot prices generating adequate gas price scenarios. We prefer this approach to methods using scenario trees or finite differences as it is independent of the underlying price process.

The financial literature on stochastic gas price models is dominated by purely stochastic approaches. The one- and two-factor models by Schwartz [12] and Schwartz and Smith [13] are general approaches applicable to many commodities, such as oil and gas. The various factors represent short- and long-term influences on the price. An important application of gas price models is the valuation of gas storage facilities. Within this context, Chen and Forsyth [7] and Boogert and de Jong [6] propose gas price models. Chen and Forsyth [7] analyze regime-switching approaches incorporating mean-reverting processes and random walks. The class of factor models is extended by Boogert and de Jong [6]. The three factors in their model represent short- and long-term fluctuations as well as the behavior of the winter–summer spread. In contrast to these models, Stoll and Wiebauer [14] propose a fundamental model with temperature as an exogenous factor. They use the temperature component as an approximation of the filling level of gas storages, which have a remarkable influence on the price.

There is a fundamental difference between the model of Stoll and Wiebauer [14] and the other models mentioned before as far as their stochastic behavior is concerned. Incorporating cumulated heating degree days over a winter as an explanatory variable leads to a seasonal effect in the variance of the prices. In this model the variance of the gas prices increases over the winter depending on the actual weather conditions and has a maximum at the end of winter. This is much more in line with the observations than the behavior of the model of Boogert and de Jong [6] where the variance of the gas price has a minimum at the end of winter as there is no effect of the winter–summer spread used there. Another major difference is the use of exogenous variables that can be observed and thus the optimal exercising decision for American-style options depends on these variables and therefore also the price of these real options will be different.

In this paper we extend the model of Stoll and Wiebauer [14] by introducing another exogenous factor to their model: the oil price. There are at least two reasons why we believe that this is useful. The main reason is that an oil price component can be considered as a proxy for the state of the world economy in the future. In contrast to other indicators, such as the gross domestic product (GDP), futures prices for oil are available on a daily basis. Furthermore, the import prices for gas in countries such as Germany are known to be oil price indexed.

Apart from the GDP or oil price there might be more candidates as an explanatory variable in the model. The most natural choice would be the forward gas price. We prefer the oil price as it gives us the chance to valuate gas derivatives that are oil price indexed, as is often the case for gas swing contracts. For the valuation of such swing contracts gas price scenarios are needed as well as corresponding oil price scenarios. This application is hardly possible with explanatory variables other than the oil price.

The rest of the paper is organized as follows. In Sect. 2 we introduce the model by Stoll and Wiebauer [14] including a short description of their model for the temperature component. In Sect. 3 we discuss the need for an oil price component in the model. The choice of the component in our model is explained. Then we fit the model to data in Sect. 4. The new model is used within a least squares Monte Carlo algorithm for valuation of gas storages and swing options in Sect. 5. The exogenous factors are included in the regression to approximate the continuation value. We finish with a short conclusion in Sect. 6.

2 A Review of the Model by Stoll and Wiebauer (2010)

Modeling the price of natural gas in Central Europe requires knowledge about the structure of supply and demand. On the supply side there are only a few sources in Central Europe, while most of the natural gas is imported from Norway and Russia. On the demand side there are mainly three classes of gas consumers: Households, industrial companies, and gas fired power plants. While households only use gas for heating purposes at low temperatures, industrial companies use gas as heating and process gas. Households and industrial companies are responsible for the major part of the total gas demand.

These two groups of consumers cause seasonalities in the gas price:

- Weekly seasonality: Many industrial companies need less gas on weekends as their operation is restricted to working days.
- Yearly seasonality: Heating gas is needed mainly in winter at low temperatures.

An adequate gas price model has to incorporate these seasonalities as well as sto-chastic deviations of these.

Stoll and Wiebauer [14] propose a model meeting these requirements and incor-porating another major influence factor: the temperature. To a certain extent the temperature dependency is already covered by the deterministic yearly seasonality. This component describes the direct influence of temperature: The lower the tem-perature, the higher the price. But the temperature influence is more complex than this. A day with average temperature of $0\,°C$ at the end of a long cold winter has a different impact on the price than a daily average of zero at the end of a "warm" winter. Similarly, a cold day at the end of a winter has a different impact on the price than a cold day at the beginning of the winter.

The different impacts are due to gas storages that are essential to cover the demand in winter. The total demand for gas is higher than the capacities of the gas pipelines from Norway and Russia. Therefore, gas providers use gas storages. These storages are filled during summer (at low prices) and emptied in winter months. At the end of a long and cold winter most gas storages will be rather empty. Therefore, additional cold days will lead to comparatively higher prices than in a normal winter.

The filling level of all gas storages in the market would be the adequate variable to model the gas price. However, these data are not available as they are private information. Therefore, we need a proxy variable for it. As the filling levels of gas storages are strongly related to the demand for gas which in turn depends on the temperature, an adequate variable can be derived from the temperature.

Stoll and Wiebauer [14] use *normalized cumulated heating degree days* to cover the influence of temperature on the gas price. They define a temperature of $15\,°C$ as the limit of heating. Any temperature below $15\,°C$ makes households as well as companies switch on their heating systems. Heating degree days are measured by $HDD_t = \max(15 - T_t, 0)$, where T_t is the average temperature of day t. As mentioned above the impact on the price depends on the number of cold days observed so far in the winter. In this context, we refer to winter as 1 October and the 181 following days till end of March. We will write $HDD_{d,w}$ for HDD_t, if t is day number d of winter w. Cumulation of heating degree days over a winter leads to a number indicating how cold the winter has been so far. Then we can define the cumulated heating degree days on the day d in winter w as

$$CHDD_{d,w} = \sum_{k=1}^{d} HDD_{k,w} \text{ for } 1 \leq d \leq 182. \tag{1}$$

The impact of cumulated heating degree days on the price depends on the comparison with a normal winter. This information is included in normalized cumulated heating degree days

$$\Lambda_{d,w} = CHDD_{d,w} - \frac{1}{w-1} \sum_{\ell=1}^{w-1} CHDD_{d,\ell} \text{ for } 1 \leq d \leq 182. \tag{2}$$

We use Λ_t instead of $\Lambda_{d,w}$ for simplicity, if t is a day in a winter. The definition of Λ_t for a summer day is described by a linear return to zero during summer. This reflects the fact that we use Λ_t as a proxy variable for filling levels of gas storages. Assuming a constant filling rate during summer we thus get the linear part of normalized cumulated heating degree days (see Fig. 1). Positive values of Λ_t describe winters colder than the average. Λ_t is included into the gas price model by a regression approach. As the seasonal components and the normalized cumulated

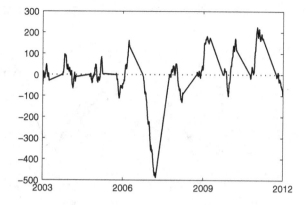

Fig. 1 Normalized cumulated heating degree days calculated based on temperature data from Eindhoven, Netherlands, for 2003–2011 (Source: Royal Netherlands Meteorological Institute)

heating degree days are linear with respect to the parameters, we can use ordinary least squares regression for parameter estimation. The complete model can be written as

$$G_t = m_t + \alpha \cdot \Lambda_t + X_t^{(G)} + Y_t^{(G)} \tag{3}$$

with the day-ahead price of gas G_t, the deterministic seasonality m_t, the normalized cumulated heating degree days Λ_t, an ARMA process $X_t^{(G)}$, and a geometric Brownian motion $Y_t^{(G)}$. For model calibration day-ahead gas prices from TTF market (Source: ICE) are used. The Dutch gas trading hub TTF offers the highest trading volumes in Central Europe. As corresponding temperature data we choose daily average temperatures from Eindhoven, Netherlands (Source: Royal Netherlands Meteorological Institute). The fit to historical prices before the crisis can be seen in Fig. 2. Outliers have been removed (see Sect. 4 for details on treatment of outliers).

3 The Oil Price Dependence of Gas Prices

The model described in Eq. (3) is capable to cover all influences on the gas price related to changes in temperature. But changes in the economic situation are not covered by that model. This was clearly observable in the economic crisis 2008/2009 (see Fig. 5). During that crisis the demand for gas by industrial companies in Central Europe was falling by more than 10 %. As a consequence the gas price rapidly decreased by more than 10 Euro per MWh.

The oil price showed a similar behavior in that period. Economic changes are the main drivers for remarkable changes in the oil price level. Short-term price movements caused by speculators or other effects cause deviations from the price level that represents the state of the world economy. Therefore, gas price changes often correspond to long-term changes in the oil price level. Such an influence can be

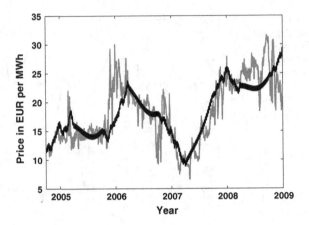

Fig. 2 $m_t + \alpha \cdot \Lambda_t$ from Eq. (3) (*black*) fitted to TTF prices from 2004–2009 (*grey*)

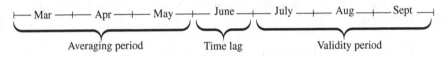

Fig. 3 In a 3-1-3 formula the price is determined by the average price of 3 months (March to May). This price is valid for July–September. The next day of price fixing is 1 October

modeled by means of a moving average of past oil prices. The averaging procedure removes short-term price movements if the averaging period is chosen sufficiently long. The result is a time series containing only the long-term trends of the oil price. Using such an oil price component in a gas price model explains the gas price movements due to changes in the economic situation. This consideration is in line with He et al. [8]. They identify cointegration between crude oil prices and a certain indicator of global economic activity.

Another important argument for the use of this oil price component is based on Central European gas markets. Countries such as Germany import gas via long-term contracts that are oil price indexed. This indexation can be described by three parameters:

1. The number of averaging months. The gas price is the average of past oil prices within a certain number of months.
2. The time lag. Possibly, there is a time lag between the months the average is taken of and the months the price is valid for.
3. The number of validity months. The price is valid for a certain number of months.

An example of a 3-1-3 formula is given in Fig. 3.

The formulas used in the gas import contracts are not known to all market participants. Theoretically, any choice of three natural numbers is possible. But from other products, like oil price indexed gas swing options, we know that some formulas are more popular than others. Examples of common formulas are 3-1-1, 3-1-3, 6-1-1,

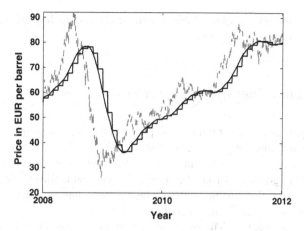

Fig. 4 The oil price (*grey*), the 6-0-1 formula (*black step function*) and the moving average of 180 days (*black*)

6-1-3, and 6-3-3. Therefore, we assume that these formulas are relevant for import contracts as well.

As there are many different import contracts with possibly different price formulas we cannot be sure that one of the mentioned formulas is responsible for the price behavior on the market. The mixture of different formulas might affect the price in the same way as a common formula or a similar one.

Evaluation of the formula leads to price jumps every time the price is fixed. The impact on the gas price will be smoother, however. The new gas price determined on a fixing day is the result of averaging a number of past oil prices. The closer to the fixing day the more prices for the averaging are known. Therefore, market participants have reliable estimations of the new import price. If the new price would be higher it would be cheaper to buy gas in advance and store it. This increases the day-ahead price prior to the fixing day and leads to a smooth transition from the old to the new price level on the day-ahead market.

This behavior of market participants leads to some smoothness of the price. In order to include this fact in a model a smoothed price formula can be used. A sophisticated smoothing approach for forward price curves is introduced by Benth et al. [3]. They assume some smoothness conditions in the knots between different price intervals. It is shown that splines of order four meet all these requirements and make sure that the result is a smooth curve. As our price formulas are step functions like forward price curves, this approach is applicable to our situation.

If the number of validity months is equal to one it is possible to use a moving average instead of a (smoothed) step function to simplify matters (see Fig. 4). This alternative is much less complex than the approach with smoothing by splines, and delivers comparable results. Therefore, the simpler method is applied in case of formulas with one validity month.

In the next section we compare various formulas regarding their ability to explain the price behavior on the gas market.

4 Model Calibration with Temperature and Oil Price

We now compare different formulas of oil prices in the regression model in order to find the one explaining the gas price best (see Fig. 5).

For the choice of the best formula we use the coefficient of determination R^2 as the measure of goodness-of-fit. We choose the reasonable formula leading to the highest value of R^2. Reasonable, in this context, means that we restrict our analysis to formulas that are equal or similar to the ones known from other oil price indexed products (compare Sect. 3). The result of this comparison is a 6-0-1 formula (see Fig. 6). Although this is not a common formula there is an explanation for it: The gas price decreased approximately six months later than the oil price in the crisis. This major price movement needs to be covered by the oil price component. As explained above we replace the step function by a moving average. Taking the moving average of 180 days is a good approximation of the 6-0-1 formula. All in all, the oil price component increases the R^2 as our measure of goodness-of-fit from 0.35 to 0.83 (see Fig. 5). Even if the new model is applied to data before the crisis the oil price component is significant. In that period the increase of R^2 is smaller but still improves the model.

These comparisons give evidence that both considerations in the previous section are valid. The included oil price component can be seen as the smoothed version of a certain formula. At the same time it can be considered as a variable describing economic influences indicated by the trends and level of the oil price.

Therefore, we model the gas price by the new model

$$G_t = m_t + \alpha_1 \Lambda_t + \alpha_2 \Psi_t + X_t^{(G)} \tag{4}$$

with Ψ_t being the oil price component. This means that the unobservable factor $Y_t^{(G)}$ in Stoll and Wiebauer [14] is replaced by the observable factor Ψ_t.

Parameter estimation of our model is based on the same data sources as the model by Stoll and Wiebauer [14]. However, we extend the period to 2011. Additionally, we need historical data for the estimation of the oil price component. Therefore, we use prices of the front month contracts of Brent crude oil traded on the Intercontinental Exchange (ICE) from 2002–2011. Using these data we can estimate all parameters applying ordinary least squares regression after removing outliers from the gas price data, G_t.

Outliers can be due to technical problems or a fire at a major gas storage. We exclude the prices on these occasions by an outlier treatment proposed by Weron [15], where values outside a range around a running median are declared to be outliers. The range is defined as three times the standard deviation. The identified outliers

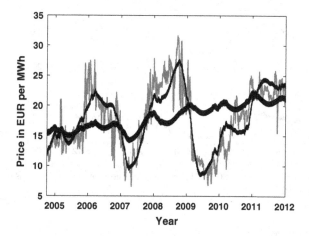

Fig. 5 The model of Stoll and Wiebauer [14] (*bold black*) and our model (*thin black line*) fitted to historical gas prices (*grey*)

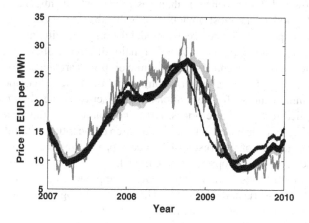

Fig. 6 Comparison of different oil price components in the model: 6-0-1 formula (*bold black*), 6-1-1 formula (*grey*) and 3-0-1 formula (*thin black*) fitted to the historical prices (*dark grey*)

are excluded in the regression. We do not remove them from our model, however, as they are still included in the estimation of the parameters of the remaining stochastic process.

Altogether, these model components give fundamental explanations for the historical day-ahead price behavior. Short-term deviations are included by a stochastic process (see Sect. 4.3). Long-term uncertainty due to the uncertain development of the oil price is included by the oil price process. Therefore, our model is able to generate reasonable scenarios for the future (see Fig. 7). We specify the stochastic models for the exogenous factors Ψ_t and Λ_t as well as the stochastic process $X_t^{(G)}$ in the following.

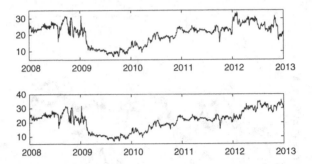

Fig. 7 The historical gas price (2008–2012) and its extensions by two realizations of the gas price process for 2012–2013

4.1 Oil Price Model

Oil prices show a different behavior than gas prices, which influences the choice of an adequate model. The most obvious fact is the absence of any seasonalities or deterministic components. Therefore, we model the oil price without a deterministic function or fundamental component. Another major difference affects the stochastic process. While the oil price and also logarithmic oil prices are not stationary the gas price is stationary after removal of seasonalities and fundamental components.

A very common model for nonstationary time series is the Brownian motion with drift applied to logarithmic prices. Drift and volatility of this process can be determined using historical data or by any estimation of the future volatility. For a stationary process, the use of an Ornstein-Uhlenbeck process or its discrete equivalent, an AR(1) process, is an appropriate simple model.

A combination of these two simple modeling approaches is given by the two-factor model by Schwartz and Smith [13]. They divide the log price into two factors: one for short-term variations and one for long-term dynamics.

$$\psi_t = \exp\left(\chi_t + \xi_t\right) \tag{5}$$

with an AR(1) process χ_t (short-term variations) and a Brownian motion ξ_t (long-term dynamics). These processes are correlated. We apply this two-factor model as it considers long- and short-term variations. The estimation of parameters in this model is more complex. The factors are not observable on the market. Following the paper by Schwartz and Smith [13] we apply the Kalman filter for parameter estimation.

The resulting process (ψ_t) is used to derive the process (Ψ_t) in Eq. (4).

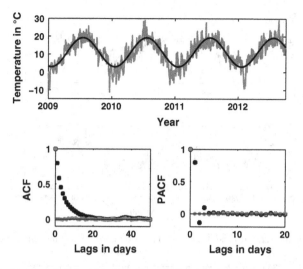

Fig. 8 *Top*: Fit of deterministic function (*black*) to the historical daily average temperature (*grey*) in Eindhoven, Netherlands. *Bottom*: Autocorrelation function (*left*) and partial autocorrelation function (*right*) of residual time series (*black*) and innovations of AR(3) process (*grey*)

4.2 Temperature Model

When modeling daily average temperature we can make use of a long history of temperature data. Here, a yearly seasonality and a linear trend can be identified. Therefore, we use a temperature model closely related to the one proposed by Benth and Benth [2].

$$T_t = a_1 + a_2t + a_3 \sin\left(\frac{2\pi t}{365.25}\right) + a_4 \cos\left(\frac{2\pi t}{365.25}\right) + X_t^{(T)} \qquad (6)$$

with $X_t^{(T)}$ being an AR(3) process. The model fit with respect to the deterministic part (ordinary least squares regression) and the AR(3) process is shown in Fig. 8. The process (T_t) (see Fig. 9) is then used to define the derived process (Λ_t) of normalized cumulated heating degree days as described in Sect. 2.

4.3 The Residual Stochastic Process

The fit of normalized cumulated heating degree days, oil price component, and deterministic components to the gas price via ordinary least squares regression (see Fig. 10) results in a residual time series. These residuals contain all unexplained, "random" deviations from the usual price behavior.

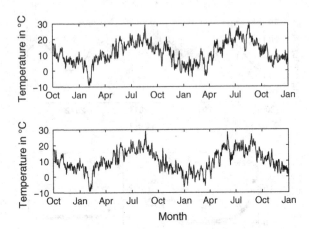

Fig. 9 Historical temperatures and its extension by two realizations of the temperature model

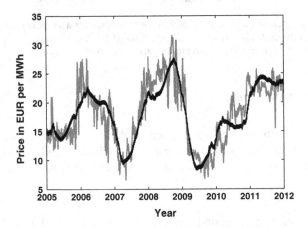

Fig. 10 Fit of deterministic function and exogenous components (*black*) to the historical gas price (*grey*)

The residuals exhibit a strong autocorrelation to the first lag. Further analysis of the partial autocorrelation function reveal an ARMA(1,2) process providing a good fit (see Fig. 11). The empirical innovations of the process show heavier tails than a normal distribution (compare Stoll and Wiebauer [14]). Therefore, we apply a distribution with heavy tails. The class of generalized hyperbolic distributions including the NIG distribution was introduced by Barndorff-Nielsen [1]. The normal-inverse Gaussian (NIG) distribution leads to a remarkably good fit (see Fig. 11).

Both the distribution of the innovations and the parameters of autoregressive processes are estimated using maximum likelihood estimation.

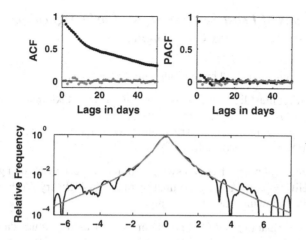

Fig. 11 *Top*: ACF (*left*) and PACF (*right*) of residual time series (*black*) and innovations of ARMA(1,2) process (*grey*). *Bottom*: Fit of NIG distribution (*grey*) to kernel density of empirical innovations (*black*)

5 Option Valuation by Least Squares Monte Carlo Including Exogenous Components

An optimal exercise of flexibility like gas storages as well as swing options is a decision under uncertainty. While the price for the next day is known, the future development of the spot prices is uncertain. Nevertheless, gas withdrawn today cannot be withdrawn on a day in the future at a possibly higher price level. The identification of an optimal trading strategy under this uncertainty is a typical problem of stochastic dynamic programming, and simulation-based approximation algorithms have been successfully applied in this area. Longstaff and Schwartz [9] introduced the least squares Monte Carlo method for the valuation of American options, Meinshausen and Hambly [10] extended the idea to swing options and Boogert and de Jong [5] applied it to the valuation of gas storages. Furthermore, Boogert and de Jong [6] found that the different components of the gas price process should be included into the regression of the least squares Monte Carlo method for the valuation of gas storages as this increases the value. While they included components that are not observable but virtual components of their price process, the price process introduced in Sect. 4 of this paper includes two exogenous and at the same time observable components. The normalized cumulated heating degree days as well as the 180 days average of the oil price can directly be observed and easily included into the exercise decision of the option that has to be done on a daily basis by a trader. The least squares Monte Carlo method including further factors is described in Sect. 5.1 and valuation examples are given in Sect. 5.2.

5.1 Extensions of Least Squares Monte Carlo Algorithm Including Exogenous Components

A gas storage is characterized by the following restrictions:

- The filling level must lie between given minimum and maximum volumes at any times $0 \le t \le T + 1$: $v_{min}(t) \le v(t) \le v_{max}(t)$
- For each day volume changes are limited by withdrawal and injection rate: $\Delta v_{min}(t, v(t)) \le \Delta v \le \Delta v_{max}(t, v(t))$

From a mathematical point of view a swing option is a special case of a gas storage. During the delivery period a daily nomination of the gas delivery for the next day is done, while the following restrictions apply:

- Daily contract quantity (DCQ): minimum as well as maximum daily volume; typical values are DCQmin 50–90 % and DCQmax 100–110 % of a given DCQ reference (where DCQ = ACQ/365)
- Annual contract quantity (ACQ): minimum as well as maximum yearly volume; typical values are ACQmin 80–90 % and ACQmax 100–110 % of a given ACQ reference

Due to these restrictions, a swing option is the same as a storage with an initial volume equal to the ACQmax of the swing

$$v_{min}(0) = v_{max}(0) = \text{ACQmax} \tag{7}$$

and the following restriction for the final volume

$$0 = v_{min}(T + 1) \le v(T + 1) \le v_{max}(T + 1) = \text{ACQmin.} \tag{8}$$

where only withdrawal is possible

$$-\text{DCQmax} = \Delta v_{min}(t, v(t)) \le \Delta v \le \Delta v_{max}(t, v(t)) = -\text{DCQmin.} \tag{9}$$

We assume that the storage is available from time $t = 0$ till time $t = T + 1$ and the holder is allowed to take an action at any discrete date $t = 1, \ldots, T$ after the spot price $S(t)$ is known. Let $v(t)$ denote the volume in storage at the start of day t and Δv the volume change during day t. In case of an injection $\Delta v > 0$, while $\Delta v < 0$ means withdrawal from the storage. The payoff on day t is

$$h(G_t, \Delta v) = \begin{cases} (-G_t - c_{WD,t}) \cdot \Delta v, & \Delta v \ge 0 \\ (-G_t - c_{IN,t}) \cdot \Delta v, & \Delta v < 0 \end{cases}. \tag{10}$$

Here $c_{WD,t}$ denotes the withdrawal costs and $c_{IN,t}$ the injection costs on day t, which can be different and may include a bid-ask spread.

Let $U(t, G_t, v(t))$ be the value of the flexibility starting at volume level $v(t)$ at time t. By $C(t, G_t, v(t), \Delta v)$ we denote the continuation value after taking an allowed action Δv from $\mathcal{D}(t, v(t))$ (the set of all admissible actions at time t if the filling level is $v(t)$). If $r(t)$ is the interest rate at time t then

$$C(t, G_t, v(t), \Delta v) = \mathbf{E}\left[e^{-r(t+1)}U(t + 1, G_{t+1}, v(t) + \Delta v)\right]. \quad (11)$$

The continuation value only depends on $v(t + 1) := v(t) + \Delta v$. Therefore, we will from now on also write $C(t, G_t, v(t + 1))$ for short. With this notation the flexibility value $U(t, G_t, v(t))$ satisfies the following dynamic program:

$$U(T + 1, G_{T+1}, v(T + 1)) = q(G_{T+1}, v(T + 1)) \quad (12)$$
$$U(t, G_t, v(t)) = \max_{\Delta v \in \mathcal{D}(t,v(t))} [h(G_t, \Delta v) + C(t, G_t, v(t), \Delta v)]$$

for all times t. In the first equation q is a possible penalty depending on the volume level at time $T + 1$ and the spot price at this time G_{T+1}.

As the continuation value cannot be determined analytically, we use the least squares Monte Carlo method to approximate the continuation value

$$C(t, G_t, v(t + 1)) \approx \sum_{l=0}^{m} \beta_{l,t}(v(t + 1)) \cdot \phi_l(G_t) \quad (13)$$

using basis functions ϕ_l. If N price scenarios are given, estimates $\hat{\beta}_{l,t}(v(t + 1))$ for the coefficients $\beta_{l,t}(v(t + 1))$ result by regression. With these coefficients an approximation $\hat{C}(t, G_t, v(t + 1))$ of the continuation value is obtained that is used to determine an approximately optimal action $\Delta v(t)$ for all volumes $v(t)$.

Moreno and Navas [11] have shown that the concrete choice of the basis functions does not have much influence on the results. For this reason we have chosen the easy to handle polynomial basis functions $\phi_l(G_t) = G_t^l$. Calculations have shown that $m = 3$ is enough to get good results. A higher number of basis functions leads to similar results.

Boogert and de Jong [6] use a multi-factor price process and include the factors of the price process into the basis used for the regression in the least squares Monte Carlo method. While their factors are unobservable, our price process (see Eq. (4)) includes two exogenous factors, which can easily be observed. We include the oil price component Ψ_t (see Sect. 3) and the temperature component Λ_t (see Sect. 2) into the regression by using

$$C(t, G_t, \Lambda_t, \Psi_t, v(t), \Delta v) \approx \sum_{l=0}^{m} \beta_{l,t} \cdot \phi_l(G_t)$$
$$+ \beta_{m+1,t}\Psi_t + \beta_{m+2,t}\Psi_t^2 + \beta_{m+3,t}\Psi_t \cdot G_t$$
$$+ \beta_{m+4,t}\Lambda_t + \beta_{m+5,t}\Lambda_t \cdot G_t. \tag{14}$$

For simplification of notation we omit to mention the explicit dependence of the parameters on the filling level $v(t+1)$ as is done in Boogert and de Jong [6] throughout the paper. Monomials of higher degree in the oil price or temperature components as well as higher mixed terms have also been examined, but do not yield better results.

5.2 Influence of Exogenous Components on Valuation Results

Gas storages and swing options are not only virtual products but are real options. This means that traders need to take exercise decisions on a daily basis. These decisions depend on all observable market information. In order to reflect this behavior in the pricing algorithm for such options we will use the least squares Monte Carlo method described above in combination with the spot price model in Sect. 4. The examples given in this section are artificial gas storages and swing options valued at two different dates, 4 July 2012 and 2 April 2013. These dates are characterized by a very different implicit volatility observed at the markets—for example for TTF the long-term volatility has significantly decreased in the 8-month period from 25 to 12 % (Source: ICE). At the same time the summer–winter spread between winter 13/14 and summer 13 has decreased from 2.40 EUR/MWh to 1.20 EUR/MWh, whereas the price level has increased from 26.15 EUR/MWh to 27.70 EUR/MWh.

The TTF market prices have been used for the valuation of a slow and a fast storage that are identical to the ones valued by Boogert and de Jong [6]. Moreover, we have also valued a flexible and an inflexible swing contract. The parameters for these storages and swings are given in Table 1. All valuations have been done using 5,000 price scenarios, which results in sufficiently convergent results.

We denote by *daily intrinsic* the value obtained if a daily price forward curve is taken and an optimal exercise is calculated (using a deterministic dynamic program). This value could be logged in immediately if each single future day could be traded as an individual forward contract. The *fair value* denotes the value resulting from the least squares Monte Carlo method, and the *extrinsic value* is the difference between fair value and daily intrinsic value. Therefore, the extrinsic value is a measure for the value of the flexibility included in the considered real option.

As can clearly be seen by comparing Tables 2 and 3 the decrease of the summer–winter spread results in a lower daily intrinsic value for the storages. In contrast to this behavior the intrinsic value of the flexible swing increases because of the higher price level in 2013 compared to 2012. Furthermore, the decrease of volatility does

Table 1 Parameters for gas storages and swing options from 1.4.2013–1.4.2014

Parameter	Slow storage	Fast storage	Inflexible swing	Flexible swing
Min volume	0 MWh	0 MWh	0 MWh	0 MWh
Max volume	100 MWh	100 MWh	438 MWh	438 MWh
Min injection	0 MWh/day	0 MWh/day	–	–
Max injection	1 MWh/day	2 MWh/day	–	–
Min withdrawal	0 MWh/day	0 MWh/day	0.6 MWh/day	0 MWh/day
Max withdrawal	1 MWh/day	5 MWh/day	1.2 MWh/day	1.2 MWh/day
Injection costs	0 EUR/MWh	0 EUR/MWh	–	–
Withdrawal costs	0 EUR/MWh	0 EUR/MWh	27 EUR/MWh	27 EUR/MWh
Start volume	0 MWh	0 MWh	438 MWh	438 MWh
Max end volume	0 MWh	0 MWh	146 MWh	146 MWh

Table 2 Results for valuation date 4 July 2012 (5,000 scenarios)

Contract	Factors in regression	Daily intrinsic	Fair value	Extrinsic value
Slow storage	Spot	360.8	382.4	21.6
	Spot & Brent	360.8	549.5	188.7
	Spot & Brent & HDD	360.8	571.2	210.4
Fast storage	Spot	517.1	561.8	44.7
	Spot & Brent	517.1	1,006.6	489.5
	Spot & Brent & HDD	517.1	1,090.1	572.9
Inflexible swing	Spot	−126.2	274.5	400.7
	Spot & Brent	−126.2	285.4	411.6
	Spot & Brent & HDD	−126.2	286.3	412.4
Flexible swing	Spot	−41.6	356.5	398.1
	Spot &Brent	−41.6	397.2	438.8
	Spot &Brent &HDD	−41.6	959.6	1,001.2

not change the extrinsic value of the two storages—very much in contrast to the swings.

For storages these findings correspond very well to the observations by Boogert and de Jong [6]. They also found that a change of volatility in the long-term component does not influence the value of gas storages—it may even decrease the value. An explanation for this behavior is that it becomes more difficult for traders to decide correctly if today's price is high or low and therefore withdrawal, injection, or no action makes most sense. Due to the decision under uncertainty about the future price development with an increased volatility, more and more wrong decisions are taken and this may decrease the value at least in case of fast storages.

The situation is completely different for swing options. With an increasing volatility their value also increases. This is not surprising as can easily be seen from looking at a special case. If the yearly restriction is not binding the swing is equivalent to a

Table 3 Results for valuation date 2, April 2013 (5,000 scenarios)

Contract	Factors in regression	Daily intrinsic	Fair value	Extrinsic value
Slow storage	Spot	227.3	309.5	82.2
	Spot & Brent	227.3	419.1	191.8
	Spot & Brent & HDD	227.3	411.7	184.4
Fast storage	Spot	353.5	593.4	240.0
	Spot & Brent	353.5	855.0	501.6
	Spot & Brent & HDD	353.5	877.0	523.5
Inflexible swing	Spot	310.0	485.2	175.2
	Spot & Brent	310.0	488.0	177.9
	Spot & Brent & HDD	310.0	471.9	161.9
Flexible swing	Spot	324.1	542.1	218.0
	Spot & Brent	324.1	558.5	234.4
	Spot & Brent & HDD	324.1	572.2	248.1

strip of European options. In this case it is well known that an increase of volatility implies an increase of the extrinsic option value under quite general assumptions on the underlying stochastic process, see e.g. Bergenthum and Rüschendorf [4].

Another important difference between swings and storages is their behavior if the exogenous components of the spot price process are included in the regression of the algorithm. For the value of storages the oil price component is much more important—in contrast to swings. For the inflexible swing both components are irrelevant, while for the flexible swing the temperature component is more important than the oil price component. For storages the oil price component is a measure for normal long-term levels. As prices revert back to this long-term level mainly defined by the oil price component, a price higher than this level is good for withdrawal while a price lower than this level is good for injection. Therefore, an inclusion in the regression is very important for the exercise decision and increases the value.

Another interesting observation is the influence of the two exogenous components on the less flexible products. While an inclusion of the oil price component increases the fair value, a further inclusion of the temperature component decreases the value slightly for valuation date 2 April 2013—but not for 4 July 2012. One important reason is that in April 2013 the end of a long and as far as heating degrees are concerned quite normal winter has just been exceeded and the linear return to zero is starting, while the winter 2011/12 has been very warm and in July the linear return with a slight gradient has half been finished.

To sum up, these results indicate that it is very important to include both exogenous components into the exercise decision for storages as well as swings, as this can significantly increase the extrinsic value.

6 Conclusion

The spot price model by Stoll and Wiebauer [14] with only temperature as an exogenous factor is not able to explain the gas price behavior during the last years. We have shown that adding an oil price component as another exogenous factor remarkably improves the model fit. It is not only a good proxy for economic influences on the price but also approximates the oil price indexation in gas import contracts on Central European gas markets. These fundamental reasons and the improvement of model fit give justification for the inclusion of the model component. The resulting simulation paths from the model are reliable. The inclusion of both exogenous factors in algorithms for the valuation of options by least squares Monte Carlo remarkably affects the valuation results.

Open Access This chapter is distributed under the terms of the Creative Commons Attribution Noncommercial License, which permits any noncommercial use, distribution, and reproduction in any medium, provided the original author(s) and source are credited.

References

1. Barndorff-Nielsen, O.E.: Hyperbolic distributions and distributions on hyperbolae. Scand. J. Stat. **5**, 151–157 (1978)
2. Benth, F.E., Benth, J.S.: The volatility of temperature and pricing of weather derivatives. Quant. Financ. **7**, 553–561 (2007)
3. Benth, F.E., Koekkebakker, S., Ollmar, F.: Extracting and applying smooth forward curves from average-based commodity contracts with seasonal variation. J. Deriv. **15**, 52–66 (2007)
4. Bergenthum, J., Rüschendorf, L.: Comparison of option prices in semimartingale models. Financ. Stochast. **10**, 222–249 (2006)
5. Boogert, A., de Jong, C.: Gas storage valuation using a Monte Carlo method. J. Deriv. **15**, 81–98 (2008)
6. Boogert, A., de Jong, C.: Gas storage valuation using a multifactor price process. J. Energy Mark. **4**, 29–52 (2011)
7. Chen, Z., Forsyth, P.A.: Implications of a regime-switching model on natural gas storage valuation and optimal operation. Quant. Financ. **10**, 159–176 (2010)
8. He, Y., Wang, S., Lai, K.K.: Global economic activity and crude oil prices: a cointegration analysis. Energy Econ. **32**, 868–876 (2010)
9. Longstaff, F.A., Schwartz, E.S.: Valuing American options by simulation: a simple least-squares approach. Rev. Financ. Stud. **14**, 113–147 (2001)
10. Meinshausen, N., Hambly, B.M.: Monte Carlo methods for the valuation of multiple-exercise options. Math. Financ. **14**, 557–583 (2004)
11. Moreno, M., Navas, J.F.: On the robustness of least-squares Monte Carlo for pricing American derivatives. Rev. Deriv. Res. **6**, 107–128 (2003)
12. Schwartz, E.S.: The stochastic behaviour of commodity prices: implications for valuation and hedging. J. Financ. **52**, 923–973 (1997)

13. Schwartz, E.S., Smith, J.E.: Short-term variations and long-term dynamics in commodity prices. Manage. Sci. **46**, 893–911 (2000)
14. Stoll, S.O., Wiebauer, K.: A spot price model for natural gas considering temperature as an exogenous factor and applications. J. Energy Mark. **3**, 113–128 (2010)
15. Weron, R.: Modeling and Forecasting Electricity Loads and Prices: A Statistical Approach. Wiley, Chichester (2006)

Copula-Specific Credit Portfolio Modeling

How the Sector Copula Affects the Tail of the Portfolio Loss Distribution

Matthias Fischer and Kevin Jakob

Abstract Traditionally, banks estimate their economic capital which has to be reserved for unexpected credit losses with individual credit portfolio models. Many of those have its roots in the CreditRisk$^+$ or in the CreditMetrics framework, which were both launched in 1997. Motivated by the current regulatory requirements, banks are required to analyze how sensitive their models (and the resulting risk figures) are with respect to the underlying assumptions. Within this context, we concentrate on the dependence structure in terms of copulas in both frameworks. By replacing the underlying copula and using other popular competitors instead, we quantify the effect on the tail, in general, and on the risk figures in specific for a hypothetical loan portfolio.

1 Introduction

After the market crash of October 1987, Value-at-Risk (VaR) became a popular management tool in financial firms. Practitioners and policy makers have invested individually in implementing and exploring a variety of new models. However, as a consequence of the financial markets turmoil around 2007/2008, the concept of VaR was exposed to fierce debates. But just a few years after the crisis, VaR is still being used albeit with greater awareness of its limitations (model risk) or in combination with scenario analysis or stress testing. In particular, banks are required to critically analyze and validate their employed VaR models which form the basis for their internal capital allocation process (ICAAP, see BaFin [1, AT.4.1]). In this context,

---

M. Fischer (✉)
Department of Statistics and Econometrics, University of Erlangen-Nuremberg,
Lange Gasse 20, 90403 Nuremberg, Germany
e-mail: Matthias.Fischer@fau.de

K. Jakob
Department of Economics, University of Augsburg,
Universitätsviertel. 16, 86159 Augsburg, Germany
e-mail: Kevin.Jakob@Student.Uni-Augsburg.de

© The Author(s) 2015   129
K. Glau et al. (eds.), *Innovations in Quantitative Risk Management*,
Springer Proceedings in Mathematics & Statistics 99,
DOI 10.1007/978-3-319-09114-3_8

the term "model validation" should be associated to the activity of assessing if the assumptions of the model are valid. Model assumptions, not computational errors, were the focus of the most common criticisms of quantitative models in the crisis. In particular, banks should be aware of the errors that can be made in the assumptions underlying their models which form one of the crucial parts of model risk, probably underestimated in the past practice of model risk management. With respect to the current regulatory requirements (see, e.g., BaFin [1] or Board of Governors of the Federal Reserve System [2]), banks are also required to quantify how sensitive their models and the resulting risk figures are if fundamental assumptions are modified.

The focus of this contribution is solely on credit risk as one of the most important risk types in the classical banking industry. Typically, the amount of economic capital which has to be reserved for credit risk is determined with a credit portfolio model. Two of the most widespread models are CreditMetrics, launched by JP Morgan (see Gupton et al. [3]) and CreditRisk$^+$, an actuarial approach proposed by Credit Suisse Financial Products (CSFP, see Wilde [4]). Shortly after their publication, Koylouglu and Hickman [5], Crouhy [6] or Gordy [7] offered a comparative anatomy of both models and described quite precisely where the models differ in functional form, distributional assumptions, and reliance on approximation formulae. Sector dependence, however, was not in the focus of these studies.

A crucial issue with credit portfolio models consists in the realistic modeling of dependencies between counterparties. Typically, all counterparties are assigned to one or more (industry/country) sectors. Consequently, high-dimensional counterparty dependence can be reduced to low(er)-dimensional sector dependence, which describes the way how sector variables are coupled together. Against this background, our focus is on the impact of different dependence structures represented in terms of copulas within credit portfolio models. Relating to Jakob and Fischer [8], we extend the analysis of the CreditRisk$^+$ model to CreditMetrics and provide comparisons between both frameworks. For this purpose, we work out the implicit and explicit sector copula of both classes in a first step and quantify the effect of exchanging the copula model on the risk figures for a hypothetical loan portfolio and a variety of recent flexible parametric copulas in a second step.

Therefore, the outline is as follows. In Sect. 2, we review the classical copula concept and briefly introduce those copulas which are used during the analysis. Section 3 summarizes and compares the underlying credit portfolio models with special emphasis on the underlying sector dependence. Finally, we empirically demonstrate the influence of different copula models on the upper tail of the loss distribution and, hence, on the risk figures for a hypothetical but realistic loan portfolio. Section 5 concludes.

2 Copulas Under Consideration

The concept of copulas dates back to Sklar [9]. In general, a copula is a multivariate distribution function on the d-dimensional unit hypercube with uniform one-dimensional margins.[1] With the help of a copula function, one can decompose an arbitrary multivariate distribution into its margins and the dependence structure. i.e., according to Sklar's Theorem, for any multivariate distribution function F on $\mathbb{R}^d$ with univariate margins F_i a unique function $C : \times_{i=1}^{d} \mathrm{Im}(F_i) \to [0, 1]$ exists, such that $F(\mathbf{x}) = C(F_1(x_1), \ldots, F_d(x_d))$ for all $\mathbf{x} \in \mathbb{R}^d$. Conversely, for arbitrary univariate distribution functions F_i and a copula C, the function F defines a valid multivariate distribution function. Because our focus is solely on the dependence structure between economic sectors, we will use Sklar's theorem in the second direction. By exchanging the copula, we can construct new multivariate distributions without affecting the margins.

Already at the beginning of this century, Li [12] incorporated the concept of copulas into the CreditMetrics model. Ebmeyer et al. [13] used a Gaussian and a t-copula within the CreditRisk$^+$ framework to model sector dependencies. Our aim is to extend these studies to a broader range of copulas and to establish a comparison between both portfolio models regarding the sensitivity of the risk figures with respect to the sector dependence. In addition to the original dependence structures, i.e., the Gaussian copula (CreditMetrics) and a specific factor copula (CreditRisk$^+$), we apply the following parametric competitors:

- **elliptical copulas**, i.e., the Gaussian copula (GC) and the t-copula (TC) (see, McNeil et al. [14]),
- **generalized hyperbolic copulas (GHC)**, implicitly defined by the family of generalized hyperbolic distributions (see Barndorff-Nielsen [15]),
- **Archimedean (AC)**, for example the Gumbel, Clayton, Joe or Frank copula and **hierarchical Archimedean copulas (HAC)** (see Savu and Trede [16], McNeil [17] or Hofert and Scherer [18]),
- **pair copula constructions (PCC)** (see Aas et al. [19]).

To estimate the unknown parameters, e.g., the dispersion matrix in case of the GC, we use the maximum likelihood (ML) approach. Other techniques, e.g., inverting Kendall's τ may be also possible. In case of the HAC and PCC, one also has to choose a suitable nesting or vine structure, [2] respectively. For this purpose, we applied the methods implemented in the R-packages "HAC" by Okhrin and Ristig [20] and "VineCopula" by Schepsmeier et al. [21], respectively. Further information about the estimation are given in Sect. 4.3. In addition, for more details about the model selection process we also refer to the mentioned articles.

---

[1] In general, we assume that the reader is already familiar with the concept of copulas as well as the most popular classes. For details, we refer to Joe [10] and Nelson [11].

[2] A vine is a directed acyclic graph, representing the decomposition sequence of a multivariate density function.

3 A Comparison Between CreditRisk$^+$ and CreditMetrics

Within this section, we shortly introduce both CreditMetrics and CreditRisk$^+$ in a comparative way to highlight the differences.

3.1 Preliminary Notes and General Remarks

CreditMetrics was developed by a group of investment banks, led by J.P. Morgan (see Gupton et al. [3]). It follows a mark to market approach and includes default risk as well as migration risk.[3] In order to ensure comparability across both models, we solely focus on the default risk. Nevertheless, in practice, migration risk is also very important and should not be neglected. CreditMetrics belongs to the class of threshold models (see McNeil et al. [14]). Here, the creditworthiness of each obligor is governed by a latent variable, which is driven by the state of the overall economy or a special sector/region as well as by an idiosyncratic factor. A default occurs if a predefined threshold, determined by the obligors' initial probability of default (PD), is exceeded.

In contrast, CreditRisk$^+$ belongs to the class of actuarial models. It was developed by the Financial Products division of Credit Suisse (see Wilde [4]). The default distribution of each counterparty is influenced by one or several factors. As in case of CreditMetrics, these factors depend on the current state of the economy as well as on idiosyncratic components. Given these values, defaults are assumed to be independent of each other.

A major difference between both models is the way how the portfolio loss distribution is achieved. Whereas in the CreditMetrics framework a Monte Carlo simulation is required to estimate the later, the same can be calculated analytically within the CreditRisk$^+$ framework. A numerically stable algorithm is described in Gundlach and Lehrbass [22, Chap. 5].

3.2 Theoretical Background

3.2.1 Model Input

We assume that for each counterparty $i = 1, \ldots, N$ the exposure at default (EAD$_i$), the loss given default (LGD$_i$) and the (unconditional) probability of default (PD$_i$) are known and not stochastic. We also assume that all business transactions of the obligors have been aggregated to a single position for each counterparty. To derive the loss distribution analytically, CreditRisk$^+$ requires the exposures to be discretized with respect to a so-called loss unit $U > 0$. The original values for EAD$_i$ and PD$_i$

---

[3] Migration risk includes the financial risk due to a change of the portfolio value caused by rating migrations (i.e., down- and upgrade).

are replaced by

$$\widetilde{\text{EAD}}_i := \max\left\{ \left\lceil \frac{\text{EAD}_i \cdot \text{LGD}_i}{U} \right\rceil, 1 \right\} \quad \text{and} \quad \widetilde{\text{PD}}_i := \frac{\text{EAD}_i \cdot \text{LGD}_i \cdot \text{PD}_i}{\widetilde{\text{EAD}}_i \cdot U}$$

respectively. The adjustment of the PDs ensures that the expected loss of the portfolio is not affected by the discretizastion. i.e., it holds:

$$\mathbb{E}\left(L\right) = \sum_{i=1}^{N} \text{EAD}_i \cdot \text{LGD}_i \cdot \text{PD}_i = \sum_{i=1}^{N} \widetilde{\text{EAD}}_i \cdot U \cdot \widetilde{\text{PD}}_i = \mathbb{E}\left(\widetilde{L}\right).$$

To simplify notation, we will omit the tilde for the discretized exposure and the PD in the following and denote them also with EAD_i and PD_i, respectively. Since the CreditMetrics model is a simulative one, such an adjustment is not necessary.

3.2.2 Sector Variables and Sector Dependencies

In order to introduce dependencies between counterparties, every obligor is mapped to one or several out of K sectors. Since the interpretations and assumptions behind the sectors variables and the corresponding counterparty specific sector weights are different, we will use an individual notation for each model. In CreditMetrics, the vector of sector variables $X = (X_1, \ldots, X_K)^T$ is assumed to follow a multivariate normal distribution. Therefore, each sector variable $X_{k=1,\ldots,K}$ has a standard normal law and the copula of $X = (X_1, \ldots, X_K)^T$ is a Gaussia one with dispersion matrix Σ.

Within CreditRisk$^+$, the sector variables S_k are assumed to follow a Gamma law with specific shape and scale parameters, such that $\mathbb{E}\left(S_k\right) = 1$ for all $k = 1, \ldots, K$. The choice of the Gamma distribution was motivated by the fact that in combination with Poisson distributed defaults, the loss distribution can be derived analytically. In order to specify the sector distributions, the sector variances σ_k^2 can be estimated from empirical data, for example, insolvency rates. In the original model of 1997, the variables S_k are also assumed to be independent of each other. In contrast, we apply the so-called CBV approach, which is an extension, published by Fischer and Dietz [23], with respect to correlated sectors. Here, each single sector variable is driven by a linear combination of $L + 1$ independent Gamma distributed variates, i.e.,

$$S_k = \overline{S}_k + \sum_{\ell=1}^{L} \gamma_{k,\ell} \hat{S}_\ell, \quad \text{for } k = 1, \ldots, K \tag{1}$$

with non-negative weights $\gamma_{k,\ell}$ for $k = 1, \ldots, K$ and $\ell = 1, \ldots, L$. The vector $\hat{S} := \left(\hat{S}_1, \ldots, \hat{S}_L\right)^T$, with $\hat{S}_\ell \sim \Gamma\left(\hat{\theta}_\ell, 1\right)$, is called common-background-vector (CBV). Besides this vector, each sector variable is also affected by an individual component $\overline{S}_k \sim \Gamma\left(\theta_k, \delta_k\right)$. Because all variables $\overline{S}_k$ and $\hat{S}_\ell$ are assumed to be independent of each other, one can reduce the CBV extension to the basic CreditRisk$^+$ model. Hence, also the CBV model can be solved analytically, too. For further details on the estimation of the Gamma parameters, we refer to Fischer and Dietz [23].

In Eq. (1), the marginal distributions of $S = (S_1, \ldots, S_K)^T$ are (in general) not Gamma anymore. An analysis of the resulting univariate distribution was established by Moschopoulos [24]. The copula of S is called a multi factor copula, which is discussed by Oh and Patton [25] in a very general way or Mai and Scherer [26].

3.2.3 Default Mechanism

In the CreditMetrics setting, a default occurs if obligor i's creditworthiness,[4] modeled by

$$A_i := \mathbf{R}_i^T \mathbf{X} + \sqrt{1 - \mathbf{R}_i^T \Sigma \mathbf{R}_i} Y_i, \tag{2}$$

falls below $\phi^{-1}(\mathrm{PD}_i)$, where ϕ^{-1} denotes the quantile function of the standard normal distribution and $Y_i \sim \mathcal{N}(0, 1)$ is independent from $\mathbf{X}$ and Y_j for $i \neq j$. The vector $\mathbf{R}_i^T \in [-1, 1]^K$, with the restriction that $\mathbf{R}_i^T \Sigma \mathbf{R}_i \leq 1$, contains the so-called factor loadings, describing the correlation between a counterparty's asset value A_i and the systemic factors X_k. Given a sector realization x of $\mathbf{X}$, the conditional PD, derived from the asset process (2) reads as

$$\mathrm{PD}_i^{\mathrm{CM}}(\mathbf{X} = x) = \phi\left[\left(\phi^{-1}(\mathrm{PD}_i) - \mathbf{R}_i^T x\right) \Big/ \sqrt{1 - \mathbf{R}_i^T \Sigma \mathbf{R}_i}\right]. \tag{3}$$

In the CreditRisk$^+$ model, the sector variables S_k are assumed to influence the conditional PD according to

$$\mathrm{PD}_i^{\mathrm{CR+}}(\mathbf{S} = s) = \mathrm{PD}_i\left(\mathbf{W}_i^T s + W_{i,0}\right) \tag{4}$$

with $\mathbf{W}_i \in [0, 1]^K$ and $W_{i,0} := \sum_{k=1}^K W_{i,k} \leq 1$. Equations (3) and (4) establish a connection between sector variables and counterparties PDs. In CreditRisk$^+$, $\mathrm{PD}_i^{\mathrm{CR+}}$ serves as intensity parameter of a Poisson distribution from which defaults are drawn independently for every counterparty. The Poisson distribution is used instead of a Bernoulli one in order to obtain a closed form expression of the loss distribution. Therefore, also multiple defaults of counterparties (especially with bad creditworthiness) are possible. This is a major drawback of the model, leading to an overestimation of the risk figures. In Sect. 4 we analyze the changes of risk figures with respect to the underlying copula. But since our focus is on relative changes, this overestimation does not influence the comparison.

---

[4] One should note, that A_i again has a standard Gaussian law. The dependence structure is described by a multi factor copula as in case of the CreditRisk$^+$- CBV model, but with a different parametrization.

4 Results on Estimated Copulas and Risk Figures

In this section the estimation results for the sector copulas are presented as well as the effect on economic capital.

4.1 Portfolio and Model Calibration

Consider a hypothetical portfolio consisting of 5,000 counterparties, each mapped to exactly one[5] out of ten industrial sectors. For reasons of simplicity, LGDs[6] are assumed to be deterministic and independent from PD. Since the absolute exposure values are chosen arbitrarily, we can assume that w.l.o.g $LGD_i = 1$ for all $i = 1,\ldots,5,000$. Because our focus is only on the relative changes of the risk figures rather than absolute values, this simplification does not restrict our results. Table 1 summarizes the number of counterparties (#CP) and exposures by industrial sectors, as well as the estimated sector parameters related to the marginal sector distributions. Although the portfolio itself is hypothetical, the distribution of exposure and counterparties across sectors might be characteristic for certain banks. Please note, that in case of CreditMetrics higher values of R_k^2 indicate a stronger dependency to systemic factors, leading to a higher risk for the specific sectors. In the CBV model

Table 1 Number of counterparties, percentage of exposures, factor loadings $\left(R_k^2, \text{CreditMetrics}\right)$ and sector variances $\left(\sigma_k^2, \text{CreditRisk}^+\right)$ by industrial sector

k	Sector	Portfolio characteristics		Sector parameters	
		# CP	EAD (%)	R_k^2	σ_k^2
1	Basic materials	16	1.7	0.070	0.42
2	Communication	5	2.5	0.045	0.29
3	Cyclical consumer goods	4,631	19.5	0.058	0.36
4	Noncyclical consumer goods	15	1.5	0.048	0.27
5	Diversified companies	28	3	0.040	0.19
6	Energy	10	4.3	0.075	0.40
7	Finance	146	45.9	0.050	0.46
8	Industry	75	11.1	0.050	0.30
9	Technology	19	1.8	0.046	0.26
10	Utilities	55	8.7	0.082	0.72

---

[5] Assigning an obligor to more than one sector would cause serious problems in the CreditMetrics framework, since, in general, the distribution of the asset value (2) is unknown if the copula of X is not Gaussian.

[6] For readers who are interested in the effect of stochastic LGDs, we refer to Gundlach and Lehrbass [22, Sect. 7] or Altman [27].

σ_k^2 represent the uncertainty about possible PD changes within the sector. Therefore, the risk related to a particular sector increases with σ_k^2.

The basis for the parameter estimation is a data pool containing monthly observations (PD estimations) from 2003 to 2012 for more than 30,000 exchange traded corporates from all over the world. The individual PD time series, derived from market data (equity prices and liabilities) via a Merton model (see Merton [28]), are aggregated on sector level via averaging. In order to take time dependencies into account, we fitted a univariate autoregressive process to every sector time series.

4.2 Parametrization of Marginal Distributions

In order to fully determine the marginal distributions, we have to specify the sector variances σ_k^2 for the CreditRisk$^+$ and the asset correlations R_k^2 for the CreditMetrics model.[7] The sector variances are estimated based on the autocovariance function of the aggregated sector time series mentioned above, which are normalized such that $\mathbb{E}(S_k) = 1$ holds, in order to ensure that the mean of the conditional PD (Eq. (4)) equals the unconditional PD. In case of the CreditMetrics model, the asset correlation parameters R_k^2 are estimated via a moment matching approach, such that the first two moments of the conditional PD in both models coincide.[8] Note, that the PD variance $\mathrm{Var}\left(\mathrm{PD}_i^{\mathrm{CM}}(X)\right)$ induced by Eq. (3) of counterparty i in sector k is given by $\Phi_2\left(\phi^{-1}(\mathrm{PD}_i), \phi^{-1}(\mathrm{PD}_i), R_k^2\right)$ whereas, in case of CreditRisk$^+$, $\mathrm{Var}\left(\mathrm{PD}_i^{\mathrm{CR}+}(S)\right)$ is simply $\mathrm{PD}_i^2\sigma_k^2$. Hence, for $k = 1, \ldots, K$ the parameter R_k^2 is chosen such that

$$\Phi_2\left(\phi^{-1}\left(\overline{\mathrm{PD}}_k\right), \phi^{-1}\left(\overline{\mathrm{PD}}_k\right), R_k^2\right) = \sigma_k^2\overline{\mathrm{PD}}_k^2,$$

where $\overline{\mathrm{PD}}_k$ denotes the mean of the time series for sector k and Φ_2 is the distribution function of the bivariate normal distribution with correlation parameter R_k^2.

4.3 Estimation of Copulas

First note that the estimations are based on the residuals of the autoregressive processes, fitted on every sector PD time series. For a more detailed discussion on this topic, we refer to Jakob and Fischer [8], for instance.

---

[7] In practice, the parametrization of both models are very different. The parameters of the CreditRisk$^+$ model are typically estimated based on default data or insolvency rates, whereas in case of the CreditMetrics model marked data are used. Using PD time series based on marked data might serve as a compromise in order to compare the results across both models.

[8] Please note that $\mathbb{E}\left(\mathrm{PD}_i^{\mathrm{CM}}(X)\right) = \mathbb{E}\left(\mathrm{PD}_i^{\mathrm{CR}+}(S)\right) = \mathrm{PD}_i$.

Copula	GC	TC	sym. GHC	GHC
Log-likelihood	634	728	8,848	13,566

Table 2 Rounded log-likelihood values for elliptical copulas and GHC

4.3.1 Elliptical and Generalized Hyperbolic Copulas

The parameters of the GC and the TC (as representatives of the elliptical copula class) are estimated via maximum likelihood using the R-Package "copulas" from Hofert et al. [29]. For the TC, we estimated 3.786 degrees of freedom indicating that a joint exceedance of high quantiles is more likely compared to the GC. Generalizing the TC, we also considered symmetric and asymmetric[9] GHC. For parameter estimation the R-package "ghyp" from Luethi and Breymann [30] was used. Please note that compared to the TC, the sym. GHC poses two more parameters due to the generalized inverse Gaussian distribution, which is used as mixing distribution for the family of generalized hyperbolic distributions and by another ten parameters because of the skewness vector in case of the asymmetric GHC. The corresponding log-likelihood values are summarized in Table 2. A standard likelihood ratio test indicates that the TC fits the data significantly better than the Gaussia one on every typical significance level. Also, the increase of the log-likelihood of the asymmetric GHC is significant to that of its symmetric counterpart. Hence, the stronger dependence between higher PDs, occurring in the asym. GHC, is significant again on every common level.

Please note that the application of the GHC in practice has several drawbacks. The estimation procedure, the MCECM (multi-cycle, expectation, conditional estimation) algorithm is much more difficult to implement and time consuming compared to estimation of GC or a TC. Furthermore, the simulation of random numbers is much more computationally intensive due to the quantile functions, which contain the modified Bessel function of the third kind, requiring methods for numerical integration.

4.3.2 (Hierarchical) Archimedean Copulas

Out of the Archimedean class, we estimated parameters for the Gumbel, Clayton, Joe, and Frank copula but only the copulas of Gumbel and Joe provided a reasonable fit to our data. Since our data represent default probabilities, the economic intuition would be that the dependence increases for higher values, i.e., in times of recession, as can be seen from the empirical data (see, Fig. 2). The Gumbel and Joe copulas exhibit a positive upper tail dependence,[10] while the lower ones are zero. Therefore, they are suitable to model this kind of asymmetric dependence. The Frank copula is

---

[9] For the symmetric GHC, we force the skewness parameter $\gamma \in \mathbb{R}^K$ to be zero for all components (notation according to Luethi and Breymann [30]).

[10] The coefficients of upper (lower) tail dependence are defined by $\lambda_U := \lim_{u \nearrow 1} \mathbb{P}\left[X_2 \geq F_2^{-1}(u) \mid X_1 \geq F_1^{-1}(u)\right]$ and $\lambda_L := \lim_{u \searrow 0} \mathbb{P}\left[X_2 \leq F_2^{-1}(u) \mid X_1 \leq F_1^{-1}(u)\right]$, respectively.

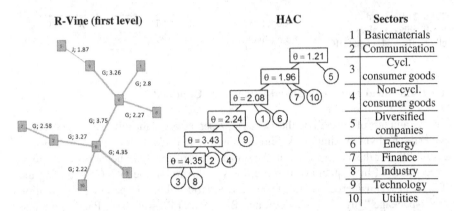

Fig. 1 First level of R-vine (with parameters of Gumbel and Joe copulas) and Hierarchical Archimedean copula (Gumbel) estimated from default data

tail independent, whereas the Clayton copula posses only a lower tail dependence. Applying goodness-of-fit tests (see Genest et al. [31]), we have to reject both copulas (Frank and Clayton) on a significance level considerably below 1 %. In addition, we also considered hierarchical Archimedean constructions. With the help of the "HAC" package from Okhrin and Ristig [20], a stepwise ML estimation procedure was used to estimate the tree of the Gumbel HAC, depicted in Fig. 1. The figure shows that the dependence parameters are in a range of 4.35 at the bottom, indicating the strongest dependence, and 1.21 at the top of the tree. For the ordinary Gumbel copula, we estimate a parameter value of 1.836, which is in the range of the HAC parameters. Since the variates selection on each level of the HAC tree is based on empirical values of Kendall's τ, the structures of the two HACs (Gumbel and Joe) coincide.

4.3.3 Pair Copula Construction (PCC)

In general, a PCC arises from a nonunique decomposition of a multivariate distribution into a product of conditional bivariate distribution, characterized by so-called vines. The estimation algorithm of a PCC in general consists of three major steps:

(I) Specification of a valid vine structure (e.g., C-, D-, or R-Vine tree),
(II) type-selection of the underlying bivariate copulas for the tree in (I) (e.g., GC or Gumbel copula),
(III) parameter estimation for the copulas, selected in (II).

Brechmann and Schepsmeier [32] describe several algorithms addressing all these issues. In particular, the specification of the vine structure is done with the help of maximum spanning trees, where on each level a tree is selected such that the sum of Kendall's τ for all pairs of variables is maximized. To determine a particular copula for the selected pairs out of a set of certain candidates, the AIC criterion is applied.

Finally, the copula parameters are estimated via ML. The corresponding steps (I)–(III) are implemented in the R-package "VineCopula" (see Schepsmeier et al. [21]), which has been used to determine a PCC for our data set. In order to allow maximum flexibility, we decided to use a R-vine, which generalizes both C- and D-vines. The candidate set for the pair copulas comprises GC, TC, Gumbel, Clayton, Frank, and Joe copula.

Analog to the HAC, the estimation algorithm of the PCC identifies sectors 3 and 8 as those with the strongest dependence. Therefore, these sectors are coupled together on the first level of the R-vine, which means that their pairwise dependence is explicitly selected to follow a Gumbel copula with $\hat{\theta} = 4.35$. In general, all except one bivariate copulas on the first level are estimated to be Gumbel with parameter values in $[1.56, 4.35]$, which is close to the HAC parameter range, see Fig. 1. Only in case of sectors 5 and 9, the Joe copula with parameter 1.87 is preferred. Again, the weakest dependence (measured by the implied value of Kendall's τ) on the first level is related to sector 5. On higher levels, all copulas out of the candidates set are selected to model conditional bivariate dependencies.

4.3.4 Parametrization of the CreditRisk [+]- CBV Copula

For the CBV model, the likelihood function is rather complex and a ML estimation is numerically not feasible. Hence, the parameters of the CBV factor copula are chosen such that the Euclidean distance between the empirical and the theoretical covariance matrix is minimal (see, e.g., Fischer and Dietz [23]).

4.3.5 Illustration for Sectors 3 and 8

Exemplarily, Fig. 2 illustrates the contour plot of the estimated copula density between sectors 3 (cycl. consumer goods) and 8 (industry) for different competitors as well as the (transformed) empirical observations. Notice that darker areas indicate higher concentration of the probability mass. In the first row, the elliptical and GHCs are displayed. Looking at the center of the unit squares, one observes that, in case of the TC and the asymmetric GHC, more probability mass is concentrated around the main diagonal as for the GC or the symmetric GHC. Since the asymmetric GHC provides a significantly better fit compared to the TC, the issue of asymmetrically distributed data seems to be more important than the absence of a positive tail dependence, at least for our data. This might be caused by the limited sample size of only 120 observations. Although the asymmetric GHC has a significantly better fit compared to the symmetric one and the skewness parameters are strictly positive, its density still looks very symmetric.

In contrast, the copula of the CBV model[11] is extremely concentrated around the main diagonal. Here, observations aside from the diagonal have a very low

---

[11] In case of the CBV copula, the density is estimated via a two dimensional kernel density estimator.

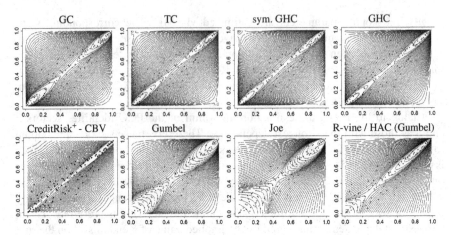

Fig. 2 Contour of the estimated copulas between sector 3 (cycl. consumer goods) and 8 (industry) together with empirical observations

probability. Please note again that the estimation procedure for this copula is different, which might explain this issue to some extend. For the ordinary Gumbel and Joe copulas, one has to choose one single parameter for all bivariate (and higher dimensional) dependencies. Therefore the estimation is always a trade-off between stronger and weaker dependencies. This leads to the effect that, in our example, the dependence in both cases seems to be rather underestimated by this copulas compared to its competitors. The HAC overcomes this drawback by using different parameters, which leads to a significantly better fit.

4.4 Effect of the Copula on the Risk Figures and the Tail of the Loss Distribution

Finally, we analyze the impact of the sector copula on the right tail and therefore on the economic capital. Since, in practice, the underlying data sets used for parametrizations of both model types are rather different and not comparable, we do not draw any comparisons between the absolute values of the risk figures across the two models. Instead, we measure the impact with the help of factors, where the risk figures of the models with the GC are normalized to one. In case of the CreditRisk[+]-CBV model, the marginal distributions of the sectors, which follow a weighted sum of Gamma distributions (see Eq. (1)), are replaced by Gamma distributed variates with the same mean and variance, for reasons of simplicity. Since this is a monotone transformation, the dependence structure is not affected. Please note that by drawing the sector realizations[12] for the CreditMetrics model, we use the survival copula,[13] because in this case higher values of the sector variates correspond to an increase rather than a decrease of obligors creditworthiness.

Table 3 summarizes all risk figures. The copulas are ordered according to the impact on the economic capital on a 99.9 % level in case of the CreditRisk[+] model.

First of all, one observes that in the CreditRisk[+] framework, the risk decreases if we switch from the original model (CBV) to another one. In both models, the GC implies the lowest risk, followed by the sym. GHC. Although both copulas are elliptically symmetric and tail independent, the risk figures differ by up to 4 %. Applying a TC, the risk rises in both models because of the positive tail dependence of $\hat{\lambda}_U = 0.69$. For the CreditMetrics model the markup is around 6 %. The highest risk arises if we use an asymmetric dependence structure, i.e., a (hierarchical) Gumbel or Joe copula, an asym. GHC, a PCC or, in case of CreditRisk[+], the factor copula induced by the CBV model. Therefore, at least for our data set and portfolio, there is an indication that the risk arising from an asymmetric dependence structure, i.e., where dependencies are higher during times of a recession, is higher compared to the risk caused by a positive tail dependence. In the CreditRisk[+] model even the economic capital in case of the HAC (Joe) copula is around 8.1 % above the amount of the model with a GC and 2 % below the basic model. In both models, the risk implied by a Joe copula is higher compared to a Gumbel copula. Since both copulas exhibit no positive lower tail dependence, whereas the upper tail dependence[14] is higher in case of the Joe copula, this observation is plausible.

As to be expected, all portfolio loss distributions exhibit a significant amount of skewness (skew) and kurtosis (kurt), measured by the third and fourth standardized moments, respectively. In addition, we calculated the right-quantile weight (RQW) for $\beta = 0.875$ which was recommended by Brys et al. [34] as a robust measure of tail weight and is defined as follows:

$$
\text{RQW}(\beta) := \frac{F_L^{-1}\left(\frac{1+\beta}{2}\right) + F_L^{-1}\left(1 - \frac{\beta}{2}\right) - 2F_L^{-1}(0.75)}{F_L^{-1}\left(\frac{1+\beta}{2}\right) - F_L^{-1}\left(1 - \frac{\beta}{2}\right)},
$$

where, in our case, F_L^{-1} denotes the quantile function of the portfolio loss distribution. First of all, it becomes obvious that the rank order observed for $EC_{99.9}$ with respect to the copula model is highly correlated to the rank order of the higher moments and of the tail weight. Secondly, all of the latter statistics derived from the CreditMetrics framework are (significantly) higher than those derived from the CreditRisk[+] framework.

---

[12] For details on the simulation of copulas in general, please refer to Mai and Scherer [33].

[13] For a random vector $\boldsymbol{u} = (u_1, \ldots, u_K)^T$ with copula C, the survival copula is defined as the copula of the vector $(1 - u_1, \ldots, 1 - u_K)^T$.

[14] The coefficients of upper tail dependence implied by the estimated parameters are 0.54 in case of the Gumbel copula and 0.66 for the Joe copula.

Table 3 Risk figures (SD: standard deviation, EC_α;ES_α: economic capital and expected shortfall on level $\alpha \in [0, 1]$ normalized to GC) and tail measures (skewness, kurtosis, and right-quantile-weight in absolute values) for different copulas in case of the CreditRisk$^+$ model (left part) and the CreditMetrics model (right part)

Copula	CreditRisk$^+$								CreditMetrics							
	SD	EC_{90}	$EC_{99,9}$	ES_{90}	$ES_{99,9}$	skew	kurt	RQW	SD	EC_{90}	$EC_{99,9}$	ES_{90}	$ES_{99,9}$	skew	kurt	RQW
CBV	1.060	1.058	1.101	1.047	1.074	1,53	6,49	0.39	–	–	–	–	–	–	–	–
GC	1.000	1.000	1.000	1.000	1.000	1,40	6,01	0.37	1.000	1.000	1.000	1.000	1.000	1,64	7,71	0.39
sGHC	1.000	0,991	1.027	1.000	1.024	1,45	6,32	0.38	1,003	0,988	1,042	1,002	1.038	1,75	8,67	0.39
TC	1.005	0.998	1.036	1,005	1.031	1,44	6,27	0.38	1.009	0,995	1,060	1,005	1.055	1,73	8,48	0.40
HAC(G)	1.001	0.988	1.053	1.008	1.039	1,50	6,57	0.39	1.006	0,978	1,095	1,009	1,073	1,83	9,18	0.41
GHC	1.010	0.997	1.057	1.011	1.047	1,56	6,86	0.38	1,007	0,992	1,090	1,009	1,068	1,87	9,40	0.41
Gumbel	0.995	0.978	1.059	1.004	1.048	1,53	6,67	0.39	1,018	0,979	1,096	1,015	1,082	1,85	9,27	0.41
PCC	1,023	1,013	1,072	1,022	1,056	1,52	6,60	0,39	1,031	1,008	1,106	1,026	1,084	1,85	9,15	0,41
Joe	1,001	0.986	1,078	1,015	1,061	1,63	7,14	0.41	1,013	0,980	1,107	1,020	1,083	1,95	9,76	0,42
HAC(J)	1,000	0.984	1,081	1,013	1,066	1,62	7,15	0.40	1,010	0,980	1,091	1,017	1,074	1,92	9,55	0,42
indep.	0.795	0.782	0.757	0.858	0.820	1,19	5,52	0.36	0.786	0,790	0,701	0,846	0.782	1,28	6,49	0.35

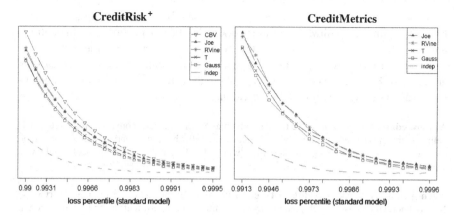

Fig. 3 Right tail of portfolio loss distribution for selected copulas

Finally, Fig. 3 exhibits the estimated densities of the portfolio loss for both models and different copulas. On the horizontal axis, the percentiles of the loss distribution of the particular standard models are displayed. The ordering of the densities confirms our results, derived from the corresponding risk figures.

5 Summary

Credit portfolio models are commonly used to estimate the future loss distribution of credit portfolios in order to derive the amount of economic capital which has to be allocated to cover unexpected losses. Therefore, capturing the (unknown) dependence between the counterparties of the portfolios or between the economic sectors to which counterparties have been assigned is a crucial issue. For this purpose, copula functions provide a flexible toolbox to specify different dependence structures.

Against this background, we analyzed the effect of different parametric copulas on the tail of the loss distribution and the risk figures for a hypothetical portfolio and for both CreditMetrics and CreditRisk$^+$, two of the most popular credit portfolio models in the financial industry. Our results indicate that the specific CreditRisk$^+$−CBV model uses a rather conservative copula. However, referring to Jakob and Fischer [8], one might come across to certain artifacts for this (implicit) copula family. In the CreditMetrics setting, the canonical assumption of a Gaussian copula allows an easy and fast implementation but also gives rise to certain drawbacks, such as the absence of a tail dependence ("extreme events occur together") or the ability to model asymmetric dependence structures for which we found evidence in the underlying data set. Replacing the Gaussian copula by alternative competitors (Student-t, Generalized hyperbolic, PCC or generalized Archimedean copulas) we could significantly improve the goodness-of-fit to the underlying PD series. As a consequence, using the

Gaussian copula might lead to an underestimation of credit risk by up to 10 % (for $EC_{99.9}$) within the CreditMetrics framework, at least for our calibration. In contrast, the CreditRisk+ model seems to be less sensitive with respect to the dependence structure, because here the markup (related to the Gaussian copula as benchmark) is around 2–4 % points lower. The question about the different behavior of both model types has to be left open for further research.

Acknowledgments We would like to thank Matthias Scherer and an anonymous referee for their helpful comments on early versions of this article. Especially the recommendation to consider a Joe copula, which proved to be a good alternative (within the class of AC) to the Gumbel copula, was very valuable.

Open Access This chapter is distributed under the terms of the Creative Commons Attribution Noncommercial License, which permits any noncommercial use, distribution, and reproduction in any medium, provided the original author(s) and source are credited.

References

1. BaFin. Erläuterung zu den MaRisk in der Fassung vom 14.12.2012, Dec 2012
2. Board of Governors of the Federal Reserve System: Supervisory guidance on model risk management, 2011. URL http://www.federalreserve.gov/bankinforeg/srletters/sr1107.htm. Letter 11-7
3. Gupton, G.M., Finger, C.C., Bhatia. M.: CreditMetrics—technical dokument, 1997. URL http://www.defaultrisk.com/_pdf6j4/creditmetrics_techdoc.pdf
4. Wilde, T.: CreditRisk+ a credit risk management framework, 1997. URL http://www.csfb.com/institutional/research/assets/creditrisk.pdf
5. Koylouglu, H.U., Hickman, A.: Reconciliable differences. RISK **11**, 56–62 (1998)
6. Crouhy, M., Galai, D., Mark, R.: A comparative analysis of current credit risk models. J. Bank. Financ. **24**, 59–117 (2000)
7. Gordy, M.B.: A comparative anatomy of credit risk models. J. Bank. Financ. **24**, 119–149 (2000)
8. Jakob, K., Fischer, M.: Quantifying the impact of different copulas in a generalized CreditRisk+ framework. Depend. Model. **2**, 1–21 (2014)
9. Sklar, A.: Fonctions de répartition à n dimensions et leurs marges. Publ. Inst. Stat. Univ. Paris **8**, 229–231 (1959)
10. Joe, H.: Multivariate Models and Dependence Concepts. Chapman & Hall/CRC, London (1997)
11. Nelson, R.B.: An Introduction to Copulas. Springer Science+Business Media Inc., New York (2006)
12. Li, D.X.: On default correlation: a copula function approach. J. Fixed Income **9**, 43–54 (2000)
13. Ebmeyer, D., Klaas, R., Quell, P.: The role of copulas in the CreditRisk+ framework. In: Copulas. Risk Books (2006)
14. McNeil, A.J., Frey, R., Embrechts, P.: Quantitative Risk Management. Princeton University Press, Princeton (2005)
15. Barndorff-Nielsen, O.E.: Exponentially decreasing distributions for the logarithm of particle size. Proc. Roy. Soc. Lond. Ser. A, Math. Phys. Sci. **353**, 401–419 (1977)
16. Savu, C., Trede, M.: Hierarchical Archimedean copulas. Quant. Financ. **10**, 295–304 (2010)
17. McNeil, A.J.: Sampling nested archimedean copulas. J. Stat. Comput. Simul. **78**, 567–581 (2008)
18. Hofert, M., Scherer, M.: CDO pricing with nested archimedean copulas. Quant. Financ. **11**(5), 775–787 (2011)

19. Aas, K., Czado, C., Frigessi, A., Bakken, H.: Pair-copula constructions of multiple dependence. Insur. Math. Econ. **44**, 182–198 (2009)
20. Okhrin, O., Ristig, A.: Hierarchical Archimedean copulae: the HAC package. Humbold Universität Berlin, Juni 2012. URL http://cran.r-project.org/web/packages/HAC/index.html
21. Schepsmeier, U., Stoeber, J., Brechmann, E.C.: VineCopula: statistical inference of vine copulas, 2013. URL http://CRAN.R-project.org/package=VineCopula. R package version 1.1-1
22. Gundlach, M., Lehrbass, F.: CreditRisk$^+$ in the Banking Industry. Springer, Berlin (2003)
23. Fischer, M., Dietz, C.: Modeling sector correlations with CreditRisk$^+$: the common background vector model. J. Credit Risk **7**, 23–43 (2011/12)
24. Moschopoulos, P.G.: The distribution of the sum of independendent gamma random variables. Ann. Inst. Stat. Math. **37**, 541–544 (1985)
25. Oh, D.H., Patton, A.J.: Modelling dependence in high dimension with factor copulas. working paper, 2013. URL http://public.econ.duke.edu/ap172/Oh_Patton_MV_factor_copula_6dec12.pdf
26. Mai, J.F., Scherer, M.: H-extendible copulas. J. Multivar. Anal. **110**, 151–160 (2012b). URL http://www.sciencedirect.com/science/article/pii/S0047259X12000802
27. Altman, E., Resti, A., Sironi A.: The link between default and recovery rates: effects on the procyclicality of regulatory capital ratios. BIS working paper, July 2002. URL http://www.bis.org/publ/work113.htm
28. Merton, R.C.: On the pricing of corporate debt: the risk structure of interest rates. J. Financ. **29**, 449–470 (1973)
29. Hofert, M., Kojadinovic, I., Mächler, M., Yan J.: Copula: multivariate dependence with copulas, r package version 0.999-5 edition, 2012. URL http://CRAN.R-project.org/package=copula
30. Luethi, D., Breymann W.: ghyp: a package on the generalized hyperbolic distribution and its special cases, 2011. URL http://CRAN.R-project.org/package=ghyp. R package version 1.5.5
31. Genest, C., Remillard, B., Beaudoin, D.: Goodness-of-fit tests for copulas: a review and a power study. Insur. Math. Econ. **44**, 199–213 (2009)
32. Brechmann, E.C., Schepsmeier, U: Modeling dependence with C- and D-vine copulas: The R-package CDVine. Technical report, Technische Universität München, 2012. URL http://www.jstatsoft.org/v52/i03/paper
33. Mai, J.F., Scherer, M.: Simulating Copulas: Stochastic Models, Sampling Algorithms, and Applications, vol. 4. World Scientific, Singapore (2012)
34. Brys, G., Hubert, M., Struyf, A.: Robust measures of tail weight. Comput. Stat. Data Anal. **50**(3), 733–759 (2006)

Implied Recovery Rates—Auctions and Models

Stephan Höcht, Matthias Kunze and Matthias Scherer

Abstract Credit spreads provide information about implied default probabilities and recovery rates. Trying to extract both parameters simultaneously from market data is challenging due to identifiability issues. We review existing default models with stochastic recovery rates and try calibrating them to observed credit spreads. We discuss the mechanisms of credit auctions and compare implied recoveries with realized auction results in the example of Allied Irish Banks (AIB).

1 Introduction

Corporate credit spreads contain the market's perception about (at least) two sources of risk: the time of default and the subsequent loss given default, respectively, the recovery rate. Default probabilities and recovery rates are unknown parameters—comparable to the volatility in the Black–Scholes model. We concern the question whether it is possible to reverse-engineer and disentangle observed credit spreads into these ingredients. Such a reverse-engineering approach translates market values into model parameters, comparable to the extraction of market implied volatilities in the Black–Scholes framework. There is growing literature in the field of implied default probabilities, whereas scientific studies on implied recoveries are sparse. Inferring implied default probabilities from market quotes of credit instruments often relies on the assumption of a fixed recovery rate of, say, $\Phi = 40\%$. Subsequently,

S. Höcht (✉)
Assenagon GmbH, Prannerstrasse 8, 80333 Munich, Germany
e-mail: stephan.hoecht@assenagon.com

M. Kunze
Assenagon Asset Management S.A., Zweigniederlassung München, Prannerstrasse 8,
80333 Munich, Germany
e-mail: matthias.kunze@assenagon.com

M. Scherer
Lehrstuhl für Finanzmathematik, Technische Universität München, Parkring 11,
85748 Garching-Hochbrück, Germany
e-mail: scherer@tum.de

© The Author(s) 2015   147
K. Glau et al. (eds.), *Innovations in Quantitative Risk Management*,
Springer Proceedings in Mathematics & Statistics 99,
DOI 10.1007/978-3-319-09114-3_9

default probabilities are chosen such that model implied credit spreads match quoted credit spreads. The assumption of fixing $\Phi = 40\%$ is close to the market-wide empirical mean (compare Altman et al. [1]), but disregards recovery risk. In many papers, the same recovery rate is assumed for all considered companies, although empirical studies suggest that recoveries are time varying (compare Altman et al. [2], Bruche and González-Aguado [3]), depend on the specific debt instrument, and vary across industry sectors (compare Altman et al. [1]). Obviously, the resulting implied default probability distribution strongly depends on the assumptions on the recovery rate. Since default probabilities and recoveries both enter theoretical spread formulas, we face a so-called identification problem. Making this more plastic, the widely known approximation via the "credit triangle" (see, e.g., Spiegeleer et al. [4, pp. 256]) suggests:

$$\text{spread } s = (1 - \Phi)\lambda \,, \tag{1}$$

where Φ is the recovery rate and λ denotes the default intensity. Obviously, for any given market spread s, the implied recovery is a function of (the assumption on) λ and vice versa. Using this simplified spread formula alone, it is clearly impossible to reverse-engineer Φ and λ simultaneously from s. As we will see, this identification problem also appears in more sophisticated credit models.

We invoke and (at least partially) answer the questions:

- Is it possible to simultaneously extract implied recovery rates and implied default probabilities (under the risk-neutral measure $\mathbb{Q}$)?
- How do implied recoveries compare to realized recoveries?[1]

We address the first question using two types of credit models, where neither the recovery rate nor the default probability distribution is fixed beforehand. As opposed to most existing approaches for the calculation of implied recoveries, both procedures only take into account prices from simultaneously traded assets. Instead of analyzing the spread of one credit instrument for different points in time, implied recoveries and default probabilities are extracted from the term structure of credit spreads. Likewise to the aforementioned implied volatility calculation, this restriction allows for an implied recovery calibration under the risk-neutral measure $\mathbb{Q}$. Analyzing the second question, both models are exemplarily calibrated to market data of Allied Irish Banks (AIB), who experienced a credit event in June 2011. Subsequently, real recovery rates were revealed and can thus be compared to their implied counterparts. In order to clarify how real recoveries are settled in today's credit markets, we start by introducing the mechanism of credit auctions.

---

[1] Here, the term realized recovery does not refer to workout recoveries but to a credit auction result. The question whether the auction procedure appropriately anticipates workout recoveries is left for future research.

2 CDS Settlement: Credit Auction

CDS are the most common and liquidly traded single-name credit derivatives—their liquidity usually even exceeds the one of the underlying bond market. In case of a credit event, the protection buyer receives a default payment, which approximates the percentage loss of a bond holder subject to this default[2] (see Schönbucher [5, preface]). This payment is referred to as loss given default (LGD). The corresponding recovery is defined as one minus the LGD. Recoveries are often quoted as rates, e.g., referring to the fraction of par the protection buyer receives, after the CDS is settled. There are mainly three types of credit events that can be distinguished:

- **Bankruptcy** A bankruptcy event occurs if the company in question faces insolvency or bankruptcy proceedings, is dissolved (other than merger, etc.), liquidated, or wound up.
- **Failure-to-pay** This occurs if the company is unable to pay back outstanding obligations in an amount at least as large as a prespecified payment requirement.
- **Restructuring** A restructuring event takes place if any clause in the company's outstanding debt is negatively altered or violated, such that it is legally binding for all debt holders. Not all types of CDS provide protection against restructuring events.

These credit events are standardized by the International Swaps and Derivatives Association (ISDA). The legally binding answer to the question, whether or not a specific credit event occurred, is given by the so-called Determinations Committees (DC).[3] CDS ISDA standard contracts as well as the responsible DCs differ among geopolitical regions. As opposed to standard European contracts, the standard North American contract does not provide protection against restructuring credit events. The differences are originated by regulatory requirements and the absence of a Chapter 11 equivalent: in order to provide capital relief from a balance sheet perspective, European contracts have to incorporate restructuring events. Our focus will be on the case of nonrestructuring credit events in what follows.

Prior to 2005, CDS were settled physically, i.e., the protection buyer received the contractually agreed notional in exchange for defaulted bonds with the same notional. Accordingly, the corresponding CDS recovery rate was the ratio of the bond's market value to its par. This procedure exhibited different shortfalls (see Haworth [6, p. 24] or Creditex and Markit [7]):

- For a protection buyer, it was necessary to own the defaulted asset. Often, this entailed an unnatural inflation of bond prices after default and became a substantial

---

[2] We will use "credit event" and "default" as synonyms. Note, however, that the terms default and credit event are sometimes distinguished in the sense that default is associated with the final liquidation procedure.

[3] More information on DCs and ISDA can be found on www.isda.org.

problem in default events, where the notional of outstanding CDS contracts exceeded the par of available bonds by multiples.[4]

- On the contrary, the protection seller was obliged to own the defaulted asset after settlement of the CDS. Thus, she or he mandatorily retained a long position with respect to the reference entity's credit risk, making it less attractive to sell protection.
- Since different bonds generally may have different prices, there was no unique settlement price and two identical CDS contracts often were settled against different recoveries, depending on the liquidity of the associated bond market.

These shortfalls were the initial motivation to alter the standard settlement procedure by introducing an auction-based method. From 2005 to 2013 auctions for the settlement of CDS and LCDS (Loan Credit Default Swaps) contracts for 112 default events were held (see Creditex and Markit [8]). On an annual basis, the number of auctions clearly peaked after the financial crisis, i.e., in 2009, where auctions for 45 default events took place. The recovery of a standard CDS contract, traded today, thus usually refers to the result of an auction, which is held subsequent to a credit event.

The auction mechanism aims at a unique and fair settlement price (recovery). It can be split into two stages: the initial bidding period and a subsequent one-sided Dutch auction. The whole process is administrated by Creditex and Markit. In the initial bidding period, each participant, i.e., each protection seller or buyer, represented by one of the bigger investment banks as their dealer, submits a two-way quote. This quote consists of a bid and an offer price for the cheapest-to-deliver bond of the reference entity together with a one-way physical settlement request. In the one-sided Dutch auction, the unique recovery for all outstanding CDS is assessed as the "fair" value of the cheapest-to-deliver bond with respect to its par.[5] Before the auction starts, a quotation amount, a maximum bid-offer spread, and the cap amount is published by ISDA. These three quantities will be explained, while passing through the auction.

2.1 Initial Biding Period

All participants submit a two-way quote together with a one-way physical settlement request. That quote refers to the price of the cheapest bond which is listed as deliverable obligation by ISDA. The request must be in the same direction as the net CDS position, e.g., participants that have net sold protection are not allowed to request delivery of an obligation. Furthermore, the two-way quote must not violate the maximum bid-offer spread. In case a dealer does not represent any outstanding CDS positions with respect to the defaulted entity, she or he is not admitted to

---

[4] Sometimes the phenomenon that some bonds were used several times for the settlement of CDS is referred to as "recycling."

[5] Restructuring events differ, since they allow for maturity specific cheapest-to-deliver bonds.

participate in the auction. Moreover, the notional of the physical settlement request is not allowed to exceed the notional of the outstanding position.

In the next step, the so-called inside market midpoint (IMM) is calculated subject to the following method:

1. Crossing quotes are canceled, i.e., in case an offer quote is smaller or equal to another bid quote, the specific bid and offer are both eliminated.[6]
2. The so-called best halves of the remaining quotes are constructed. The best bid half refers to the (rounded up) upper half of the remaining bid quotes. Accordingly, the best offer half contains the same number of lowest non-canceled offer quotes.
3. The IMM is defined as the average of all quotes in those best halves.

Any participant, whose bid and ask price are both violating the IMM has to pay an adjustment amount.[7] This penalty is supposed to assure that the IMM reflects the underlying bond market in an appropriate way.[8] The initial bidding period is concluded by calculating the net open interest, i.e., the netted notional of physical settlement requests, which is simply carried out by aggregation. In case this amount is zero, the IMM is fixed as the auction result and consequently as the recovery for all CDS, which were supposed to be settled via the auction. Otherwise, the IMM serves as a benchmark for the second part of the auction procedure.

To illustrate this first step, we consider the failure-to-pay event of AIB on June 21, 2011. Two auctions were held, one for senior and one for subordinated CDS referring to AIB. We only consider the senior auction. Table 1 displays the submitted two-way quotes from all 14 participants. For the calculation of the IMM, the reported bid quotes are arranged in descending order, whereas the offers start from the lowest quote.

The first quotes from Nomura (bid) and Citigroup (offer) are canceled out, since the corresponding bid exceeds the offer. Note that this cancelation does not entail a settlement, both quotes are merely neglected with regard to the IMM calculation. Therefore, 13 bid and offer quotes remain and the best halves are the seven highest bid and lowest offer quotes, which are emphasized in Table 1. The IMM is calculated via averaging over these quotes and rounding to one eighth, yielding an IMM of 71.375. The maximum bid-offer spread was 2.50 %-points and the quotation amount was EUR 2 MM. In Table 2, the corresponding physical settlement requests are reported.

As the aggregated notional from bid quotes exceeds the aggregated notional from offer quotes, the auction type is "to buy". Since there is netted demand for the cheapest-to-deliver senior bond, initial offers falling below the IMM are considered

---

[6] Note that they are not settled, but only not taken into account for the calculation of the IMM.

[7] The term "violating" refers to both quotes falling below the IMM (auction is "to buy") or exceeding the IMM (auction is "to sell"), respectively.

[8] Suppose the net open interest is "to sell", i.e., there is a surplus on the seller side. If a participant submits a bid exceeding the IMM, he or she is considered off-market, since prices are supposed to go down and not up. Then the corresponding participant has to pay the prefixed quotation amount times the difference between the IMM and his or her bid. The penalty works vice versa for off-market offers if the open interest is "to buy".

Table 1 Dealer inside market quotes for the first stage of the auction of senior AIB CDS (see Creditex and Markit [8]). Published with the kind permission of ?Creditex Group Inc. and Markit Group Limited 2013. All rights reserved

Dealer	Bid	Offer	Dealer
Nomura Int. PLC	72.00	70.50	Citigroup Global Markets Ltd.
Goldman Sachs Int.	**71.00**	**71.50**	**Société Générale**
Bank of America N.A.	**70.50**	**72.00**	**Credit Suisse Int.**
Barclays Bank PLC	**70.50**	**72.00**	**Deutsche Bank AG**
BNP Paribas	**70.50**	**72.00**	**JPMorgan Chase Bank N.A.**
HSBC Bank PLC	**70.50**	**72.25**	**Morgan Stanley &Co. Int. PLC**
The Royal Bank of Scotland PLC	**70.50**	**72.50**	**UBS AG**
Deutsche Bank AG	**70.00**	**73.00**	**Bank of America N.A.**
UBS AG	70.00	73.00	Barclays Bank PLC
Morgan Stanley &Co. Int. PLC	69.75	73.00	BNP Paribas
Credit Suisse Int.	69.50	73.00	HSBC Bank PLC
JPMorgan Chase Bank N.A.	69.50	73.00	The Royal Bank of Scotland PLC
Société Générale	69.00	73.50	Goldman Sachs Int.
Citigroup Global Markets Ltd.	68.00	74.50	Nomura Int. PLC
Resulting IMM	71.375		

All quotes are reported in %

Table 2 Physical settlement requests for the first stage of the auction of AIB (see Creditex and Markit [8]). Published with the kind permission of ?Creditex Group Inc. and Markit Group Limited 2013. All rights reserved

Dealer	Type	Size in EUR MM
BNP Paribas	Offer	48.00
Credit Suisse Int.	Offer	43.90
Morgan Stanley &Co. Int. PLC	Offer	11.80
Barclays Bank PLC	Bid	30.00
JPMorgan Chase Bank N.A.	Bid	52.00
Nomura Int. PLC	Bid	7.75
UBS AG	Bid	16,00
Total (net)	"To buy"	2.05

off-market and the corresponding dealers have to pay an adjustment amount. In Table 1, only Citigroup's offer of 70.50 is considered off-market. The difference to the IMM is 0.875. Using the quotation amount as notional, the resulting adjustment amount is EUR 17, 500. The second part of the auction aims at satisfying the net physical settlement request of EUR 2.05 MM demand.

2.2 Dutch Auction

This second step is designed as a one-sided Dutch auction, i.e., only quotes in the opposite direction of the net open interest are allowed. In case the net open interest is "to sell", dealers are only allowed to submit bid limit orders and vice versa. For the senior CDS auction of AIB, the net physical settlement request is "to buy" and thus only offer limit orders are allowed. As opposed to the first stage of the auction, there is no restriction with respect to the size of the submitted orders, regardless of the initial settlement request. In order to prevent manipulations, particularly in case of a low net open interest, the prefixed cap amount, which is usually half of the maximum bid-offer spread, imposes a further restriction on the possible limit orders. In case the auction is "to sell", orders are bounded from above by the IMM plus the cap amount and vice versa if the net open interest is "to buy".

In addition to these new limit orders, the appropriate side from the initial two-way quotes from the first stage of the auction are carried over to the second stage—as long as the order does not violate the IMM. All quotes, which are carried over, are determined to have the same size, i.e., the prespecified quotation amount, which was already used to assess the adjustment amount.

Now, all submitted and carried over limit orders are filled, until the net open interest is matched. In case the auction is "to sell", i.e., there is a surplus of bond offerings, the bid limit orders are processed in descending order, starting from the highest quote. Analogously, if the auction is "to buy", offer quotes are filled, starting from the lowest quote. The unique auction price corresponds to the last quote which was at least partially filled. Furthermore, the result may not exceed 100 %.[9]

Reconsider the credit event auction for outstanding senior AIB CDS. Both, carried over offer quotes (first) as well as offers from the second stage (second) of the auction are reported in Table 3.

Recalling that the net physical settlement request was EUR 2.05 MM, we observe that the first two orders were partially filled. The associated limit orders were 70.125 %, which is consequently fixed as the final auction result, i.e., all outstanding senior CDS for AIB were settled subject to a recovery rate of 70.125 %. Following an auction, all protection buyers, who decided to settle their contracts physically beforehand, are obliged to deliver one of the deliverable obligations in exchange for par. Naturally, they are interested in choosing the cheapest among all possible deliverables. Thus, in case of a default, protection buyers are long a cheapest-to-delivery option (compare, e.g., Schönbucher [5, p. 36]), enhancing the position of a protection buyer. Details about the value of that option can be found in Haworth [6, pp.30–32] and Jankowitsch et al. [9].

---

[9] For Northern Rock Asset Management, the European DC resolved that a restructuring credit event occurred on December, 15, 2011. Two auctions took place on February, 2, 2012 and the first one theoretically would have led to an auction result of 104.25 %. Consequently, the recovery was fixed at 100 % (compare Creditex and Markit [8]).

Table 3 Limit orders for the senior auction of AIB (see Creditex and Markit [8]). Published with the kind permission of ?Creditex Group Inc. and Markit Group Limited 2013. All rights reserved

Dealer	Type	Quote (%)	Size (EUR MM)	Aggregated size (USD MM)
JPMorgan Chase Bank N.A.	**Second**	**70.125**	**2.05**	**2.05**
Barclays Bank PLC	**Second**	**70.125**	**2.05**	**4.10**
Credit Suisse Int.	Second	70.25	2.05	6.15
BNP Paribas	Second	70.25	1.00	7.15
BNP Paribas	Second	70.375	1.05	8.20
Citigroup Global Markets Ltd.	First	71.375	2.00	10.20
			...	
Nomura Int. PLC	Second	75	2.00	42.25

2.3 Summary of the Auction Procedure

The auction-based settlement of CDS is designed to approximate the loss of the cheapest-to-deliver bond. The term "CDS auction" might thus be misleading, since it is an auction, where the market value of the cheapest from a set of bonds is assessed. Consequently, the recovery rate of a CDS contract is the market value of this bond divided by its par.

In the above example, JPMorgan's and Barclays' orders were the only ones filled. Both dealers had a considerable physical settlement request of EUR 52 MM and EUR 30 MM, respectively, possibly reflecting a long CDS position. By submitting the lowest possible quote for a notional of EUR 2.05 MM each, both dealers stretched the recovery to the possible maximum. In case, both parties indeed represented large long CDS positions, they profited from the low open interest. Moreover, the final auction result was below the IMM. Thus, if one dealer would have quoted the final auction result already in the first step, she or he would have been considered off-market and consequently penalized.

Another problem appeared during a restructuring credit event of SNS bank, where senior and subordinated CDS were settled in the same auction. Due to government intervention, subordinated bond holders experienced a full write-down ("bail-in") before the auction. Thus, there were no more subordinated deliverables and senior and subordinated CDS had the very same recovery (either 95.5 or 85.5 %, depending on the maturity of the CDS), contradicting the connection between the subordinated bond holder's loss and the subordinated CDS recovery. Another case for a counterintuitive auction result concerned the settlement of CDS referring to Fannie Mae or Freddy Mac, where subordinated contracts recovered above senior. Moreover, as the determination committees and dealers are big investment banks, there might be conflicts of interest when determining whether a credit event occurred or not.

These are reasons for an ongoing discussion about whether this one-sided auction design is fair or not (compare Du and Zhu [10] for the proposal of an alternative auction design). Currently, ISDA is working on a further supplement to the credit derivative definitions, involving among others the introduction of a new credit event as a solution to what happened with subordinated SNS CDS.

3 Examples of Implied Recovery Models

As explained above, the recovery of a CDS, $\Phi_\tau \in [0, 1]$, refers to the result of an auction which is held after a credit event at time τ and is designed to approximate the relative "left-over" for a bond holder. Before a default event and the following auction takes place this recovery is unknown. One way to assess this quantity for nondefaulted securities is to reverse-engineer implied recoveries from market CDS quotes. Any basic pricing approach for the "fair" spread s_T of a CDS with maturity $T > 0$ is of the form

$$s_T = \mathbb{E}_{\mathbb{Q}}[f(\tau, \Phi_\tau)]. \tag{2}$$

I.e., the spread is the risk-neutral expectation of a function of the default time (or default probability, respectively) and the recovery rate in case of default. Specifying τ and Φ_τ, two models are revisited and calibrated by minimizing the root mean squared error (RMSE) between $\mathbb{E}_{\mathbb{Q}}[f(\tau, \Phi_\tau)]$ and market spreads over a term structure of CDS spreads.

3.1 Cox–Ingersoll–Ross Type Reduced-Form Model

This reduced-form model resembles the one presented in Jaskowski and McAleer [11], although applied in a different context. All reduced-form models are based on the same principle. The time of a credit event τ is the first jump of a stochastic counting process $Z = \{Z_t\}_{t\geq0} \in \mathbb{N}_0$, i.e., $\tau = \inf\{t \geq 0 : Z_t > 0\}$. In this case Z will be a Cox-Process governed by a Cox–Ingersoll–Ross type intensity process λ, i.e.,

$$d\lambda_t = \kappa(\theta - \lambda_t)dt + \sigma\sqrt{\lambda_t}dW_t, \quad \lambda_0 > 0.$$

The recovery in this model is defined as an exponential function of the intensity process, i.e.,

$$\Phi_\tau := ae^{-\frac{1}{\tau}\int_0^\tau \lambda_t dt},$$

where $a \in [0, 1]$ is referred to as the recovery parameter. A default in a period of high expected distress, e.g., in an economic downturn, entails lower recoveries

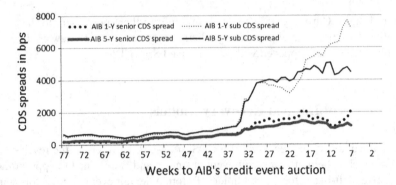

Fig. 1 Weekly average spreads for AIB senior and subordinated CDS with 1 and 5 years maturity. The spreads represent two whole term structures, which are used to calibrate the presented implied recovery approaches in every displayed week independently

and vice versa. Comparable choices for modeling recoveries can be found, e.g., in Madan et al. [12], Das and Hanouna [13], Höcht and Zagst [14], or Jaskowski and McAleer [11]. Since the model will be calibrated to one CDS spread curve, one has to be restrictive concerning the amount of free model parameters in the recovery model. Using this model, the risk-neutral spread $s_T(\kappa, \theta, \sigma, \lambda_0, a)$ has an integral-free representation. The resulting risk-neutral parameters and subsequently the risk-neutral implied recovery and probability of default are determined by minimizing the RMSE:

$$(\kappa^*, \theta^*, \sigma^*, \lambda_0^*, a^*) := \operatorname{argmin} \sqrt{\frac{1}{|\mathbb{I}|} \sum_{T \in \mathbb{I}} \left(s_T^M - s_T(\kappa, \theta, \sigma, \lambda_0, a)\right)^2}, \qquad (3)$$

where $\mathbb{I}$ is the set of maturities with observable market quotes for CDS spreads s_T^M. In case senior as well as subordinated CDS are available for a certain defaultable entity, two different recovery parameters a_{sen} and a_{sub} are used, while the intensity parameters are the same for both seniorities. This reflects the fact that in case of a credit event both CDS types are settled, although usually in different auctions.[10] In this case, the optimization in Eq. (3) is simply carried out by matching senior and subordinated spreads simultaneously. For the calibration, we reconsider the example of AIB. Figure 1 exemplarily shows weekly average quotes for AIB senior and subordinated CDS spreads with maturities 1 and 5 years.

Approaching the time of default, a spread widening and inversion of both senior and subordinated term structures can be observed. Calibrating the introduced Cox-Ingersoll-Ross model to AIB CDS quotes for each week independently for several maturities leads to the resulting implied recoveries and 5-year default probabilities shown in Fig. 2.

---

[10] In the current version of the upcoming ISDA supplement, subordinated CDS may also settle without effecting senior CDS. However, so far either both or none settles.

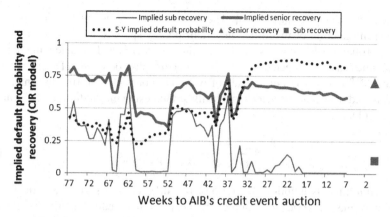

Fig. 2 Weekly calibration results for the CIR model applied to CDS spreads of AIB before its default in June 2011

Implied senior and subordinated recoveries and implied default probabilities vary substantially over time. One reason is that term structure shapes and general spread regimes also vary unusually strong from week to week, since AIB is in distress. Furthermore, there are co-movements of the 5-year implied default probability and the implied recoveries. This is caused by the fact that a (recovery) and θ (long-term default intensity) have a similar effect on long term CDS spreads. Assuming $\lambda_t \equiv \theta$ for all $t > t^* > 0$, the fair long term spread can be approximated via

$$s_T \approx c_0 + (1 - ae^{-\theta})\theta, \quad \text{for all } T > t^*, \tag{4}$$

where $c_0 \in \mathbb{R}$ is constant. Hence, using the above approximation for a given spread s_T, the optimal recovery parameter a^* can be seen as a function of the long term default intensity, denoted as $a^*(\theta)$. This entails the existence of a continuum of parameter values $(\kappa^*, \theta, \sigma^*, \lambda_0^*, a^*(\theta))$, $\theta > 0$, which all generate a comparable long term spread and thus a similar RMSE. Consequently, a minor variation in the quoted spreads might cause a substantial change in the resulting optimal parameters and thus in the implied recovery and implied probability of default. This is referred to as identification problem.

The following section contains a framework to circumvent this identification problem.

3.2 Pure Recovery Model

Two CDS contracts with the same reference entity and maturity, but differently ranked reference obligations, face the same default probabilities, but different recoveries. The general idea of the "pure recovery model" goes back to Unal et al. [15] and

Schläfer and Uhrig-Homburg [16]. The approach makes use of this fact by considering the fraction of two differently ranked CDS spreads, which is then free of default probabilities. Hence, spread ratios are considered and modeled and default probabilities can be neglected. A comparable approach is outlined in Doshi [17]. Let s^{sen} and s^{sub} denote the fair spreads of two CDS contracts referring to senior and subordinated debt. The basic idea can be illustrated using the credit triangle formula from Eq. (1), i.e.,

$$\frac{s^{sen}}{s^{sub}} \approx \frac{(1 - \Phi^{sen})\lambda}{(1 - \Phi^{sub})\lambda} = \frac{1 - \Phi^{sen}}{1 - \Phi^{sub}}. \tag{5}$$

Under simplified assumptions the ratio of two different types of CDS spreads is a function of the recoveries Φ^{sen} and Φ^{sub}. In case of the credit triangle formula, for instance, the underlying assumptions include independence of λ and Φ. The crucial point is to find a suitable and sophisticated model, such that this fraction again only contains recovery information. Implied recoveries are then extracted by calibrating fractions of senior and subordinated spreads. We propose a model that allows for time variation in Φ but no dependence on the default time τ.

In a first step, a company-wide recovery rate X_T is defined, i.e., a recovery for the whole company in case of a default until T, where T_{max} is the maximum of all instruments' maturities which should be captured by the model. Suppose $\mu_0 \in (0, 1)$, $\mu_1 \in (-1, 1)$, and $\mu_0 + \mu_1 \in (0, 1)$. Furthermore, let $v \in (0, 1)$. For a certain maturity $T_{max} > T > 0$, X_T is assumed to be Beta-distributed with the following expectation and variance:

$$\mathbb{E}_{\mathbb{Q}}[X_T] = \mu(T) := \mu_0 + \mu_1\sqrt{T/T_{max}}, \tag{6}$$

$$\mathbb{V}\mathrm{ar}_{\mathbb{Q}}[X_T] = \sigma^2(T) := v[\mu(T) - \mu(T)^2]. \tag{7}$$

The Beta distribution is a popular choice for modeling stochastic recovery rates, since it allows for an U-shaped density on [0, 1] that is empirically confirmed for recovery rates. The above parameter restrictions assure that a Beta distribution with $\mathbb{E}_{\mathbb{Q}}[X_T]$ and $\mathbb{V}\mathrm{ar}_{\mathbb{Q}}[X_T]$ as above actually exists. The square-root specification allows for a higher differentiation between maturity specific recoveries near $T = 0$, a phenomenon which is also widely reflected in CDS market term structures. Overall, this company-wide recovery distribution varies in time without depending on τ. In a second step, the seniority specific recoveries Φ_T^{sen} and Φ_T^{sub} are defined as functions of X_T. In legal terms, such a relation is established via a pecking order, defined by the Absolute Priority Rule (APR): In case of a default event, any class of debt with a lower priority than another will only be repaid if all higher ranked debt is repaid in full. Furthermore, all claimants of the same seniority will recover simultaneously, i.e., they receive the same proportion of their par value. Let d_{sec}, d_{sen}, and d_{sub} denote the proportions of secured, senior unsecured, and subordinated unsecured debt, respectively, on the balance sheet of a company at default, such that $d_{sub} + d_{sen} + d_{sec} = 1$. Figure 3 illustrates the APR.

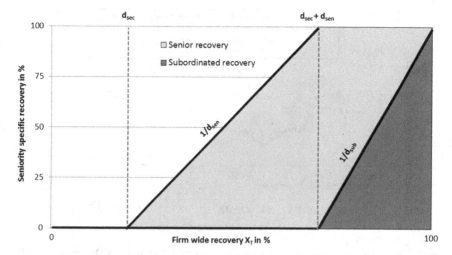

Fig. 3 Absolute priority rule: seniority specific recoveries depend on the stochastic firm-wide recovery and the debt structure of the company

The parameters d_{sub}, d_{sen}, and d_{sec} determine, which proportion of X_T is assigned to senior and subordinated debt holders if a default occurs. Motivated by the linkage of bonds and CDS in the auction mechanism, Φ_T^{sen} and Φ_T^{sub} are also assumed to be the appropriate CDS recoveries. Note, however, that in practice, APR violations often occur and are widely examined (see, e.g., Betker [18] and Eberhart and Weiss [19]). Using the APR rule, a general spread representation as in Eq. (2) as well as independence of Φ and τ, the recoveries are deterministic functions of the company-wide recovery X_T and the fraction of senior to subordinated CDS spreads is given by

$$
\frac{s_T^{sen}}{s_T^{sub}} = \frac{1 - \int_{d_{sec}}^{d_{sec}+d_{sen}} \frac{x-d_{sec}}{d_{sen}} f_{p_T,q_T}(x)dx - \int_{d_{sec}+d_{sen}}^{1} f_{p_T,q_T}(x)dx}{1 - \int_{d_{sec}+d_{sen}}^{1} \frac{x-(d_{sec}+d_{sen})}{d_{sub}} f_{p_T,q_T}(x)dx}, \tag{8}
$$

where $f_{p_T,q_T}(x)$ denotes the density of a Beta(p_T, q_T)-distributed random variable. The variables p_T and q_T are linked to the parameters μ_0, μ_1, and v via Eqs. (6) and (7) and the first two moments of the Beta distribution. They are calibrated using the above formula, whereas the balance sheet parameters d_{sec}, d_{sen}, and d_{sub} are directly taken from quarterly reports. Instead of calibrating a single-spread curve, the calibration is carried out by matching theoretical fractions $s_T^{sen}/s_T^{sub}(\mu_0, \mu_1, v)$ in Eq. (8) for a set of several maturities to their market counterparts $s_T^{M,sen}/s_T^{M,sub}$, i.e.

$$
(\mu_0^*, \mu_1^*, v^*) := \arg\min \sqrt{\frac{1}{|\mathbb{I}|} \sum_{T \in \mathbb{I}} \left(s_T^{M,sen}/s_T^{M,sub} - s_T^{sen}/s_T^{sub}(\mu_0, \mu_1, v) \right)^2}.
$$

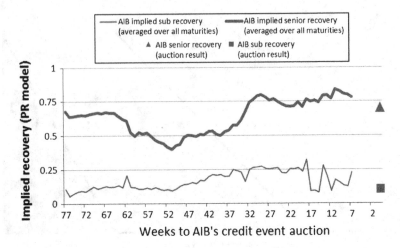

Fig. 4 Weekly calibration results for the pure recovery model applied to CDS spreads of AIB before its default in June 2011

The resulting risk-neutral implied distribution of the company-wide recovery can be translated into risk-neutral seniority specific recovery distributions by applying the APR rule. Furthermore, we could proceed to use this implied recovery result and extract implied default probabilities in a second step.

Calibrating the pure recovery model to senior and subordinated spreads from AIB (see Fig. 1) before its default yields implied recoveries for senior and sub debt, averaged over all maturities as displayed in Fig. 4.

As opposed to the Cox–Ingersoll–Ross model, the resulting recoveries do not exhibit sudden jumps, but are more stable over time. Only during the last weeks before default (weeks 17 to 7), particularly the subordinated recovery fluctuates. However, this is related to the significant movements of the market spreads and not originated by an identification problem among the parameters. Moreover, both senior and subordinated recoveries are in line with the later auction results, at least with respect to their proportional relation.

4 Conclusion and Outlook

Extracting implied recoveries and implied default probabilities in a risk-neutral setting tends to generate instable parameter estimates. The identification problem among long-term default probabilities and recovery rates is not limited to the presented CIR model, but can also be observed, e.g., in jump-to-default equity models such as the one proposed in Das and Hanouna [13]. We illustrated one way to circumvent the problem by reducing the calibrated expression to a form, where only recovery-related parameters appear. This is possible by considering instruments with different seniorities, such as senior and subordinated CDS.[11] Furthermore, the extracted risk-neutral recoveries are more in line with the observed final auction results. Generally, further instruments, e.g., loans or the recently more popular contingent convertibles could be used in a similar way.

Acknowledgments The authors would like to thank Michael Hünseler, Rudi Zagst, and an anonymous referee for their valuable remarks on earlier versions of the manuscript.

Open Access This chapter is distributed under the terms of the Creative Commons Attribution Noncommercial License, which permits any noncommercial use, distribution, and reproduction in any medium, provided the original author(s) and source are credited.

References

1. Altman, E.I., Kishore, I., Vellore, M.: Almost everything you wanted to know about recoveries on defaulted bonds. Financ. Anal. J. **52**(6), 57–64 (1996)
2. Altman, E.I., Brady, B., Resti, A., Sironi, A.: The link between default and recovery rate: theory, empirical evidence and implications. J. Bus. **78**(6), 2203–2227 (2005)
3. Bruche, M., González-Aguado, C.: Recovery rates, default probabilities and the credit cycle. J. Bank. Financ. **34**(4), 754–764 (2010)
4. De Spiegeleer, J., Van Hulle, C., Schoutens, W.: The Handbook of Hybrid Securities. Wiley Finance Series. Wiley (2014)
5. Schönbucher, P.J.: Credit Derivatives Pricing Models. Wiley Finance Series. Wiley, Canada (2003)
6. Haworth, H.: A guide to credit event and auctions (2011). http://www.credit-suisse.com/researchandanalytics
7. Creditex and Markit. Credit event auction primer (2010). http://www.creditfixings.com
8. Creditex and Markit. Credit event fixings (2013). http://www.creditfixings.com
9. Jankowitsch, R., Pullirsch, R., Veža, T.: The delivery option in credit default swaps. J. Bank. Financ. **32**(7), 1269–1285 (2008)
10. Du, S., Zhu, H.: Are CDS auctions biased? Working Paper, Stanford University (2011)
11. Jaskowski, M., McAleer, M.: Estimating implied recovery rates from the term structure of CDS spreads. Tinbergen Institute Discussion Papers 13–005/III, Tinbergen Institute (2012)
12. Madan, D.B., Bakshi, G.S., Zhang F.X.: Understanding the role of recovery in default risk models: empirical comparisons and implied recovery rates. FDIC CFR Working Paper No. 06; EFA 2004 Maastricht Meetings Paper No. 3584; FEDS Working Paper; AFA 20004 Meetings, 2006
13. Das, S., Hanouna, P.: Implied recovery. J. Econ. Dyn. Control **33**(11), 1837–1857 (2009)
14. Höcht, S., Zagst, R.: Pricing credit derivatives under stochastic recovery in a hybrid model. Appl. Stoch. Models Bus. Ind. **26**(3), 254–276 (2010)
15. Unal, H., Madan, D., Güntay, L.: Pricing the risk of recovery in default with absolute priority rule violation. J. Bank. Financ. **27**(6), 1001–1218 (2003)
16. Schläfer T.S., Uhrig-Homburg, M.: Estimating market-implied recovery rates from credit default swap premia. University of Karlsruhe Working Paper, 2009
17. Doshi, H.: The term structure of recovery rates. McGill University Working Paper (2011)

---

[11] Note that there is an ongoing discussion regarding the admission of credit events, which are only binding for subordinated CDS. Such a possibility is explicitly excluded in the proposed pure recovery model.

18. Betker, B.L.: Management's incentives, equity's bargaining power and deviations from absolute priority in Chapter 11 bankruptcies. J. Bus. **68**(2), 161–183 (1995)
19. Eberhart, A.C., Weiss, L.A.: The importance of deviations form the absolute priority rule in Chapter 11 bankruptcy proceedings. Financ. Manag. **27**(4), 106–110 (1998)

Upside and Downside Risk Exposures of Currency Carry Trades via Tail Dependence

Matthew Ames, Gareth W. Peters, Guillaume Bagnarosa and Ioannis Kosmidis

Abstract Currency carry trade is the investment strategy that involves selling low interest rate currencies in order to purchase higher interest rate currencies, thus profiting from the interest rate differentials. This is a well known financial puzzle to explain, since assuming foreign exchange risk is uninhibited and the markets have rational risk-neutral investors, then one would not expect profits from such strategies. That is, according to uncovered interest rate parity (UIP), changes in the related exchange rates should offset the potential to profit from such interest rate differentials. However, it has been shown empirically, that investors can earn profits on average by borrowing in a country with a lower interest rate, exchanging for foreign currency, and investing in a foreign country with a higher interest rate, whilst allowing for any losses from exchanging back to their domestic currency at maturity.

This paper explores the financial risk that trading strategies seeking to exploit a violation of the UIP condition are exposed to with respect to multivariate tail dependence present in both the funding and investment currency baskets. It will outline in what contexts these portfolio risk exposures will benefit accumulated portfolio returns and under what conditions such tail exposures will reduce portfolio returns.

M. Ames (✉)
Department of Statistical Science, University College London, London, UK
e-mail: m.ames.12@ucl.ac.uk

G.W. Peters
Department of Statistical Science, University College London, London, UK

G.W. Peters
Commonwealth Scientific and Industrial Research Organisation, Canberra, Australia

G.W. Peters
Oxford-Man Institute, Oxford University, Oxford, UK

G. Bagnarosa
Department of Computer Science, ESC Rennes School of Business, University College London, London, UK

I. Kosmidis
Department of Statistical Science, University College London, London, UK

© The Author(s) 2015   163
K. Glau et al. (eds.), *Innovations in Quantitative Risk Management*,
Springer Proceedings in Mathematics & Statistics 99,
DOI 10.1007/978-3-319-09114-3_10

Keywords Currency carry trade · Multivariate tail dependence · Forward premium puzzle · Mixture models · Generalised archimedean copula

1 Currency Carry Trade and Uncovered Interest Rate Parity

One of the most robust puzzles in finance still to be satisfactorily explained is the uncovered interest rate parity puzzle and the associated excess average returns of currency carry trade strategies. Such trading strategies are popular approaches which involve constructing portfolios by selling low interest rate currencies in order to buy higher interest rate currencies, thus profiting from the interest rate differentials. The presence of such profit opportunities, pointed out by [2, 10, 15] and more recently by [5–7, 20, 21, 23], violates the fundamental relationship of uncovered interest rate parity (UIP). The UIP refers to the parity condition in which exposure to foreign exchange risk, with unanticipated changes in exchange rates, is uninhibited and therefore if one assumes rational risk-neutral investors, then changes in the exchange rates should offset the potential to profit from the interest rate differentials between high interest rate (investment) currencies and low interest rate (funding) currencies. We can more formally write this relation by assuming that the forward price, F_t^T, is a martingale under the risk-neutral probability $\mathbb{Q}$ ([24]):

$$E_{\mathbb{Q}}\left[\frac{S_T}{S_t}\bigg|\mathscr{F}_t\right] = \frac{F_t^T}{S_t} = e^{(r_t-r_t^\star)(T-t)}. \tag{1}$$

The UIP Eq. (1) thus states that under the risk-neutral probability, the expected variation of the exchange rate S_t should equal the differential between the interest rate of the two associated countries, denoted by, respectively, r_t and $r_t^\star$. The currency carry trade strategy investigated in this paper aims at exploiting violations of the UIP relation by investing a certain amount in a basket of high interest rate currencies (the long basket), while funding it through a basket of low interest rate currencies (the short basket). When the UIP holds, then given foreign exchange market equilibrium, no profit should arise on average from this strategy, however, such opportunities are routinely observed and exploited by large volume trading strategies.

In this paper, we build on the existing literature by studying a stochastic feature of the joint tail behaviours of the currencies within each of the long and the short baskets, which form the carry trade. We aim to explore to what extent one can attribute the excess average returns with regard to compensation for exposure to tail risk, for example either dramatic depreciations in the value of the high interest rate currencies or dramatic appreciations in the value of the low interest rate currencies in times of high market volatility.

We postulate that such analyses should also benefit from consideration not only of the marginal behaviours of the processes under study, in this case the exchange rates of currencies in a portfolio, but also a rigorous analysis of the joint dependence features of such relationships. We investigate such joint relationships in light of the UIP condition. To achieve this, we study the probability of joint extreme movements in the funding and investment currency baskets and interpret these extremal tail probabilities as relative risk exposures of adverse and beneficial joint currency movements, which would affect the portfolio returns. This allows us to obtain a relative contribution to the exposure of the portfolio profit decomposed in terms of the downside and upside risks that are contributed from such tail dependence features in each currency basket. We argue that the analysis of the carry trade is better informed by jointly modelling the multivariate behaviour of the marginal processes of currency baskets accounting for potential multivariate extremes, whilst still incorporating heavy tailed relationships studied in marginal processes.

We fit mixture copula models to vectors of daily exchange rate log returns between 1989 and 2014 for both the investment and funding currency baskets making up the carry trade portfolio. The method and the dataset considered for the construction of the respective funding and investing currencies baskets are thoroughly described in [1]. The currency compositions of the funding and investment baskets are varying daily over time as a function of the interest rate differential processes for each currency relative to the USD.

Our analysis concludes that the appealing high return profile of a carry portfolio is not only compensating the tail thickness of each individual component probability distribution but also the fact that extreme returns tend to occur simultaneously and lead to a portfolio particularly sensitive to the risk of what is known as drawdown. Furthermore, we also demonstrate that high interest rate currency baskets and low interest rate currency baskets can display periods during which the tail dependence gets inverted, demonstrating when periods of construction of the aforementioned carry positions are being undertaken by investors.

2 Interpreting Tail Dependence as Financial Risk Exposure in Carry Trade Portfolios

In order to fully understand the tail risks of joint exchange rate movements present when one invests in a carry trade strategy, we can look at both the downside extremal tail exposure and the upside extremal tail exposure within the funding and investment baskets that comprise the carry portfolio. The downside tail exposure can be seen as the crash risk of the basket, i.e. the risk that one will suffer large joint losses from each of the currencies in the basket. These losses would be the result of joint appreciations of the currencies that one is short in the low interest rate basket and/or joint depreciations of the currencies that one is long in the high interest rate basket.

Definition 1 (*Downside Tail Risk Exposure in Carry Trade Portfolios*) Consider the investment currency (long) basket with n-exchange rates relative to base currency, on day t, with currency log returns $(X_t^{(1)}, X_t^{(2)}, \ldots, X_t^{(d)})$. Then, the downside tail exposure risk for the carry trade will be defined as the conditional probability of adverse currency movements in the long basket, corresponding to its upper tail dependence (a loss for a long position results from a forward exchange rate increase), given by,

$$\lambda_{\mathcal{U}}^{(i)}(u) := \Pr\left(X_t^{(i)} > F_i^{-1}(u)|X_t^{(1)} > F_1^{-1}(u), \ldots, X_t^{(i-1)} > F_{i-1}^{-1}(u), X_t^{(i+1)} > F_{i+1}^{-1}(u), \ldots, X_t^{(d)} > F_d^{-1}(u)\right)$$
(2)

for a currency of interest $i \in \{1, 2, \ldots, d\}$ where F_i is the marginal distribution for the asset i. Conversely, the downside tail exposure for the funding (short) basket with d currencies will be defined as the conditional probability of adverse currency movement in the short basket (a loss for a short position results from a forward exchange rate decrease), given by

$$\lambda_{\mathcal{L}}^{(i)}(u) := \Pr\left(X_t^{(i)} < F_i^{-1}(u)|X_t^{(1)} < F_1^{-1}(u), \ldots, X_t^{(i-1)} < F_{i-1}^{-1}(u), X_t^{(i+1)} < F_{i+1}^{-1}(u), \ldots, X_t^{(d)} < F_d^{-1}(u)\right).$$
(3)

In general, then a basket's upside or downside risk exposure would be quantified by the probability of a loss (or gain) arising from an appreciation or depreciation jointly of magnitude u and the dollar cost associated to a given loss/gain of this magnitude. The standard approach in economics would be to associate say a linear cost function in u to such a probability of loss to get say the downside risk exposure in dollars according to $E_i(u) = C(F_i^{-1}(u)) \times \lambda_{\mathcal{U}}(u)$, which will be a function of the level u. As $\lambda_{\mathcal{U}}$ becomes independent of the marginals, i.e. as $u \to 0$ or $u \to 1$, $C_{\mathcal{U}}$ also becomes independent of the marginals.

Conversely, we will also define the upside tail exposure that will contribute to profitable returns in the carry trade strategy when extreme movements that are in favour of the carry position held. These would correspond to precisely the probabilities discussed above applied in the opposite direction. That is the upside risk exposure in the funding (short) basket is given by Eq. (2) and the upside risk exposure in the investment (long) basket is given by Eq. (3). That is the upside tail exposure of the carry trade strategy is defined to be the risk that one will earn large joint profits from each of the currencies in the basket. These profits would be the result of joint depreciations of the currencies that one is short in the low interest rate basket and/or joint appreciations of the currencies that one is long in the high interest rate basket.

Remark 1 In a basket with d currencies, $d \geq 2$, if one considers capturing the upside and downside financial risk exposures from a model-based calculation of these extreme probabilities, then if the parametric model is exchangeable, such as an Archimedean copula, then swapping currency i in Eqs. (2) and (3) with another currency from the basket, say j will not alter the downside or upside risk exposures. If they are not exchangeable, then one can consider upside and downside risks for each individual currency in the carry trade portfolio.

We thus consider these tail upside and downside exposures of the carry trade strategy as features that can show that even though average profits may be made from the violation of UIP, it comes at significant tail exposure.

We can formalise the notion of the dependence behaviour in the extremes of the multivariate distribution through the concept of tail dependence, limiting behaviour of Eqs. (2) and (3), as $u \uparrow 1$ and $u \downarrow 0$ asymptotically. The interpretation of such quantities is then directly relevant to assessing the chance of large adverse movements in multiple currencies which could potentially increase the risk associated with currency carry trade strategies significantly, compared to risk measures which only consider the marginal behaviour in each individual currency. Under certain statistical dependence models, these extreme upside and downside tail exposures can be obtained analytically. We develop a flexible copula mixture example that has such properties below.

3 Generalised Archimedean Copula Models for Currency Exchange Rate Baskets

In order to study the joint tail dependence in the investment or funding basket, we consider an overall tail dependence analysis which is parametric model based, obtained by using flexible mixtures of Archimedean copula components. Such a model approach is reasonable since typically the number of currencies in each of the long basket (investment currencies) and the short basket (funding currencies) is 4 or 5.

In addition, these models have the advantage that they produce asymmetric dependence relationships in the upper tails and the lower tails in the multivariate model. We consider three models; two Archimedean mixture models and one outer power transformed Clayton copula. The mixture models considered are the Clayton-Gumbel mixture and the Clayton-Frank-Gumbel mixture, where the Frank component allows for periods of no tail dependence within the basket as well as negative dependence. We fit these copula models to each of the long and short baskets separately.

Definition 2 (*Mixture Copula*) A mixture copula is a linear weighted combination of copulae of the form:

$$C_M(\mathbf{u}; \boldsymbol{\theta}) = \sum_{i=1}^{N} \lambda_i C_i(\mathbf{u}; \boldsymbol{\theta}_i),$$ (4)

where $0 \leq \lambda_i \leq 1 \ \forall i \in \{1, \ldots, N\}$ and $\sum_{i=1}^{N} \lambda_i = 1$.

Definition 3 (*Archimedean Copula*) A d-dimensional copula C is called Archimedean if it can be represented by the form:

$$C(\mathbf{u}) = \psi\{\psi^{-1}(u_1) + \cdots + \psi^{-1}(u_d)\} \ \forall \mathbf{u} \in [0, 1]^d,$$ (5)

where ψ is an Archimedean generator satisfying the conditions given in [22]. $\psi^{-1} : [0, 1] \rightarrow [0, \infty)$ is the inverse generator with $\psi^{-1}(0) = \inf\{t : \psi(t) = 0\}$.

In the following section, we consider two stages to estimate the multivariate basket returns, first the estimation of suitable heavy tailed marginal models for the currency exchange rates (relative to USD), followed by the estimation of the dependence structure of the multivariate model composed of multiple exchange rates in currency baskets for long and short positions.

Once the parametric Archimedean mixture copula model has been fitted to a basket of currencies, it is possible to obtain the upper and lower tail dependence coefficients, via closed form expressions for the class of mixture copula models and outer power transform models we consider. The tail dependence expressions for many common bivariate copulae can be found in [25]. This concept was recently extended to the multivariate setting by [9].

Definition 4 (*Generalised Archimedean Tail Dependence Coefficient*) Let $X = (X_1, \ldots, X_d)^T$ be an d-dimensional random vector with distribution $C(F_1(X_1), \ldots, F_d(X_d))$, where C is an Archimedean copula and $F_1, \ldots, F_d$ are the marginal distributions. The coefficients of upper and lower tail dependence are defined respectively as:

$$
\begin{aligned}
\lambda_{\mathcal{U}}^{1,\ldots,h|h+1,\ldots,d} &= \lim_{u \to 1-} P\left(X_1 > F_1^{-1}(u), \ldots, X_h > F_h^{-1}(u) | X_{h+1} > F_{h+1}^{-1}(u), \ldots, X_d > F_d^{-1}(u)\right) \\
&= \lim_{t \to 0+} \frac{\sum_{i=1}^{d} \left(\binom{d}{d-i} i (-1)^i \left[\psi'(it)\right]\right)}{\sum_{i=1}^{d-h} \left(\binom{d-h}{d-h-i} i (-1)^i \left[\psi'(it)\right]\right)},
\end{aligned}
\tag{6}
$$

$$
\begin{aligned}
\lambda_{\mathcal{L}}^{1,\ldots,h|h+1,\ldots,d} &= \lim_{u \to 0+} P\left(X_1 < F_1^{-1}(u), \ldots, X_h < F_h^{-1}(u) | X_{h+1} < F_{h+1}^{-1}(u), \ldots, X_d < F_d^{-1}(u)\right) \\
&= \lim_{t \to \infty} \frac{d}{d-h} \frac{\psi'(dt)}{\psi'((d-h)t)}
\end{aligned}
\tag{7}
$$

for the model dependence function 'generator' $\psi(\cdot)$ and its inverse function.

In [9], the analogous form of the generalised multivariate upper and lower tail dependence coefficients for outer power transformed Clayton copula models is provided. The derivation of Eqs. (6) and (7) for the outer power case follows from [12], i.e. the composition of a completely monotone function with a non-negative function that has a completely monotone derivative is again completely monotone. The densities for the outer power Clayton copula can be found in [1].

In the above definitions of model-based parametric upper and lower tail dependence, one gets the estimates of joint extreme deviations in the whole currency basket. It will often be useful in practice to understand which pairs of currencies within a given currency basket contribute significantly to the downside or upside risks of the overall currency basket. In the class of Archimedean-based mixtures we consider, the feature of exchangeability precludes decompositions of the total basket downside and upside risks into individual currency specific components. To be precise, we aim to perform a decomposition of say the downside risk of the funding basket into contributions from each pair of currencies in the basket, we will do this via a simple linear projection onto particular subsets of currencies in the portfolio that are of interest, which leads, for example to the following expression:

$$\mathbb{E}\left[\hat{\lambda}_{\mathscr{U}}^{i|1,2,\dots,i-1,i+1,\dots,d}\,\middle|\,\hat{\lambda}_{\mathscr{U}}^{2|1},\hat{\lambda}_{\mathscr{U}}^{3|1},\hat{\lambda}_{\mathscr{U}}^{3|2},\dots,\hat{\lambda}_{\mathscr{U}}^{d|d-1}\right]=\alpha_0+\sum_{i\neq j}^{d}\alpha_{ij}\hat{\lambda}_{\mathscr{U}}^{i|j}, \qquad (8)$$

where $\hat{\lambda}_{\mathscr{U}}^{i|1,2,\dots,i-1,i+1,\dots,d}$ is a random variable since it is based on parameters of the mixture copula model which are themselves functions of the data and therefore random variables. Such a simple linear projection will then allow one to interpret directly the marginal linear contributions to the upside or downside risk exposure of the basket obtained from the model, according to particular pairs of currencies in the basket by considering the coefficients α_{ij}, i.e. the projection weights. To perform this analysis, we need estimates of the pairwise tail dependence in the upside and downside risk exposures $\hat{\lambda}_{\mathscr{U}}^{i|j}$ and $\hat{\lambda}_{\mathscr{L}}^{i|j}$ for each pair of currencies $i,\,j\in\{1,2,\dots,d\}$. We obtain this through non-parametric (model-free) estimators, see [8].

Definition 5 Non-Parametric Pairwise Estimator of Upper Tail Dependence (Extreme Exposure)

$$\hat{\lambda}_{\mathscr{U}}=2-\min\left[2,\frac{\log\hat{C}_n\left(\frac{n-k}{n},\frac{n-k}{n}\right)}{\log(\frac{n-k}{n})}\right]\qquad k=1,2,\dots,n-1, \qquad (9)$$

where $\hat{C}_n(u_1,u_2)=\frac{1}{n}\sum_{i=1}^{n}\mathbf{1}\left(\frac{R_{1i}}{n}\leq u_1,\frac{R_{2i}}{n}\leq u_2\right)$ and R_{ji} is the rank of the variable in its marginal dimension that makes up the pseudo data.

In order to form a robust estimator of the upper tail dependence, a median of the estimates obtained from setting k as the 1st, 2nd, ..., 20th percentile values was used. Similarly, k was set to the 80th, 81st, ..., 99th percentiles for the lower tail dependence.

4 Currency Basket Model Estimations via Inference Function for the Margins

The inference function for margins (IFM) technique introduced in [17] provides a computationally faster method for estimating parameters than Full Maximum Likelihood, i.e. simultaneously maximising all model parameters and produces in many cases a more stable likelihood estimation procedure. This two-stage estimation procedure was studied with regard to the asymptotic relative efficiency compared with maximum likelihood estimation in [16] and in [14]. It can be shown that the IFM estimator is consistent under weak regularity conditions.

In modelling parametrically the marginal features of the log return forward exchange rates, we wanted flexibility to capture a broad range of skew-kurtosis relationships as well as potential for sub-exponential heavy tailed features. In addition, we wished to keep the models to a selection which is efficient to perform inference and easily interpretable. We consider a flexible three parameter model for the marginal distributions given by the Log-Generalised Gamma distribution (l.g.g.d.), see details in [19], where Y has a l.g.g.d. if $Y = \log(X)$ such that X has a g.g.d. The density of Y is given by

$$f_Y(y; k, u, b) = \frac{1}{b\Gamma(k)} \exp\left[k\left(\frac{y-u}{b}\right) - \exp\left(\frac{y-u}{b}\right) \right], \qquad (10)$$

with $u = \log(\alpha)$, $b = \beta^{-1}$ and the support of the l.g.g.d. distribution is $y \in \mathbb{R}$.

This flexible three-parameter model admits the LogNormal model as a limiting case (as $k \to \infty$). In addition, the g.g.d. also includes the exponential model ($\beta = k = 1$), the Weibull distribution ($k = 1$) and the Gamma distribution ($\beta = 1$).

As an alternative to the l.g.g.d. model, we also consider a time series approach to modelling the marginals, given by the GARCH(p,q) model, as described in [3, 4], and characterised by the error variance:

$$\sigma^2 = \alpha_0 + \sum_{i=1}^{q} \alpha_i \varepsilon_{t-i}^2 + \sum_{i=1}^{p} \beta_i \sigma_{t-i}^2. \qquad (11)$$

4.1 Stage 1: Fitting the Marginal Distributions via MLE

The estimation for the three model parameters in the l.g.g.d. can be challenging due to the fact that a wide range of model parameters, especially for k, can produce similar resulting density shapes (see discussions in [19]). To overcome this complication and to make the estimation efficient, it is proposed to utilise a combination of profile likelihood methods over a grid of values for k and perform profile likelihood based MLE estimation for each value of k, over the other two parameters b and u. The differentiation of the profile likelihood for a given value of k produces the system of two equations:

$$\exp(\tilde{\mu}) = \left[\frac{1}{n} \sum_{i=1}^{n} \exp\left(\frac{y_i}{\tilde{\sigma}\sqrt{k}}\right) \right]^{\tilde{\sigma}\sqrt{k}} ; \quad \frac{\sum_{i=1}^{n} y_i \exp\left(\frac{y_i}{\tilde{\sigma}\sqrt{k}}\right)}{\sum_{i=1}^{n} \exp\left(\frac{y_i}{\tilde{\sigma}\sqrt{k}}\right)} - \bar{y} - \frac{\tilde{\sigma}}{\sqrt{k}} = 0,$$
$$\qquad (12)$$

where n is the number of observations, $y_i = \log x_i$, $\tilde{\sigma} = b/\sqrt{k}$ and $\tilde{\mu} = u + b \log k$. The second equation is solved directly via a simple root search to give an estimation for $\tilde{\sigma}$ and then substitution into the first equation results in an estimate for $\tilde{\mu}$. Note, for each value of k we select in the grid, we get the pair of parameter estimates $\tilde{\mu}$

and $\tilde{\sigma}$, which can then be plugged back into the profile likelihood to make it purely a function of k, with the estimator for k then selected as the one with the maximum likelihood score. As a comparison, we also fit the GARCH(1,1) model using the MATLAB MFEtoolbox using the default settings.

4.2 Stage 2: Fitting the Mixture Copula via MLE

In order to fit the copula model, the parameters are estimated using maximum likelihood on the data after conditioning on the selected marginal distribution models and their corresponding estimated parameters obtained in Stage 1. These models are utilised to transform the data using the CDF function with the l.g.g.d. MLE parameters $(\hat{k}, \hat{u}$ and $\hat{b})$ or using the conditional variances to obtain standardised residuals for the GARCH model. Therefore, in this second stage of MLE estimation, we aim to estimate either the one parameter mixture of CFG components with parameters $\underline{\theta} = (\rho_{\text{clayton}}, \rho_{\text{frank}}, \rho_{\text{gumbel}}, \lambda_{\text{clayton}}, \lambda_{\text{frank}}, \lambda_{\text{gumbel}})$, the one parameter mixture of CG components with parameters $\underline{\theta} = (\rho_{\text{clayton}}, \rho_{\text{gumbel}}, \lambda_{\text{clayton}}, \lambda_{\text{gumbel}})$ or the two parameter outer power transformed Clayton with parameters $\underline{\theta} = (\rho_{\text{clayton}}, \beta_{\text{clayton}})$. The log likelihood expression for the mixture copula models, is given generically by:

$$l(\underline{\theta}) = \sum_{i=1}^{n} \log c(F_1(X_{i1}; \hat{\mu}_1, \hat{\sigma}_1), \ldots, F_d(X_{id}; \hat{\mu}_d, \hat{\sigma}_d)) + \sum_{i=1}^{n} \sum_{j=1}^{d} \log f_j(X_{ij}; \hat{\mu}_j, \hat{\sigma}_j).$$

(13)

This optimization is achieved via a gradient descent iterative algorithm which was found to be quite robust given the likelihood surfaces considered in these models with the real data. Alternative estimation procedures such as expectation-maximisation were not found to be required.

5 Exchange Rate Multivariate Data Description and Currency Portfolio Construction

In our study, we fit copula models to the high interest rate basket and the low interest rate basket updated for each day in the period 02/01/1989 to 29/01/2014 using log return forward exchange rates at one month maturities for data covering both the previous 6 months and previous year as a sliding window analysis on each trading day in this period.

Our empirical analysis consists of daily exchange rate data for a set of 34 currency exchange rates relative to the USD, as in [23]. The currencies analysed included: Australia (AUD), Brazil (BRL), Canada (CAD), Croatia (HRK), Cyprus (CYP), Czech Republic (CZK), Egypt (EGP), Euro area (EUR), Greece (GRD), Hungary (HUF), Iceland (ISK), India (INR), Indonesia (IDR), Israel (ILS), Japan (JPY),

Malaysia (MYR), Mexico (MXN), New Zealand (NZD), Norway (NOK), Philippines (PHP), Poland (PLN), Russia (RUB), Singapore (SGD), Slovakia (SKK), Slovenia (SIT), South Africa (ZAR), South Korea (KRW), Sweden (SEK), Switzerland (CHF), Taiwan (TWD), Thailand (THB), Turkey (TRY), Ukraine (UAH) and the United Kingdom (GBP).

We have considered daily settlement prices for each currency exchange rate as well as the daily settlement price for the associated 1 month forward contract. We utilise the same dataset (albeit starting in 1989 rather than 1983 and running up until January 2014) as studied in [20, 23] in order to replicate their portfolio returns without tail dependence risk adjustments. Due to differing market closing days, e.g. national holidays, there was missing data for a couple of currencies and for a small number of days. For missing prices, the previous day's closing prices were retained.

As was demonstrated in Eq. (1), the differential of interest rates between two countries can be estimated through the ratio of the forward contract price and the spot price, see [18] who show this holds empirically on a daily basis. Accordingly, instead of considering the differential of risk-free rates between the reference and the foreign countries, we build our respective baskets of currencies with respect to the ratio of the forward and the spot prices for each currency. On a daily basis, we compute this ratio for each of the d currencies (available in the dataset on that day) and then build five baskets. The first basket gathers the $d/5$ currencies with the highest positive differential of interest rate with the US dollar. These currencies are thus representing the 'investment' currencies, through which we invest the money to benefit from the currency carry trade. The last basket will gather the $d/5$ currencies with the highest negative differential (or at least the lowest differential) of interest rate. These currencies are thus representing the 'financing' currencies, through which we borrow the money to build the currency carry trade.

Given this classification, we investigate then the joint distribution of each group of currencies to understand the impact of the currency carry trade, embodied by the differential of interest rates, on currencies returns. In our analysis, we concentrate on the high interest rate basket (investment currencies) and the low interest rate basket (funding currencies), since typically when implementing a carry trade strategy one would go short the low interest rate basket and go long the high interest rate basket.

6 Results and Discussion

In order to model the marginal exchange rate log returns, we considered two approaches. First, we fit Log Generalised Gamma models to each of the 34 currencies considered in the analysis, updating the fits for every trading day based on a 6 month sliding window. A time series approach was also considered to fit the marginals, as is popular in much of the recent copula literature, see for example [4], using GARCH(1,1) models for the 6-month sliding data windows. In each case we are assuming approximate local stationarity over these short 6 month time frames.

Table 1 Average AIC for the Generalised Gamma (GG) and the GARCH(1,1) for the four most frequent currencies in the high interest rate and the low interest rate baskets over the 2001–2014 data period split into two chunks, i.e. 6 years

		01–07		07–14	
Investment	Currency	GG	GARCH	GG	GARCH
	TRY	356.9 (3.5)	341.1 (21.7)	358.7 (3.0)	349.1 (16.8)
	MXN	360.0 (1.2)	357.04 (3.8)	358.6 (4.0)	344.5 (28.1)
	ZAR	358.7 (3.0)	353.5 (11.4)	358.0 (6.1)	352.8 (12.2)
	BRL	359.0 (2.8)	341.6 (19.4)	360.0 (2.1)	341.6 (23.2)
Funding	JPY	361.2 (0.9)	356.5 (7.2)	356.9 (6.8)	355.0 (7.0)
	CHF	360.8 (1.4)	359.1 (2.9)	358.6 (7.4)	355.4 (8.8)
	SGD	360.0 (2.7)	356.8 (5.7)	360.0 (2.6)	353.7 (7.5)
	TWD	358.7 (6.2)	347.0 (16.4)	359.1 (5.8)	348.5 (13.2)

Standard deviations are shown in parentheses. Similar performance was seen between 1989 and 2001

A summary of the marginal model selection can be seen in Table 1, which shows the average AIC scores for the four most frequent currencies in the high interest rate and the low interest rate baskets over the data period. Whilst the AIC for the GARCH(1,1) model is consistently lower than the respective AIC for the Generalised Gamma, the standard errors are sufficiently large for there to be no clear favourite between the two models.

However, when we consider the model selection of the copula in combination with the marginal model, we observe lower AIC scores for copula models fitted on the pseudo-data resulting from using Generalised Gamma margins than using GARCH(1,1) margins. This is the case for all three copula models under consideration in the paper. Figure 1 shows the AIC differences when using the Clayton-Frank-Gumbel copula in combination with the two choices of marginal for the high interest rate and the low interest rate basket, respectively. Over the entire data period, the mean difference between the AIC scores for the CFG model with Generalised Gamma versus GARCH(1,1) marginals for the high interest rate basket is 12.3 and for the low interest rate basket is 3.6 in favour of the Generalised Gamma.

Thus, it is clear that the Generalised Gamma model is the best model in our copula modelling context and so is used in the remainder of the analysis. We now consider the goodness-of-fit of the three copula models applied to the high interest rate basket and low interest rate basket pseudo data. We used a scoring via the AIC between the three component mixture CFG model versus the two component mixture CG model versus the two parameter OpC model. One could also use the Copula-Information-Criterion (CIC), see [13] for details.

The results are presented for this comparison in Fig. 2, which shows the differentials between AIC for CFG versus CG and CFG versus OpC for each of the high interest rate and the low interest rate currency baskets. We can see it is not unreasonable to consider the CFG model for this analysis, since over the entire data period, the mean difference between the AIC scores for the CFG and the CG models

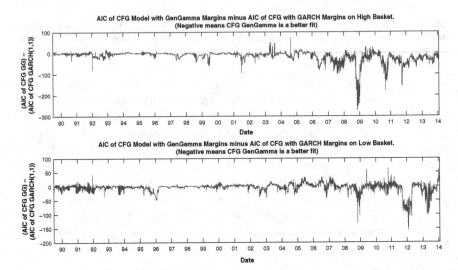

Fig. 1 Comparison of AIC for Clayton-Frank-Gumbel model fit on the pseudo-data resulting from generalised gamma versus GARCH(1,1) margins. The high interest rate basket is shown in the *upper panel* and the low interest rate basket is shown in the *lower panel*

for the high interest rate basket is 1.33 and for the low interest rate basket is 1.62 in favour of the CFG.

However, from Fig. 2, we can see that during the 2008 credit crisis period, the CFG model is performing much better. The CFG copula model provides a much better fit when compared to the OpC model, as shown by the mean difference between the AIC scores of 9.58 for the high interest rate basket and 9.53 for the low interest rate basket. Similarly, the CFG model performs markedly better than the OpC model during the 2008 credit crisis period.

6.1 Tail Dependence Results

Below, we will examine the time-varying parameters of the maximum likelihood fits of this mixture CFG copula model. Here, we shall focus on the strength of dependence present in the currency baskets, given the particular copula structures in the mixture, which is considered as tail upside/downside exposure of a carry trade over time. Figure 3 shows the time-varying upper and lower tail dependence, i.e. the extreme upside and downside risk exposures for the carry trade basket, present in the high interest rate basket under the CFG copula fit and the OpC copula fit. Similarly, Fig. 4 shows this for the low interest rate basket.

Remark 2 (Model Risk and its Influence on Upside and Downside Risk Exposure) In fitting the OpC model, we note that independent of the strength of true tail dependence

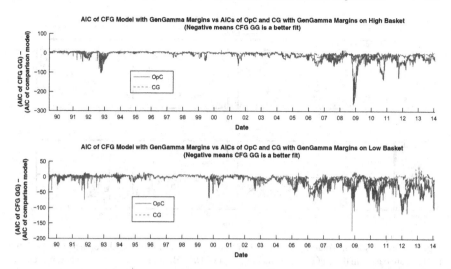

Fig. 2 Comparison of AIC for Clayton-Frank-Gumbel model with Clayton-Gumbel and outer power clayton models on high and low interest rate baskets with generalised gamma margins. The high interest rate basket is shown in the *upper panel* and the low interest rate basket is shown in the *lower panel*

in the multivariate distribution, the upper tail dependence coefficient $\lambda_{\mathscr{U}}$ for this model strictly increases with dimension very rapidly. Therefore, when fitting the OpC model, if the basket size becomes greater than bivariate, i.e. from 1999 onwards, the upper tail dependence estimates become very large (even for outer power parameter values very close to $\beta = 1$). This lack of flexibility in the OpC model only becomes

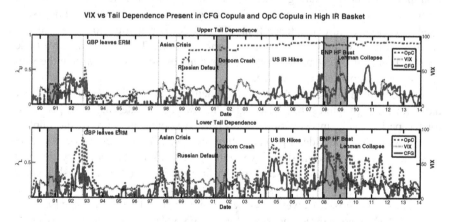

Fig. 3 Comparison of Volatility Index (VIX) with upper and lower tail dependence of the high interest rate basket in the CFG copula and OpC copula. US NBER recession periods are represented by the shaded *grey* zones. Some key crisis dates across the time period are labelled

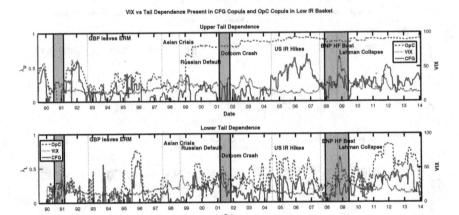

Fig. 4 Comparison of Volatility Index (VIX) with *upper* and *lower tail* dependence of the low interest rate basket in the CFG copula and OpC copula. US NBER recession periods are represented by the shaded *grey* zones. Some key crisis dates across the time period are labelled

apparent in baskets of dimension greater than 2, but is also evident in the AIC scores in Fig. 2. Here, we see an interesting interplay between the model risk associated to the dependence structure being fit and the resulting interpreted upside or downside financial risk exposures for the currency baskets.

Focusing on the tail dependence estimate produced from the CFG copula fits, we can see that there are indeed periods of heightened upper and lower tail dependence in the high interest rate and the low interest rate baskets. There is a noticeable increase in upper tail dependence in the high interest rate basket at times of global market volatility. Specifically, during late 2007, i.e. the global financial crisis, there is a sharp peak in upper tail dependence. Preceding this, there is an extended period of heightened lower tail dependence from 2004 to 2007, which could tie in with the building of the leveraged carry trade portfolio positions. This period of carry trade construction is also very noticeable in the low interest rate basket through the very high levels of upper tail dependence.

We compare in Figs. 3 and 4 the tail dependence plotted against the VIX volatility index for the high interest rate basket and the low interest rate basket, respectively, for the period under investigation. The VIX is a popular measure of the implied volatility of S&P 500 index options—often referred to as the *fear index*. As such, it is one measure of the market's expectations of stock market volatility over the next 30 days. We can clearly see here that in the high interest rate basket, there are upper tail dependence peaks at times when there is an elevated VIX index, particularly post-crisis. However, we would not expect the two to match exactly since the VIX is not a direct measure of global FX volatility. We can thus conclude that investors' risk aversion clearly plays an important role in the tail behaviour. This conclusion corroborates recent literature regarding the skewness and the kurtosis features characterising the currency carry trade portfolios [5, 11, 23].

6.2 Pairwise Decomposition of Basket Tail Dependence

In order to examine the contribution of each pair of currencies to the overall n-dimensional basket tail dependence, we calculated the corresponding non-parametric pairwise tail dependencies for each pair of currencies. In Fig. 5, we can see the average upper and lower non-parametric tail dependence for each pair of currencies during the 2008 credit crisis, with the 3 currencies most frequently in the high interest rate and the low interest rate baskets labelled accordingly. The lower triangle represents the non-parametric pairwise lower tail dependence and the upper triangle represents the non-parametric pairwise upper tail dependence.

If one was trying to optimise their currency portfolio with respect to the tail risk exposures, i.e. to minimise negative tail risk exposure and maximise positive tail risk exposure, then one would sell short currencies with high upper tail dependence and low lower tail dependence, whilst buying currencies with low upper tail dependence and high lower tail dependence.

Similarly, in Fig. 6 we see the pairwise non-parametric tail dependencies averaged over the last 12 months (01/02/2013 to 29/01/2014). Comparing this heat map to the heat map during the 2008 credit crisis (Fig. 5), we notice that in general there are lower values of tail dependence amongst the currency pairs.

We performed linear regression of the pairwise non-parametric tail dependence on the respective basket tail dependence for the days, during the period (01/02/2013 to 29/01/2014), on which the 3 currencies all appeared in the basket (224 out of 250 for the lower interest rate basket and 223 out of 250 for the high interest rate basket). The regression coefficients and R^2 values can be seen in Table 2. We can

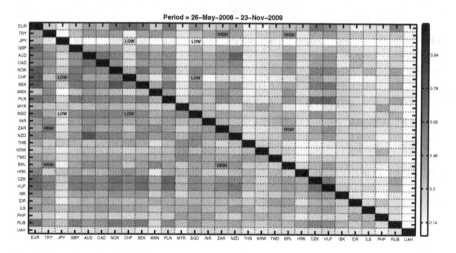

Fig. 5 Heat map showing the strength of non-parametric tail dependence between each pair of currencies averaged over the 2008 credit crisis period. *Lower* tail dependence is shown in the *lower triangle* and *upper tail* dependence is shown in the *upper triangle*. The 3 currencies most frequently in the high interest rate and the low interest rate baskets are labelled

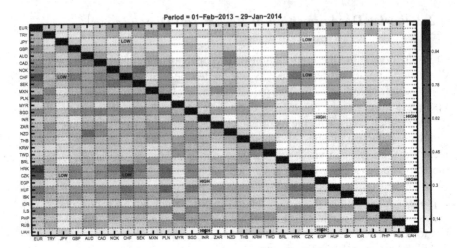

Fig. 6 Heat map showing the strength of non-parametric tail dependence between each pair of currencies averaged over the last 12 months (01/02/2013–29/01/2014). *Lower tail* dependence is shown in the *lower triangle* and *upper tail* dependence is shown in the *upper triangle*. The 3 currencies most frequently in the high interest rate and the low interest rate baskets are labelled

interpret this as the relative contribution of each of the 3 currency pairs to the overall basket tail dependence. We note that for the low interest rate lower tail dependence and for the high interest rate upper tail dependence, there is a significant degree of cointegration between the currency pair covariates and hence we might be able to use a single covariate due to the presence of a common stochastic trend.

Table 2 Pairwise non-parametric tail dependence, during the period 01/02/2013 to 29/01/2014, regressed on respective basket tail dependence (standard errors are shown in parentheses)

Low IR Basket	Constant	CHF JPY	CZK CHF	CZK JPY	R^2
Upper TD	0.22 (0.01)	0.02 (0.03)	0.18 (0.02)	0.38 (0.05)	0.57
Lower TD	0.71 (0.17)	−0.62 (0.25)	−0.38 (0.26)	0.23 (0.32)	0.28
High IR Basket	Constant	EGP INR	UAH EGP	UAH INR	R^2
Upper TD	0.07 (0.01)	−0.06 (0.33)	0.59 (0.08)	2.37 (0.42)	0.4
Lower TD	0.1 (0.02)	0.56 (0.05)	0.44 (0.08)	−0.4 (0.07)	0.44

The 3 currencies most frequently in the respective baskets are used as independent variables

6.3 Understanding the Tail Exposure Associated with the Carry Trade and Its Role in the UIP Puzzle

As was discussed in Sect. 2, the tail exposures associated with a currency carry trade strategy can be broken down into the upside and downside tail exposures within each of the long and short carry trade baskets. The downside relative exposure adjusted returns are obtained by multiplying the monthly portfolio returns by one minus the upper and the lower tail dependence present, respectively, in the high interest rate basket and the low interest rate basket at the corresponding dates. The upside relative exposure adjusted returns are obtained by multiplying the monthly portfolio returns by one plus the lower and upper tail dependence present, respectively, in the high interest rate basket and the low interest rate basket at the corresponding dates. Note that we refer to these as relative exposure adjustments only for the tail exposures since we do not quantify a market price per unit of tail risk. However, this is still informative as it shows a decomposition of the relative exposures from the long and short baskets with regard to extreme events.

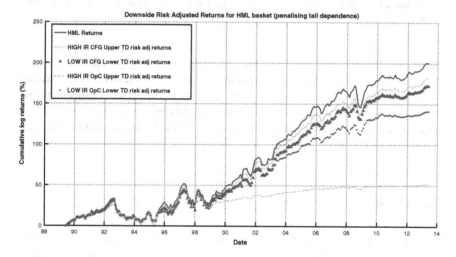

Fig. 7 Cumulative log returns of the carry trade portfolio (HML = High interest rate basket minus low interest rate basket). Downside exposure adjusted cumulative log returns using *upper/lower tail* dependence in the high/low interest rate basket for the CFG copula and the OpC copula are shown for comparison

As can be seen in Fig. 7, the relative adjustment to the absolute cumulative returns for each type of downside exposure is greatest for the low interest rate basket, except under the OpC model, but this is due to the very poor fit of this model to baskets containing more than 2 currencies which we see transfers to financial risk exposures. This is interesting because intuitively one would expect the high interest rate basket to be the largest source of tail exposure. However, one should be careful when

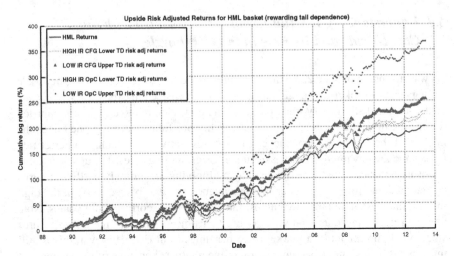

Fig. 8 Cumulative log returns of the carry trade portfolio (HML = High interest rate basket minus low interest rate basket). Upside exposure adjusted cumulative log returns using *lower/upper tail dependence* in the high/low interest rate basket for the CFG copula and the OpC copula are shown for comparison

interpreting this plot, since we are looking at the extremal tail exposure. The analysis may change if one considered the intermediate tail risk exposure, where the marginal effects become significant. Similarly, Fig. 8 shows the relative adjustment to the absolute cumulative returns for each type of upside exposure is greatest for the low interest rate basket. The same interpretation as for the downside relative exposure adjustments can be made here for upside relative exposure adjustments.

7 Conclusion

In this paper, we have shown that the positive and negative multivariate tail risk exposures present in currency carry trade baskets are additional factors needing careful consideration when one constructs a carry portfolio. Ignoring these exposures leads to a perceived risk return profile that is not reflective of the true nature of such a strategy. In terms of marginal model selection, it was shown that one is indifferent between the log Generalised Gamma model and the frequently used GARCH(1,1) model. However, in combination with the three different Archimedean copula models considered in this paper, the log Generalised Gamma marginals provided a better overall model fit.

Open Access This chapter is distributed under the terms of the Creative Commons Attribution Noncommercial License, which permits any noncommercial use, distribution, and reproduction in any medium, provided the original author(s) and source are credited.

References

1. Ames, M., Bagnarosa, G., Peters, G.W.: Reinvestigating the Uncovered Interest Rate Parity Puzzle via Analysis of Multivariate Tail Dependence in Currency Carry Trades. arXiv:1303.4314 (2013)
2. Backus, D.K., Foresi, S., Telmer, C.I.: Affine term structure models and the forward premium anomaly. J. Financ. **56**(1), 279–304 (2001)
3. Bollerslev, T.: Generalized autoregressive conditional heteroskedasticity. J. Econom. **31**(3), 307–327 (1986)
4. Brechmann, E.C., Czado, C.: Risk management with high-dimensional vine copulas: an analysis of the Euro Stoxx 50. Stat. Risk Model. **30**(4), 307–342 (2012)
5. Brunnermeier, M.K., Nagel, S., Pedersen, L.H.: Carry trades and currency crashes. Working paper 14473, National Bureau of Economic Research, November 2008
6. Burnside, C., Eichenbaum, M., Kleshchelski, I., Rebelo, S.: Do peso problems explain the returns to the carry trade? Rev. Financ. Stud. **24**(3), 853–891 (2011)
7. Christiansen, C., Ranaldo, A., Söderlind, P.: The time-varying systematic risk of carry trade strategies. J. Financ. Quant. Anal. **46**(04), 1107–1125 (2011)
8. Cruz, M., Peters, G., Shevchenko, P.: Handbook on Operational Risk. Wiley, New York (2013)
9. De Luca, G., Rivieccio G.: Multivariate tail dependence coefficients for archimedean copulae. Advanced Statistical Methods for the Analysis of Large Data-Sets, p. 287 (2012)
10. Fama, E.F.: Forward and spot exchange rates. J. Monet. Econ. **14**(3), 319–338 (1984)
11. Farhi, E., Gabaix, X.: Rare disasters and exchange rates. Working paper 13805, National Bureau of Economic Research, February 2008
12. Feller, W.: An Introduction to Probability Theory and Its Applications, vol. 2. Wiley, New York (1971)
13. Grønneberg, S.: The copula information criterion and its implications for the maximum pseudo-likelihood estimator. Dependence Modelling: The Vine Copula Handbook, World Scientific Books, pp. 113–138 (2010)
14. Hafner, C.M., Manner, H.: Dynamic stochastic copula models: estimation, inference and applications. J. Appl. Econom. **27**(2), 269–295 (2010)
15. Hansen, L.P., Hodrick, R.J.: Forward exchange rates as optimal predictors of future spot rates: an econometric analysis. J. Polit. Econ. 829–853 (1980)
16. Joe, H.: Asymptotic efficiency of the two-stage estimation method for copula-based models. J. Multivar. Anal. **94**(2), 401–419 (2005)
17. Joe, H., Xu, J.J.: The Estimation Method of Inference Functions for Margins for Multivariate Models. Technical report, Technical Report 166, Department of Statistics, University of British Columbia (1996)
18. Juhl, T., Miles, W., Weidenmier, M.D.: Covered interest arbitrage: then versus now. Economica **73**(290), 341–352 (2006)
19. Lawless, J.F.: Inference in the generalized gamma and log gamma distributions. Technometrics **22**(3), 409–419 (1980)
20. Lustig, H., Roussanov, N., Verdelhan, A.: Common risk factors in currency markets. Rev. Financ. Stud. **24**(11), 3731–3777 (2011)
21. Lustig, H., Verdelhan, A.: The cross section of foreign currency risk premia and consumption growth risk. Am. Econ. Rev. **97**(1), 89–117 (2007)
22. McNeil, A.J., Nešlehová, J.: Multivariate archimedean copulas, d-monotone functions and L1-norm symmetric distributions. Ann. Stat. **37**(5B), 3059–3097 (2009)
23. Menkhoff, L., Sarno, L., Schmeling, M., Schrimpf, A.: Carry trades and global foreign exchange volatility. J. Financ. **67**(2), 681–718 (2012)
24. Musiela, M., Rutkowski, M.: Martingale Methods in Financial Modelling. Springer, Berlin (2011)
25. Nelsen, R.B.: An Introduction to Copulas. Springer, New York (2006)

Part III
Insurance Risk
and Asset Management

Participating Life Insurance Contracts under Risk Based Solvency Frameworks: How to Increase Capital Efficiency by Product Design

Andreas Reuß, Jochen Ruß and Jochen Wieland

Abstract Traditional participating life insurance contracts with year-to-year (cliquet-style) guarantees have come under pressure in the current situation of low interest rates and volatile capital markets, in particular when priced in a market consistent valuation framework. In addition, such guarantees lead to rather high capital requirements under risk-based solvency frameworks such as Solvency II or the Swiss Solvency Test (SST). We introduce several alternative product designs and analyze their impact on the insurer's financial situation. We also introduce a measure for Capital Efficiency that considers both, profits and capital requirements, and compare the results of the innovative products to the traditional product design with respect to Capital Efficiency in a market consistent valuation model.

Keywords Capital efficiency · Participating life insurance · Embedded options · Interest rate guarantees · Market consistent valuation · Risk based capital requirements · Solvency II · SST

1 Introduction

Traditional participating life insurance products play a major role in old-age provision in Continental Europe and in many other countries. These products typically come with a guaranteed benefit at maturity, which is calculated using some guaranteed minimum interest rate. Furthermore, the policyholders receive an annual surplus participation that depends on the performance of the insurer's assets. With the so-

A. Reuß
Institut für Finanz- und Aktuarwissenschaften, Lise-Meitner-Straße 14, 89081 Ulm, Germany
e-mail: a.reuss@ifa-ulm.de

J. Ruß · J. Wieland (✉)
Institut für Finanz- und Aktuarwissenschaften, Ulm University, Lise-Meitner-Straße 14,
89081 Ulm, Germany
e-mail: j.russ@ifa-ulm.de

J. Wieland
e-mail: j.wieland@ifa-ulm.de

© The Author(s) 2015
K. Glau et al. (eds.), *Innovations in Quantitative Risk Management*,
Springer Proceedings in Mathematics & Statistics 99,
DOI 10.1007/978-3-319-09114-3_11

185

called cliquet-style guarantees, once such surplus has been assigned to the policy at the end of the year, it increases the guaranteed benefit based on the same guaranteed minimum interest rate. This product design can create significant financial risk.

Briys and de Varenne [8] were among the first to analyze the impact of interest rate guarantees on the insurer's risk exposure. However, they considered a simple point-to-point guarantee where surplus (if any) is credited at maturity only. The financial risks of cliquet-style guarantee products have later been investigated, e.g., by Grosen and Jorgensen [17]. They introduce the "average interest principle", where the insurer aims to smooth future bonus distributions by using a bonus reserve as an additional buffer besides the policy reserve (the client's account). Besides valuing the contract they also calculate default probabilities (however, under the risk-neutral probability measure $\mathbb{Q}$). Grosen et al. [19] extend the model of Grosen and Jorgensen [17], and introduce mortality risk. Grosen and Jorgensen [18] modify the model used by Briys and de Varenne [8] by incorporating a regulatory constraint for the insurer's assets and analyzing the consequences for the insurer's risk policy. Mitersen and Persson [23] analyze a different cliquet-style guarantee framework with the so-called terminal bonuses, whereas Bauer et al. [4] specifically investigate the valuation of participating contracts under the German regulatory framework.

While all this work focuses on the risk-neutral valuation of life insurance contracts (sometimes referred to as "financial approach"), Kling et al. [20, 21] concentrate on the risk a contract imposes on the insurer (sometimes referred to as "actuarial approach") by means of shortfall probabilities under the real-world probability measure $\mathbb{P}$.

Barbarin and Devolder [3] introduce a methodology that allows for combining the financial and actuarial approach. They consider a contract similar to Briys and de Varenne [8] with a point-to-point guarantee and terminal surplus participation. To integrate both approaches, they use a two-step method of pricing life insurance contracts: First, they determine a guaranteed interest rate such that certain regulatory requirements are satisfied, using value at risk and expected shortfall risk measures. Second, to obtain fair contracts, they use risk-neutral valuation and adjust the participation in terminal surplus accordingly. Based on this methodology, Gatzert and Kling [14] investigate parameter combinations that yield fair contracts and analyze the risk implied by fair contracts for various contract designs. Gatzert [13] extends this approach by introducing the concept of "risk pricing" using the "fair value of default" to determine contracts with the same risk exposure. Graf et al. [16] (also building on Barbarin and Devolder [3]) derive the risk minimizing asset allocation for fair contracts using different risk measures like the shortfall probability or the relative expected shortfall.

Under risk-based solvency frameworks such as Solvency II or the Swiss Solvency Test (SST), the risk analysis of interest rate guarantees becomes even more important. Under these frameworks, capital requirement is derived from a market consistent valuation considering the insurer's risk. This risk is particularly high for long term contracts with a year-to-year guarantee based on a fixed (i.e., not path dependent) guaranteed interest rate. Measuring and analyzing the financial risk in relation to the required capital, and analyzing new risk figures such as the Time Value of Options and Guarantees (TVOG) is a relatively new aspect, which gains importance with new solvability frameworks, e.g., the largest German insurance company (Allianz) announced in a press conference on June 25, 2013[1] the introduction of a new participating life insurance product that (among other features) fundamentally modifies the type of interest rate guarantee (similar to what we propose in the remainder of this paper). It was stressed that the TVOG is significantly reduced for the new product. Also, it was mentioned that the increase of the TVOG resulting from an interest rate shock (i.e., the solvency capital requirement for interest rate risk) is reduced by 80 % when compared to the previous product. This is consistent with the findings of this paper.

The aim of this paper is a comprehensive risk analysis of different contract designs for participating life insurance products. Currently, there is an ongoing discussion, whether and how models assessing the insurer's risk should be modified to reduce the capital requirements (e.g., by applying an "ultimate forward rate" set by the regulator). We will in contrast analyze how (for a given model) the insurer's risk, and hence capital requirement can be influenced by product design. Since traditional cliquet-style participating life insurance products lead to very high capital requirements, we will introduce alternative contract designs with modified types of guarantees, which reduce the insurer's risk and profit volatility, and therefore also the capital requirements under risk-based solvency frameworks. In order to compare different product designs from an insurer's perspective, we develop and discuss the concept of Capital Efficiency, which relates profit to capital requirements.[2] We identify the key drivers of Capital Efficiency, which are then used in our analyses to assess different product designs.

The remainder of this paper is structured as follows:

In Sect. 2, we present three considered contract designs that all come with the same level of guaranteed maturity benefit but with different types of guarantee:

- Traditional product: a traditional contract with a cliquet-style guarantee based on a guaranteed interest rate > 0.
- Alternative product 1: a contract with the same guaranteed maturity benefit, which is, however, valid only at maturity; additionally, there is a 0 % year-to-year guarantee on the account value meaning that the account value cannot decrease from one year to the next.
- Alternative product 2: a contract with the same guaranteed maturity benefit that is, however, valid only at maturity; there is no year-to-year guarantee on the account value meaning that the account value may decrease in some years.

---

[1] Cf. [1], particularly slide D24.

[2] Of course, there already exist other well-established measures linking profit to required capital, such as the return on risk-adjusted capital (RORAC). However, they may not be suitable to assess products with long-term guarantees since they consider the required capital on a one-year basis only. To the best of our knowledge there is no common measure similar to what we define as Capital Efficiency that relates the profitability of an insurance contract to the risk it generates, and hence capital it requires over the whole contract term.

On top of the different types of guarantees, all three products include a surplus participation depending on the insurer's return on assets. Our model is based on the surplus participation requirements given by German regulation. That means in particular that each year at least 90 % of the (book value) investment return has to be distributed to the policyholders.

To illustrate the mechanics, we will first analyze the different products under different deterministic scenarios. This shows the differences in product design and how they affect the insurer's risk.

In Sect. 3, we introduce our stochastic model, which is based on a standard financial market model: The stock return and short rate processes are modeled using a correlated Black-Scholes and Vasicek model.[3] We then describe how the evolution of the insurance portfolio and the insurer's balance sheet are simulated in our asset-liability-model. The considered asset allocation consists of bonds with different maturities and stocks. The model also incorporates management rules as well as typical intertemporal risk sharing mechanisms (e.g., building and dissolving unrealized gains and losses), which are an integral part of participating contracts in many countries and should therefore not be neglected.

Furthermore, we introduce a measure for Capital Efficiency based on currently discussed solvency regulations such as the Solvency II framework. We also propose a more tractable measure for an assessment of the key drivers of Capital Efficiency.

In Sect. 4, we present the numerical results. We show that the alternative products are significantly more capital efficient: financial risk, and therefore also capital requirement is significantly reduced, although in most scenarios all products provide the same maturity benefit to the policyholder.[4] We observe that the typical "asymmetry", i.e., particularly the heavy left tail of the insurer's profit distribution is reduced by the modified products. This leads to a significant reduction of both, the TVOG and the solvency capital requirement for interest rate risk.

Section 5 concludes and provides an outlook for further research.

2 Considered Products

In this section, we describe the three different considered contract designs. Note that for the sake of simplicity, we assume that in case of death in year t, always only the current account value AV_t (defined below) is paid at the end of year t. This allows us to ignore mortality for the calculation of premiums and actuarial reserves.

---

[3] The correlated Black-Scholes and Vasicek model is applied in Zaglauer and Bauer [29] and Bauer et al. [5] in a similar way.

[4] Note: In scenarios where the products' maturity benefits do differ, the difference is limited since the guaranteed maturity benefit (which is the same for all three products) is a lower bound for the maturity benefit.

2.1 The Traditional Product

First, we consider a traditional participating life insurance contract with a cliquet-style guarantee. It provides a guaranteed benefit G at maturity T based on annual premium payments P. The pricing is based on a constant guaranteed interest rate i and reflects annual charges c_t. The actuarial principle of equivalence[5] yields

$$\sum_{t=0}^{T-1} (P - c_t) \cdot (1+i)^{T-t} = G. \tag{1}$$

During the lifetime of the contract, the insurer has to build up sufficient (prospective) actuarial reserves AR_t for the guaranteed benefit based on the same constant interest rate i:

$$AR_t = G \cdot \left(\frac{1}{1+i}\right)^{T-t} - \sum_{k=t}^{T-1} (P - c_k) \cdot \left(\frac{1}{1+i}\right)^{k-t}. \tag{2}$$

The development of the actuarial reserves is then given by:

$$AR_t = (AR_{t-1} + P - c_{t-1}) \cdot (1+i).$$

Traditional participating life insurance contracts typically include an annual surplus participation that depends on the performance of the insurer's assets. For example, German regulation requires that at least a "minimum participation" of $p = 90\%$ of the (local GAAP book value) earnings on the insurer's assets has to be credited to the policyholders' accounts. For the traditional product, any surplus assigned to a contract immediately increases the guaranteed benefit based on the same interest rate i. More precisely, the surplus s_t is credited to a bonus reserve account BR_t (where $BR_0 = 0$) and the interest rate i will also apply each year on the bonus reserve:

$$BR_t = BR_{t-1} \cdot (1+i) + s_t.$$

The client's account value AV_t consists of the sum of the actuarial reserve AR_t and the bonus reserve BR_t; the maturity benefit is equal to AV_T.

As a consequence, each year at least the rate i has to be credited to the contracts. The resulting optionality is often referred to as asymmetry: If the asset return is above i, a large part (e.g., $p = 90\%$) of the return is credited to the client as a surplus and the shareholders receive only a small portion (e.g., $1 - p = 10\%$) of the return. If, on the other hand, the asset returns are below i, then 100% of the shortfall has to be compensated by the shareholder. Additionally, if the insurer distributes a high surplus, this increases the insurer's future risk since the rate i has to be credited also to this surplus amount in subsequent years. Such products constitute a significant

---

[5] For the equivalence principle, see e.g., Saxer [25], Wolthuis [28].

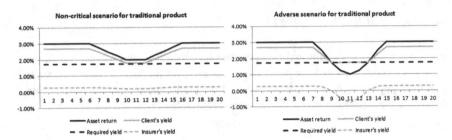

Fig. 1 Two illustrative deterministic scenarios for the traditional product: asset returns and yield distribution

financial risk to the insurance company, in particular in a framework of low interest rates and volatile capital markets.[6]

The mechanics of this year-to-year guarantee are illustrated in Fig. 1 for two illustrative deterministic scenarios. We consider a traditional policy with term to maturity $T = 20$ years and a guaranteed benefit of $G = €20,000$. Following the current situation in Germany, we let $i = 1.75\%$ and assume a surplus participation rate of $p = 90\%$ on the asset returns.

The first scenario is not critical for the insurer. The asset return (which is here arbitrarily assumed for illustrative purposes) starts at 3 %, then over time drops to 2 % and increases back to 3 % where the x axis shows the policy year. The chart shows this asset return, the "client's yield" (i.e., the interest credited to the client's account including surplus), the "required yield" (which is defined as the minimum rate that has to be credited to the client's account), and the insurer's yield (which is the portion of the surplus that goes to the shareholder). Obviously, in this simple example, the client's yield always amounts to 90 % of the asset return and the insurer's yield always amounts to 10 % of the asset return. By definition, for this contract design, the required yield is constant and always coincides with $i = 1.75\%$.

In the second scenario, we let the asset return drop all the way down to 1 %. Whenever 90 % of the asset return would be less than the required yield, the insurer has to credit the required yield to the account value. This happens at the shareholder's expense, i.e., the insurer's yield is reduced and even becomes negative. This means that a shortfall occurs and the insurer has to provide additional funds.

It is worthwhile noting that in this traditional product design, the interest rate i plays three different roles:

- pricing interest rate i_p used for determining the ratio between the premium and the guaranteed maturity benefit,
- reserving interest rate i_r, i.e., technical interest rate used for the calculation of the prospective actuarial reserves,
- year-to-year minimum guaranteed interest rate i_g, i.e., a minimum return on the account value.

---

[6] This was also a key result of the QIS5 final report preparing for Solvency II, cf. [2, 11].

2.2 Alternative Products

We will now introduce two alternative product designs, which are based on the idea to allow different values for the pricing rate, the reserving rate and the year-to-year minimum guaranteed interest rate on the account value. So Formulas 1 and 2 translate to the following formulae for the relation between the annual premium, the guaranteed benefit and the actuarial reserves:

$$\sum_{t=0}^{T-1} (P - c_t) \cdot (1 + i_p)^{T-t} = G$$

$$AR_t = G \cdot \left(\frac{1}{1+i_r} \right)^{T-t} - \sum_{k=t}^{T-1} (P - c_k) \cdot \left(\frac{1}{1+i_r} \right)^{k-t}.$$

Note, that in the first years of the contract, negative values for AR_t are possible in case of $i_p < i_r$, which implies a "financial buffer" at the beginning of the contract. The year-to-year minimum guaranteed interest rate i_g is not relevant for the formulae above, but it is simply a restriction for the development of the client's account, i.e.,

$$AV_t \geq (AV_{t-1} + P - c_{t-1}) \cdot (1 + i_g),$$

where $AV_0 = \max \{AR_0, 0\}$ is the initial account value of the contract.

The crucial difference between such new participating products and traditional participating products is that the guaranteed maturity benefit is not explicitly increased during the lifetime of the contract (but, of course, an increase in the account value combined with the year-to-year minimum guaranteed interest rate can implicitly increase the maturity guarantee).

In this setting, the prospective reserve AR_t is only a minimum reserve for the guaranteed maturity benefit: The insurer has to make sure that the account value does not fall below this minimum reserve. This results in a "required yield" explained below. Under "normal" circumstances the account value (which is also the surrender value) exceeds the minimum reserve. Therefore, the technical reserve (under local GAAP), which may not be below the surrender value, coincides with the account value.

The required yield on the account value in year t is equal to

$$z_t = \max \left\{ \frac{\max \{AR_t, 0\}}{AV_{t-1} + P - c_{t-1}} - 1, i_g \right\}. \tag{3}$$

The left part of (3) assures that the account value is nonnegative and never lower than the actuarial reserve. The required yield decreases if the bonus reserve (which is included in AV_{t-1}) increases.

The surplus participation rules remain unchanged: the policyholder's share p (e.g., 90 %) of the asset return is credited to the policyholders (but not less than z_t). Hence, as long as the policyholder's share is always above the technical interest rate used in the traditional product, there is no difference between the traditional and the alternative product designs.

Obviously, only combinations fulfilling $i_g \leq i_p \leq i_r$ result in suitable products: If the first inequality is violated, then the year-to-year minimum guaranteed interest rate results in a higher (implicitly) guaranteed maturity benefit than the (explicit) guarantee resulting from the pricing rate. If the second inequality is violated then at $t = 0$, additional reserves (exceeding the first premium) are required.

In what follows, we will consider two concrete alternative contract designs. Obviously, the choice of i_g fundamentally changes the mechanics of the guarantee embedded in the product (or the "type" of guarantee), whereas the choice of i_p changes the level of the guarantee. Since the focus of this paper is on the effect of the different guarantee mechanisms, we use a pricing rate that coincides with the technical rate of the traditional product. Hence, the guaranteed maturity benefit remains unchanged. Since the legally prescribed maximum value for the reserving rate also coincides with the technical rate of the traditional product, we get $i_p = i_r = 1.75\%$ for both considered alternative designs.

In our alternative product 1, we set $i_g = 0\%$ (0 % year-to-year guarantee) and for alternative 2 we set $i_g = -100\%$ (no year-to-year guarantee). In order to illustrate the mechanics of the alternative products, Figs. 2 and 3 show the two scenarios from Fig. 1 for both alternative products. In the first scenario (shown on the left), the required yield z_t on the account value gradually decreases for both alternative contract designs since the bonus reserve acts as some kind of buffer (as described above). For alternative 1, the required yield can of course not fall below $i_g = 0\%$, while for the alternative 2 it even becomes negative after some years.

The adverse scenario on the right shows that the required yield rises again after years with low asset returns since the buffer is reduced. However, contrary to the traditional product, the asset return stays above the required level and no shortfall occurs.

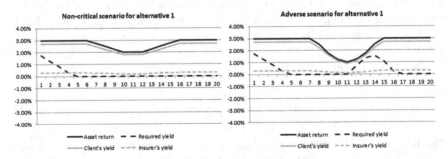

Fig. 2 Two illustrative deterministic scenarios for alternative 1 product: asset returns and yield distribution

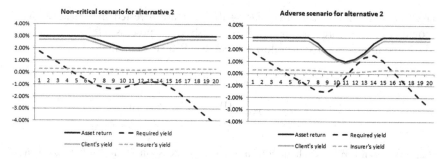

Fig. 3 Two illustrative deterministic scenarios for alternative 2 product: asset returns and yield distribution

From a policyholder's perspective, both alternative contract designs provide the same maturity benefit as the traditional contract design in the first scenario since the client's yield is always above 1.75 %. In the second scenario, however, the maturity benefit is slightly lower for both alternative contract designs since (part of) the buffer built up in years 1 to 8 can be used to avoid a shortfall. In this scenario, the two alternative products coincide, since the client's yield is always positive.

Even if scenarios where the products differ appear (or are) unlikely, the modification has a significant impact on the insurer's solvency requirements since the financial risks particularly in adverse scenarios are a key driver for the solvency capital requirement. This will be considered in a stochastic framework in the following sections.

3 Stochastic Modeling and Analyzed Key Figures

Since surplus participation is typically based on local GAAP book values (in particular in Continental Europe), we use a stochastic balance sheet and cash flow projection model for the analysis of the product designs presented in the previous section. The model includes management rules concerning asset allocation, reinvestment strategy, handling of unrealized gains and losses and surplus distribution. Since the focus of the paper is on the valuation of future profits and capital requirements we will introduce the model under a risk-neutral measure. Similar models have been used (also in a real-world framework) in Kling et al. [20, 21] and Graf et al. [16].

3.1 The Financial Market Model

We assume that the insurer's assets are invested in coupon bonds and stocks. We treat both assets as risky assets in a risk-neutral, frictionless and continuous financial market. Additionally, cash flows during the year are invested in a riskless bank account (until assets are reallocated). We let the short rate process r_t follow a Vasicek[7] model, and the stock price S_t follow a geometric Brownian motion:

$$dr_t = \kappa \left(\theta - r_t \right) dt + \sigma_r dW_t^{(1)} \text{ and}$$

$$\frac{dS_t}{S_t} = r_t dt + \rho \sigma_S dW_t^{(1)} + \sqrt{1 - \rho^2} \sigma_S dW_t^{(2)},$$

where $W_t^{(1)}$ and $W_t^{(2)}$ each denote a Wiener process on some probability space $(\Omega, \mathscr{F}, \mathbb{F}, \mathbb{Q})$ with a risk-neutral measure $\mathbb{Q}$ and the natural filtration $\mathbb{F} = \mathscr{F}_t = \sigma\left(\left(W_s^{(1)}, W_s^{(2)} \right), s < t \right)$. The parameters $\kappa, \theta, \sigma_r, \sigma_S$ and ρ are deterministic and constant. For the purpose of performing Monte Carlo simulations, the stochastic differential equations can be solved to

$$S_t = S_{t-1} \cdot \exp \left(\int_{t-1}^{t} r_u du - \frac{\sigma_S^2}{2} + \int_{t-1}^{t} \rho \sigma_S dW_u^{(1)} + \int_{t-1}^{t} \sqrt{1 - \rho^2} \sigma_S dW_u^{(2)} \right) \text{ and}$$

$$r_t = e^{-\kappa} \cdot r_{t-1} + \theta \left(1 - e^{-\kappa} \right) + \int_{t-1}^{t} \sigma_r \cdot e^{-\kappa(t-u)} dW_u^{(1)},$$

where $S_0 = 1$ and the initial short rate r_0 is a deterministic parameter. Then, the bank account is given by $B_t = \exp \left(\int_0^t r_u du \right)$. It can be shown that the four (stochastic) integrals in the formulae above follow a joint normal distribution.[8] Monte Carlo paths are calculated using random realizations of this multidimensional distribution. The discretely compounded yield curve at time t is then given by[9]

$$r_t(s) =$$

$$\exp \left[\frac{1}{s} \left(\frac{1 - e^{-\kappa s}}{\kappa} r_t + \left(s - \frac{1 - e^{-\kappa s}}{\kappa} \right) \cdot \left(\theta - \frac{\sigma_r^2}{2\kappa^2} \right) + \left(\frac{1 - e^{-\kappa s}}{\kappa} \right)^2 \frac{\sigma_r^2}{4\kappa} \right) \right] - 1$$

for any time t and term $s > 0$. Based on the yield curves, we calculate par yields that determine the coupon rates of the considered coupon bonds.

---

[7] Cf. [27].

[8] Cf. Zaglauer and Bauer [29]. A comprehensive explanation of this property is included in Bergmann [6].

[9] See Seyboth [26] as well as Branger and Schlag [7].

Table 1 Balance sheet at time t

Assets	Liabilities
BV_t^S	X_t
BV_t^B	AV_t

3.2 The Asset-Liability Model

The insurer's simplified balance sheet at time t is given by Table 1. Since our analysis is performed for a specific portfolio of insurance contracts on a stand-alone basis, there is no explicit allowance for shareholders' equity or other reserves on the liability side. Rather, X_t denotes the shareholders' profit or loss in year t, with corresponding cash flow at the beginning of the next year. Together with AV_t as defined in Sect. 2, this constitutes the liability side of our balance sheet.

In our projection of the assets and insurance contracts, incoming cash flows (premium payments at the beginning of the year, coupon payments and repayment of nominal at the end of the year) and outgoing cash flows (expenses at the beginning of the year and benefit payments at the end of the year) occur. In each simulation path, cash flows occurring at the beginning of the year are invested in a bank account. At the end of the year, the market values of the stocks and coupon bonds are derived and the asset allocation is readjusted according to a rebalancing strategy with a constant stock ratio q based on market values. Conversely, $(1 - q)$ is invested in bonds and any money on the bank account is withdrawn and invested in the portfolio consisting of stocks and bonds.

If additional bonds need to be bought in the process of rebalancing, the corresponding amount is invested in coupon bonds yielding at par with term M. However, toward the end of the projection, when the insurance contracts' remaining term is less than M years, we invest in bonds with a term that coincides with the longest remaining term of the insurance contracts. If bonds need to be sold, they are sold proportionally to the market values of the different bonds in the existing portfolio.

With respect to accounting, we use book-value accounting rules following German GAAP, which may result in unrealized gains or losses (UGL): Coupon bonds are considered as held to maturity and their book value BV_t^B is always given by their nominal amounts (irrespective if the market value is higher or lower). In contrast, for the book value of the stocks BV_t^S, the insurer has some discretion.

Of course, interest rate movements as well as the rebalancing will cause fluctuations with respect to the UGL of bonds. Also, the rebalancing may lead to the realization of UGL of stocks. In addition, we assume an additional management rule with respect to UGL of stocks: We assume that the insurer wants to create rather stable book value returns (and hence surplus distributions) in order to signal stability to the market. We, therefore, assume that a ratio d_{pos} of the UGL of stocks is realized annually if unrealized gains exist and a ratio d_{neg} of the UGL is realized annually if unrealized losses exist. In particular, $d_{neg} = 100\%$ has to be chosen in a legal framework where unrealized losses on stocks are not possible.

Based on this model, the total asset return on a book value basis can be calculated in each simulation path each year as the sum of coupon payments from bonds, interest payments on the bank account, and the realization of UGL. The split between policyholders and shareholders is driven by the minimum participation parameter p explained in Sect. 2. If the cumulative required yield on the account values of all policyholders is larger than this share, there is no surplus for the policyholders, and exactly the respective required yield z_t is credited to every account. Otherwise, surplus is credited, which amounts to the difference between the policyholders' share of the asset return and the cumulative required yield. Following the typical practice, e.g., in Germany, we assume that this surplus is distributed among the policyholders such that all policyholders receive the same client's yield (defined by the required yield plus surplus rate), if possible. To achieve that, we apply an algorithm that sorts the accounts by required yield, i.e., $\left(z_t^{(1)}, \ldots, z_t^{(k)} \right)$, $k \in \mathbb{N}$ in ascending order. First, all contracts receive their respective required yield. Then, the available surplus is distributed: Starting with the contract(s) with the lowest required yield $z_t^{(1)}$, the algorithm distributes the available surplus to all these contracts until the gap to the next required yield $z_t^{(2)}$ is filled. Then, all the contracts with a required yield lower or equal to $z_t^{(2)}$ receive an equal amount of (relative) surplus until the gap to $z_t^{(3)}$ is filled, etc. This is continued until the entire surplus is distributed. The result is that all contracts receive the same client's yield if this unique client's yield exceeds the required yield of all contracts. Otherwise, there exists a threshold z^* such that all contracts with a required yield above z^* receive exactly their required yield (and no surplus) and all contracts with a required yield below z^* receive z^* (i.e., they receive some surplus).

From this, the insurer's profit X_t results as the difference between the total asset return and the amount credited to all policyholder accounts. If the profit is negative, a shortfall has occurred, which we assume to be compensated by a corresponding capital inflow (e.g., from the insurer's shareholders) at the beginning of the next year.[10] Balance sheet and cash flows are projected over τ years until all policies that are in force at time zero have matured.

3.3 Key Drivers for Capital Efficiency

The term Capital Efficiency is frequently used in an intuitive sense, in particular among practitioners, to describe the feasibility, profitability, capital requirement, and riskiness of products under risk-based solvency frameworks. However, to the best of our knowledge, no formal definition of this term exists. Nevertheless, it seems obvious that capital requirement alone is not a suitable figure for managing a

---

[10] We do not consider the shareholders' default put option resulting from their limited liability, which is in line with both, Solvency II valuation standards and the Market Consistent Embedded Value framework (MCEV), cf. e.g., [5] or [10], Sect. 5.3.4.

product portfolio from an insurer's perspective. Rather, capital requirement and the resulting cost of capital should be considered in relation to profitability.

Therefore, a suitable measure of Capital Efficiency could be some ratio of profitability and capital requirement, e.g., based on the distribution of the random variable

$$\frac{\sum_{t=1}^{\tau} \frac{X_t}{B_t}}{\sum_{t=1}^{\tau} \frac{RC_{t-1} \cdot CoC_t}{B_t}}. \tag{4}$$

The numerator represents the present value of the insurer's future profits, whereas the denominator is equal to the present value of future cost of capital: RC_t denotes the required capital at time t under some risk-based solvency framework, i.e., the amount of shareholders' equity needed to support the business in force. The cost of capital is derived by applying the cost of capital rate CoC_t for year t on the required capital at the beginning of this year.[11] In practical applications, however, the distribution of this ratio might not be easy to calculate. Therefore, moments of this distribution, a separate analysis of (moments of) the numerator and the denominator or even just an analysis of key drivers for that ratio could create some insight.

In this spirit, we will use a Monte Carlo framework to calculate the following key figures using the model described above:

A typical market consistent measure for the insurer's profitability is the expected present value of future profits (PVFP),[12] which corresponds to the expected value of the numerator in (4). The PVFP is estimated based on Monte Carlo simulations:

$$\text{PVFP} = \frac{1}{N} \sum_{n=1}^{N} \sum_{t=1}^{\tau} \frac{X_t^{(n)}}{B_t^{(n)}} = \frac{1}{N} \sum_{n=1}^{N} \text{PVFP}^{(n)},$$

where N is the number of scenarios, $X_t^{(n)}$ denotes the insurer's profit/loss in year t in scenario n, $B_t^{(n)}$ is the value of the bank account after t years in scenario n, and hence $\text{PVFP}^{(n)}$ is the present value of future profits in scenario n.

In addition, the degree of asymmetry of the shareholder's cash flows can be characterized by the distribution of $\text{PVFP}^{(n)}$ over all scenarios[13] and by the time value of options and guarantees (TVOG). Under the MCEV framework,[14] the latter is defined by

$$\text{TVOG} = \text{PVFP}_{CE} - \text{PVFP}$$

---

[11] This approach is similar to the calculation of the cost of residual nonhedgeable risk as introduced in the MCEV Principles in [9], although RC_t reflects the total capital requirement including hedgeable risks.

[12] The concept of PVFP is introduced as part of the MCEV Principles in [9].

[13] Note that this is a distribution under the risk-neutral measure and has to be interpreted carefully. However, it can be useful for explaining differences between products regarding PVFP and TVOG.

[14] Cf. [9].

Table 2 Product parameters I

	Traditional Product (%)	Alternative 1 (%)	Alternative 2 (%)
i_p, i_r	1.75	1.75	1.75
i_g	1.75	0	-100

where $\text{PVFP}_{CE} = \sum_{t=1}^{\tau} \frac{X_t^{(CE)}}{B_t^{(CE)}}$ is the present value of future profits in the so-called "certainty equivalent" (CE) scenario. This deterministic scenario reflects the expected development of the capital market under the risk-neutral measure. It can be derived from the initial yield curve $r_0(s)$ based on the assumption that all assets earn the forward rate implied by the initial yield curve.[15] The TVOG is also used as an indicator for capital requirement under risk-based solvency frameworks.

Comparing the PVFP for two different interest rate levels—one that we call basic level and a significantly lower one that we call stress level—provides another important key figure for interest rate risk and capital requirements. In the standard formula[16] of the Solvency II framework

$$\Delta \text{PVFP} = \text{PVFP(basic)} - \text{PVFP(stress)}$$

determines the contribution of the respective product to the solvency capital requirement for interest rate risk (SCR_{int}). Therefore, we also focus on this figure which primarily drives the denominator in (4).

4 Results

4.1 Assumptions

The stochastic valuation model described in the previous section is applied to a portfolio of participating contracts. For simplicity, we assume that all policyholders are 40 years old at inception of the contract and mortality is based on the German standard mortality table (DAV 2008 T). We do not consider surrender. Furthermore, we assume annual charges c_t that are typical in the German market consisting of annual administration charges $\beta \cdot P$ throughout the contract's lifetime, and acquisition charges $\alpha \cdot T \cdot P$, which are equally distributed over the first 5 years of the contract. Hence, $c_t = \beta \cdot P + \alpha \frac{T \cdot P}{5} \mathbb{1}_{t \in \{0,\dots,4\}}$. Furthermore, we assume that expenses coincide with the charges. Product parameters are given in Tables 2 and 3.

Stochastic projections are performed for a portfolio that was built up in the past 20 years (i.e., before $t = 0$) based on 1,000 new policies per year. Hence, we have a

---

[15] Cf. Oechslin et al. [24].

[16] A description of the current version of the standard formula can be found in [12].

Table 3 Product parameters II

$G(\text{€})$	T (years)	$P(\text{€})$	β (%)	α (%)
20,000	20	896.89	3	4

portfolio at the beginning of the projections with remaining time to maturity between 1 year and 19 years (i.e., $\tau = 19$ years).[17] For each contract, the account value at $t = 0$ is derived from a projection in a deterministic scenario. In this deterministic scenario, we use a flat yield curve of 3.0% (consistent with the mean reversion parameter θ of the stochastic model after $t = 0$), and parameters for management rules described below. In line with the valuation approach under Solvency II and MCEV, we do not consider new business.

The book value of the asset portfolio at $t = 0$ coincides with the book value of liabilities. We assume a stock ratio of $q = 5\%$ with unrealized gains on stocks at $t = 0$ equal to 10% of the book value of stocks. The coupon bond portfolio consists of bonds with a uniform coupon of 3.0% where the time to maturity is equally split between 1 year and $M = 10$ years.

Capital market parameters for the basic and stress projections are shown in Table 4. The parameters $\kappa, \sigma_r, \sigma_S$ and ρ are directly adopted from Graf et al. [16]. The parameters θ and r_0 are chosen such that they are more in line with the current low interest rate level. The capital market stress corresponds to an immediate drop of interest rates by 100 basis points.

The parameters for the management rules are given in Table 5 and are consistent with current regulation and practice in the German insurance market.

For all projections, the number of scenarios is $N = 5,000$. Further analyses showed that this allows for a sufficiently precise estimation of the relevant figures.[18]

Table 4 Capital market parameters

	r_0 (%)	θ (%)	κ (%)	σ_r (%)	σ_S (%)	ρ (%)
Basic	2.5	3.0	30.0	2.0	20.0	15.0
Stress	1.5	2.0				

---

[17] Note that due to mortality before $t = 0$, the number of contracts for the different remaining times to maturity is not the same.

[18] In order to reduce variance in the sample an antithetic path selection of the random numbers is applied, cf. e.g., Glasserman [15].

Table 5 Parameters for management rules

q (%)	M (years)	d_{pos} (%)	d_{neg} (%)	p (%)
5	10	20	100	90

4.2 Comparison of Product Designs

In Table 6, the PVFP and the TVOG for the base case are compared for the three products. All results are displayed as a percentage of the present value of future premium income from the portfolio. For alternative 1, the PVFP increases from 3.63 to 4.24 %, i.e., by 0.61 percentage points (pp), compared to the traditional contract design (which corresponds to a 17 % increase of profitability). This means that this product with a "maturity only" guarantee and an additional guarantee that the account value will not decrease is, as expected, more profitable than the product with a traditional year-to-year (cliquet-style) guarantee. This difference is mainly caused by the different degree of asymmetry of the shareholders' cash flows which is characterized by the TVOG. Since $PVFP_{CE}$ amounts to 4.26 % for all products in the base case, the difference of TVOG between the traditional product and alternative 1 is also 0.61 pp. This corresponds to a TVOG reduction of more than 90 % for alternative 1, which shows that the risk resulting from the interest rate guarantee is much lower for the modified product.

Compared to this, the differences between alternative 1 and alternative 2 are almost negligible. The additional increase of the PVFP is only 0.01 pp, which is due to a slightly lower TVOG compared to alternative 1. This shows that the fact that the account value may decrease in some years in alternative 2 does not provide a material additional risk reduction.

Additional insights can be obtained by analyzing the distribution of $PVFP^{(n)}$ (see Fig. 4)[19]: For the traditional contract design, the distribution is highly asymmetric with a strong left tail and a significant risk of negative shareholder cash flows (on a present value basis). In contrast, both alternative contract designs exhibit an almost symmetric distribution of shareholder cash flows which explains the low TVOG. Hence, the new products result in a significantly more stable profit perspective for the shareholders, while for the traditional product the shareholder is exposed to significantly higher shortfall risk.

Ultimately, the results described above can be traced back to differences in the required yield. While for the traditional product, by definition, the required yield always amounts to 1.75 %, it is equal to 0 % in most scenarios for the alternative 1 product. Only in the most adverse scenarios, the required yield rises toward 1.75 %.[20] For the alternative 2 product, it is even frequently negative.

---

[19] Cf. Footnote 13.

[20] Note that here, the required yield in the first projection year reflects the financial buffer available for the considered portfolio of existing contracts at $t = 0$. This is different from the illustrations in Sect. 2, which consider individual contracts from inception to maturity.

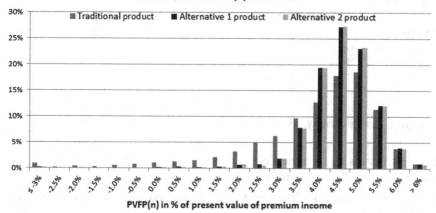

Fig. 4 Histogram of PVFP$^{(n)}$ in base case

Table 6 PVFP and TVOG for base case (as percentage of the present value of premium income)

	Traditional product (%)	Alternative 1 (%)	Alternative 2 (%)
PVFP	3.63	4.24	4.25
TVOG	0.63	0.02	0.01

Apart from the higher profitability, the alternative contract designs also result in a lower capital requirement for interest rate risk. This is illustrated in Table 7, which displays the PVFP under the interest rate stress and the difference to the basic level. Compared to the basic level, the PVFP for the traditional product decreases by 75%, which corresponds to an SCR$_{int}$ of 2.73% of the present value of future premium income. In contrast, the PVFP decreases by only around 40% for the alternative contract designs and thus the capital requirement is only 1.66 and 1.65%, respectively.

We have seen that a change in the type of guarantee results in a significant increase of the PVFP. Further analyses show that a traditional product with guaranteed interest rate $i = 0.9\%$ instead of 1.75% would have the same PVFP (i.e., 4.25%) as the alternative contract designs with $i_p = 1.75\%$. Hence, although changing only the type of guarantee and leaving the level of guarantee intact might be perceived as a rather small product modification by the policyholder, it has the same effect on the insurer's profitability as reducing the level of guarantee by a significant amount.

Furthermore, our results indicate that even in an adverse capital market situation the alternative product designs may still provide an acceptable level of profitability: The profitability of the modified products if interest rates were 50 basis points lower roughly coincides with the profitability of the traditional product in the base case.

Table 7 PVFP for stress level and PVFP difference between basic and stress level

	Traditional product (%)	Alternative 1 (%)	Alternative 2 (%)
PVFP(basic)	3.63	4.24	4.25
PVFP(stress)	0.90	2.58	2.60
ΔPVFP	2.73	1.66	1.65

4.3 Sensitivity Analyses

In order to assess the robustness of the results presented in the previous section, we investigate three different sensitivities:

1. Interest rate sensitivity: The long-term average θ and initial rate r_0 in Table 4 are replaced by $\theta = 2.0\,\%$, $r_0 = 1.5\,\%$ for the basic level, and $\theta = 1.0\,\%$, $r_0 = 0.5\,\%$ for the stress level.
2. Stock ratio sensitivity: The stock ratio is set to $q = 10\,\%$ instead of 5 %.
3. Initial buffer sensitivity: The initial bonus reserve $BR_t = AV_t - AR_t$ is doubled for all contracts.[21]

The results are given in Table 8.

Interest rate sensitivity If the assumed basic interest rate level is lowered by 100 basis points, the PVFP decreases and the TVOG increases significantly for all products. In particular, the alternative contract designs now also exhibit a significant TVOG. This shows that in an adverse capital market situation, also the guarantees embedded in the alternative contract designs can lead to a significant risk for the shareholder and an asymmetric distribution of profits as illustrated in Fig. 5. Nevertheless, the alternative contract designs are still much more profitable and less volatile than the traditional contract design and the changes in PVFP/TVOG are much less pronounced than for the traditional product: while the TVOG rises from 0.63 to 2.13 %, i.e., by 1.50 pp for the traditional product, it rises by only 0.76 pp (from 0.02 to 0.78 %) for alternative 1.

As expected, an additional interest rate stress now results in a larger SCR_{int}. For all product designs, the PVFP after stress is negative and the capital requirement increases significantly. However, as in the base case (cf. Table 7), the SCR_{int} for the traditional product is more than one percentage point larger than for the new products.

Stock ratio sensitivity The stock ratio sensitivity also leads to a decrease of PVFP and an increase of TVOG for all products. Again, the effect on the PVFP of the traditional product is much stronger: The profit is about cut in half (from 3.63 to 1.80 %), while for the alternative 1 product the reduction is much smaller (from 4.24 to 3.83 %), and even smaller for alternative 2 (from 4.25 to 3.99 %). It is noteworthy that with a larger stock ratio of $q = 10\,\%$ the difference between the two alternative

---

[21] The initial book and market values of the assets are increased proportionally to cover this additional reserve.

Table 8 PVFP, TVOG, PVFP under interest rate stress and ΔPVFP for base case and all sensitivities

Base case	Traditional product (%)	Alternative 1 (%)	Alternative 2 (%)
PVFP	3.63	4.24	4.25
TVOG	0.63	0.02	0.01
PVFP(stress)	0.90	2.58	2.60
ΔPVFP	2.73	1.66	1.65
Interest rate sensitivity			
PVFP	0.90	2.58	2.60
TVOG	2.13	0.78	0.76
PVFP(stress)	−4.66	−1.81	−1.76
ΔPVFP	5.56	4.39	4.36
Stock ratio sensitivity			
PVFP	1.80	3.83	3.99
TVOG	2.45	0.43	0.26
PVFP(stress)	−1.43	1.65	1.92
ΔPVFP	3.23	2.18	2.07
Initial buffer sensitivity			
PVFP	3.74	4.39	4.39
TVOG	0.64	<0.01	<0.01
PVFP(stress)	1.02	2.87	2.91
ΔPVFP	2.72	1.52	1.48

products becomes more pronounced, which is reflected by the differences of the TVOG. Alternative 2 has a lower shortfall risk than alternative 1 since the account value may decrease in some years as long as the account value does not fall below the minimum reserve for the maturity guarantee. Hence, we can conclude that the guarantee that the account value may not decrease becomes more risky if asset returns exhibit a higher volatility.

The results for the stressed PVFPs under the stock ratio sensitivity are in line with these results: First, the traditional product requires even more solvency capital: The SCR_{int} is half a percentage point larger than in the base case (3.23 % compared to 2.73 %), and it is also more than one percentage point larger than for the alternative products with 10 % stocks (2.18/2.07 %). Second, the interest rate stress shows a more substantial difference between the two different alternative products. While the difference of the SCR_{int} between alternative 1 and 2 was 0.01 % in the base case, it is now 0.11 %.

Initial buffer sensitivity If the initial buffer is increased, we observe a slight increase of the PVFP for all products. However, there are remarkable differences for the effect on TVOG between the traditional and the alternative products: While for the traditional product the TVOG remains approximately the same, for the alternative products it is essentially reduced to zero. This strongly supports our product

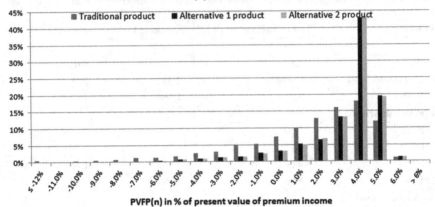

Fig. 5 Histogram of $PVFP^{(n)}$ for interest rate sensitivity (-100 basis points)

motivation in Sect. 2: For the alternative products, larger surpluses from previous years reduce risk in future years.[22] Furthermore, the stressed PVFPs imply that the decrease of capital requirement is significantly larger for the alternative products: 0.14 % reduction (from 1.66 to 1.52 %) for alternative 1 and 0.17 % reduction (from 1.65 to 1.48 %) for alternative 2, compared to just 0.01 % reduction for the traditional product.

4.4 Reduction in the Level of Guarantee

So far we have only considered contracts with a different type of guarantee. We will now analyze contracts with a lower level of guarantee, i.e., products where $i_p < i_r$. If we apply a pricing rate of $i_p = 1.25$ % instead of 1.75 %, the annual premium required to achieve the same guaranteed maturity benefit rises by approx. 5.4 %, which results in an additional initial buffer for this contract design. For the sake of comparison, we also calculate the results for the traditional product with a lower guaranteed interest rate $i = 1.25$ %. The respective portfolios at $t = 0$ are derived using the assumptions described in Sect. 4.1.

The results are presented in Table 9. We can see that the PVFP is further increased and the TVOG is very close to 0 for the modified alternative products, which implies an almost symmetric distribution of the PVFP. The TVOG can even become slightly negative due to the additional buffer in all scenarios. Although the risk situation for the traditional product is also improved significantly due to the lower guarantee, the

---

[22] From this, we can conclude that if such alternative products had been sold in the past, the risk situation of the life insurance industry would be significantly better today in spite of the rather high nominal maturity guarantees for products sold in the past.

Table 9 PVFP, TVOG, PVFP under interest rate stress and ΔPVFP for the alternative products with lower pricing rate

	Traditional product (%)	Alternative 1 (%)	Alternative 2 (%)	Traditional $i = 1.25$ (%)	Alternative 1 $i_p = 1.25$ (%)	Alternative 2 $i_p = 1.25$ (%)
PVFP	3.63	4.24	4.25	4.12	4.31	4.31
TVOG	0.63	0.02	0.01	0.14	−0.05	−0.05
PVFP (stress)	0.90	2.58	2.60	2.43	3.28	3.32
ΔPVFP	2.73	1.66	1.65	1.69	1.03	0.99

alternative products can still preserve their advantages. A more remarkable effect can be seen for the SCR_{int}, which amounts to 1.03 and 0.99 % for the alternative products 1 and 2, respectively, compared to 1.69 % for the traditional product. Hence, the buffer leads to a significant additional reduction of solvency capital requirements for the alternative products meaning that these are less affected by interest rate risk.

5 Conclusion and Outlook

In this paper, we have analyzed different product designs for traditional participating life insurance contracts with a guaranteed maturity benefit. A particular focus of our analysis was on the impact of product design on capital requirements under risk-based solvency frameworks such as Solvency II and on the insurer's profitability.

We have performed a market consistent valuation of the different products and have analyzed the key drivers of Capital Efficiency, particularly the value of the embedded options and guarantees and the insurer's profitability.

As expected, our results confirm that products with a typical year-to-year guarantee are rather risky for the insurer, and hence result in a rather high capital requirement. Our proposed product modifications significantly enhance Capital Efficiency, reduce the insurer's risk, and increase profitability. Although the design of the modified products makes sure that the policyholder receives less than with the traditional product only in extreme scenarios, these products still provide a massive relief for the insurer since extreme scenarios drive the capital requirements under Solvency II and SST.

It is particularly noteworthy that starting from a standard product where the guaranteed maturity benefit is based on an interest rate of 1.75 %, changing the type of the guarantee to our modified products (but leaving the level of guarantee intact) has the same impact on profitability as reducing the level of guarantee to an interest rate of 0.9 % and not modifying the type of guarantee. Furthermore, it is remarkable that the reduction of SCR_{int} from the traditional to the alternative contract design is very robust throughout our base case as well as all sensitivities and always amounts to slightly above one percentage point.

We would like to stress that the product design approach presented in this paper is not model arbitrage (hiding risks in "places the model cannot see"), but a real reduction of economic risks. In our opinion, such concepts can be highly relevant in practice if modified products keep the product features that are perceived and desired by the policyholder, preserve the benefits of intertemporal risk sharing, and do away with those options and guarantees of which policyholders often do not even know they exist. Similar modifications are also possible for many other old age provision products like dynamic hybrid products[23] or annuity payout products. Therefore, we expect that the importance of "risk management by product design" will increase. This is particularly the case since—whenever the same pool of assets is used to back new and old products—new capital efficient products might even help reduce the risk resulting from an "old" book of business by reducing the required yield of the pool of assets.

We, therefore, feel that there is room for additional research: It would be interesting to analyze similar product modifications for the annuity payout phase. Also—since many insurers have sold the traditional product in the past—an analysis of a change in new business strategy might be worthwhile: How would an insurer's risk and profitability change and how would the modified products interact with the existing business if the insurer has an existing (traditional) book of business in place and starts selling modified products today?

Another interesting question is how the insurer's optimal strategic asset allocation changes if modified products are being sold: If typical criteria for determining an optimal asset allocation are given (e.g., maximizing profitability under the restriction that some shortfall probability or expected shortfall is not exceeded), then the c.p. lower risk of the modified products might allow for a more risky asset allocation, and hence also higher expected profitability for the insurer and higher expected surplus for the policyholder. So, if this dimension is also considered, the policyholder would be compensated for the fact that he receives a weaker type of guarantee.

Finally, our analysis so far has disregarded the demand side. If some insurers keep selling the traditional product type, there should be little demand for the alternative product designs with reduced guarantees unless they provide some additional benefits. Therefore, the insurer might share the reduced cost of capital with the policyholder, also resulting in higher expected benefits in the alternative product designs.

Since traditional participating life insurance products play a major role in old-age provision in many countries and since these products have come under strong pressure in the current interest environment and under risk-based solvency frameworks, the concept of Capital Efficiency and the analysis of different product designs should be of high significance for insurers, researchers, and regulators to identify sustainable life insurance products. In particular, we would hope that legislators and regulators would embrace sustainable product designs where the insurer's risk is significantly reduced, but key product features as perceived and requested by policyholders are still present.

---

[23] Cf. Kochanski and Karnarski [22].

Open Access This chapter is distributed under the terms of the Creative Commons Attribution Noncommercial License, which permits any noncommercial use, distribution, and reproduction in any medium, provided the original author(s) and source are credited.

References

1. Allianz. Presentation Allianz Capital Markets Day, 2013. Available at https://www.allianz.com/v_1372138505000/media/investor_relations/en/conferences/capital_markets_days/documents/2013_allianz_cmd.pdf
2. BaFin. Ergebnisse der fünften quantitativen Auswirkungsstudie zu Solvency II (QIS 5), 2011. Available at http://www.bafin.de/SharedDocs/Downloads/DE/Versicherer_Pensionsfonds/QIS/dl_qis5_ergebnisse_bericht_bafin.pdf?__blob=publicationFile&v=8
3. Barbarin, J., Devolder, P.: Risk measure and fair valuation of an investment guarantee in life insurance. Insur.: Math. Econ. **37**(2), 297–323 (2005)
4. Bauer, D., Kiesel, R., Kling, A., Ruß, J.: Risk-neutral valuation of participating life insurance contracts. Insur.: Math. Econ. **39**(2), 171–183 (2006)
5. Bauer, D., Reuß, A., Singer, D.: On the calculation of solvency capital requirement based on nested simulations. ASTIN Bull. **42**(2), 453–499 (2012)
6. Bergmann, D.: Nested Simulations in Life Insurance. PhD thesis, University of Ulm (2011)
7. Branger, N., Schlag, C.: Zinsderivate. Modelle und Bewertung, Berlin (2004)
8. Briys, E., de Varenne, F.: On the risk of insurance liabilities: debunking some common pitfalls. J. Risk Insur. **64**(4), 637–694 (1997)
9. CFO-Forum. Market Consistent Embedded Value Principles, 2009. Available at http://www.cfoforum.nl/downloads/MCEV_Principles_and_Guidance_October_2009.pdf
10. DAV. DAV Fachgrundsatz zum Market Consistent Embedded Value. Köln (2011)
11. EIOPA. EIOPA Report on the fifth Quantitative Impact Study (QIS5) for Solvency II, 2011. Available at http://eiopa.europa.eu/fileadmin/tx_dam/files/publications/reports/QIS5_Report_Final.pdf
12. EIOPA. Technical Specifications on the Long Term Guarantee Assessment, 2013. Available at https://eiopa.europa.eu/consultations/qis/insurance/long-term-guarantees-assessment/technical-specifications/index.html
13. Gatzert, N.: Asset management and surplus distribution strategies in life insurance: an examination with respect to risk pricing and risk measurement. Insur.: Math. Econ. **42**(2), 839–849 (2008)
14. Gatzert, N., Kling, A.: Analysis of participating life insurance contracts: a unification approach. J. Risk Insur. **74**(3), 547–570 (2007)
15. Glasserman, P.: Monte Carlo Methods in Financial Engineering. Springer, New York (1994)
16. Graf, S., Kling, A., Ruß, J.: Risk analysis and valuation of life insurance contracts: combining actuarial and financial approaches. Insur.: Math. Econ. **49**(1), 115–125 (2011)
17. Grosen, A., Jorgensen, P.: Fair valuation of life insurance liabilities: the impact of interest rate guarantees, surrender options, and bonus policies. Insur.: Math. Econ. **26**(1), 37–57 (2000)
18. Grosen, A., Jorgensen, P.: Life insurance liabilities at market value: an analysis of insolvency risk, bonus policy, and regulatory intervention rules in a barrier option framework. J. Risk Insur. **69**(1), 63–91 (2002)
19. Grosen, A., Jensen, B., Jorgensen, P.: A finite difference approach to the valuation of path dependent life insurance liabilities. Geneva Pap. Risk Insur. Theory **26**, 57–84 (2001)
20. Kling, A., Richter, A., Ruß, J.: The impact of surplus distribution on the risk exposure of with profit life insurance policies including interest rate guarantees. J. Risk Insur. **74**(3), 571–589 (2007)
21. Kling, A., Richter, A., Ruß, J.: The interaction of guarantees, surplus distribution, and asset allocation in with-profit life insurance policies. Insur.: Math. Econ. **40**(1), 164–178 (2007)

22. Kochanski, M., Karnarski, B.: Solvency capital requirement for hybrid products. Eur. Actuar. J. **1**(2), 173–198 (2011)
23. Mitersen, K., Persson, S.-A.: Guaranteed investment contracts: distributed and undistributed excess return. Scand. Actuar. J. **103**(4), 257–279 (2003)
24. Oechslin, J., Aubry, O., Aellig, M.: Replicating embedded options. Life Pensions pp. 47–52 (2007)
25. Saxer, W.: Versicherungsmathematik. Springer, Berlin (1955)
26. Seyboth, M.: Der Market Consistent Appraisal Value und seine Anwendung im Rahmen der wertorientierten Steuerung von Lebensversicherungsunternehmen. PhD thesis, University of Ulm (2011)
27. Vasicek, O.: An equilibrium characterization of the term structure. J. Financ. Econ. **5**(2), 177–188 (1977)
28. Wolthuis, H.: Life Insurance Mathematics. CAIRE, Brussels (1994)
29. Zaglauer, K., Bauer, D.: Risk-neutral valuation of participating life insurance contracts in a stochastic interest rate environment. Insur.: Math. Econ. **43**(1), 29–40 (2008)

Reducing Surrender Incentives Through Fee Structure in Variable Annuities

Carole Bernard and Anne MacKay

Abstract In this chapter, we study the effect of the fee structure of a variable annuity on the embedded surrender option. We compare the standard fee structure offered in the industry (fees set as a fixed percentage of the variable annuity account) with periodic fees set as a fixed, deterministic amount. Surrender charges are also taken into account. Under fairly general conditions on the premium payments, surrender charges and fee schedules, we identify the situation when it is never optimal for the policyholder to surrender. Solving partial differential equations using finite difference methods, we present numerical examples that highlight the effect of a combination of surrender charges and deterministic fees in reducing the value of the surrender option and raising the optimal surrender boundary.

1 Introduction

A variable annuity (VA) is a unit-linked insurance product, which guarantees a certain amount at some future dates. Usually, the policyholder pays an initial premium for the contract. This premium is invested in a mutual fund chosen by the policyholder. There are different kinds of VAs defined by the type of guarantees embedded in the contract (for more details see Hardy [9]). In this paper, we focus on a variable annuity contract that pays the maximum of the mutual fund value and a guaranteed amount at maturity. This type of VA is referred to as a guaranteed minimum accumulation benefit (GMAB) (see Bauer et al. [1]).

Typically, the fee that covers the management of the VA and embedded financial guarantees is set as a constant percentage of the VA account and withdrawn directly from it at regular intervals. When the account value is high, the financial guarantee is worth very little, but the fee is still being paid as the same percentage. Thus, it represents an incentive for the policyholder to surrender the contract and take the

C. Bernard · A. MacKay (✉)
University of Waterloo, Waterloo, ON, Canada
e-mail: anne.mackay@math.ethz.ch

C. Bernard
e-mail: c3bernar@uwaterloo.ca

© The Author(s) 2015
K. Glau et al. (eds.), *Innovations in Quantitative Risk Management*,
Springer Proceedings in Mathematics & Statistics 99,
DOI 10.1007/978-3-319-09114-3_12

44

amount accumulated in the account. Such surrenders represent an important risk for VA issuers as the expenses linked to the sale of the policy are typically reimbursed through the fees collected throughout the duration of the contract. As exposed by Kling et al. [11], unexpected surrenders also compromise the efficiency of dynamic hedging strategies.

There are various ways to reduce the incentive to surrender a VA contract with guarantees. For example, insurance companies usually impose surrender charges, which reduce the amount available at surrender. Milevsky and Salisbury [13] argue that these charges are necessary for VA contracts to be both hedgeable and marketable. The design of VA benefits can also discourage policyholders from surrendering. Kling et al. [11] discuss for example the impact of ratchet options (possibility to reset the maturity guarantee as the fund value increases) to convince policyholders to keep the VA alive. Yet another way to reduce the incentive to surrender can be to modify the way fees are paid from the VA account. As explained above, the typical constant percentage fee structure leads to a mismatch between the fee paid and the value of the financial guarantee, which can discourage the policyholder from staying in the contract.[1] By reducing the fee paid when the value of the financial guarantee is low, it is possible to reduce the value of the real option to surrender embedded in a VA. The new fee structure can take different forms. For example, Bernard et al. [2] suggest to set a certain account value above which no fee will be paid. This is shown to modify the rational policyholder's surrender incentive. In this paper, we explore another fee structure so that part of (or all) the fee is paid as a deterministic periodic amount. The intuition behind this fee structure is that the amount will represent a lower percentage of the account value as the value of the financial guarantee decreases. This will affect the surrender incentive, and reduce the additional value created by the possibility to surrender the contract.

To explore the effect of the deterministic fee amount on the surrender incentive, we consider a VA with a simple GMAB. We assume that the total fee withdrawn from the VA account throughout the term of the contract is set as the sum of a fixed percentage c of the account value, and a deterministic, pre-determined amount p_t at time t (in other words, the deterministic amount does not need to be constant).[2] Our paper constitutes a significant extension of the results obtained on the optimal surrender strategy for a fee set as a fixed percentage of the fund [4], since the deterministic fee structure increases the complexity of the dynamics of the VA account value. For this reason, we need to resort to PDE methods to obtain the optimal surrender strategy

---

[1] Specifically, the policyholder has the option to surrender the contract and to receive a "surrender benefit", which can be more valuable than the contract itself. This additional value, as well as the optimal surrender strategy, is explored and quantified by Bernard, MacKay, and Muehlbeyer in [4] in the case when the fees are paid as a percentage of the underlying fund.

[2] Note that the deterministic amount component of the fee can be interpreted as a variable percentage of the account value F_t. In fact, let ρ denote the percentage of the fund value that yields the same fee amount as the deterministic amount p_t. Then, ρ is a function of time and of the fund value F_t, and can be computed as $\rho(t, F_t) = p_t/F_t$. Then, $\rho(t, F_t)F_t = p_t$ is the fee paid at time t.

when a portion of the fee is set as a deterministic amount. This paper also extends the work done on state-dependent fee structures, since Bernard et al. [2] do not quantify the reduction in the surrender incentive resulting from the new fee structure.

Throughout the paper, our main goal is to investigate the impact of the deterministic fee amount on the value of the surrender option. In Sect. 2, we describe the model and the VA contract. Section 3 introduces a theoretical result and discusses the valuation of the surrender option. Numerical examples are presented in Sects. 4 and 5 concludes.

2 Assumptions and Model

Consider a market with a bank account yielding a constant risk-free rate r and an index evolving as in the Black-Scholes model so that

$$\frac{dS_t}{S_t} = r\,dt + \sigma\,dW_t,$$

under the risk-neutral measure $\mathbb{Q}$, where $\sigma > 0$ is the constant instantaneous volatility of the index. Let $\mathscr{F}_t$ be the natural filtration associated with the Brownian motion W_t.

In this paper, we use a Black-Scholes setting since its simplicity allows us to compute prices explicitly, and thus to study the surrender incentive precisely. More realistic market models could be considered, but resorting to Monte Carlo methods or more advanced numerical methods would be required. Since the focus of this paper is on the surrender incentive, we believe that the Black-Scholes model's approximation of market dynamics is sufficient to provide insight on the effect of the deterministic amount fee structure.

2.1 Variable Annuity

We consider a VA contract with an underlying fund fully invested in the index S. At time t, we assume that the fee paid is the sum of a constant percentage $c \geq 0$ of the account value and a deterministic amount p_t. Setting $p_t = 0$, we will find back the results commonly used in the literature with the fee being only paid as a percentage of the fund (see for example [4]).

The motivation to study periodic deterministic fees is that the surrender incentives when the fees are paid as a fixed percentage of the fund are larger than when the fees are set as a deterministic amount. This will be illustrated via numerical examples in Sect. 4.

We further assume that the investment of the policyholder is P_0 at time 0, and that regular additional premiums a_t are paid at time t. Additional contributions are common in variable annuities but they are regularly neglected in the literature and most academic research focuses on the single premium case as it is simpler. When additional contributions can be made to the account throughout time, VAs are called Flexible Premiums Variable Annuities (FPVAs). Chi and Lin [7] provide examples of such VAs where the policyholder is given the choice between a single premium and a periodic monthly payment in addition to some initial lump sum. Analytical formulae for the value of such contracts can be found in [8, 10]. In the first part of this chapter, we show how flexible premium payments influence the surrender value.

We assume that all premiums paid at 0 and at later times t are invested in the fund. All fees (percentage or fixed fees) are taken from the fund. We need to model the dynamics of the fund. Our approach is inspired by Chi and Lin [7]. For the sake of simplicity, we assume that all cash flows happen in continuous time, so that a fixed payment of A at time 1 (say, end of the year) is similar to a payment made continuously over the interval $[0, 1]$. Due to the presence of a risk-free rate r, an amount paid at time T equal to A is equivalent to an instantaneous contribution of $a_t\,dt$ at any time $t \in (0, 1]$ so that the annual amount paid per year is $A = \int_0^1 a_t e^{r(1-t)}dt$. By abuse of notation, if a_t is constant over the year, we will write that a_t is the annual rate of contribution per year (although there is no compounding effect).

Specifically, the dynamics of the fund can be written as follows

$$dF_t = (r - c)F_t dt + \sigma F_t dW_t + a_t dt - p_t dt$$

with $F_0 = P_0$, and where F_t denotes the value of the fund at time t, a_t is the annual rate of contributions, c is the annual rate of fees, and p_t is the annual amount of fee to pay for the options. Similarly as [7] it is straightforward to show that

$$F_t = F_0 e^{(r-c-\frac{\sigma^2}{2})t+\sigma W_t} + \int_0^t (a_s - p_s)e^{(r-c-\frac{\sigma^2}{2})(t-s)+\sigma(W_t-W_s)}ds, \quad t \geq 0,$$

that is

$$F_t = S_t e^{-ct} + \int_0^t (a_s - p_s)e^{-c(t-s)}\frac{S_t}{S_s}ds, \tag{1}$$

in particular $P_0 = F_0 = S_0$. To simplify the notation, we will write

$$F_t = S_t e^{-ct} + \int_0^t b_s e^{-c(t-s)}\frac{S_t}{S_s}ds, \tag{2}$$

where $b_s = a_s - p_s$ can take values in $\mathbb{R}$. While in the case of regular contributions, b_s is typically positive, it can also be negative, for example in the single premium case, or if the regular premiums are very low. We will split b_s into contributions a_s and deterministic fees p_s when it is needed for the interpretation of the results.

This formulation can be seen as an extension of the case studied in [7], where it is assumed that a constant contribution parameter $a_t = a$ for all t and there is no periodic fees, so that $p_t = 0$. It is clear from (2) that the fund value becomes path-dependent and involves a continuous arithmetic average. Without loss of generality, let $F_0 = S_0$.

2.2 Benefits

We assume that there is a guaranteed minimum accumulation rate $g < r$ on all the contributions of the policyholder until time t so that the accumulated guaranteed benefit G_t at time t has dynamics

$$dG_t = gG_t dt + a_t dt$$

where $G_0 = P_0$ at time 0. Thus, at time t the guaranteed amount G_t can be expressed as

$$G_t = P_0 e^{gt} + \int_0^t a_s e^{g(t-s)} ds.$$

When the annual rate of contribution is constant ($a_t = a$), the guaranteed value can be simplified to

$$G_t = P_0 e^{gt} + a \left(\frac{e^{gt} - 1}{g} \mathbb{1}_{\{g>0\}} + t \mathbb{1}_{\{g=0\}} \right).$$

Chi and Lin [7] develop techniques to price and hedge the guarantee at time t. Using their numerical approach it is possible to estimate the fair fee for the European VA (Proposition 3 in their paper).

As in [4, 13], we assume that the policyholder has the option to surrender the policy at any time t and to receive a surrender benefit at surrender time equal to

$$(1 - \kappa_t) F_t$$

where κ_t is a penalty percentage charged for surrendering at time t. As presented for instance in [3, 13] or [15], a standard surrender penalty is decreasing over time. Typical VAs sold in the US have a surrender charge period. In general, the maximum surrender charge is around 8 % of the account value and decreases during the surrender charge period. A typical example is New York Life's Premier Variable Annuity [14], for which the surrender charge starts at 8 % in the first contract year, decreases by 1 % per year to reach 2 % in year 7. From year 8 on, there is no penalty on surrender. In another example, "the surrender charge is 7 % during the first Contract Year and decreases by 1 % each subsequent Contract Year. No surrender charge is deducted for surrenders occurring in Contract Years 8 and later" [17].

3 Valuation of the Surrender Option

In this section, we discuss the valuation of the variable annuity contract with maturity benefit and surrender option.[3] We first present a sufficient condition to eliminate the possibility of optimal surrender. We then explain how we evaluate the value of the surrender option using partial differential equations (PDEs). We consider a variable annuity contract with maturity benefit only, which can be surrendered. We choose to ignore the death benefits that are typically added to that type of contract since our goal is to analyze the effect of the fee structure on the value of the surrender option.

3.1 Notation and Optimal Surrender Decision

We denote by $v(t, F_t)$ and $V(t, F_t)$ the value of the contract without and with surrender option, respectively. In this paper, we ignore death benefits and assume that the policyholder survives to maturity.[4] Thus, the value of the contract without the surrender option is simply the risk-neutral expectation of the payoff at maturity, conditional on the filtration up to time t.

$$v(t, F_t) = E[e^{-r(T-t)} \max(G_T, F_T)|\mathscr{F}_t] \tag{3}$$

We assume that the difference between the value of the maturity benefit and the full contract is only attributable to the surrender option, which we denote by $e(t, F_t)$. Then, we have the following decomposition.

$$V(t, F_t) = v(t, F_t) + e(t, F_t) \tag{4}$$

The value of the contract with surrender option is calculated assuming that the policyholder surrenders optimally. This means that the contract is surrendered as soon as its value drops below the value of the surrender benefit. To express the total value of the variable annuity contract, we must introduce further notation. We denote by $\mathscr{T}_t$ the set of all stopping times τ greater than t and bounded by T. Then, we can express the continuation value of the VA contract as

---

[3] In this paper, we quantify the value added by the possibility for the policyholder to surrender his policy. We call it the surrender option, as in [13]. It is not a guarantee that can be added to the variable annuity, but rather a real option created by the fact that the contract can be surrendered.

[4] See [2] for instance for a treatment on how to incorporate mortality benefits.

$$V^*(t, F_t) = \sup_{\tau \in \mathscr{T}_t} E[e^{-r(\tau-t)}\psi(\tau, F_\tau)],$$

where

$$\psi(t, x) = \begin{cases} (1 - \kappa_t)x, & \text{if } t \in (0, T) \\ \max(G_T, x), & \text{if } t = T \end{cases}$$

is the payoff of the contract at surrender or maturity. Finally, we let $\mathscr{S}_t$ be the optimal surrender region at time $t \in [0, T]$. The optimal surrender region is given by the fund values for which the surrender benefit is worth more than the VA contract if the policyholder continues to hold it for at least a small amount of time. Mathematically speaking, it is defined by

$$\mathscr{S}_t = \{F_t : V^*(t, F_t) \leqslant \psi(t, F_t)\}.$$

The complement of the optimal surrender region $\mathscr{S}_t$ will be referred to as the continuation region. We also define B_t, the optimal surrender boundary at time t, by

$$B_t = \inf_{F_t \in [0,\infty)} \{F_t \in \mathscr{S}_t\}.$$

3.2 Theoretical Result on Optimal Surrender Behavior

According to (2) the account value F_t can be written as follows at time t

$$F_t = e^{-ct}S_t + \int_0^t b_s e^{-c(t-s)} \frac{S_t}{S_s} ds, \quad t \geqslant 0,$$

and at time $t + dt$, it is equal to

$$F_{t+dt} = e^{-c(t+dt)}S_{t+dt} + \int_0^{t+dt} b_s e^{-c(t+dt-s)} \frac{S_{t+dt}}{S_s} ds.$$

Proposition 3.1 (Sufficient condition for no surrender) *For a fixed time $t \in [0, T]$, a sufficient condition to eliminate the surrender incentive at time 't' is given by*

$$(\kappa_t' + (1 - \kappa_t)c)F_t < b_t(1 - \kappa_t), \tag{5}$$

where $\kappa'_t = \partial\kappa_t/\partial t$. Here, are some special cases of interest:

- When $a_t = p_t = 0$ (no periodic investment, no periodic fee) and $\kappa_t = 1-e^{-\kappa(T-t)}$ (situation considered by [4]) then $b_t = 0$ and (5) becomes

$$\kappa > c.$$

- When $a_t = 0$ (no periodic investment, i.e., a single lump sum paid at time 0), then $b_t = -p_t \leq 0$. Assume that $p_t > 0$ so that $b_t < 0$ thus

  - If $\kappa'_t + (1-\kappa_t)c > 0$ (for example if κ is constant), then the condition can never be satisfied and no conclusion can be drawn.
  - If $\kappa'_t + (1-\kappa_t)c < 0$ then it is not optimal to surrender when

$$F_t > \frac{-p_t(1-\kappa_t)}{\kappa'_t + (1-\kappa_t)c}.$$

When $\kappa_t = \kappa$ and $b_t = b$ are constant over time, condition (5) can be rewritten as

$$F_t < \frac{b(1-\kappa)}{c(1-\kappa)} = \frac{b}{c}.$$

Remark 3.1 Proposition 3.1 shows that in the absence of periodic fees and investment, an insurer can easily ensure that it is never optimal to surrender by choosing a surrender charge equal to $1 - e^{-\kappa t}$ at time t, with a penalty parameter κ higher than the percentage fee c. Proposition 3.1 shows that it is also possible to eliminate the surrender incentive when there are periodic fees and investment opportunities, but the conditions are more complicated.

Proof Consider a time t at which it is optimal to surrender. This implies that for any time interval of length $dt > 0$, it is better to surrender at time t than to wait until time $t + dt$. In other words, the surrender benefit at time t must be at least equal to the expected discounted value of the contract at time $t + dt$, and in particular larger than the surrender benefit at time $t + dt$. Thus

$$(1 - \kappa_t)F_t \geq E[e^{-rdt}(1 - \kappa_{t+dt})F_{t+dt} | \mathscr{F}_t]$$

Using the martingale property for the discounted stock price S_t and the independence of increments for the Brownian motion, we know that $E[S_{t+dt}e^{-rdt}] = S_t$ and $E\left[\frac{S_{t+dt}}{S_t} \middle| \mathscr{F}_t\right] = E\left[\frac{S_{t+dt}}{S_t}\right] = e^{rdt}$ thus

$$E[e^{-rdt}F_{t+dt}|\mathscr{F}_t] = e^{-c(t+dt)}S_t + \int_0^t b_s e^{-c(t+dt-s)}\frac{S_t}{S_s}ds$$

$$+ \int_t^{t+dt} b_s e^{-c(t+dt-s)}e^{-rdt} E\left[\frac{S_{t+dt}}{S_s}\right]ds,$$

$$= e^{-c(t+dt)}S_t + \int_0^t b_s e^{-c(t+dt-s)}\frac{S_t}{S_s}ds + \int_t^{t+dt} b_s e^{-c(t+dt-s)}ds,$$

$$= e^{-cdt}F_t + e^{-cdt}\int_t^{t+dt} b_s e^{-c(t-s)}ds. \tag{6}$$

Thus

$$(1-\kappa_t)F_t \geq (1-\kappa_{t+dt})\left(e^{-cdt}F_t + e^{-cdt}\int_t^{t+dt} b_s e^{-c(t-s)}ds\right)$$

We then use $\kappa_{t+dt} = \kappa_t + \kappa'_t dt + o(dt)$, $e^{-cdt} = 1 - cdt + o(dt)$ and $\int_t^{t+dt} b_s e^{-c(t-s)}\,ds = b_t dt + o(dt)$ to obtain

$$(1-\kappa_t)F_t \geq (1-\kappa_t - \kappa'_t dt)((1-cdt)F_t + (1-cdt)b_t dt) + j(dt),$$

which can be further simplified into

$$(\kappa'_t + (1-\kappa_t)c)F_t dt \geq b_t(1-\kappa_t)dt + j(dt). \tag{7}$$

where the function $j(dt)$ is $o(dt)$. Since this holds for any $dt > 0$, we can divide (7) by dt and take the limit as $dt \to 0$. Then, we get that if it is optimal to surrender the contract at time t, then

$$(\kappa'_t + (1-\kappa_t)c)F_t \geq b_t(1-\kappa_t).$$

It follows that if $(\kappa'_t + (1-\kappa_t)c)F_t < b_t(1-\kappa_t)$, it is not optimal to surrender the contract at t.   □

3.3 Valuation of the Surrender Option Using PDEs

To evaluate the surrender option $e(t, F_t)$, we subtract the value of the maturity benefit from the value of the VA contract. These values can be compared to American and European options, respectively, since the guarantee in the former is only triggered when the contract expires, while the latter can be exercised at any time before maturity.

From now on, we assume that the deterministic fee p_t is constant over time, so that $p_t = p$ for any time t. We also assume that the policyholder makes no contribution after the initial premium (so that $a_t = 0$ for any t).

It is well-known[5] that the value of a European contingent claim on the fund value F_t follows the following PDE:

$$\frac{\partial v}{\partial t} + \frac{1}{2}\frac{\partial^2 v}{\partial F_t^2} F_t^2 \sigma^2 + \frac{\partial v}{\partial F_t}(F_t(r-c) - p) - rv = 0. \tag{8}$$

Note that Eq. (8) is very similar to the Black-Scholes equation for a contingent claim on a stock that pays dividends (here, the constant fee c represents the dividends), with the addition of the term $\frac{\partial v}{\partial F_t} p$ resulting from the presence of a deterministic fee. Since it represents the contract described in Sect. 2, Eq. (8) is subject to the following conditions:

$$v(T, F_T) = \max(G_T, F_T)$$
$$\lim_{F_t \to 0} v(t, F_t) = G_T e^{-r(T-t)}.$$

The last condition results from the fact that when the fund value is very low, the guarantee is certain to be triggered. When $F_t \to \infty$, the problem is unbounded. However, we have the following asymptotic behavior:

$$\lim_{F_t \to \infty} v(t, F_t) = E_t[F_T e^{-r(T-t)}], \tag{9}$$

which stems from the value of the guarantee approaching 0 for very high fund values. We will use this asymptotic result to solve the PDE numerically, when truncating the grid of values for F_t. The expectation in (9) is easily calculated and is given in the proof of Proposition 3.1.

As it is the case for the American put option,[6] the VA contract with surrender option gives rise to a free boundary problem. In the continuation region, $V^*(t, F_t)$ follows Eq. (8), the same equation as for the contract without surrender option. However, in the optimal surrender region, the value of the contract with surrender is the value of the surrender benefit:

$$V^*(t, F_t) = \psi(t, F_t), \quad t \in [0, T], F_t \in \mathscr{S}_t. \tag{10}$$

For the contract with surrender, the PDE to solve is thus subject to the following conditions:

---

[5] See, for example [5, Sect. 7.3].
[6] See, for example [6].

$$V^*(T, F_T) = \max(G_T, F_T)$$

$$\lim_{F_t \to 0} V^*(t, F_t) = G_T e^{-r(T-t)}$$

$$\lim_{F_t \to B_t} V^*(t, F_t) = \psi(t, B_t).$$

$$\lim_{F_t \to B_t} \frac{\partial}{\partial F_t} V^*(t, F_t) = 1 - \kappa_t.$$

For any time $t \in [0, T]$, the value of the VA with surrender is given by

$$V(t, F_t) = \max(V^*(t, F_t), \psi(t, F_t)).$$

This free boundary problem is solved in Sect. 4 using numerical methods.

4 Numerical Example

To price the VA using a PDE approach, we modify Eq. (8) to express it in terms of $x_t = \log F_t$. We discretize the resulting equation over a rectangular grid with time steps $dt = 0.0001$ ($dt = 0.0002$ for $T = 15$) and $dx = \sigma\sqrt{3dt}$ (following suggestions by Racicot and Théoret [16]), from 0 to T in t and from 0 to $\log 450$ in x. We use an explicit scheme with central difference in x and in x^2.

Throughout this section, we assume that the contract is priced so that only the maturity benefit is covered. In other words, we set c and p such that

$$P_0 = v(t, F_t), \tag{11}$$

where P_0 denotes the initial premium paid by the policyholder. In this section, when the fee is set in the manner, we call it the *fair fee*, even if it does not cover the full value of the contract. We set the fee in this manner to calculate the value added by the possibility to surrender.

4.1 Numerical Results

We now consider variable annuities with the maturity benefit described in Sect. 2. We assume that the initial premium $P_0 = 100$, that there are no periodic premium ($a_s = a = 0$), that the deterministic fee is constant ($p_t = p$) and that the guaranteed roll-up rate is $g = 0$. We further assume that the surrender charge, if any, is of the form $\kappa_t = 1 - e^{\kappa(T-t)}$, and that $r = 0.03$ and $\sigma = 0.2$.

For contracts with and without surrender charge and with maturity 5, 10 and 15 years, the results are presented in Table 1. In each case, the fee levels c and p are chosen such that $P_0 = v(t, F_t)$. As a percentage of the initial premium, the fair fee

Table 1 Value of the surrender option in 5-, 10- and 15-year variable annuity contracts for various fee structures and surrender charges

T = 5				T = 10				T = 15			
		Surrender option				Surrender option				Surrender option	
Fee		κ		Fee		κ		Fee		κ	
c (%)	p	0%	0.5%	c (%)	p	0%	0.5%	c (%)	p	0%	0.4%
0.00	4.150	3.09	2.09	0.00	2.032	3.07	1.02	0.00	1.259	2.76	0.23
1.00	2.971	3.32	2.33	0.50	1.387	3.50	1.46	0.30	0.842	3.30	0.77
2.00	1.796	3.56	2.57	1.00	0.744	3.92	1.89	0.60	0.427	3.84	0.84
3.53	0.000	3.92	2.94	1.58	0.000	4.43	2.39	0.91	0.000	4.40	1.86

For the 15-year contract, we lowered the surrender charge parameter to $\kappa = 0.4\%$ to ensure that the optimal surrender boundary is always finite

when it is paid as a deterministic amount is higher than the fair constant percentage fee. In fact, for high fund values, the deterministic fee is lower than the amount paid when the fee is set as a constant percentage. But when the fund value is low, the deterministic fee represents a larger proportion of the fund compared to the constant percentage fee. This higher proportion drags the fund value down and increases the option value. The effect of each fee structure on the amount collected by the insurer can explain the difference between the fair fixed percentage and deterministic fees.

The results in Table 1 show that when the fee is set as a fixed amount, the value of the surrender option is always lower than when the fee is expressed as a percentage of the fund. When a mix of both types of fees is applied, the value of the surrender option decreases as the fee set as a percentage of the fund decreases. When the fee is deterministic, a lower percentage of the fund is paid out when the fund value is high. Consequently, the fee paid by the policyholder is lower when the guarantee is worth less, reducing the surrender incentive. This explains why the value of the surrender option is lower for deterministic fees. This result can be observed both with and without surrender charges. However, surrender charges decrease the value of the surrender option, as expected. The effect of using a deterministic amount fee, instead of a fixed percentage, is even more noticeable when a surrender charge is added. A lower surrender option value means that the possibility to surrender adds less value to the contract. In other words, if the contract is priced assuming that policyholders do not surrender, unexpected surrenders will result in a smaller loss, on average.

Figure 1 shows the optimal surrender boundaries for the fee structures presented in Table 1 for 10-year contracts. As expected, the optimal boundaries are higher when there is a surrender charge. Those charges are put in place in part to discourage policyholders from surrendering early. The boundaries are also less sensitive to the fee structure when there is a surrender charge. In fact, when there is a surrender charge, setting the fee as a fixed amount leads to a higher optimal boundary during most of the contract. This highlights the advantage of the fixed amount fee structure combined with surrender charges. Without those charges, the fixed fee amount could lead to more surrenders. We also note that the limiting case $p = 0$ corresponds to

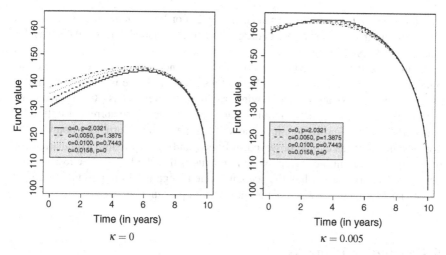

Fig. 1 Optimal surrender boundary when $T = 10$

the situation when fees are paid as a percentage of the fund. The optimal boundary obtained using the PDE approach in this paper coincides with the optimal boundary derived in [4] by solving an integral equation numerically.

Table 1 also shows the effect of the maturity combined with the fee structure on the surrender option. For all maturities, setting the fee as a fixed amount instead of a fixed percentage has a significant effect on the value of the surrender option. This effect is amplified for longer maturities. As for the 10-year contract, combining the fixed amount fee with a surrender charge further reduces the value of the surrender option, especially when $T = 15$. The optimal surrender boundaries for different fee

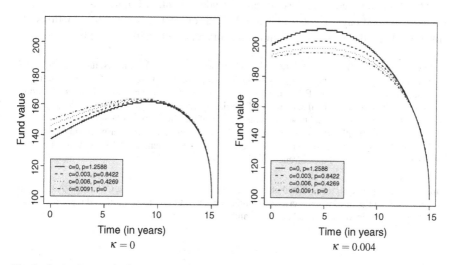

Fig. 2 Optimal surrender boundary when $T = 15$

structures when $T = 15$ are presented in Fig. 2. For longer maturities such as this one, the combination of surrender charges and deterministic fee raises the surrender boundary more significantly.

In all cases, the decrease in the value of the surrender option caused by the combination of a deterministic amount fee and a surrender charge is significant. In our example with a 15-year contract, moving from a fee entirely set as a fixed percentage to a fee set as a deterministic amount reduces the value of the surrender option by over 85 %. This is surprising since the shift in the optimal surrender boundary is not as significant (as can be observed in Figs. 1 and 2). A possible explanation for the sharp decrease in the surrender option value is that the fee income lost when a policyholder surrenders when the account value is high is less important, relatively to the value of the guarantee, than in the constant percentage fee case.

5 Concluding Remarks

In this chapter, the maturity guarantee fees are paid during the term of the contract as a series of deterministic amounts instead of a percentage of the fund, which is more common in the industry. We give a sufficient condition that allows the elimination of optimal surrender incentives for variable annuity contracts with fairly general fee structures. We also show how deterministic fees and surrender charges affect the value of the surrender option and the optimal surrender boundary. In particular, we highlight the efficiency of combining deterministic fees and exponential surrender charges in decreasing the value of the surrender option. In fact, although the optimal surrender boundary remains at a similar level, a fee set as a deterministic amount reduces the value of the surrender option, which makes the contract less risky for the insurer. This result also suggests that the state-dependent fee suggested in [2] could also be efficient in reducing the optimal surrender incentive. Future work could focus on more general payouts (see for example [12] for ratchet and lookback options [4] for Asian benefits) in more general market models, and include death benefits.

Acknowledgments Both authors gratefully acknowledge support from the Natural Sciences and Engineering Research Council of Canada and from the Society of Actuaries Center of Actuarial Excellence Research Grant. C. Bernard thanks the Humboldt Research Foundation and the hospitality of the chair of mathematical statistics of Technische Universität München where the paper was completed. A. MacKay also acknowledges the support of the Hickman scholarship of the Society of Actuaries. We would like to thank Mikhail Krayzler for inspiring this chapter by raising a question at the Risk Management Reloaded conference in Munich in September 2013 about the impact of fixed deterministic fees on the surrender boundary.

Open Access This chapter is distributed under the terms of the Creative Commons Attribution Noncommercial License, which permits any noncommercial use, distribution, and reproduction in any medium, provided the original author(s) and source are credited.

References

1. Bauer, D.K., Kling, A., Russ, J.: A universal pricing framework for guaranteed minimum benefits in variable annuities. ASTIN Bull. **38**(2), 621 (2008)
2. Bernard, C., Hardy, M., MacKay, A.: State-dependent fees for variable annuity guarantees. ASTIN Bull. (2014) forthcoming
3. Bernard, C., Lemieux, C.: Fast simulation of equity-linked life insurance contracts with a surrender option. In: Proceedings of the 40th Conference on Winter Simulation, pp. 444–452. Winter Simulation Conference (2008)
4. Bernard, C., MacKay, A., Muehlbeyer, M.: Optimal surrender policy for variable annuity guarantees. Insur. Math. Econ. **55**(C), 116–128 (2014)
5. Björk, T.: Arbitrage Theory in Continuous Time. Oxford University Press, Oxford (2004)
6. Carr, P., Jarrow, R., Myneni, R.: Alternative characterizations of American put options. Math. Financ. **2**(2), 87–106 (1992)
7. Chi, Y., Lin, S.X.: Are flexible premium variable annuities underpriced? ASTIN Bull. (2013) forthcoming
8. Costabile, M.: Analytical valuation of periodical premiums for equity-linked policies with minimum guarantee. Insur. Math. Econ. **53**(3), 597–600 (2013)
9. Hardy, M.R.: Investment Guarantees: Modelling and Risk Management for Equity-Linked Life Insurance. Wiley, New York (2003)
10. Hürlimann, W.: Analytical pricing of the unit-linked endowment with guarantees and periodic premiums. ASTIN Bull. **40**(2), 631 (2010)
11. Kling, A., Ruez, F., Ruß, J.: The impact of policyholder behavior on pricing, hedging, and hedge efficiency of withdrawal benefit guarantees in variable annuities. Eur. Actuar. J. (2014) forthcoming
12. Krayzler, M., Zagst, R., Brunner, B.: Closed-form solutions for guaranteed minimum accumulation benefits. Working Paper available at SSRN: http://ssrn.com/abstract=2425801(2013)
13. Milevsky, M.A., Salisbury, T.S.: The real option to lapse a variable annuity: can surrender charges complete the market. In: Conference Proceedings of the 11th Annual International AFIR Colloquium (2001)
14. New York Life: "Premier Variable Annuity", fact sheet, http://www.newyorklife.com/nyl-internet/file-types/NYL-Premier-VA-Fact-Sheet-No-CA-NY.pdf (2014). Accessed 10 May 2014
15. Palmer, B.: Equity-indexed annuities: fundamental concepts and issues, working Paper (2006)
16. Racicot, F.-E., Théoret, R.: Finance computationnelle et gestion des risques. Presses de l'Université du Quebec (2006)
17. Thrivent financial: flexible premium deferred variable annuity, prospectus, https://www.thrivent.com/insurance/autodeploy/Thrivent_VA_Prospectus.pdf (2014). Accessed 12 May 2014

A Variational Approach for Mean-Variance-Optimal Deterministic Consumption and Investment

Marcus C. Christiansen

Abstract A significant number of life insurance contracts are based on deterministic investment strategies—this justifies to restrict the set of admissible controls to deterministic controls. Optimal deterministic controls can be identified by Hamilton-Jacobi-Bellman techniques, but for the corresponding partial differential equations only numerical solutions are available and so the general existence of optimal controls is unclear. We present a non-constructive existence result and derive necessary characterizations for optimal controls by using a Pontryagin maximum principle. Furthermore, based on the variational idea of the Pontryagin maximum principle, we derive a numerical optimization algorithm for the calculation of optimal controls.

1 Introduction

Among many other applications, individual investment strategies arise in pension saving contracts, see for example Cairns [6]. While in dynamic optimal consumption-investment problems one typically aims to find an optimal control from the set of adapted processes, in insurance practice quite a number of contracts rely on deterministic investment strategies. Deterministic investment and consumption strategies have the advantage that they are easier to organize in asset management, that they make future consumption predictable, and that they are easier to communicate. From a mathematical point of view, deterministic control avoids unwanted features of stochastic control such as diffusive consumption, satisfaction points and consistency problems. For further arguments and a detailed comparison of stochastic versus deterministic control see also Menkens [17].

The present paper is motivated by Christiansen and Steffensen [9], where mean-variance-optimal deterministic consumption and investment is discussed in a Black-Scholes market. Sufficient conditions for optimal strategies are derived from a Hamilton-Jacobi-Bellman approach, but only numerical solutions and no analytical

M.C. Christiansen (✉)
University of Ulm, 89069 Ulm, Germany
e-mail: marcus.christiansen@uni-ulm.de

© The Author(s) 2015   225
K. Glau et al. (eds.), *Innovations in Quantitative Risk Management*,
Springer Proceedings in Mathematics & Statistics 99,
DOI 10.1007/978-3-319-09114-3_13

solutions are given. That means that the general existence of solutions remains unclear. We fill that gap, allowing for a slightly more general model with non-constant Black-Scholes market parameters. By applying a Pontryagin maximum principle, we additionally verify that the sufficient conditions of Christiansen and Steffensen [9] for optimal controls are actually necessary. Furthermore, we present an alternative numerical algorithm for the calculation of optimal controls. Therefore, we make use of the variational idea behind the Pontryagin maximum principle. In a first step, we define generalized gradients for our objective function, which, in a second step, allows us to construct a gradient ascent method.

Mean-variance investment is a true classic since the seminal work by Markowitz [16]. Since then various authors have improved and extended the results, see for example Korn and Trautmann [12], Korn [13], Zhou and Li [18], Basak and Chabakauri [3], Kryger and Steffensen [15], Kronborg and Steffensen [14], Alp and Korn [1], Björk, Murgoci and Zhou [5] and others.

Deterministic optimal control is fundamental in Herzog et al. [11] and Geering et al. [10]. But apart from other differences, they disregard income and consumption and focus on the pure portfolio problem without cash flows. Bäuerle and Rieder [2] study optimal investment for both, adapted stochastic strategies and deterministic strategies. They discuss various objectives including mean-variance objectives under constraints. In the present paper, we discuss an unconstrained mean-variance-objective and we also control for consumption.

The paper is structured as follows. In Sect. 2, we set up a basic model framework and specify the optimal consumption and investment problem that we discuss here. In Sect. 3, we present an existence result for the optimal control. Section 4 derives necessary conditions for optimal controls by applying a Pontryagin maximum principle. Section 5 defines and calculates generalized gradients for the objective, which helps to set up a numerical optimization algorithm in Sect. 6. In Sect. 7 we illustrate the numerical algorithm.

2 The Mean-Variance-Optimal Deterministic Consumption and Investment Problem

Let $B([0, T])$ denote the space of bounded Borel-measurable functions, equipped with the uniform norm $\| \cdot \|_\infty$. On some finite time interval $[0, T]$, we assume that we have a continuous income with nonnegative rate $a \in B([0, T])$ and a continuous consumption with nonnegative rate $c \in B([0, T])$. Let $C([0, T])$ bet the set of continuous functions on $[0, T]$. The positive initial wealth x_0 and the stochastic wealth $X(t)$ at $t > 0$ is distributed between a bank account with risk-free interest rate $r \in C([0, T])$ and a stock or stock fund with price process

$$dS(t) = S(t)\alpha(t)dt + S(t)\sigma(t)dW(t), \quad S(0) = 1,$$

where $\alpha(t) > r(t) \geq 0$, $\sigma(t) > 0$ and $\alpha, \sigma \in C([0, T])$. We write $\pi(t)$ for the proportion of the total value invested in stocks and call it the investment strategy. The wealth process $X(t)$ is assumed to be self-financing. Thus, it satisfies

$$dX(t) = X(t)(r(t) + (\alpha(t) - r(t))\pi(t))dt + (a(t) - c(t))dt + X(t)\sigma(t)\pi(t)dW(t) \tag{1}$$

with initial value $X(0) = x_0$ and has the explicit representation

$$X(t) = x_0 e^{\int_0^t dU} + \int_0^t (a(s) - c(s)) e^{\int_s^t dU} ds, \tag{2}$$

where

$$dU(t) = (r(t) + (\alpha(t) - r(t))\pi(t) - \frac{1}{2}\sigma(t)^2\pi(t)^2)dt$$
$$+ \sigma(t)\pi(t)dW(t), \quad U(0) = 0. \tag{3}$$

It is important to note that the process $(X(t))_{t \geq 0}$ depends on the choice of the investment strategy $(\pi(t))_{t \geq 0}$ and the consumption rate $(c(t))_{t \geq 0}$. In order to make that dependence more visible, we will also write $X = X^{(\pi,c)}$. For some arbitrary but fixed risk aversion parameter $\gamma > 0$ of the investor, we define the risk measure

$$MV_\gamma[\cdot] := E[\cdot] - \gamma Var[\cdot].$$

We aim to maximize the functional

$$G(\pi, c) := MV_\gamma \left[\int_0^T e^{-\rho s} c(s)ds + e^{-\rho T} X^{(\pi,c)}(T) \right] \tag{4}$$

with respect to the investment strategy π and the consumption rate c. The parameter $\rho \geq 0$ describes the preference for consuming today instead of tomorrow.

3 Existence of Optimal Deterministic Control Functions

In Christiansen and Steffensen [9], where a Hamilton-Jacobi-Bellman approach is used, the existence of optimal control functions is related to the existence of solutions for the Hamilton-Jacobi-Bellman partial differential equation. However, only numerical solutions are available, so the general existence of solutions is unclear. Here, we fill that gap by giving an existence result for optimal deterministic control functions. The proof needs rather weak assumptions, but it is not constructive.

Theorem 1 *Let* $G : D \to (-\infty, \infty)$ *be defined by* (4) *for*

$$D := \left\{ (\pi, c) \in B([0, T]) \times B([0, T]) : \underline{c}(t) \leq c(t) \leq \overline{c}(t), \quad t \in [0, T] \right\}$$

with lower and upper consumption bounds $\underline{c}, \overline{c} \in B([0, T])$. *Then, the functional* G *is continuous and has a finite upper bound.*

Proof We first show that $MV_\gamma[X(T)] = MV_\gamma[X^{(\pi,c)}(T)]$ has a finite upper bound that does not depend on (π, c). Defining the stochastic process

$$Y(t) := X(t) - \gamma X(t)^2 + \gamma E[X(t)]^2, \quad t \in [0, T], \tag{5}$$

we have $MV_\gamma[X(T)] = E[Y(T)]$. So it suffices to show that $E[Y(T)]$ has a finite upper bound that does not depend on (π, c). Since the quadratic variation process of X satisfies $d[X](t) = X(t)^2 \sigma(t)^2 \pi(t)^2 dt$, from Ito's Lemma we get that

$$dE[X(t)] = E[X(t)] \left\{ r(t) + (\alpha(t) - r(t))\pi(t) \right\} dt + (a(t) - c(t)) dt, \tag{6}$$

$$dE[X(t)^2] = 2E[X(t)^2] \left\{ r(t) + (\alpha(t) - r(t))\pi(t) \right\} dt + E[X(t)](a(t) - c(t)) dt \\ + E[X(t)^2]\sigma(t)^2 \pi(t)^2 dt. \tag{7}$$

Hence, the expectation function of Y solves the differential equation

$$dE[Y(t)] = E[X(t)] \left\{ r(t) + (\alpha(t) - r(t))\pi(t) \right\} dt + (a(t) - c(t)) dt \\ - \gamma E[X(t)^2]\sigma(t)^2 \pi(t)^2 dt - 2\gamma \left(E[X(t)^2] - E[X(t)]^2 \right) \tag{8} \\ \times \left\{ r(t) + (\alpha(t) - r(t))\pi(t) \right\} dt.$$

The right hand side of (8) is maximal with respect to $\pi(t)$ for

$$\pi^*(t) = \frac{1}{2\gamma} \frac{\alpha(t) - r(t)}{\sigma(t)^2} \frac{E[X(t)] - 2\gamma \, \text{Var}[X(t)]}{E[X(t)^2]}. \tag{9}$$

Plugging (9) into (8) and rearranging terms yields

$$dE[Y(t)] \leq r(t) \left(E[X(t)] - 2\gamma Var[X(t)] \right) dt + (a(t) - c(t)) dt \\ + \frac{1}{4\gamma} \frac{(\alpha(t) - r(t))^2}{\sigma(t)^2} \frac{\left(E[X(t)] - 2\gamma \, \text{Var}[X(t)] \right)^2}{E[X(t)^2]} dt. \tag{10}$$

Recall that we assumed $\gamma > 0$ and $\sigma(t) > 0$, so the first and second denominator are never zero. If the third denominator $E[X(t)^2]$ is zero, we implicitly get $E[X(t)] = 0$, and (10) is still true by defining $0/0 := 0$. The first line on the right hand side of (10) has an upper bound of

$$r(t) E[Y(t)]dt + (a(t) - c(t))dt.$$

With the help of the equality

$$\big(E[X(t)] - 2\gamma \, \mathrm{Var}[X(t)]\big)^2 = E[X(t)]^2 - 4\gamma \, \mathrm{Var}[X(t)] \, E[Y(t)]$$

and the inequalities $(E[X(t)])^2 \le E[X(t)^2]$ and $\mathrm{Var}[X(t)] \le E[X(t)^2]$, we can show that the second line on the right hand side of (10) has an upper bound of

$$\frac{1}{4\gamma} \frac{(\alpha(t) - r(t))^2}{\sigma(t)^2} \Big(1 + 4\gamma \, |E[Y(t)]|\Big) dt.$$

All in all, we obtain

$$dE[Y(t)] \le \big(C_1|E[Y(t)]| + C_2\big)dt, \quad t \in [0, T] \tag{11}$$

for some finite positive constants C_1 and C_2, since the functions $r(t), a(t), \alpha(t)$ are uniformly bounded on $[0, T]$, since $-c(t) \le -\underline{c}(t)$ for a uniformly bounded function $\underline{c}$, and since the positive and continuous function $\sigma(t)$ has a uniform lower bound greater than zero. Thus, we have $E[Y(t)] \le g(t)$ for $g(t)$ defined by the differential equation

$$dg(t) = \big(C_1|g(t)| + C_2\big)dt, \quad g(0) = Y(0) = x_0 > 0. \tag{12}$$

This differential equation for $g(t)$ has a unique solution, which is bounded on $[0, T]$ and does not depend on the choice of (π, c). Hence, also $MV_\gamma[X(T)] = E[Y(T)]$ has a finite upper bound that does not depend on the choice of (π, c). The same is true for the functional (4), since

$$G(\pi, c) \le \int_0^T e^{-\rho s} \overline{c}(s) ds + e^{-\rho T} MV_{\gamma e^{-\rho T}} [X(T)].$$

Now we show the continuity of the functional G. Suppose that $(\pi_n, c_n)_{n \ge 1}$ is an arbitrary but fixed sequence in D that converges to (π_0, c_0) with respect to the supremum norm. Since D is a Banach space, the limit (π_0, c_0) is also an element of D. Let $X_n(t) := X^{(\pi_n, c_n)}(t)$ for all t. As the sequence $(\pi_n, c_n)_{n \ge 1}$ is convergent and within D, the absolutes $|\pi_n(t)|$ and $|c_n(t)|$ have finite upper bounds, uniformly in n and uniformly in t. Therefore, analogously to inequality (11), from Eq. (6) we get that

$$dE[X_n(t)] \le \big(C_3 |E[X_n(t)]| + C_4\big)dt, \quad t \in [0, T], \quad n = 0, 1, 2, \ldots$$

for some positive finite constants C_3 and C_4. Arguing analogously to (12), we obtain that $E[X_n(t)] \le f(t)$ for some bounded function $f(t)$. Using similar arguments for $-E[X_n(t)]$, we get that also the absolute $|E[X_n(t)]|$ is uniformly bounded in n and in t. Applying Eq. (7), we obtain

$$dE[X_n(t)^2] = 2E[X_n(t)^2]\{r(t) + (\alpha(t) - r(t))\pi_n(t)\}dt$$
$$+ E[X_n(t)](a(t) - c_n(t))dt + E[X_n(t)^2]\sigma(t)^2\pi_n(t)^2 dt.$$

Using the uniform boundedness of $|E[X_n(t)]|$, $|\pi_n(t)|$ and $|c_n(t)|$, we can conclude that

$$dE[X_n(t)^2] \leq (C_5 E[X_n(t)^2] + C_6)dt, \quad t \in [0, T], \quad n = 0, 1, 2, \ldots$$

for some positive finite constants C_5 and C_6. Hence, arguing analogously to above, the value $E[X_n(t)^2]$ is uniformly bounded in n and in t. Let $Y_n(t)$ be the process according to definition (5) but with X_n instead of X. Using (8) and the uniform boundedness of $|E[X_n(t)]|$, $E[X_n(t)^2]$, $|\pi_n(t)|$ and $|c_n(t)|$, we can show that

$$dE[Y_0(t) - Y_n(t)] \leq \left(C_7 \sup_{t \in [0,T]} |\pi_0(t) - \pi_n(t)| + \sup_{t \in [0,T]} |c_0(t) - c_n(t)| \right) dt, \quad t \in [0, T]$$

for some positive finite constant C_7. Thus, we get

$$E[Y_0(T) - Y_n(T)] \leq T C_7 \sup_{t \in [0,T]} |\pi_0(t) - \pi_n(t)| + T \sup_{t \in [0,T]} |c_0(t) - c_n(t)|,$$

where we used that $Y_0(0) - Y_n(0) = x_0 - x_0 = 0$. Arguing similarly for $-E[Y_0(t) - Y_n(t)]$, we can conclude that

$$|G(\pi_0, c_0) - G(\pi_n, c_n)|$$

$$= \left| \int_0^T e^{-\rho s}(c_0(s) - c_n(s))ds + e^{-\rho T} MV_{\gamma e^{-\rho T}}[X_0(T)] - e^{-\rho T} MV_{\gamma e^{-\rho T}}[X_n(T)] \right|$$

$$\leq T \sup_{t \in [0,T]} |c_0(t) - c_n(t)| + e^{-\rho T} \left| E[\tilde{Y}_0(T) - \tilde{Y}_n(T)] \right|$$

$$\leq T C_8 \sup_{t \in [0,T]} |\pi_0(t) - \pi_n(t)| + 2T \sup_{t \in [0,T]} |c_0(t) - c_n(t)|$$

for some finite constant C_8, where the processes $\tilde{Y}_0(t)$ and $\tilde{Y}_n(t)$ are defined as above but with γ replaced by $\gamma e^{-\rho T}$. Since we assumed that $(\pi_n, c_n)_{n \geq 1}$ converges in supremum norm, we obtain that $G(\pi_n, c_0)$ converges to $G(\pi_0, c_0)$, i.e. the functional G is continuous.

As G has a finite upper bound on the domain D, the supremum

$$\sup_{(\pi,c)\in D} G(\pi, c)$$

indeed exists. Since G is continuous and D is a Banach space, we can conclude that on each compact subset K of D there exists a pair (π^*, c^*) for which

$$G(\pi^*, c^*) = \sup_{(\pi,c)\in K} G(\pi, c). \tag{13}$$

4 A Pontryagin Maximum Principle

Christiansen and Steffensen [9] identify characterizing equations for optimal investment and consumption rate by using a Hamilton-Jacobi-Bellman approach. Here, we show that those characterizing equations are indeed necessary by using a Pontryagin maximum principle (cf. Bertsekas [4]).

Defining the moment functions

$$
\begin{aligned}
m_i(t) &= E[(X(t))^i], \qquad i = 1, 2, \\
p_i(t) &= E[(\int_t^T (a(s) - c(s)) e^{\int_s^T dU} ds)^i], \qquad i = 1, 2, \\
n_i(t) &= E[e^{i\int_t^T dU}], \qquad i = 1, 2, \\
k(t) &= E[e^{\int_t^T dU} \int_t^T (a(s) - c(s)) e^{\int_s^T dU} ds],
\end{aligned}
\tag{14}
$$

as in Christiansen and Steffensen [9], we can represent the objective function $G(\pi, c)$ by

$$
\begin{aligned}
G(\pi, c) = \int_0^T e^{-\rho s} c(s) ds &+ e^{-\rho T} \left(m_1(t) n_1(t) + p_1(t) \right) \\
&- \gamma e^{-2\rho T} \left(m_2(t) n_2(t) + 2 m_1(t) k(t) + p_2(t) - (m_1(t) n_1(t) + p_1(t))^2 \right)
\end{aligned}
\tag{15}
$$

for any t in $[0, T]$. Simple calculations give us that

$$
\begin{aligned}
\tfrac{d}{dt} m_1(t) &= (r(t) + (\alpha(t) - r(t))\pi(t)) m_1(t) + (a(t) - c(t)), \\
\tfrac{d}{dt} m_2(t) &= \left(2r(t) + 2(\alpha(t) - r(t))\pi(t) + \pi(t)^2 \sigma(t)^2 \right) m_2(t) + 2(a(t) - c(t)) m_1(t).
\end{aligned}
\tag{16}
$$

Similarly to m_1 and m_2, also n_1, n_2, p_1, p_2, and k solve a system of ordinary differential equations but with terminal instead of initial conditions, see Christiansen and Steffensen [9].

Theorem 2 *Let (π^*, c^*) be an optimal control in the sense of (13), and let $m_i^*(t)$, $p_i^*(t)$, $n_i^*(t)$, $i = 1, 2$, and $k^*(t)$ be the corresponding moment functions according to (14). Then, we have necessarily*

$$\pi^*(t) = \frac{\alpha(t) - r(t)}{\sigma(t)^2}\left(\frac{e^{\rho T} m_1^*(t) n_1^*(t) - 2\gamma m_1^*(t)\left(k^*(t) - n_1^*(t) m_1^*(T)\right)}{2\gamma m_2^*(t) n_2^*(t)} - 1\right),$$

(17)

$$c^*(t) = \begin{cases} \bar{c}(t) & \text{if } e^{\rho(T-t)} - n_1^*(t) + 2\gamma e^{-\rho T}\left(m_1^*(t) n_2^*(t) + k^*(t) - n_1^*(t) m_1^*(T)\right) > 0 \\ \underline{c}(t) & \text{else.} \end{cases}$$

(18)

Proof With (π^*, c^*) being an optimal control, we define local alternatives by

$$(\pi^\varepsilon(t), c^\varepsilon(t)) = \begin{cases} (\pi^*(t), c^*(t)) + (h(t), l(t)) & \text{for } t \in (t_0 - \varepsilon, t_0] \\ (\pi^*(t), c^*(t)) & \text{else} \end{cases}$$

for continuous functions h and l. As $G(\pi^*, c^*)$ is maximal, by applying (15) for $t = t_0$ we obtain that

$$G(\pi^*, c^*) - G(\pi^\varepsilon, c^\varepsilon) = -\int_{t_0-\varepsilon}^{t_0} e^{-\rho s} l(s)\,ds + e^{-\rho T}\left(m_1^*(t_0) - m_1^\varepsilon(t_0)\right) n_1^*(t_0)$$
$$- \gamma e^{-2\rho T}\Big\{\left(m_2^*(t_0) - m_2^\varepsilon(t_0)\right) n_2^*(t_0) + 2\left(m_1^*(t_0) - m_1^\varepsilon(t_0)\right) k^*(t_0)$$
$$- \left(m_1^*(t_0)^2 - m_1^\varepsilon(t_0)^2\right) n_1^*(t_0)^2 - 2\big(m_1^*(t_0)$$
$$- m_1^\varepsilon(t_0)\big) n_1^*(t_0) p_1^*(t_0)\Big\}$$
$$= -\int_{t_0-\varepsilon}^{t_0} e^{-\rho s} l(s)\,ds + \left(m_1^*(t_0) - m_1^\varepsilon(t_0)\right)$$
$$\times \left\{e^{-\rho T} n_1^*(t_0) - 2\gamma e^{-2\rho T}\left(k^*(t_0) - n_1^*(t_0) p_1^*(t_0)\right)\right\}$$
$$- \left(m_1^*(t_0)^2 - m_1^\varepsilon(t_0)^2\right)\gamma e^{-2\rho T} n_1^*(t_0)^2$$
$$- \left(m_2^*(t_0) - m_2^\varepsilon(t_0)\right)\gamma e^{-2\rho T} n_2^*(t_0)$$

(19)

must be nonnegative. Equation (16) implies that

$$m_1^*(t_0) - m_1^\varepsilon(t_0) = \int_{t_0-\varepsilon}^{t_0}\left(\frac{d}{ds}m_1^*(s) - \frac{d}{ds}m_1^\varepsilon(s)\right)ds$$
$$= \int_{t_0-\varepsilon}^{t_0}\left(\left\{(r(s) + (\alpha(s) - r(s))\pi^*(s))m_1^*(s)\right.\right.$$

$$- \left(r(s) + (\alpha(s) - r(s))\pi^\varepsilon(s)\right)m_1^\varepsilon(s)\bigg\} + l(s)\bigg)ds,$$

since $\|m_1^* - m_1^\varepsilon\| \to 0$ for $\varepsilon \to 0$. Moreover, since we have $m_1^*(t) \to m_1^*(t_0)$, $r(t) \to r(t_0), \alpha(t) \to \alpha(t_0), \sigma(t) \to \sigma(t_0)$ for $t \to t_0$, we get that

$$m_1^*(t_0) - m_1^\varepsilon(t_0) = -(\alpha(t_0) - r(t_0))\, m_1^*(t_0) \int_{t_0-\varepsilon}^{t_0} h(s)ds + \int_{t_0-\varepsilon}^{t_0} l(s)ds + o(\varepsilon).$$

$$(20)$$

For the squared functions we use

$$m_1^*(t_0)^2 - m_1^\varepsilon(t_0)^2 = \left(m_1^*(t_0) - m_1^\varepsilon(t_0)\right)\left(m_1^*(t_0) + m_1^\varepsilon(t_0)\right)$$
$$= 2m_1^*(t_0)\left(m_1^*(t_0) - m_1^\varepsilon(t_0)\right) - \left(m_1^*(t_0) - m_1^\varepsilon(t_0)\right)^2$$

and then apply the asymptotic formula (20), which leads to

$$m_1^*(t_0)^2 - m_1^\varepsilon(t_0)^2 = -2(\alpha(t_0) - r(t_0))\, m_1^*(t_0)^2 \int_{t_0-\varepsilon}^{t_0} h(s)ds$$

$$+ 2m_1^*(t_0) \int_{t_0-\varepsilon}^{t_0} l(s)ds + o(\varepsilon).$$

Similarly, we can show that

$$m_2^*(t_0) - m_2^\varepsilon(t_0) = \bigg\{ -2(\alpha(t_0) - r(t_0))\, m_2^*(t_0)$$

$$- 2\pi^*(t_0)\sigma(t_0)^2 m_2^*(t_0)\bigg\} \int_{t_0-\varepsilon}^{t_0} h(s)ds$$

$$+ 2m_1^*(t_0) \int_{t_0-\varepsilon}^{t_0} l(s)ds + o(\varepsilon).$$

$$(21)$$

Plugging Eq. (21) into Eq. (19) and rearranging, we get

$$o(\varepsilon) \leq \int_{t_0-\varepsilon}^{t_0} l(s)ds \bigg(-e^{\rho t_0} + e^{-\rho T}n_1^*(t_0) - 2\gamma e^{-2\rho T}\big\{m_1^*(t_0)n_2^*(t_0) + k^*(t_0)$$

$$- n_1^*(t_0)\big(n_1^*(t_0)m_1^*(t_0) + p_1^*(t_0)\big)\big\}\bigg)$$

$$+ \int_{t_0-\varepsilon}^{t_0} h(s) \, ds \, (\alpha(t_0) - r(t_0)) \Big(-m_1^*(t_0)n_1^*(t_0)e^{-\rho T} - 2\gamma e^{-2\rho T} m_1^*(t_0)$$

$$\times \Big\{ -k^*(t_0) + n_1^*(t_0)\big(n_1^*(t_0)m_1^*(t_0) + p_1^*(t_0)\big)\Big\}\Big)$$

$$- 2\gamma e^{-2\rho T} m_2^*(t_0)n_2^*(t_0)\Big\{ -1 - \pi^*(t_0)\frac{\sigma(t_0)^2}{\alpha(t_0) - r(t_0)}\Big\}$$

for all continuous functions l and h. Note that $n_1^*(t_0)m_1^*(t_0) + p_1^*(t_0) = m_1^*(T)$. Consequently, we must have that the sign of $l(t_0)$ equals the sign of

$$-e^{\rho t_0} + e^{-\rho T} n_1^*(t_0) - 2\gamma e^{-2\rho T} \Big(m_1^*(t_0)n_2^*(t_0) + k^*(t_0) - n_1^*(t_0)m_1^*(T)\Big),$$

which means that (18) holds, and we have necessarily that

$$0 = -m_1^*(t_0)n_1^*(t_0)e^{-\rho T} - 2\gamma e^{-2\rho T} m_2^*(t_0)n_2^*(t_0)\Big(-1 - \pi^*(t_0)\frac{\sigma(t_0)^2}{\alpha(t_0)-r(t_0)}\Big)$$
$$-2\gamma e^{-2\rho T} m_1^*(t_0)\Big(-k^*(t_0) + n_1^*(t_0)m_1^*(T)\Big),$$

$$\tag{22}$$

which means that (17) is satisfied.

Recalling that $n_1^*(t_0)m_1^*(t_0) + p_1^*(t_0) = m_1^*(T)$, we observe that Eqs. (17) and (18) are equal to Eqs. (19) and (20) in Christiansen and Steffensen [9], which means that the latter equations are not only sufficient but also necessary.

5 Generalized Gradients for the Objective

For differentiable functions on the Euclidean space, a popular method to find maxima is to use the gradient ascent method. We want to follow that variational concept, however our objective is a mapping on a functional space. Therefore, we first need to discuss the definition and calculation of proper gradient functions.

Theorem 3 *Let* $(\pi, c) \in D$ *for* D *as defined in Theorem 1. For each pair of continuous functions* (h, l) *on* $[0, T]$, *we have*

$$\lim_{\delta \to 0} \frac{G(\pi + \delta h, c + \delta l) - G(\pi, c)}{\delta} = \int_0^T h(s)(\nabla_\pi G(\pi, c))(s)ds$$

$$+ \int_0^T l(s)(\nabla_c G(\pi, c))(s)ds$$

with

$$(\nabla_\pi G(\pi, c))(s) = (\alpha(s) - r(s))\Big(m_1(s)n_1(s)e^{-\rho T} - 2\gamma e^{-2\rho T}m_2(s)n_2(s)$$
$$\times \Big(1 + \pi(s)\frac{\sigma(s)^2}{\alpha(s) - r(s)}\Big) - 2\gamma e^{-2\rho T}m_1(s)$$
$$\times \Big(k(s) - n_1(s)m_1(T)\Big)\Big)$$

and

$$(\nabla_c G(\pi, c))(s) = e^{\rho s} - e^{-\rho T}n_1(s)$$
$$+ 2\gamma e^{-2\rho T}\Big(m_1(s)n_2(s) + k(s) - n_1(s)m_1(T)\Big).$$

The limit

$$\lim_{\delta \to 0} \frac{G(\pi + \delta h, c + \delta l) - G(\pi, c)}{\delta} = \frac{d}{d\delta}\Big|_{\delta=0} G(\pi + \delta h, c + \delta l)$$

is the so-called Gateaux derivative (or directional derivative) of the functional G at (π, c) in direction (h, l). Following Christiansen [7], we interpret the two-dimensional function $(\nabla_\pi G(\pi, c), \nabla_\pi G(\pi, c))$ as the gradient of G at (π, c).

Proof (Proof of Theorem 3) In the proof of Theorem 2 we already implicitly showed that

$$G(\pi, c) - G(\pi + h\mathbf{1}_{(t_0-\varepsilon,t_0]}, c + l\mathbf{1}_{(t_0-\varepsilon,t_0]}) = -(\nabla_\pi G(\pi, c))(t_0) \int_{t_0-\varepsilon}^{t_0} h(s)ds$$

$$- (\nabla_c G(\pi, c))(t_0) \int_{t_0-\varepsilon}^{t_0} l(s)ds + o(\varepsilon)$$

for all $t_0 \in [0, T]$, $(\pi, c) \in D$, and $h, l \in C([0, T])$. Defining an equidistant decomposition of the interval $[0, T]$ by

$$\tau_i := \frac{i}{n}T, \quad i = 0, \ldots, n,$$

we can rewrite the difference $G(\pi + \delta h, c + \delta l) - G(\pi, c)$ to

$$G(\pi + \delta h, c + \delta l) - G(\pi, c)$$

$$= \sum_{i=1}^{n} \left(G(\pi + \delta h 1_{[0,\tau_i]}, c + \delta l 1_{[0,\tau_i]}) - G(\pi + \delta h 1_{[0,\tau_{i-1}]}, c + \delta l 1_{[0,\tau_{i-1}]}) \right)$$

$$= \delta \sum_{i=1}^{n} (\nabla_\pi G(\pi + \delta h 1_{[0,\tau_{i-1}]}, c + \delta l 1_{[0,\tau_{i-1}]}))(\tau_i) \int_{\tau_{i-1}}^{\tau_i} h(s) ds$$

$$+ \delta \sum_{i=1}^{n} (\nabla_c G(\pi + \delta h 1_{[0,\tau_{i-1}]}, c + \delta l 1_{[0,\tau_{i-1}]}))(\tau_i) \int_{\tau_{i-1}}^{\tau_i} l(s) ds + \sum_{i=1}^{n} o(T/n)$$

for all $0 < \delta \leq 1$. The moments p_1, p_2, n_1, n_2, k, interpreted as mappings of (π, c) from the domain $B([0, T])^2$ with L_2-norm into the codomain $C([0, T])$ with supremum norm, are continuous. Hence, the gradient functions on the right hand side of the last equation are continuous with respect to the parameters τ_{i-1} and τ_i. Thus, for $n \to \infty$ we obtain

$$\frac{G(\pi + \delta h, c + \delta l) - G(\pi, c)}{\delta} = \int_{0}^{T} (\nabla_\pi G(\pi + \delta h 1_{[0,t]}, c + \delta l 1_{[0,s]}))(s) h(s) ds$$

$$+ \int_{0}^{T} (\nabla_c G(\pi + \delta h 1_{[0,s]}, c + \delta l 1_{[0,s]}))(s) l(s) ds.$$

Since the moment functions p_1, p_2, n_1, n_2, k (interpreted as mappings of (π, c) from the domain $B([0, T])^2$ with supremum-norm into the codomain $C([0, T])$ with supremum norm) are even uniformly continuous, the above gradient functions are uniformly continuous with respect to parameter δ. Thus, for $\delta \to 0$ we end up with the statement of the theorem.

6 Numerical Optimization by a Gradient Ascent Method

With the help of the gradient function $(\nabla_\pi G(\pi, c), \nabla_\pi G(\pi, c))$ of the objective $G(\pi, c)$, we can construct a gradient ascent method. A similar approach is also used in Christiansen [8].

Algorithm

1. Choose a starting control $(\pi^{(0)}, c^{(0)})$.
2. Calculate a new scenario by using the iteration

Fig. 1 Sequence of investment rates $\pi^{(i)}$, $i = 0, \ldots, 40$ calculated by the gradient ascent method. The higher the number i the darker the color of the corresponding graph

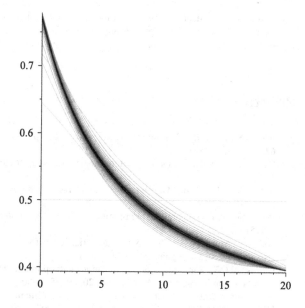

$$\left(\pi^{(i+1)}, c^{(i+1)}\right) := \left(\pi^{(i)}, c^{(i)}\right) + K \left(\nabla_\pi G(\pi^{(i)}, c^{(i)}), \nabla_\pi G(\pi^{(i)}, c^{(i)})\right)$$

where $K > 0$ is some step size that has to be chosen. If $c^{(i+1)}$ is above or below the bounds $\bar{c}$ and $\underline{c}$, we cut it off at the bounds.

3. Repeat step 2 until $\left|G(\pi^{(i+1)}, c^{(i+1)}) - G(\pi^{(i)}, c^{(i)})\right|$ is below some error tolerance.

7 Numerical Example

Here, we demonstrate the gradient ascent method of the previous section with a numerical example. For simplicity, we fix the consumption rate c and only control the investment rate π. We take the same parameters as in Christiansen and Steffensen [9] in order to have comparable results: For the Black-Scholes market we assume that $r = 0.04$, $\alpha = 0.06$ and $\sigma = 0.2$. The time horizon is set to $T = 20$, the initial wealth is $x_0 = 200$, and the savings rate is $a(t) - c(t) = 100 - 80 = 20$. The preference parameter of consuming today instead tomorrow is set to $\rho = 0.1$, and the risk aversion parameter is set to $\gamma = 0.003$.

Starting from $\pi^{(0)} = 0.5$, Fig. 1 shows the converging series of investment rates $\pi^{(i)}$, $i = 0, \ldots, 40$ for $K = 0.2$. The last iteration step $\pi^{(40)}$ perfectly fits the corresponding numerical result in Christiansen and Steffensen [9].

Open Access This chapter is distributed under the terms of the Creative Commons Attribution Noncommercial License, which permits any noncommercial use, distribution, and reproduction in any medium, provided the original author(s) and source are credited.

References

1. Alp, Ö.S., Korn, R.: Continuous-time mean-variance portfolios: a comparison. Optimization **62**, 961–973 (2013)
2. Bäuerle, N., Rieder, U.: Optimal deterministic investment strategies for insurers. Risks **1**, 101–118 (2013)
3. Basak, S., Chabakauri, G.: Dynamic mean-variance asset allocation. Rev. Financ. Stud. **23**, 2970–3016 (2010)
4. Bertsekas, D.P.: Dynamic Programming and Optimal Control. Athena Scientific, Belmont (1995)
5. Björk, T., Murgoci, A., Zhou, X.: Mean-variance portfolio optimization with state dependent risk aversion. Math. Financ. **24**, 1–24 (2014)
6. Cairns, A.J.G.: Some notes on the dynamics and optimal control of stochastic pension fund models in continuous time. ASTIN Bull. **20**, 19–55 (2000)
7. Christiansen, M.C.: A sensitivity analysis concept for life insurance with respect to a valuation basis of infinite dimension. Insur.: Math. Econ. **42**, 680–690 (2008)
8. Christiansen, M.C.: Making use of netting effects when composing life insurance contracts. Eur. Actuar. J. **1**(Suppl 1), 47–60 (2011)
9. Christiansen, M.C., Steffensen, M.: Deterministic mean-variance-optimal consumption and investment. Stochastics **85**, 620–636 (2013)
10. Geering, H.P., Herzog, F., Dondi, G.: Stochastic optimal control with applications in financial engineering. In: Chinchuluun, A., Pardalos, P.M., Enkhbat, R., Tseveendorj, I. (eds.) Optimization and Optimal Control: Theory and Applications, pp. 375–408. Springer, Berlin (2010)
11. Herzog, F., Dondi, G., Geering, H.P.: Stochastic model predictive control and portfolio optimization. Int. J. Theor. Appl. Financ. **10**, 231–244 (2007)
12. Korn, R., Trautmann, S.: Continuous-time portfolio optimization under terminal wealth constraints. Z. für Op. Res. **42**, 69–92 (1995)
13. Korn, R.: Some applications of L^2-hedging with a non-negative wealth process. Appl. Math. Financ. **4**, 64–79 (1997)
14. Kronborg, M.T., Steffensen, M.: Inconsistent investment and consumption problems. Available at SSRN: http://ssrn.com/abstract=1794174 (2011)
15. Kryger, E.M., Steffensen, M.: Some solvable portfolio problems with quadratic and collective objectives. Available at SSRN: http://ssrn.com/abstract=1577265 (2010)
16. Markowitz, H.M.: Portfolio selection. J. Financ. **7**, 77–91 (1952)
17. Menkens, O.: Worst-case scenario portfolio optimization given the probability of a crash. In: Glau, K. et al. (eds.) Innovations in Quantitative Risk Management, Springer Proceedings in Mathematics & Statistics, vol. 99 (2015)
18. Zhou, X., Li, D.: Continuous-time mean-variance portfolio selection: a stochastic LQ framework. Appl. Math. Optim. **42**, 19–33 (2000)

Risk Control in Asset Management: Motives and Concepts

Thomas Dangl, Otto Randl and Josef Zechner

Abstract In traditional portfolio theory, risk management is limited to the choice of the relative weights of the riskless asset and a diversified basket of risky securities, respectively. Yet in industry, risk management represents a central aspect of asset management, with distinct responsibilities and organizational structures. We identify frictions that lead to increased importance of risk management and describe three major challenges to be met by the risk manager. First, we derive a framework to determine a portfolio position's marginal risk contribution and to decide on optimal portfolio weights of active managers. Second, we survey methods to control downside risk and unwanted risks since investors frequently have nonstandard preferences, which make them seek protection against excessive losses. Third, we point out that quantitative portfolio management usually requires the selection and parametrization of stylized models of financial markets. We, therefore, discuss risk management approaches to deal with parameter uncertainty, such as shrinkage procedures or resampling procedures, and techniques of dealing with model uncertainty via methods of Bayesian model averaging.

We thank Victor DeMiguel, Matthias Scherer, Neal Stoughton, Raman Uppal, Arne Westerkamp, Rudi Zagst and an anonymous referee for helpful comments.

T. Dangl
Vienna University of Technology, Vienna, Austria
e-mail: thomas.dangl@tuwien.ac.at

O. Randl · J. Zechner(✉)
WU Vienna University of Economics and Business, Vienna, Austria
e-mail: otto.randl@wu.ac.at

J. Zechner
e-mail: josef.zechner@wu.ac.at

© The Author(s) 2015

239

K. Glau et al. (eds.), *Innovations in Quantitative Risk Management*,
Springer Proceedings in Mathematics & Statistics 99,
DOI 10.1007/978-3-319-09114-3_14

1 Introduction

In traditional portfolio theory the scope for risk management is limited. Wilson [63] showed that in the absence of frictions the consumption allocation of each agent in an efficient equilibrium satisfies a linear sharing rule as long as agents have equi-cautious HARA utilities. This implies that investors are indifferent between the universe of securities and having access to only two appropriately defined portfolio positions, a result that is usually referred to as the Two-Fund Separation Theorem. If a riskless asset exists, then these two portfolios can be identified as the riskless asset and the tangency portfolio. Risk management in this traditional portfolio theory is, therefore, trivial: the portfolio manager only needs to choose the optimal location on the line that combines the riskless asset with the tangency portfolio, i.e., on the capital market line. Risk management is thus equivalent to choosing the relative weights that should be given to the tangency portfolio and to the riskless asset, respectively.

In a more realistic model that allows for frictions, risk management in asset management becomes a much more central and complex component of asset management. First, a world with costly information acquisition will feature informational asymmetries regarding the return moments, as analyzed in the seminal paper by Grossman and Stiglitz [29]. In this setup, investors generally do not hold the same portfolio of risky assets and the two fund separation theorem brakes down (see, e.g., Admati [1]). We will refer to such portfolios as active portfolios. In such a setup, risk management differs from the simple structure described above for the traditional portfolio theory. Second, frictions such as costly information acquisition frequently require delegated portfolio management, whereby an investor transfers decision power to a portfolio manager. This gives rise to principal-agent conflicts that may be mitigated by risk monitoring and portfolio risk control. Third, investors may have nonstandard objective functions. For example, the investor may exhibit large costs if the end-of-period portfolio value falls below a critical level. This may be the case, for example, because investors are subject to their own principal-agent conflicts. Alternatively, investors may be faced with model risk, and thus be unable to derive probability distributions over possible portfolio outcomes. In such a setting investors may have nonstandard preferences, such as ambiguity aversion. We will now discuss each of these deviations from the classical frictionless paradigm and analyze how it affects portfolio risk management.

2 Risk Management for Active Portfolios

If the optimal portfolio differs from the market portfolio, portfolio risk management becomes a much more complicated and important task for the portfolio manager. For active portfolios individual positions' risk contributions are no longer fully determined by their exposures to systematic risk factors that affect the overall market portfolio. A position's contribution to overall portfolio risk must not be measured by the sensitivity to the systematic risk factors, but instead by the sensitivity to the investor's portfolio return. For active portfolios the manager must, therefore, correctly measure each asset's risk-contribution to the overall portfolio risk and ensure that it corresponds to the expected return contribution of the asset. We will now derive a simple framework that a portfolio manager may use to achieve this.

We consider an investor who wishes to maximize his expected utility, $E[\tilde{u}]$. In this section, we consider the case where the investor exhibits constant absolute risk aversion with the coefficient of absolute risk aversion denoted by Γ. In the following derivations, we borrow ideas from Sharpe [61] and assume for convenience that investment returns and their dispersions are small relative to initial wealth, V_0. Thus, we can approximate $\Gamma \cong \gamma / V_0$ with γ denoting the investor's relative risk aversion. This allows for easy translation of the results into the context of later sections, where we focus on relative risk aversion.[1] An expected-utility maximizer with constant absolute risk aversion solves

$$\max_{w} E[\tilde{u}] = \max_{w} E[-\exp(-\Gamma(V_0(1 + w'\tilde{r})))] = \max_{w} E[-\exp(-\gamma(1 + w'\tilde{r}))],$$

(1)

where w represents the $(N \times 1)$ vector of portfolio weights and $\tilde{r}$ is the $(N \times 1)$ vector of securities returns. We make standard assumptions of mean-variance analysis, and denote μ^e as the $(N \times 1)$ vector of securities' expected returns in excess of the risk free rate r_f, $\sigma_p^2(w)$ the portfolio's return variance given weights w, and Σ the covariance matrix of excess returns. $\text{MR} = 2\Sigma w$ constitutes the vector of marginal risk contributions resulting from a marginal increase in portfolio weight of the respective asset, i.e., $\text{MR} = \partial(\sigma_p^2)/\partial w$, financed against the riskless asset. For each asset i in the portfolio we must, therefore, have

$$\mu_i^e = \gamma e_i' \Sigma w = \frac{1}{2} \gamma \text{MR}_i,$$

(2)

which implies

$$\frac{\mu_i^e}{\text{MR}_i} = \frac{\mu_j^e}{\text{MR}_j} = \frac{1}{2}\gamma \quad \forall i, j.$$

(3)

These results show the fundamental difference between risk management for active and passive portfolios. While in the traditional world of portfolio theory, each asset's risk contribution was easily measured by a constant (vector of) beta coefficient(s) to the systematic risk factor(s), the active investor must measure a security's risk contribution by the sensitivity of the asset to the specific portfolio return, expressed by $2e_i'\Sigma w$. This expression makes clear that each position's marginal risk contribution depends not only on the covariance matrix Σ, but also on the portfolio weights, i.e., the chosen vector w. It actually converges to the portfolio variance, σ_p^2, as the security's weight approaches one. In the case of active portfolios, these weights are likely to change over time, and so will each position's marginal risk

---

[1] See, e.g., Pennacchi [55] for more details on this assumption.

contribution. The portfolio manager can no longer observe a position's relevant risk characteristics from readily available data providers such as the stock's beta reported by Bloomberg, but must calculate the marginal risk contributions based on the portfolio characteristics. As shown in Eq. (3), a major responsibility of the portfolio risk manager now is to ensure that the ratios of securities' expected excess returns over their marginal risk contribution are equated.

2.1 Factor Structure and Portfolio Risk

A prevalent model of investment management in practice features a CIO who decides on the portfolio's asset allocation and on the allocation between passively or actively managed mandates within each asset class. The actual management of the positions within each asset class is then delegated to external managers. In the following we provide a consistent framework within which such a problem can be analyzed. We hereby assume a linear return generating process so that the vector of asset excess returns, r^e can be written as

$$r^e = \alpha + Bf^e + \epsilon, \tag{4}$$

where

- r^e is the $(N \times 1)$ vector of fund or manager returns in excess of the risk free return
- B is a $(N \times K)$ matrix that denotes the exposure of each of the N assets to the K return factors
- f^e is a $(K \times 1)$ vector of factor excess returns and
- ϵ is the error term (independent of f^e).

Let Σ_f denote the covariance matrix of factor excess returns and Ω the covariance matrix of residuals, ϵ. Then, the covariance matrix of managers' excess returns Σ is given by

$$
\begin{aligned}
\Sigma &= E(r^e r^{e\prime}) \\
&= E([Bf^e + \epsilon][Bf^e + \epsilon]') \\
&= E(Bf^e f^{e\prime} B' + \epsilon\epsilon') \\
&= BE(f^e f^{e\prime})B' + E(\epsilon\epsilon') \\
&= B\Sigma_f B' + \Omega.
\end{aligned}
$$

Let w denote the $N \times 1$ vector of weights assigned to managers by the CIO, then the portfolio excess return r_p^e is given by

$$r_p^e = w'r^e.$$

If e_i is the ith column of the $(N \times N)$ identity matrix then

$$
\begin{aligned}
\mathrm{Cov}(r_i^e, r_p^e) &= \mathrm{Cov}(e_i' Bf^e + e_i'\epsilon, \ w'Bf^e + w'\epsilon) \\
&= \mathrm{Cov}(e_i' Bf^e, \ w'Bf^e) + \mathrm{Cov}(e_i'\epsilon, \ w'\epsilon) \\
&= e_i' B\Sigma_f B'w + e_i'\Omega w.
\end{aligned}
$$

The beta of manager i's return with respect to the portfolio is then

$$
\tilde{\beta}_i = \frac{e_i' B\Sigma_f B'w + e_i'\Omega w}{w'(B\Sigma_f B' + \Omega)w}.
$$

Thus, we have an orthogonal decomposition of the vector of betas, $\tilde{\beta}$, into a part that is due to factor exposure, $\tilde{\beta}^S$, and a part that is due to the residuals of active managers (tracking error), $\tilde{\beta}^I$

$$
\tilde{\beta} = \frac{B\Sigma_f B'w + \Omega w}{w'(B\Sigma_f B' + \Omega)w} = \underbrace{\frac{B\Sigma_f B'w}{w'(B\Sigma_f B' + \Omega)w}}_{\tilde{\beta}^S} + \underbrace{\frac{\Omega w}{w'(B\Sigma_f B' + \Omega)w}}_{\tilde{\beta}^I}.
$$

We can now determine the beta of a pure factor excess return f_k^e to the portfolio. With e_k^F denoting the kth column of the $(K \times K)$ identity matrix, the covariance between the factor excess return and the portfolio excess return is

$$
\begin{aligned}
\mathrm{Cov}(f_k^e, r_p^e) &= \mathrm{Cov}(e_k^{F'} f^e, \ w'Bf^e + w'\epsilon) \\
&= e_k^{F'} \Sigma_f B'w.
\end{aligned}
$$

The vector of pure factor betas, $\tilde{\beta}^F$, to the portfolio is therefore

$$
\tilde{\beta}^F = \frac{\Sigma_f B'w}{w'(B\Sigma_f B' + \Omega)w}
$$

We thus have $\tilde{\beta}^S = B\tilde{\beta}^F$. Consequently, a position's beta to the portfolio can be written as

$$
\tilde{\beta} = B\tilde{\beta}^F + \beta^I.
$$

i.e., we can decompose the position's beta into the exposure-weighted betas of the pure factor returns plus the beta of the position's residual return.

Next we can derive the vector of marginal risk contributions of the portfolio positions. Given the factor structure above, the effect of a small change in portfolio weights, w, on portfolio risk, σ_p^2 is given by MR:

$$\frac{1}{2}\text{MR} = \frac{1}{2}\frac{\partial}{\partial w}w'\Sigma w = \Sigma w$$
$$= \sigma_p^2\tilde{\beta} = \sigma_p^2(B\tilde{\beta}^F + \tilde{\beta}^I)$$

Thus, an individual portfolio position i's marginal risk contribution, MR_i, is given by

$$\frac{1}{2}\text{MR}_i = \frac{1}{2}\frac{\partial}{\partial w_i}w'\Sigma w$$
$$= e_i'\Sigma w = \sigma_p^2\tilde{\beta}_i$$
$$= \sigma_p^2(B\tilde{\beta}_i^F + \tilde{\beta}_i^I). \tag{5}$$

2.2 Allocation to Active and Passive Funds

One important objective of risk control in a world with active investment strategies is to ensure that an active portfolio manager's contribution to the portfolio return justifies his idiosyncratic risk or "tracking error". If this is not the case, then it is better to replace the active manager with a passive position that only provides a pure factor exposures but no idiosyncratic risks. To analyze this question we define ν^e as the vector of expected excess returns of the factor-portfolios and assume without loss of generality $\nu^e > 0$. Then, the vector of expected portfolio excess returns can be written as

$$\mu^e = E(\alpha + Bf^e + \epsilon) = \alpha + B\nu^e. \tag{6}$$

The first order optimality condition (3) states that the portfolio weight assigned to manager i should not be reduced as long as it holds that:

$$\frac{\mu_i^e}{\text{MR}_i} \geq \frac{\mu_j^e}{\text{MR}_j}, \quad \forall j.$$

Substituting marginal risk contribution from (5) and expected return from (6) into the above relation, we conclude that a manager i with $\text{MR}_i > 0$ justifies her portfolio weight relative to a pure factor investment in factor k iff

$$\frac{\mu_i^e}{E(f_k^e)} = \frac{e_i'BE(f^e) + \alpha_i}{E(f_k^e)} = \frac{e_i'B\nu^e + \alpha_i}{\nu_k^e} \geq \frac{\tilde{\beta}_i}{\tilde{\beta}_k^F} = \frac{e_i'B\tilde{\beta}^F + e_i'\tilde{\beta}^I}{\tilde{\beta}_k^F}.$$

Consider the case where asset manager i has exposure only to factor k, denoted by $B_{i,k}$. Then, this manager justifies her capital allocation iff

$$\frac{B_{i,k}\nu_k^e + \alpha_i}{\nu_k^e} \geq \frac{B_{i,k}\tilde{\beta}_k^F + \tilde{\beta}_i^I}{\tilde{\beta}_k^F}$$

$$B_{i,k} + \frac{\alpha_i}{\nu_k^e} \geq B_{i,k} + \frac{\tilde{\beta}_i^I}{\tilde{\beta}_k^F}$$

$$\alpha_i \geq \frac{\tilde{\beta}_i^I}{\tilde{\beta}_k^F}\nu_k^e.$$

Note that in general this condition depends on the portfolio weight. For sufficiently small weights w_i, manager i's tracking error risk will be "non-systematic" in the portfolio context, i.e., $\tilde{\beta}_i^I = 0$. However, as manager i's weight in the portfolio increases, his tracking error becomes "systematic" in the portfolio context. Therefore, the manager's hurdle rate increases with the portfolio weight. This is illustrated in Example 1.

Example 1 Consider the special case where there is only one single factor and a portfolio, which consists of a passive factor-investment and a single active fund. The portfolio weight of the passive investment is denoted by w_1 and that of the active fund by w_2. The active fund is assumed to have a beta with respect to the factor denoted by β and idiosyncratic volatility of σ_I.[2]

The covariance of factor returns is then a simple scalar equal to the factor return variance, the matrix of factor exposures B has dimension (2×1) and the idiosyncratic covariance matrix is (2×2)

$$w = \begin{pmatrix} 1 - w_2 \\ w_2 \end{pmatrix}, \quad \Sigma_f = \sigma_\nu^2, \quad B = \begin{pmatrix} 1 \\ \beta \end{pmatrix}, \quad \Omega = \begin{pmatrix} 0 & 0 \\ 0 & \sigma_I^2 \end{pmatrix}.$$

The usual assumption $\nu^e > 0$, $\sigma_I^2 > 0$ applies. The hurdle to be met by the alpha of the active fund is accordingly given by

$$\alpha \geq H(w_2) = \frac{\tilde{\beta}_2^I}{\tilde{\beta}^F}\nu^e = \frac{\sigma_I^2 w_2}{\sigma_\nu^2(1 - (1 - \beta)w_2)}\nu^e.$$

The derivative of this hurdle with respect to the weight of the active fund w_2 is

$$\frac{dH}{dw_2} = \frac{\sigma_I^2}{\sigma_\nu^2}\nu^e\frac{1}{(1 - (1 - \beta)w_2)^2} > 0,$$

i.e., the hurdle $H(w_2)$ has a strictly positive slope, thus, the higher the portfolio weight w_2 of an active fund, the higher is the required α it must deliver. This is so because with low portfolio weight, the active fund's idiosyncratic volatility is almost orthogonal to

---

[2] Note that β is the linear exposure of the fund to the factor. It is a constant and independent of portfolio weights. In contrast, betas of portfolio constituents relative to the portfolio, $\tilde{\beta}^F$ and $\tilde{\beta}^I$, depend on weights.

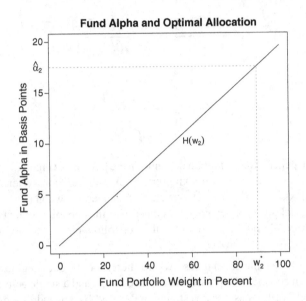

Fig. 1 Minimum alpha justifying portfolio weights

the portfolio return, and so its contribution to the overall portfolio risk is low. When in contrast the active fund has a high portfolio weight, its idiosyncratic volatility already co-determines the portfolio return and is—in the portfolio's context—a systematic component. The marginal risk contribution of the fund is then larger and consequently demands a higher compensation, translating into an upward-sloping α-hurdle.

Take as an example JPMorgan Funds—Highbridge US STEEP, an open-end fund incorporated in Luxembourg that has exposure primarily to U.S. companies, through the use of derivatives. Using monthly data from 12/2008 to 12/2013 (data source: Bloomberg), we estimate

$$\hat{\Sigma}_f = \hat{\sigma}_\nu^2 = 0.002069, \quad \hat{B} = \begin{pmatrix} 1 \\ 0.9821 \end{pmatrix}, \quad \hat{\Omega} = \begin{pmatrix} 0 & 0 \\ 0 & 0.000303 \end{pmatrix}.$$

Furthermore, we use the historical average of the market risk premium $\bar{\nu} = 0.013127$, and the fund's estimated alpha $\hat{\alpha} = 0.001751$. The optimal allocation is the vector of weights w^* such that the marginal excess return divided by the marginal risk contribution is equal for both assets in the portfolio. The increasing relationship between alpha and optimal fund weight is illustrated in Fig. 1. At the estimated alpha of 17.51 basis points, the optimal weights are given by

$$w^* = \begin{pmatrix} 0.1029 \\ 0.8971 \end{pmatrix}.$$

3 Dealing with Investors Downside-Risk Aversion

When discussing investor's utility optimization in Sect. 2, we referred to literature showing that under fairly general assumptions optimal static sharing rules are linear in the investment's payoff, i.e., optimal risk sharing implies holding a certain fraction of a risky investment rather than negotiating contracts with nonlinear payoffs. In a dynamic context, Merton [51] derives an optimal savings-consumption rule that is also in accordance with this finding. Consider a continuous-time framework with a single risky and a riskless asset, where the investor can change the allocation w_t to the risky asset over time. When the risky asset follows a geometric Brownian motion with drift μ and volatility σ, and utility exhibits a constant relative risk aversion γ, then the optimal allocation to the risky asset is constant over time and can be described as $w_t = \mu^e/(\gamma\sigma^2)$. This means with constant investment opportunities (μ and σ constant over time) investors keep the proportions of the risky and risk-free assets in the portfolio unchanged over time. To keep weights constant, portfolio rebalancing requires buying the risky asset when it decreases in value and selling it with increasing prices.

While these theoretical results suggest that an investor should not avoid exposure to risky investments even after sharp draw-downs of her portfolio's value, financial intermediaries face strong demand for products that provide portfolio insurance. That is, investors seem to have considerable downside-risk aversion. Rebalancing to constant portfolio weights is in clear contrast to portfolio insurance strategies, where the allocation to the risky asset has to be decreased if it falls in value, and the risky asset will be purchased in response to price increases. Perold and Sharpe [56] note that these opposing rebalancing rules lead to different shapes of strategy payoff curves. Buying stocks as they fall (as in the Merton model) leads to concave payoff curves. Such strategies do well in flat but oscillating markets, as assets are bought cheaply and sold at higher prices. However, in persistent downmarkets losses are aggravated from buying ever more stocks as they fall. Portfolio insurance rebalancing rules prescribe the opposite: selling stocks as they fall. This limits the impact of persistent down markets on the final portfolio value and at the same time keeps the potential of upmarkets intact, leading to a convex payoff profile. Yet if markets turn out flat but oscillating, convex strategies perform poorly.

3.1 Portfolio Insurance

In this paper, we define portfolio insurance as a dynamic investment strategy that is designed to limit downside risk. The variants of portfolio insurance are, therefore, popular examples of convex strategies. The widespread use of portfolio insurance strategies among both individual and institutional investors indicates that not all market participants are equally capable of bearing the downside risk associated with their average holding of risky assets. Individual investors might be subject to habit formation or recognition of subsistence levels that define a minimum level of wealth required. For corporations, limited debt capacity makes it impossible to benefit from profitable investment projects if wealth falls below a critical value. Furthermore, kinks in the utility function could originate in agency problems, e.g., career concerns of portfolio managers, who see fund flows and pay respond in an asymmetric way to performance. In the literature on portfolio insurance, Leland [47] has stated the prevalence of convex over concave strategies for an investor whose risk aversion decreases in wealth more rapidly than for the representative agent. Alternatively, portfolio insurance strategies should be demanded by investors with average risk tolerance, but above average return expectations. Leland argues that insured strategies allow such an optimistic investor to more fully exploit positive alpha situations through greater levels of risky investment, while still keeping risk within manageable bounds.

Brennan and Solanki [14] contrast this analysis and derive a formal condition for optimality of an option like payoff that is typical for portfolio insurance. It can be shown that a payoff function where the investor receives the maximum of the reference portfolio's value and a guaranteed amount is optimal only under the stringent conditions of a zero risk premium and linear utility for wealth levels in excess of the guaranteed amount. Similarly, Benninga and Blume [9] argue that in complete markets utility functions consistent with optimality of portfolio insurance would have to exhibit unrealistic features, like unbounded risk aversion at some wealth level. However, they make the point that portfolio insurance can be optimal if markets are not complete. An extreme example of market incompleteness in this context, which makes portfolio insurance attractive, is the impossibility for an investor to allocate funds into the risk-free asset. Grossman and Vila [30] discuss portfolio insurance in complete markets, noting that the solution of an investor's constrained portfolio optimization problem (subject to a minimum wealth constraint $V_T > K$) can be characterized by the solution of the unconstrained problem plus a put option with exercise price K. More recently, Dichtl and Drobetz [19] provide empirical evidence that portfolio insurance is consistent with prospect theory, introduced by Kahneman and Tversky [41]. Loss-averse investors seem to use a reference point to evaluate portfolio gains and losses. They experience an asymmetric response to increasing versus decreasing wealth, in being more sensitive to losses than to gains. In addition, risk aversion also depends on the current wealth level relative to the reference point. The model by Gomes [27] shows that the optimal dynamic strategy followed by loss-averse investors can be consistent with portfolio insurance.[3]

---

[3] It is interesting to study the potential effects of portfolio insurance on the aggregate market. As our focus is the perspective of a risk-manager who does not take into account such market-wide effects of his actions, we do not cover this literature. We refer the interested reader to Leland and Rubinstein [46], Brennan and Schwartz [13], Grossman and Zhou [32] and Basak [6] as a starting point.

3.2 Popular Portfolio Insurance Strategies

The main portfolio insurance strategies used in practice are stop-loss strategies, option-based portfolio insurance, constant proportion protfolio insurance, ratcheting strategies with adjustments to the minimum wealth target, and value-at-risk based portfolio insurance.

3.2.1 Stop-Loss Strategies

The simplest dynamic strategy for an investor to limit downside risk is to protect his investment using a stop-loss strategy. In this case, the investor sets a minimum wealth target or floor F_T, that must be exceeded by the portfolio value V_T at the investment horizon T. He then monitors if the current value of the portfolio V_t exceeds the present value of the floor $\exp(-r_f(T-t))F_T$, where r_f is the riskless rate of interest. When the portfolio value reaches the present value of the floor, the investor sells the risky and buys the riskfree asset. While this strategy has the benefit of simplicity, there are several disadvantages. First, due to discreteness of trading or illiquidity of assets, the transaction price might be undesirably far below the price triggering portfolio reallocation. Second, once the allocation has switched into the riskfree asset the portfolio will grow deterministically at the riskfree rate, making it impossible to even partially participate in a possible recovery in the price of the risky asset.

3.2.2 Option-Based Portfolio Insurance

Brennan and Schwartz [12] and Leland [47] describe that portfolio insurance can be implemented in two eqivalent ways: (1) holding the reference portfolio plus a put option, or (2) holding the riskfree asset plus a call option. When splitting his portfolio into a position S_0 in the risky asset and P_0 in a protective put option at time $t = 0$, the investor has to take into account the purchase price of the option when setting the exercise price K, solving $(S_0 + P_0(K)) \cdot (F_T/V_0) = K$ for K. The ratio F_T/V_0 is the minimum wealth target expressed as a fraction of initial wealth. If such an option is available on the market it can be purchased and no further action is needed over the investment horizon; alternatively such an option can be synthetically replicated as popularized by Rubinstein and Leland [58]. Again, the risky asset will be bought on price increases and sold on falling prices, but in contrast to the stop-loss strategy, changes in the portfolio allocation will now be implemented smoothly. Even after a fall in the risky asset's price there is scope to partially participate in an eventual recovery as long as Delta is strictly positive. Toward the end of the investment horizon, Delta will generally be very close to either zero or one, potentially leading to undesired portfolio switching if the risky asset fluctuates around the present value of the exercise price.

3.2.3 Constant Proportion Portfolio Insurance

In order to provide a simpler alternative to the option replication approach described above, Black and Jones [10] propose CPPI for equity portfolios. Black and Perold [11] describe properties of CPPI and propose a kinked utility function for which CPPI is the optimal strategy. Implementation of CPPI starts with calculation of the *cushion* $C_t = V_t - F_t$, which is the amount by which the current portfolio value V_t exceeds the present value of the minimum wealth target $(F_t = \exp(-r_f(T - t))F_T)$. Thus, the cushion can be interpreted as the risk capital available at time t. The exposure E_t to the risky asset is determined as a constant multiple m of the cushion C_t, while the remainder is invested risk free. To avoid excessive leverage, exposure will typically be determined subject to the constraint of a maximum leverage ratio l, hence $E_t = \min\{m \cdot C_t, l \cdot V_t\}$. If the portfolio is monitored in continuous time, the portfolio value at time T cannot fall below F_T. However, discrete trading in combination with sudden price jumps could lead to a breach of the minimum wealth target (gap risk).

3.2.4 Ratcheting Strategies

The portfolio insurance strategies discussed so far limit the potential shortfall from the start of the investment period to its end, frequently a calendar year. But investors may also be concerned with losing unrealized profits that have been earned within the year. Estep and Kritzman [23] propose a technique called TIPP (time invariant portfolio insurance) as a simple way of achieving (partial) protection of interim gains in addition to the protection offered by CPPI. Their methodology adjusts the floor F_t used to calculate the cushion C_t over time. The TIPP floor is set as the maximum of last period's floor and a fraction k of the current portfolio value: $F_t = \max(F_{t-1}, kV_t)$. This method of *ratcheting the floor up* is time invariant in the sense that the notion of a target date T is lost. However, if the percentage protection is required with respect to a specific target date, the method can be easily adjusted by setting a target date floor F_T proportional to current portfolio value V_t, which is then discounted. Grossman and Zhou [31] provide a formal analysis of portfolio insurance with a rolling floor, while Brennan and Schwartz [13] characterize a complete class of time-invariant portfolio insurance strategies, where asset allocation is allowed to depend on current portfolio value, but is independent of time.

3.2.5 Value-at-Risk-Based Portfolio Insurance

In a broader context, Value-at-Risk (VaR) has emerged as a standard for measurement and management of financial market risk. VaR has to be specified with confidence a and horizon Δt and is the loss amount that will be exceeded only with probability $(1 - a)$ over the time span Δt. It is, therefore, a natural measure to control portfolio drawdown risk. The typical definition of VaR assumes that over the time horizon no adjustments are made to the portfolio. Yet, if under adverse market movements risk reducing transactions are implemented, VaR is likely to overestimate actual losses, making portfolio insurance even more effective. On the other hand, poor estimation of the return distribution will lead to bad quality of the VaR estimate. Herold et al. [35, 36] describe a VaR-based method for controlling shortfall risk. The allocation to the risky asset is chosen such that the VaR equals the prespecified minimum return. They note that their method can be seen as a generalized version of CPPI with a dynamic multiplier $m_t = 1/(\Phi^{-1}(a)\sqrt{\Delta t}\sigma_t)$, where $\Phi^{-1}(a)$ is the a-percentile of the standard normal distribution, and σ_t is the volatility of the reference portfolio. Typically, market volatility increases when markets crash, leading to a more pronounced reduction of the allocation to the risky asset as both the cushion and the multiplier shrink. This offers the potential advantage of VaR-based risk control that if markets calm, the allocation to the risky asset will increase again, allowing the portfolio to benefit from a recovery. Basak and Shapiro [7] take a critical view on VaR-based risk management: Strictly interpreting VaR as a risk quantile, managers could be inclined to deliberately assume extreme risks if they are not penalized for the severity of losses that occur with a probability less than $1 - a$. However, in a portfolio insurance context this could be easily fixed, e.g., by restrictions on assuming tail risks.

3.3 Performance Comparison

Benninga [8] uses Monte Carlo simulation techniques to compare stop-loss, OBPI, and CPPI. Surprisingly, he finds that stop-loss dominates with respect to terminal wealth and Sharpe ratio. Dybvig [21] considers asset allocation and portfolio payouts in the context of endowment management. If payouts are not allowed to decrease, CPPI exhibits more desirable properties than constant mix strategies. Balder et al. [4] analyze risks associated with implementation of CPPI under discrete-time trading and transaction costs. Zagst and Kraus [64] compare OBPI and CPPI with respect to stochastic dominance. Taking into account that implied volatility—which is relevant for OBPI—is usually higher than realized volatilty—relevant for CPPI—they find that under specific parametrizations CPPI dominates. Recently, Dockner [20] compares buy-and-hold, OBPI and CPPI concluding that there does not exist a clear ranking of the alternatives. Dichtl and Drobetz [19] consider prospect theory (Kahneman and Tversky [41]) as framework to evaluate portfolio insurance strategies. They use a twofold methodological approach: Monte Carlo simulation and historical simulation with data for the German stock market. Within the behavioral finance context chosen, their findings provide clear support for the justification of downside protection strategies. Interestingly, in their study stop-loss, OBPI and CPPI turn out attractive while the high protection level of TIPP associated with opportunity costs in terms of reduced upside potential turns out to be suboptimal. Finally, they recommend to implement CPPI aggressively by using the highest multiplier m consistent with tolerance for overnight or gap risk.

Example 2 In 4 out of the 18 calendar years from 1995 to 2013, the S&P 500 total return index lost more than 5 %. For investors with limited risk capacity it was not helpful that these losses happened three times in a row (2000, 2001, and 2002), or were severe (2008). The following example illustrates how simple versions of common techniques to control downside risk have performed over these 18 years. We assume investment opportunities in the S&P 500 index and a risk-free asset, an investment horizon equal to the calendar year, and a frictionless market (no transaction costs). Each calendar year the investment starts with a January 1st portfolio value of 100. Rebalancing is possible with daily frequency. For the portfolio insurance strategies investigated, the desired minimum wealth is given with 95, and free parameters are set in a way to make the strategies comparable, by ensuring equal equity allocations at portfolio start. This is achieved by resetting the multiples *m* for CPPI and TIPP each January 1st according to the Delta of the OBPI strategy. Similarly, the VaR confidence level is set to achieve this same equity proportion at the start of the calendar year. OBPI Delta also governs the initial equity portion of the buy-and-hold portfolio. Table 1 reports the main results, and Fig. 2 summarizes the distribution of year-end portfolio values in a box plot.

The achieved minimum wealth levels show that for CPPI, TIPP, OBPI, and VaR-based portfolio insurance even in the worst year the desired minimum wealth has been missed just slightly, while in the case of the stop loss strategy there is a considerable gap. This can be partly explained by the simple setup of the eample (e.g., rebalancing using daily closing prices only, while in practice intraday decision-making and trading will happen). But a possibly large gap between desired and achieved minimum wealth is also systematic of stop loss strategies because of the mechanics of stop-loss orders. The moment the stop limit is reached, a market order to sell the entire portfolio is executed. The trading price, therefore, can and frequently will be lower than the limit. This can pose considerable problems in highly volatile and illiquid market environments. Option replication comes next in missing desired wealth protection.

Table 1 Portfolio insurance strategies

	Mean	Median	SD all	SD lower	Min	Max	Turnover
Long only	110.92	113.91	19.29	15.55	63.91	137.59	0.00
Buy & hold	108.32	107.15	12.52	9.18	79.99	132.11	0.00
Stop loss	108.77	105.77	16.09	6.52	89.13	137.59	0.42
CPPI	107.71	104.77	12.52	2.82	94.62	136.89	4.58
TIPP	105.31	104.20	7.45	3.29	94.63	122.75	1.03
OBPI	108.50	105.21	12.43	4.29	95.00	135.58	3.63
Option repl.	108.84	107.07	11.93	4.72	92.04	132.59	3.64
VaR	108.21	104.15	13.21	2.64	94.79	137.59	8.16

Comparison of portfolio insurance strategies, annual horizon, S&P 500, 1995–2013. We report end-of-year wealth levels per investment of 100 (mean, median, min, max); standard deviations calculated both over the whole sample (SD all) and for the subsample where the annual S&P 500 total return is below its mean (SD lower); turnover in the annual turnover ratio

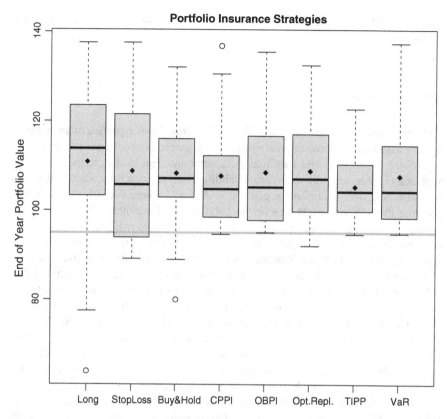

Fig. 2 Comparison of portfolio insurance strategies, annual horizon, S&P 500, 1995–2013. For each strategy, the *shaded area* indicates the observations from the 25th to the 75th percentile, the median is shown as the *line* across the box and the mean as a *diamond* within the box. The whiskers denote the lowest datum still within 1.5 interquartile range of the lower quartile, and the highest datum still within 1.5 interquartile range of the upper quartile. If there are more extreme observations they are shown separately by a *circle*. The semitransparent *horizontal line* indicates the desired minimum wealth level

In the example, this might be due the simplified setup, where the exercise price of the option to be replicated is determined only once per year (at year start), and then daily Delta is calculated for this option and used for allocation into the risky and the riskless asset. In practice, new information on volatility and the level of interest rates will also lead to a reset of the strike used for calculation of the Delta. Another observation is that the standard deviation of annual returns is lowest for TIPP, which comes at the price of the lowest average return. If the cross-sectional standard deviation is computed only for the years with below-average S&P 500 returns, it is lowest for VaR-based risk control. For all methods shown, practical implementation will typically use higher levels of sophistication. For example, trading filters will be applied to avoid adjusting portfolios as frequently as in the example leading to high turnover values.

3.4 Other Risks

In the previous discussion, shortfall risk was seen from the perspective of an investor holding assets only. However, many institutional investors simultaneously optimize a portfolio of assets A and liabilities L. Sharpe and Tint [62] describe a flexible approach to systematically incorporate liabilities into pension fund asset allocation, by optimizing over a surplus measure $S = A - kL$, where $k \in [0, 1]$ is a factor denoting the relative weight attached to liabilities. In the context of asset liability management, Ang et al. [2] analyze the effect of downside risk aversion, and offer an explanation why risk aversion tends to be high when the value of the assets approaches the value of the liabilities. Ang et al. [2] specify the objective function of the fund as mean-variance over asset returns plus a downside risk penalty on the liability shortfall that is proportional to the value of an option to exchange the optimal portfolio for the random value of the liabilities. An investor following their advice tends to be more risk averse than a portfolio manager implementing the Sharpe and Tint [62] model. For very high funding ratios, the impact of downside risk on risk taking, and therefore the asset allocation of the pension fund manager is small. For deeply underfunded plans, the value of the option is also relatively insensitive to changes in volatility, again leading to a small impact on asset allocation. The effect of liabilities on asset allocation is strongest when the portfolio value is close to the value of liabilities. In this case, lower volatility reduces the value of the exchange option, leading to a smaller penalty.

Another hedging motive arises if investors wish to bear only specific risks. This might be due to specialization of the investor in a certain asset class, making it desirable to hedge against risks not primarily driving the returns of this asset class. A popular example is currency risk, which has been recently analyzed by Campbell et al. [15] who find full currency hedging to be optimal for a variance-minimizing bond investor, but discuss the potential for overall risk reduction from keeping foreign exchange exposure partly unhedged in the case of equity portfolios.

4 Parameter Uncertainty and Model Uncertainty

Quantitative portfolio management builds on optimization output of stylized models, which (i) need to be carefully chosen to capture relevant features of the market framework and (ii) must be calibrated and parameterized. These choices, model selection, as well as model calibration, bear the risk of misspecification, which might have severely negative consequences on the desired out-of-sample properties of the portfolio. Thus, a main application of risk management in asset management is controlling the risk inherent in model specification and parameter selection. In this section, we distinguish between parameter uncertainty and model uncertainty in the following way. With *parameter uncertainty* we refer to the case where we know the structure of the data generating process that lies behind the observed set of data, but the parameters of the process must be empirically determined.[4] Finite data history is the only limiting factor, which prevents us from deriving the true values of the model parameters. Under the assumption of the null hypothesis, we can derive the joint distribution of the estimated parameters relative to the true values, and finally the joint predictive distribution of asset returns under full consideration of estimation problems. Thus, we can treat parameter uncertainty simply as an additional source of variability in returns. It is noncontroversial to assume that a decision-maker does not distinguish between uncertainty in returns caused by the general variability of returns and uncertainty that has its origin in estimation problems, and hence the portfolio optimization paradigm is not affected.

In contrast, with *model uncertainty* we refer to the case where a decision-maker is not sure, which model is the correct formulation that describes the underlying dynamics of asset returns. In such a case, it is generally not possible to specify probabilities for the models considered as feasible. Thus, model uncertainty increases uncertainty about asset returns, but we are not able to state a definite probability distribution of returns, which incorporates model uncertainty. That is, model uncertainty is a prototypical case of Knightian uncertainty, referring to Knight [42], where it is not possible to characterize the uncertain entity (in our case the asset return) by means of a probability distribution. Consequently, model uncertainty fundamentally changes the decision-making framework and we have to make assumptions regarding a decision-maker's preferences concerning situations of ambiguity.

4.1 Parameter Uncertainty

The most obvious estimation problem in a traditional minimum-variance portfolio optimization task arises when determining the covariance structure of asset returns. This is so because estimates of the sample covariance matrix turn out to be weakly conditioned in general and—as soon as the number of assets is larger than the number of periods considered in the return history —the sample covariance matrix is singular by construction.

Example 3 Consider as a broad asset universe the S&P 500 with $N = 500$ constituents. It is common practice to estimate the covariance structure of stock returns from two years of weekly returns. The argument for a restriction of the history to $T = 104$ weeks is a reaction to the fact that there is apparently some time-variation

---

[4] We assume in general, that the model has a structure, which ensures that parameters are identifiable. For example, it is assumed that log-returns are normally distributed, but mean and variance must be estimated from observed data.

in the covariance structure, which the estimate is able to capture only if one restricts the used history.[5]

Let r denote the $(T \times N)$ matrix containing weekly returns, then the sample covariance matrix $\hat{\Sigma}_S$ is determined by

$$\hat{\Sigma}_S = \frac{1}{T-1} r' M r, \tag{7}$$

where the symmetric and idempotent matrix M is the residual maker with respect to a regression onto a constant,

$$M = \mathbb{I} - \mathbf{1} (\mathbf{1}' \mathbf{1})^{-1} \mathbf{1}',$$

with $\mathbb{I}$ the $(T \times T)$ identity matrix and $\mathbf{1}$ a column vector containing T times the constant 1.

In the assumed setup, the sample covariance matrix is singular by construction. This is so because from (7) it follows that the rank of $\hat{\Sigma}_S$ is bounded from above by $\min\{N, T - 1\}$.[6] And even in the case where the number of return observations per asset exceeds the number of assets $(T > N + 1)$ the sample covariance matrix is weakly determined, hence, subject to large estimation errors since one has to estimate $N(N + 1)/2$ elements of $\hat{\Sigma}_S$ from $T \cdot N$ observations.

Since a simple Markowitz optimization, see Markowitz [49], needs to invert the covariance matrix, matrix singularity prohibits any attempt of advanced portfolio optimization, and is thus the most evident estimation problem in portfolio management. Elton and Gruber [22] is an early contribution, which proposes the use of structural estimators of the covariance matrix. Jobson and Korkie [38] provide a rigorous analysis of the small sample properties of estimates of the covariance structure of returns.

Less evident are the problems caused by errors in the estimates of return expectations, whereas it turns out that they are economically much more critical. Jorion [39] shows in the context of international equity portfolio selection that the errors in the estimates of return expectations have a severe impact on the out-of-sample

---

[5] Such an approach is typical for dealing with inadequate model specification. The formal estimate is based on the assumption that the covariance structure is stable. Since data show that the covariance structure is not stable, an ad-hoc adaptation—the limitation of the data history—is used to capture the recent covariance structure. The optimal amount of historical data that should be used cannot be derived within the model, but must be roughly calibrated to some measure of goodness-of-fit, which balances estimation error against timely response to time variations.

[6] The residual maker M has at most rank $T - 1$ because it generates residuals from a projection onto a one-dimensional subspace of $\mathbb{R}^T$. Since r has at most rank N, we have

$$\text{rank}(\hat{\Sigma}_S) \leq \min\{N, T - 1\}.$$

For example, the sample covariance matrix estimated from two years of weekly returns of the 500 constituents of the S&P 500 (104 observations per stock) has at most rank 103. Hence, it is not positive definite and not invertible, because at least 397 of its 500 eigenvalues are exactly equal 0.

performance of optimized portfolios. He further shows that the Bayes-Stein shrinkage approach introduced in Jorion [40] helps mitigate errors and at the same time improves out-of-sample properties of the portfolio.

Structural Estimators Means and covariances of asset returns are the most basic inputs into a portfolio optimization model. However, estimation errors in further model parameters like some measure of risk aversion, speed of reversion to long-term averages, etc., must be estimated from empirical data and are, thus, equally likely inflicted with estimation errors. While sample estimates of distribution means, (co-)variances and higher moments are generally unbiased and efficient, they tend to be noisy. This can be improved by imposing some sort of structure on the estimated parameters. Such structural estimates are less prone to estimation errors at the expense of ignoring part of the information inherent in the observed data sample. When determining the covariance structure of asset returns, Elton and Gruber [22] analyze a set of different structural assumptions, e.g., what they call the single index model (assuming that the pairwise covariance of asset returns is only generated by the assets individual correlation to a market index), the mean model (pairwise correlations between assets are assumed constant across the asset universe), and models that assume that the correlation structure of asset returns is determined by within industry averages or across industry averages or by a (small) number of principal components of the sample covariance matrix. They show that especially the particularly restrictive estimates (single index model and mean model) deliver forecasts of future correlation that are more accurate than the simple historical sample estimates.[7]

Shrinkage Estimators When determining model parameters θ, it is very popular to apply some shrinkage approach. This approach aims to combine the advantages of a sample estimate $\hat{\theta}_S$ (pure reliance on sample data) and a structural estimate $\hat{\theta}_{struct}$ (robustness) by computing some sort of weighted average[8]

$$\hat{\theta} = \lambda\hat{\theta}_S + (1 - \lambda)\hat{\theta}_{struct}.$$

While practitioners often use ad hoc weighting schemes, the literature provides a powerful Bayesian interpretation of shrinkage, which allows for the computation of *optimal* weights. In this Bayesian view, the structural estimator serves as the prior, which anchors the location of model parameters θ and the sample estimate acts as the conditioning signal. Bayes' rule then gives a stringent advice of how to combine prior and signal in order to compute the updated posterior that is used as an input for the portfolio optimization. The abovementioned Bayes-Stein shrinkage used in Jorion [39, 40] focuses on estimates of the expected returns. In the context of covariance estimation, an early contribution is Frost and Savarino [25]. More recently, Ledoit and Wolf [43] determine a more general Bayesian framework to optimize the shrinkage intensity, in which the authors explicitly correct for the fact

---

[7] See, e.g., Dangl and Kashofer [18] for an overview of structural estimates of the covariance structure of large equity portfolios—including shrinkage estimates.

[8] Shrinkage is usually a multivariate concept, i.e., λ is in general not a fixed scalar, but it depends on the observed data in some nonlinear fashion.

that the prior (i.e., the structural estimate of the covariance structure) as well as the updating information (i.e., the sample covariance matrix) are determined from the same data. Consequently, errors in these two inputs are not independent and the Bayesian estimate must control for the interdependence.[9]

Weight Restrictions A commonly observed reaction to parameter uncertainty in portfolio management is imposing ad hoc restrictions on portfolio weights. That is, the discretion of a portfolio optimizer is limited by maximum as well as minimum constraints on the weights of portfolio constituents.[10] In sample, weight restrictions clearly reduce portfolio performance (as measured by the objective function used in the optimization approach).[11] Nevertheless, out of sample studies show, that in many cases weight restrictions improve the risk-return trade-off of portfolios. Jagannathan and Ma [37] provide evidence why weight restrictions might be an efficient response to estimation errors in the covariance structure. Analyzing minimum-variance portfolios they show that binding long only constraints are equivalent to shrinking extreme covariance estimates toward more moderate levels.

Robust Optimization A more systematic approach to parameter uncertainty than weight restrictions is robust optimization. After determining the uncertainty set S for the relevant parameter vector p, robust portfolio optimization is usually formulated as a max-min problem where the vector w of portfolio weights solves the equation

$$w \in \text{argmax}_w \{\min_{p \in S} f(w; p)\},$$

with $f(w; p)$ being the planner's objective function that she seeks to maximize. This is a conservative or worst-case approach, which in many real-world applications shows favorable out-of-sample properties (see Fabozzi et al. [24], or for more details on robust and convex optimization problems and its applications in finance see Lobo et al. [48]). Provided a distribution of the parameters is available, the rather extreme max-min approach could be relaxed by applying convex risk measures. In the context of derivatives pricing, Bannoer and Scherer [5] develop the concept of risk-capturing functionals and exemplify risk averse pricing using an average Value-at-Risk measure.

Resampling A different approach to deal with parameter uncertainty in asset management is resampling. This technique does not attempt to produce more robust parameter estimates or to build a portfolio-optimization model, which directly regards parameter uncertainty in portfolio optimization. Resampling is a simulation-based approach that was first described in the portfolio-optimization context by Michaud [52] and exists in different specifications. It takes the sample estimates of mean

---

[9] See also Ledoit and Wolf [44, 45] for more on shrinkage estimates of the covariance structure.

[10] Weight restrictions are frequently part of regulatory measures targeting the fund industry aimed to control the risk characteristics of investment funds.

[11] Green and Hollifield [28] argue that in the apparent presence of a strong factor structure in the cross section of equity returns, mean-variance optimal portfolios should take large short positions in selected assets. Hence, a restriction to a long-only portfolio is expected to negatively influence portfolio performance.

returns as well as of the covariance matrix and resamples a number R of return 'histories' (where R is typically between 1,000 and 10,000). From each of these return histories, an estimate of the vector of mean returns as well as of the covariance matrix is derived. These estimates form the ingredients to calculate R different versions of the mean-variance frontier. Resampling approaches differ in the set of restrictions used to determine the mean-variance frontiers and in the way how the frontiers are averaged to get the definite portfolio weights. Some authors criticize that the unconditionally optimal portfolio does not simply follow from an average over R vectors of conditionally optimal portfolio weights (see, e.g., Scherer [59] or Markowitz and Usmen [50]), others point out that the ad-hoc approach of resampling could be improved by using a Bayesian approach (see, e.g., Scherer [60], or Harvey et al. [33, 34]). Despite the critique, all those studies appreciate the out-of-sample characteristics of resampled portfolios.

Example 4 This simple example builds on Example 1 which discusses the optimal weight of an active fund relative to a passive factor investment. An index-investment in the S&P 500 serves as the passive factor investment and an active fund with the constituents of the S&P 500 as its investment universe is the delegated active investment strategy. In Example 1 we take a history of five years of monthly log-returns (60 observations) to estimate mean returns as well as the covariance structure and the alpha, which the fund generates relative to the passive investment. We use these estimates to conclude that the optimal portfolio weight of the fund should be roughly 90 % and only 10 % of wealth should be held as a passive investment.

Being concerned about the quality of our parameter estimation that feeds into the optimization, we first examine the regression, which was performed to come up with these estimates. Assuming that log-returns are normally distributed, we conclude from the regression in Example 1 that our best estimates of the parameters α, β and ν are

$$\hat{\alpha} = 17.51 \, \text{bp/month}, \quad \hat{\beta} = 0.9821, \quad \hat{\nu} = 131.27 \, \text{bp/month},$$

and that the estimation errors are t-distributed with a standard deviation[12]

$$\sigma_{58}(\hat{\alpha}) = 23.40 \, \text{bp/month}, \quad \sigma_{58}(\hat{\beta}) = 0.0498, \quad \sigma_{59}(\hat{\nu}) = 454.91 \, \text{bp/month}.$$

Furthermore, estimation errors in $\hat{\alpha}$ and $\hat{\beta}$ are negatively correlated with a correlation coefficient $\rho = -27.93 \, \%$ and errors in the estimate of the market risk premium $\hat{\nu}$ are uncorrelated to the errors in $\hat{\alpha}$ and $\hat{\beta}$.

A statistician would now conclude that neither the fund's α nor the factor's risk premium ν is significantly different from zero, and thus an investor should seek exposure to none of the two. Another approach is to extend the optimization problem and include parameter uncertainty as an additional source of variability in the final outcome.

---

[12] Subscripts denote degrees of freedom.

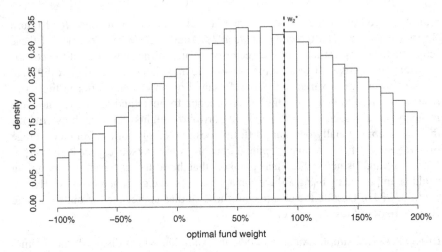

Fig. 3 Distribution of optimal portfolio weight in the interval [−100 %, 200 %] of the active investment over 100,000 resampled histories. Approximately 29 % of weights lie outside the stated interval

In contrast to a full consideration of parameter uncertainty, we use a resampling approach, which addresses this issue in a more ad hoc manner. We take the empirical estimates as the true moments of the joint distribution of factor returns and active returns, and resample 100,000 histories.[13] Then, we perform the optimization discussed in Example 1 on each of the simulated histories. Figure 3 illustrates the distribution of optimal active weights across theses 100,000 histories. Given the null hypothesis that returns are normally distributed with the estimated moments, resampling gives a good and reliable overview of the joint distribution of model parameters we estimate and—finally—an overview of the distribution of optimal weights. We can conclude that in the present setup, optimal active weights are not well determined since the estimation of the optimization model from only 60 observations per time series is too noisy to get a well-determined outcome. While resampling generates a good picture of the overall effects of parameter uncertainty, it provides no natural advice for the optimal portfolio decision beyond this illustrative insight.[14]

---

[13] This is the simplest version of resampling, mostly used in portfolio optimization. Given the null hypothesis that returns are normally distributed, we know that the empirical estimates of distribution moments are t-distributed around the true parameters, see Jobson and Korkie [38] for a detailed derivation of the small sample properties of these estimates. Thus, a more advanced approach samples for each of the histories, first the model parameters from their joint distribution, and then—given the selected moments—the history of normally distributed returns. Harvey et al. [33] is an example that uses advanced resampling to compare Bayesian inference with simple resampling.

[14] Some authors do propose schemes how to generate portfolio decisions from the cross section of the simulation results, see, e.g., Michaud and Michaud [53]. These schemes are, however, criticized by other authors for not being well-founded in decision theory, e.g., Markowitz and Usmen [50] and others mentioned in the text above.

Finally, a study that perfectly illustrates the strong implications of parameter uncertainty on optimal portfolio decisions is Pastor and Stambaugh [54]. The authors question the paradigm that due to mean reverting returns, stocks are less risky in the long run than over short horizons. This proposition is true if we know the parameters of the underlying mean reverting process with certainty. Pastor and Stambaugh [54] show that as soon as we properly regard estimation errors in model parameters, additional uncertainty from estimation errors dominates the variance reduction due to mean reversion, and thus they provide strong evidence against time diversification in equity returns.

4.2 Model Uncertainty

Qualitatively different from dealing with parameter uncertainty is the issue of model uncertainty. Since it is not at all clear what the exact characteristics of the data-generating process, which underlies asset returns are, it is not obvious which attributes a model must feature in order to capture all economically relevant effects of the portfolio selection process. Hence, every model of optimal portfolio choice bears the risk of being misspecified. In Sect. 4.1 we already mention the fact that traditional portfolio models assume that mean returns and the covariance structure of returns are constant over time. This is in contrast to empirical evidence that the moments of the return distribution are time varying. Limiting the history, which is used to estimate distribution parameters, is a frequently used procedure to get a more actual estimate. The correct length of historical data that shall be used is, however, only rarely determined in a systematic manner.

Bayesian Model Averaging A systematic approach to estimation under model uncertainty is Bayesian model averaging. It builds on the concept of a Bayesian decision-maker that has a prior about the probability weights of competing models that are constructed to predict relevant variables (e.g., asset returns) one period ahead. Observed returns are then used to determine posterior probability weights for each of the models considered applying Bayes rule.[15] Each of the competing models generates a predictive density for the next period's return. After observing the return, models which have assigned a high likelihood to the observed value (compared to others) experience an upward revision of their probability weight. In contrast, models that have assigned a low likelihood to the observed value experience a downward revision of their weight. Finally, the overall predictive density is calculated as a probability-weighted sum of all models' predictive densities. This Bayesian model averaging is an elegant way to approach a problem of model uncertainty to transform it into a standard portfolio problem to find the optimal risk-return trade-off under the derived predictive return distribution. This approach can, however, only be applied

---

[15] The posterior probability that a certain model is the correct model is proportional to the product of the model's prior probability weight and the realized likelihood of the observed return.

under the assumption that the decision-maker has a single prior and that she shows no aversion against the ambiguity inherent in the model uncertainty.[16]

Raftery et al. [57] provide the technical details of Bayesian model averaging and Avramov [3], Cremers [16], and Dangl and Halling [17] are applications to return prediction. Bayesian model averaging treats model uncertainty just as an additional source of variation. The predictive density for next period's returns becomes more disperse the higher the uncertainty about models, which differ in their prediction. The optimal portfolio selection is then unchanged, but regards the additional contribution to uncertainty.

Ambiguity Aversion If it is not possible to explicitly assess the probability that a certain model correctly mirrors the portfolio selection problem and investors are averse to this form of ambiguity, alternative portfolio selection approaches are needed. Garlappi et al. [26] develop a portfolio selection approach for investors who have multiple priors over return expectations and show ambiguity aversion. The authors prove that the portfolio selection problem of such an ambiguity-averse investor can be formulated by imposing two modifications to the standard mean-variance model, (i) an additional constraint that guarantees that the expected return lies in a specified confidence region (the way how multiple priors are modeled) and (ii) an additional minimization over all expected returns that conform to the priors (mirroring ambiguity aversion). This model gives an intuitive illustration of the fact that ambiguity averse investors show explicit desire for robustness.

5 Conclusion

The asset management industry has substantial influence on financial markets and on the welfare of many citizens. Increasingly, citizens are saving for retirement via delegated portfolio managers such as pension funds or mutual funds. In many cases there are multiple layers of delegation. It is, therefore, crucial for the welfare of modern societies that portfolio managers manage and control their portfolio risks. This article provides an eagle's perspective on risk management in asset management.

In traditional portfolio theory, the scope for risk control in portfolio management is limited. Risk management is essentially equivalent to determining the fraction of capital that the manager invests in a broadly and well diversified basket of risky securities. Thus, the "risk manager" only needs to find the optimal location on the securities market line. By contrast, in a more realistic model of the world that accounts for frictions, risk management becomes a central and important module in asset management that is frequently separate from other divisions of an asset manager. We identify several major frictions that require risk management that goes beyond choosing the weight of the riskless asset in the portfolio. First, in a world with costly information acquisition, investors do not hold the same mix of risky assets. This

---

[16] As explained in the introduction to this section, ambiguity aversion refers to preferences that express discomfort with uncertainty in the sense of Knight [42].

requires measuring a position's risk contribution relative to the specific portfolio. Thus, risk management requires constant measurement of each portfolio position's marginal risk contribution and comparing it to its marginal return contribution. This article derives a framework to calculate the marginal risk contributions and to decide on optimal portfolio weights of active managers.

In many realistic instances, investors have nonstandard preferences, which make them particularly sensitive to downside risks. We, therefore, review the main portfolio insurance concepts to achieve protection against downside risk. Stop-loss strategies, option-based portfolio insurance, constant proportion portfolio insurance, ratcheting strategies, and value-at-risk-based portfolio insurance. Using data for the S&P 500 since 1995 we simulate these alternative risk management concepts and demonstrate their risk and return characteristics.

Finally, we point out that quantitative portfolio management usually builds on the output from rather stylized models, which must be chosen to capture the relevant market environment, and which must be calibrated and parameterized. Both these choices, i.e., model selection and model calibration, contain the risk of misspecification, and thus the risk of negative effects on out-of-sample portfolio performance. We survey and discuss risk management approaches to deal with parameter uncertainty, such as shrinkage procedures or resampling procedures. Qualitatively different from parameter uncertainty is the effect of model uncertainty. Different ways of dealing with model uncertainty via methods of Bayesian model averaging and the consideration of ambiguity aversion are, therefore, surveyed and discussed.

The increased risk during the financial crisis and the following sovereign debt crisis has lead to a substantially increased focus on risk control in the asset management industry. At the same time these market episodes have also demonstrated the limitations of risk management in asset management. For example that volatile markets without strong trends make existing downside protection strategies very expensive for investors. Furthermore, risk management concepts for long-term investors are still in their infancy. Scenario-based approaches, possibly combined with min-max strategies may be more useful in this context than standard risk management tools.

Open Access This chapter is distributed under the terms of the Creative Commons Attribution Noncommercial License, which permits any noncommercial use, distribution, and reproduction in any medium, provided the original author(s) and source are credited.

References

1. Admati, A.: A noisy rational expectations equilibrium for multi-asset securities markets. Econometrica **53**(3), 629–657 (1985)
2. Ang, A., Chen, B., Sundaresan, S.: Liability-driven investment with downside risk. J. Portf. Manag. **40**(1), 71–87 (2013)
3. Avramov, D.: Stock return predictability and model uncertainty. J. Financ. Econ. **64**, 423–458 (2002)

4. Balder, S., Brandl, M., Mahayni, A.: Effectiveness of CPPI strategies under discrete-time trading. J. Econ. Dyn. Control **33**(1), 204–220 (2009)
5. Bannör, K.F., Scherer, M.: Capturing parameter risk with convex risk measures. Eur. Actuar. J. **3**(1), 97–132 (2013)
6. Basak, S.: A comparative study of portfolio insurance. J. Econ. Dyn. Control **26**, 1217–1241 (2002)
7. Basak, S., Shapiro, A.: Value-at-risk-based risk management: optimal policies and asset prices. Rev. Financ. Stud. **14**(2), 371–405 (2001)
8. Benninga, S.: Comparing portfolio insurance strategies. Finanzmarkt Portf. Manag. **4**(1), 20–30 (1990)
9. Benninga, S., Blume, M.E.: On the optimality of portfolio insurance. J. Financ. **40**(5), 1341–1352 (1985)
10. Black, F., Jones, R.: Simplifying portfolio insurance. J. Portf. Manag. **14**(1), 48–51 (1987)
11. Black, F., Perold, A.F.: Theory of constant proportion portfolio insurance. J. Econ. Dyn. Control **16**(3–4), 403–426 (1992)
12. Brennan, M.J., Schwartz, E.S.: The pricing of equity-linked life insurance policies with an asset value guarantee. J. Financ. Econ. **3**, 195–213 (1976)
13. Brennan, M.J., Schwartz, E.S.: Time-invariant portfolio insurance strategies. J. Financ. **43**(2), 283–299 (1988)
14. Brennan, M.J., Solanki, R.: Optimal portfolio insurance. J. Financ. Quant. Anal. **16**(3), 279–300 (1981)
15. Campbell, J.Y., Medeiros, K.S.-D., Viceira, L.M.: Global currency hedging. J. Financ. **65**(1), 87–121 (2010)
16. Cremers, M.K.J.: Stock return predictability: a Bayesian model selection perspective. Rev. Financ. Stud. **15**, 1223–1249 (2002)
17. Dangl, T., Halling, M.: Predictive regressions with time-varying coefficients. J. Financ. Econ. **106**, 157–181 (2012)
18. Dangl, T., Kashofer, M.: Minimum-variance stock picking—a shift in preferences for minimum-variance portfolio constituents. Working paper (2013)
19. Dichtl, H., Drobetz, W.: Portfolio insurance and prospect theory investors: popularity and optimal design of capital protected financial products. J. Bank. Financ. **35**(7), 1683–1697 (2011)
20. Dockner, E.: Sind Finanzprodukte mit Kapitalgarantie eine attraktive Anlageform? In: Frick, R., Gantenbein, P., Reichling, P. (eds.) Asset Management, pp. 271–284. Haupt, Bern Stuttgart Wien (2012)
21. Dybvig, P.H.: Using asset allocation to protect spending. Financ. Anal. J. **55**(1), 49–62 (1999)
22. Elton, E.J., Gruber, M.J.: Estimating the dependence structure of share prices-implications for portfolio selection. J. Financ. **28**(5), 1203–1232 (1973)
23. Estep, T., Kritzman, M.: TIPP: insurance without complexity. J. Portf. Manag. **14**(4), 38–42 (1988)
24. Fabozzi, F.J., Kolm, P.N., Pachamanova, D., Focardi, S.M.: Robust Portfolio Optimization and Management. Wiley, Hoboken (2007)
25. Frost, P.A., Savarino, J.E.: An empirical Bayes approach to efficient portfolio selection. J. Financ. Quant. Anal. **21**(3), 293–305 (1986)
26. Garlappi, L., Uppal, R., Wang, T.: Portfolio selection with parameter and model uncertainty: a multi-prior approach. Rev. Financ. Stud. **20**(1), 41–81 (2007)
27. Gomes, F.J.: Portfolio choice and trading volume with loss-averse investors. J. Bus. **78**(2), 675–706 (2005)
28. Green, R.C., Hollifield, B.: When will mean-variance efficient portfolios be well diversified? J. Financ. **47**(5), 1785–1809 (1992)
29. Grossman, S.J., Stiglitz, J.E.: On the impossibility of informationally efficient markets. Am. Econ. Rev. **70**(3), 393–408 (1980)
30. Grossman, S.J., Vila, J.-L.: Portfolio insurance in complete markets: a note. J. Bus. **62**(4), 473–476 (1989)

31. Grossman, S.J., Zhou, Z.: Optimal investment strategies for controlling drawdowns. Math. Financ. **3**(3), 241–276 (1993)
32. Grossman, S.J., Zhou, Z.: Equilibrium analysis of portfolio insurance. J. Financ. **51**(4), 1379–1403 (1996)
33. Harvey, C.R., Liechty, J.C., Liechty, M.W.: Bayes vs. resampling: a rematch. J. Invest. Manag. **6**(1), 1–17 (2008)
34. Harvey, C.R., Liechty, J.C., Liechty, M.W., Müller, P.: Portfolio selection with higher moments. Quant. Financ. **10**(5), 469–485 (2010)
35. Herold, U., Maurer, R., Purschaker, N.: Total return fixed-income portfolio management. A risk-based dynamic strategy. J. Portf. Manag. Spring **31**, 32–43 (2005)
36. Herold, U., Maurer, R., Stamos, M., Vo, H.T.: Total return strategies for multi-asset portfolios: dynamically managing portfolio risk. J. Portf. Manag. **33**(2), 60–76 (2007)
37. Jagannathan, R., Ma, T.: Risk reduction in large portfolios: why imposing the wrong constraints helps. J. Financ. **58**(4), 1651–1684 (2003)
38. Jobson, J., Korkie, B.: Estimation for Markowitz efficient portfolios. J. Am. Stat. Assoc. **75**(371), 544–554 (1980)
39. Jorion, P.: International portfolio diversification with estimation risk. J. Bus. **58**(3), 259–278 (1985)
40. Jorion, P.: Bayes-Stein estimation for portfolio analysis. J. Financ. Quant. Anal. **21**(3), 279–292 (1986)
41. Kahneman, D., Tversky, A.: Prospect theory: an analysis of decision under risk. Econometrica **47**(2), 263–291 (1979)
42. Knight, F.H.: Risk, uncertainty and profit. Sentry press, Reprinted 1956 (1921)
43. Ledoit, O., Wolf, M.: Improved estimation of the covariance matrix of stock returns with an application to portfolio selection. J. Empir. Financ. **10**, 603–621 (2003)
44. Ledoit, O., Wolf, M.: Honey, I shrunk the sample covariance matrix. J. Portf. Manag. **30**(4), 110–119 (2004a)
45. Ledoit, O., Wolf, M.: A well-conditioned estimator for large-dimensional covariance matrices. J. Multivar. Anal. **88**, 365–411 (2004b)
46. Leland, H., Rubinstein, M.: Comments on the market crash: six months after. J. Econ. Perspect. **2**(3), 45–50 (1988)
47. Leland, H.E.: Who should buy portfolio insurance? J. Financ. **35**(2), 581–594 (1980)
48. Lobo, M.S., Vandenberghe, L., Boyd, S., Lebret, H.: Applications of second-order cone programming. Linear Algebra Appl. **284**(1–2), 193–228 (1998)
49. Markowitz, H.: Portfolio selection. J. Financ. **7**, 77–91 (1952)
50. Markowitz, H., Usmen, N.: Resampled frontiers versus diffuse Bayes: an experiment. J. Invest. Manag. **4**(1), 9–25 (2003)
51. Merton, R.: Optimum consumption and portfolio rules in a continuous-time model. J. Econ. Theory **3**(4), 373–413 (1971)
52. Michaud, R.O.: Efficient Asset Management. Oxford University Press, Oxford (1998)
53. Michaud, R.O., Michaud, R.O.: Efficient Asset Management, 2nd edn. Oxford University Press, Oxford (2008)
54. Pastor, L., Stambaugh, R.F.: Are stocks really less volatile in the long run? J. Financ. **67**(2), 431–477 (2012)
55. Pennacchi, G.: Theory of Asset Pricing. The Addison-Wesley series in finance. Addison-Wesley, Boston (2008)
56. Perold, A.F., Sharpe, W.F.: Dynamic strategies for asset allocation. Financ. Anal. J. **44**(1), 16–27 (1988)
57. Raftery, A.E., Madigan, D., Hoeting, J.A.: Bayesian model averaging for linear regression models. J. Am. Stat. Assoc. **92**, 179–191 (1997)
58. Rubinstein, M., Leland, H.E.: Replicating options with positions in stock and cash. Financ. Anal. J. **37**(4), 63–72 (1981)
59. Scherer, B.: Portfolio resampling: review and critique. Financ. Anal. J. **58**(6), 98–109 (2002)

60. Scherer, B.: A note on the out-of-sample performance of resampled efficiency. J. Asset Manag. **7**(3/4), 170–178 (2006)
61. Sharpe, W.F.: Decentralized investment management. J. Financ. **36**(2), 217–234 (1981)
62. Sharpe, W.F., Tint, L.G.: Liabilities—a new approach. J. Portf. Manag. **16**(2), 5–10 (1990)
63. Wilson, R.: The theory of syndicates. Econometrica **36**(1), 119–132 (1968)
64. Zagst, R., Kraus, J.: Stochastic dominance of portfolio insurance strategies—OBPI versus CPPI. Ann. Op. Res. **185**(1), 75–103 (2011)

Worst-Case Scenario Portfolio Optimization Given the Probability of a Crash

Olaf Menkens

Abstract Korn and Wilmott [9] introduced the worst-case scenario portfolio problem. Although Korn and Wilmott assume that the probability of a crash occurring is unknown, this paper analyzes how the worst-case scenario portfolio problem is affected if the probability of a crash occurring is known. The result is that the additional information of the known probability is not used in the worst-case scenario. This leads to a q-quantile approach (instead of a worst case), which is a value at risk-style approach in the optimal portfolio problem with respect to the potential crash. Finally, it will be shown that—under suitable conditions—every stochastic portfolio strategy has at least one superior deterministic portfolio strategy within this approach.

1 Introduction

Portfolio optimization in continuous time goes back to Merton [17]. Merton assumes that the investor has two investment opportunities; one risk-free asset (bond) and one risky asset (stock) with dynamics given by

$$\mathrm{d}P_{0,0}(t) = P_{0,0}(t)\, r_0\, \mathrm{d}t, \qquad P_{0,0}(0) = 1, \qquad \text{"bond"}$$
$$\mathrm{d}P_{0,1}(t) = P_{0,1}(t)\,[\mu_0\, \mathrm{d}t + \sigma_0\, \mathrm{d}W_0(t)], \qquad P_{0,1}(0) = p_1, \qquad \text{"stock"}$$

with constant market coefficients μ_0, r_0, $\sigma_0 > 0$, and where W_0 is a Brownian motion on a complete probability space $(\Omega, \mathscr{F}, P)$. Finally, X_0^π denotes the *wealth process* of the investor given the portfolio strategy π (which denotes the fraction invested in the risky asset). More specifically, the wealth process satisfies

$$\mathrm{d}X_0^\pi(t) = X_0^\pi(t)\,[(r_0 + \pi(t)\,[\mu_0 - r_0])\, \mathrm{d}t + \pi(t)\sigma_0\, \mathrm{d}W_0(t)],$$
$$X_0^\pi(0) = x.$$

O. Menkens (✉)
School of Mathematical Sciences, Dublin City University, Dublin 9,
Glasnevin, Ireland
e-mail: olaf.menkens@dcu.ie

© The Author(s) 2015
K. Glau et al. (eds.), *Innovations in Quantitative Risk Management*,
Springer Proceedings in Mathematics & Statistics 99,
DOI 10.1007/978-3-319-09114-3_15

Assuming that the utility function $U(x)$ of the investor is given by $U(x) = \ln(x)$, one can define the *performance function* for an arbitrary admissible portfolio strategy $\pi(t)$ by

$$\mathscr{J}_0(t, x, \pi) := \mathbb{E}\left[\ln\left(X_0^{\pi, t, x}(T)\right)\right] = \ln(x) + \mathbb{E}\left[\int_t^T \left[\Psi_0 - \frac{\sigma_0^2}{2}(\pi(s) - \pi_0^*)^2\right] ds\right]. \quad (1)$$

Here,

$$\Psi_0 := r_0 + \frac{1}{2}\left(\frac{\mu_0 - r_0}{\sigma_0}\right)^2 = r_0 + \frac{\sigma_0^2}{2}(\pi_0^*)^2 \quad \text{and} \quad \pi_0^* := \frac{\mu_0 - r_0}{\sigma_0^2}$$

will be called the *utility growth potential* or *earning potential* and the *optimal portfolio strategy* or *Merton fraction*, respectively. Using this, the *portfolio optimization problem* in the Merton case (that is without taking possible jumps into account) is given by

$$\sup_{\pi(\cdot) \in A_0(x)} \mathscr{J}_0(t, x, \pi) =: v_0(t, x) \quad [= \ln(x) + \Psi_0(T - t)], \quad (2)$$

where v_0 is known as the *value function* in the Merton case. From Eq. (1), it is clear that π_0^* maximizes $\mathscr{J}_0$. Hence, it is the optimal portfolio strategy for Eq. (2).

Merton's model has the disadvantage that it cannot model jumps in the price of the risky asset. Therefore, Aase [1] extended Merton's model to allow for jumps in the risky asset. In the simplest case, the dynamics of the risky asset changes to

$$dP_J(t) = P_J(t)[\mu_0\, dt + \sigma_0\, dW_0(t) - k\, dN(t)],$$

where N is a Poisson process with intensity $\lambda > 0$ on $(\Omega, \mathscr{F}, P)$ and $k > 0$ is the crash or jump size. In this setting, the performance function is given by

$$\mathscr{J}_J(t, x, \pi) = \ln(x) + \mathbb{E}\left[\int_t^T \left[\Psi_0 - \frac{\sigma_0^2}{2}(\pi(s) - \pi_0^*)^2 - \ln(1 - \pi(s)k)\lambda\right] ds\right].$$

Using this, the optimal portfolio strategy can be computed to

$$\pi_J^* = \frac{1}{2}\left(\pi_0^* + \frac{1}{k}\right) - \sqrt{\frac{1}{4}\left(\pi_0^* - \frac{1}{k}\right)^2 + \frac{\lambda}{\sigma_0^2}}.$$

Figure 1 shows the fraction invested in the risky asset in Merton's (solid line) and Aase's model for various λ (all the other lines). The dashed line below the solid line is the case where $\frac{1}{\lambda} = 50$, that is the investor expects on average one crash within 50 years. By comparison, the lowest line (the dash–dotted line) is the case

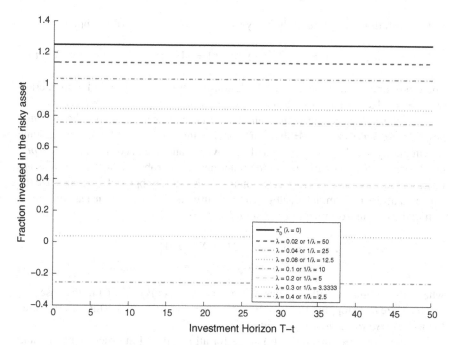

Fig. 1 Examples of Merton's optimal portfolio strategies. This figure is plotted with $\pi_0^* = 1.25$, $\sigma_0 = 0.25$, $r = 0.05$, $k = 0.25$, and $T = 50$. This implies that $\lambda_0 = \frac{\sigma_0^2 \pi_0^*}{k} = 0.3125$, $\Psi_0 \approx 0.098828$, and $\frac{1}{k^*} = 4$

where $\frac{1}{\lambda} = 2.5$, that is the investor expects on average one crash within 2.5 years. Note, however, that the fraction invested in the risky asset is negative in this case, meaning that the optimal strategy is that the investor goes short in the risky asset. This strategy is very risky because the probability that the investor will go bankrupt is strictly positive. This can also be observed in practice where several hedge funds went bankrupt which were betting on a crash in the way described above.

Therefore, let us consider an ansatz which overcomes this problem.

1.1 Alternative Ansatz of Korn and Wilmott

The ansatz made by Korn and Wilmott [9] is to distinguish between *normal times* and *crash time*. In normal times, the same set up as in Merton's model is used. At the crash time, the price of the risky asset falls by a factor of $k \in [k_*, k^*]$ (with $0 \le k_* \le k^* < 1$). This implies that the wealth process $X_0^\pi(t)$ just before the crash time $\tau-$ satisfies

$$X_0^\pi(\tau-) = \underbrace{[1 - \pi(\tau)] X_0^\pi(\tau-)}_{\text{bond investment}} + \underbrace{\pi(\tau) X_0^\pi(\tau-)}_{\text{stock investment}}.$$

At the crash time, the price of the risky asset drops by a factor of k, implying

$$[1 - \pi(\tau)] X_0^\pi(\tau-) + \pi(\tau)X_0^\pi(\tau-)[1-k] = [1 - \pi(\tau)k] X_0^\pi(\tau-) = X_0^\pi(\tau).$$

Therefore, one has a straightforward relationship of the wealth right before a crash with the wealth right after a crash.

The main disadvantage of this ansatz is that one needs to know the *maximal possible number of crashes M* that can happen at most—in the following, we assume for simplicity that $M = 1$ if not stated otherwise—and one needs to know the *worst crash size k^** that can happen. On the other hand, no probabilistic assumptions are made on the crash time or crash size. Therefore, Merton's approach, to maximize the expected utility of terminal wealth, cannot be used in this context. Instead the aim is to find the best uniform *worst-case bound*, e.g. solve

$$\sup_{\pi(\cdot)\in A(x)} \inf_{\substack{0\leq\tau\leq T \\ k\in K}} \mathbb{E}\left[\ln\left(X^\pi(T)\right)\right], \tag{3}$$

where the terminal wealth satisfies $X^\pi(T) = (1 - \pi(\tau)k) X_0^\pi(T)$ in the case of a crash of size k at stopping time τ. Moreover, $K = \{0\} \cup [k_*, k^*]$. This will be called the *worst-case scenario portfolio problem*.

Note that one requires that $\pi(t) < \frac{1}{k^*}$ for all $t \in [0, T]$ in order to avoid bankruptcy. The *value function* to the above problem is defined via

$$v_c(t, x) := \sup_{\pi(\cdot)\in A(t,x)} \inf_{\substack{t\leq\tau\leq T, \\ k\in K}} \mathbb{E}\left[\ln\left(X^{\pi,t,x}(T)\right)\right]. \tag{4}$$

Observe that this optimization problem can be interpreted as a stochastic differential game (see Korn and Steffensen [12]), where the investor tries to maximize her expected utility of terminal wealth while the counterparty (the market or nature) tries to hit the investor as badly as possible by triggering a crash. The control of the investor is π, the fraction of wealth invested into the risky asset, while the control of the counterparty is the crash time τ and the crash size k. Figure 2 depicts the optimization problem. For each control choice (that is portfolio strategy) of the investor (e.g., π_2), the investor calculates the expected utility of terminal wealth for all possible control choices (that is (τ, k)) of the counterparty (which is the dotted line for the strategy π_2). Then the worst-case scenario is determined for each portfolio strategy (e.g., $\left(\tau^{(\pi_2)}, k^{(\pi_2)}\right)$ for the strategy π_2). Afterwards, the expected utility of terminal wealth for this worst-case scenario is calculated and denoted by $\mathscr{WC}(\pi_2)$. The last step is to find the strategy which maximizes the worst-case scenario function $\mathscr{WC}(\cdot)$. For the three examples given in Fig. 2, this would be π_3. Notice that π_3 is special in that all choices (τ, k) lead to the same worst-case scenario, that is $\mathscr{WC}(\pi_3)$ is independent of the scenario (τ, k).

Observe that Aase [1] would fix k, model the crash time via a Poisson distribution, and maximize the expected utility of terminal wealth. Whereas, by comparison, the

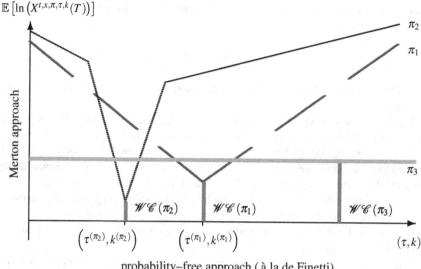

Fig. 2 Schematic interpretation of the worst-case scenario optimization

worst-case scenario optimization method uses a probability-free approach on crash time and size.

Apparently, it is quite cumbersome to determine the optimal portfolio strategy as described above. Instead consider the following approach. Define as v_1 the value function as in Eq. (2), except that the subscript 1 indicates that this is the value function in the Merton case after a crash has happened (and where the market parameters might change—see Sect. 2 for details). To that end, a portfolio strategy $\hat{\pi} \geq 0$ determined via the equation

$$\mathscr{J}_0\left(t, x, \hat{\pi}\right) = v_1\left(t, x\left(1 - \hat{\pi}(t)k^*\right)\right) \quad \text{for all} \quad t \in [0, T] \tag{5}$$

will be called a *crash indifference strategy*. This is, because the investor gets the same expected utility of terminal wealth if either no crash happens (left-hand side) or a crash of the worst-case size k^* happens (right-hand side). It is straightforward to verify (see Korn and Menkens [10]) that there exists a unique crash indifference strategy $\hat{\pi}$, which is given by the solution of the differential equation

$$\hat{\pi}'(t) = \frac{\sigma_0^2}{2}\left(\hat{\pi}(t) - \frac{1}{k^*}\right)\left(\hat{\pi}(t) - \pi_0^*\right)^2, \tag{6}$$

with $\quad \hat{\pi}(T) = 0.$ $\tag{7}$

This crash indifference strategy is bounded by $0 \leq \hat{\pi} \leq \min\{\pi_0^*, \frac{1}{k^*}\}$. It can be shown (see Korn and Wilmott [9] or Korn and Menkens [10]) that the optimal portfolio strategy for an investor, who wants to maximize her worst-case scenario portfolio problem, is given by

$$\bar{\pi}(t) := \min\left\{\hat{\pi}(t), \pi_0^*\right\} \qquad \text{for all } t \in [0, T]. \tag{8}$$

$\bar{\pi}$ will be named the *optimal crash hedging strategy* or *optimal worst-case scenario strategy*.

Figure 3 shows the optimal worst-case scenario strategies of Korn/Wilmott if at most one (solid line), two (dashed line), or three (dash–dotted line) crashes can happen. Assuming that the investor has an initial investment horizon of $T = 50$ and expects to see at most three crashes, a optimal worst-case scenario investor would use the portfolio strategy $\hat{\pi}_3(t)$ until she observes a first crash, say at time τ_1. After having observed a crash, the investor would switch to the strategy $\hat{\pi}_2(t)$, since the investor expects to see at most two further crashes in the remaining investment horizon $T - \tau_1$; and so on. Finally, if the investor expects to observe no further crash, she will switch to the Merton fraction π_0^*.

The worst case scenario strategies are now compared to the optimal portfolio strategy in Aase's model, where $\lambda(t) = \frac{1}{T-t}$ (see dotted line in Fig. 3), that is the investor expects to see on average one crash over his remaining investment horizon $T - t$. Clearly, setting λ in this way is somewhat unrealistic. Nevertheless, this extreme example is used to point out several disadvantages of the expected utility

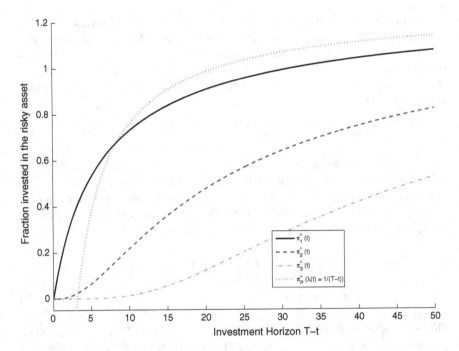

Fig. 3 Examples of worst-case optimal portfolio strategies

approach of Aase. First, considering a λ which changes over time and depends on the investment horizon of the investor, leads not only to a time-changing optimal strategy $\pi_P(t)$, but also to a price dynamics of the risky asset which depends on the investment horizon of the investor. Hence, any two investors with different investment horizons would work with different price dynamics of the risky assets. Second, as the investor approaches the investment horizon, $\lambda(t) \to \infty$ (that is a crash happens almost surely), thus, $\pi_P(t) \to -\infty$, which would lead to big losses on short-term investment horizons if no crash happens. Of course, these losses would average out with the gains made if a crash happens remembering that the assumption is that—on average—every second scenario would observe at least one crash. This is the effect of averaging the crash out in an expected utility sense (compared to the worst-case approach of Korn/Wilmott).

Basically, it would be possible to cut off $\pi_p(t)$ at zero, that is, one would not allow for short-selling. This would imply to cut off $\lambda(t)$ at $\frac{\mu_0 - r_0}{k}$ and there is no economic interpretation why this should be done (except that short-selling might not be allowed). Finally, note that it is also possible to set λ such that one expects to see at least one crash with probability q (e.g., $q = 5\%$), however this would not remedy the two disadvantages mentioned above.

Why is the worst-case scenario approach more suitable than the standard expected utility approach in the presence of jumps? The standard expected utility approach will average out the impact of the jumps over all possible scenarios. With other words, the corresponding optimal strategy will offer protection only on average over all possible scenarios, which will be good as long as either no jump or just a small jump happens. However, if a large jumps happens, the protection is negligible. By comparison, the worst-case scenario approach will offer full protection from a jump up to the worst-case jump size assumed.

The situation can be compared to the case of buying liability insurance. The standard utility approach would look at the average of all possible claim sizes (say e.g., 100,000 EUR)—and its optimal strategy would be to buy liability insurance with a cover of 100,000 EUR. However, the usual advice is to buy liability insurance with a cover which is as high as possible—this solution corresponds to the worst-case scenario approach. With other words, the aim is to insure the rare large jumps. This observation is supported by the fact that many insurances offer retention (which excludes small jumps from the insurance).

1.2 Literature Review

To the best of our knowledge, Hua and Wilmott [8] were the first to consider the worst-case scenario approach in a binomial model to price derivatives. Korn and Wilmott [9] were the first to apply this approach to portfolio optimization, and Korn and Menkens [10] developed a stochastic control framework for this approach, while Korn and Steffensen [12] considered this approach as a stochastic differential game. Korn and Menkens [10] and Menkens [16] looked at changing market coefficients after a crash. Seifried [22] evolved a martingale approach for the worst-case scenario. Moreover, the worst-case scenario approach has been applied to the optimal investment problem of an insurance company (see Korn [11]) and to optimize reinsurance for an insurance company (see Korn et al. [13]). Korn et al. [13] show also in their setting that the worst-case scenario approach has a negative diversification effect. Furthermore, both portfolio optimization under proportional transaction costs (see Belak et al. [4]) and the infinite time consumption problem (see Desmettre et al. [7]) have been studied in a worst-case scenario optimization setting. Mönnig [18] applies the worst-case scenario approach in a stochastic target setting to compute option prices. Finally, Belak et al. [2, 3] allow for a random number of crashes, while Menkens [15] analyzes the costs and benefits of using the worst-case scenario approach.

Notice that there is a different worst-case scenario optimization problem which is also known as *Wald's Maximin approach* (see Wald [23, 24]). The following quotation is taken from Wald [23, p. 279]:

> A problem of statistical inference may be interpreted as a zero sum two person game as follows: Player 1 is Nature and player 2 is the statistician. [...] The outcome $K[\theta, w(E)]$ of the game is the risk $r[\theta|w(E)]$ of the statistician. Clearly, the statistician wishes to minimize $r[\theta|w(E)]$. Of course, we cannot say that Nature wants to maximize $r[\theta|w(E)]$. However, if the statistician is in complete ignorance as to Nature's choice, it is perhaps not unreasonable to base the theory of a proper choice of $w(E)$ on the assumption that Nature wants to maximize $r[\theta|w(E)]$.

This is a well-known concept in decision theory and is also known as *robust optimization* (see e.g., Bertsimas et al. [5] or Rustem and Howe [21] and the references therein). However, while the ansatz is the same, it is usually assumed that the parameters (in our case r_0, μ_0, and σ_0) are unknown within certain boundaries. Therefore, this is a parameter uncertainty problem which is solved using a worst-case scenario approach—instead of using perturbation analysis. Observe that this usually involves optimization procedures done by a computer. Finally, note that the optimal strategies can be computed directly only in the special case that only μ_0 is uncertain (see Mataramvura and Øksendal [14], Øksendal and Sulem [19], or Pelsser [20] for a recent application in an insurance setting).

By comparison, the worst-case scenario approach considered in this paper is taking (possibly external) shocks/jumps/crashes into account—and not parameter uncertainty. While the original idea and the wording are similar or even the same, it is clear that the worst-case scenario approach of Korn and Wilmott [9] is different from the robust optimization approach in decision theory.

The remainder of this paper is organized as follows. Section 2 introduces the set up of the model which will be considered; and Sect. 3 solves the optimization problem if the probability of a potential crash is known. As a consequence, the q-quantile crash hedging strategy will be developed in Sect. 4. Section 5 gives examples of the q-quantile crash hedging strategy, while Sect. 6 shows that stochastic portfolio strategies are always inferior to their corresponding deterministic portfolio strategies. Finally, Sect. 7 concludes.

2 Setup of the Model

Let us work with the model introduced above and let us make the following refinements. First, it has been tacitly assumed that the investor is able to realize that the crash has happened. Thus, let us model its occurrence via a $\mathscr{F}$—stopping time τ. To model the fact that the investor is able to realize that a jump of the stock price has happened it is supposed that the investor's decisions are adapted to the P-augmentation $\{\mathscr{F}_t\}$ of the filtration generated by the Brownian motion $W(t)$. The difficulty of this approach is to determine the optimal strategy after a crash because the starting point is random, however Seifried [22] solved this problem.

Let us further suppose that the market conditions change after a possible crash. Let therefore k (with $k \in [k_*, k^*]$) be the arbitrary size of a crash at time τ. The price of the bond and the risky asset after a crash of size k happened at time τ is assumed to be

$$dP_{1,0}(t) = P_{1,0}(t) r_1 \, dt \, , \qquad\qquad P_{1,0}(\tau) = P_{0,0}(\tau) \, , \qquad (9)$$
$$dP_{1,1}(t) = P_{1,1}(t) [\mu_1 \, dt + \sigma_1 \, dW(t)] \, , \qquad P_{1,1}(\tau) = (1 - k) P_{0,1}(\tau) \, , \quad (10)$$

with constant market coefficients r_1, μ_1, and $\sigma_1 > 0$ after a possible crash of size k at time τ. That is, this is the same market model as before the crash except that the market parameters are allowed to change after a crash has happened.

It is important to keep in mind that the investor does *not* know for certain that a crash will occur—the investor only thinks that it is possible. An investor who knows that a crash will happen within the time horizon $[0, T]$ has additional information and is therefore an insider. The set of possible crash heights of the insider is indeed $K_I := [k_*, k^*]$, while the set of possible crash heights of the investor who thinks that a crash is possible is $K := \{0\} \cup [k_*, k^*]$. In this paper, only the portfolio problem of the investor, who thinks a crash is possible, is considered.

For simplicity, the initial market will also be called market 0, while the market after a crash will be called market 1. In order to set up the model, the following definitions are needed.

Definition 1 (i) For $i = 0, 1$, let $A_i(s, x)$ be the *set of admissible portfolio processes* $\pi(t)$ corresponding to an initial capital of $x > 0$ at time s, i.e., $\{\mathscr{F}_t, s \leq t \leq T\}$-progressively measurable processes such that

(a) the *wealth equation* in market i in the usual crash-free setting

$$dX_i^{\pi,s,x}(t) = X_i^{\pi,s,x}(t) [(r_i + \pi(t) [\mu_i - r_i]) \, dt + \pi(t) \sigma_i dW_i(t)] \, , \quad (11)$$
$$X_i^{\pi,s,x}(s) = x \qquad (12)$$

has a unique non-negative solution $X_i^{\pi,s,x}(t)$ and satisfies

$$\int_s^T \left[\pi(t) X_i^{\pi,s,x}(t)\right]^2 dt < \infty \quad P\text{-a.s.}, \tag{13}$$

i.e. $X_i^{\pi,s,x}(t)$ is the *wealth process in market i* in the crash-free world, which uses the portfolio strategy π and starts at time s with initial wealth x.

Furthermore, $X_i^{\pi}(t) := X_i^{\pi,0,x}(t)$ will be used as an abbreviation.
(b) $\pi(t)$ has left-continuous paths with right limits.

(ii) the corresponding *wealth process $X^{\pi}(t)$ in the crash model*, defined as

$$X^{\pi}(t) = \begin{cases} X_0^{\pi}(t) & \text{for } s \leq t < \tau \\ [1 - \pi(\tau)k] X_1^{\pi,\tau,X_0^{\pi}(\tau)}(t) & \text{for } t \geq \tau \geq s, \end{cases} \tag{14}$$

given the occurrence of a jump of height k at time τ, is strictly positive. Thereby, it is assumed that the crash time τ is a $\mathscr{F}$-stopping time. The set of admissible portfolio strategies is obviously given by $A_0(s, x)$ as long as no crash happens. After a crash at time τ, the set is given by $A_1(\tau, x)$, which is defined scenario-wise, that is via $A_1(\tau(\omega), x)$ for all $\omega \in \Omega$. Hence,

$$A(s, x) := \left\{ \phi(t) \text{ with } t \in [s, T] : \phi|_{[s,\tau]} \in A_0(s, x)|_{[s,\tau]} \text{ and } \phi|_{[\tau,T]} \in A_1(\tau, x) \right\}.$$

(iii) $A(x)$ is used as an abbreviation for $A(0, x)$.

Finally, it is clear how to extend the definitions given above only for $i = 0$ to $i = 1$. Simply, replace the zeros by ones.

3 Optimal Portfolios Given the Probability of a Crash

In this section, let us suppose that the investor knows the probability of a crash occurring. Let p, with $p \in [0, 1]$, be the probability that a crash can happen (but must not necessarily happen)[1]. Note that the following argument holds also for time-dependent p (that is $p(t)$), however to simplify the notation, it is assumed that p is constant. In this situation, the optimization problem can be split up into two problems (crash can occur, no crash happens) which have to be solved simultaneously. To that end define for $p \in [0, 1]$

---

[1] Observe that the important information is that no crash will happen with a probability of at least $1 - p$. If one would say that a crash will happen with probability p, the investor would become an insider with an adjusted optimization problem as described in Sect. 2, p. 9. However, this insider approach would make the discussion way more difficult. Therefore, to simplify the discussion, the approach of no crash happens/a crash can happen is taken here.

$$\mathbb{E}_p\left[\ln\left(X^{\pi,t,x}(T)\right)\right] := p\underbrace{\mathbb{E}\left[\ln\left(X^{\pi,t,x}(T)\right)\right]}_{\text{A crash can occur.}} + (1-p)\underbrace{\mathbb{E}\left[\ln\left(X_0^{\pi,t,x}(T)\right)\right]}_{\text{No crash happens.}}.$$

Using this definition, the optimization problem can be written as

$$\sup_{\pi(\cdot)\in A(t,x)} \inf_{\substack{t\leq\tau\leq T,\\ k\in K}} \mathbb{E}_p\left[\ln\left(X^{\pi,t,x}(T)\right)\right]$$

$$= \sup_{\pi(\cdot)\in A(t,x)} \left\{ p\left\{ \inf_{\substack{t\leq\tau\leq T,\\ k\in K}} \mathbb{E}\left[\ln\left(X^{\pi,t,x}(T)\right)\right] \right\} + (1-p)\mathbb{E}\left[\ln\left(X_0^{\pi,t,x}(T)\right)\right] \right\}$$

$$= \sup_{\pi(\cdot)\in A(t,x)} \left\{ p\cdot\left\{ \inf_{\substack{t\leq\tau\leq T,\\ k\in K}} v_1\left(\tau, X_0^{\pi,t,x}(\tau)\left(1-\pi(\tau)k\right)\right) \right\} + (1-p)\,\mathscr{J}_0\left(t,x,\pi\right) \right\}.$$

$$(15)$$

Observe that the two extremes, $p \in \{0, 1\}$ are straightforward to solve:

(A) $p = 1$:

$$\sup_{\pi(\cdot)\in A(t,x)} \inf_{\substack{t\leq\tau\leq T,\\ k\in K}} \mathbb{E}_1\left[\ln\left(X^{\pi,t,x}(T)\right)\right] = \sup_{\pi(\cdot)\in A(t,x)} \inf_{\substack{t\leq\tau\leq T,\\ k\in K}} \mathbb{E}\left[\ln\left(X^{\pi,t,x}(T)\right)\right].$$

Thus, this is the original worst-case scenario portfolio problem. The solution is already known.

(B) $p = 0$:

$$\sup_{\pi(\cdot)\in A(t,x)} \inf_{\substack{t\leq\tau\leq T,\\ k\in K}} \mathbb{E}_0\left[\ln\left(X^{\pi,t,x}(T)\right)\right] = \sup_{\pi(\cdot)\in A_0(t,x)} \mathbb{E}\left[\ln\left(X_0^{\pi,t,x}(T)\right)\right],$$

which is the classical optimal portfolio problem of Merton. The solution is well known and is given in our notation (compare with Eq. (2)) by π_0^*.

Let us now consider the case $p \in (0, 1)$. Denoting the optimal crash hedging strategy in this situation by $\hat{\pi}_p$, Eq. (15) can be rewritten as

$$\mathscr{J}_0\left(t,x,\hat{\pi}_p\right) = p\cdot v_1\left(t, x\left(1-\hat{\pi}_p(t)k^*\right)\right) + (1-p)\,\mathscr{J}_0\left(t,x,\hat{\pi}_p\right)$$
$$\Longleftrightarrow \quad \mathscr{J}_0\left(t,x,\hat{\pi}_p\right) = v_1\left(t, x\left(1-\hat{\pi}_p(t)k^*\right)\right),$$

where the last equation is obtained from the first equation by solving the first equation for $\mathscr{J}_0$. Since the latter equation is the indifference Eq. (5) in this setting, which leads to the same ODE and boundary condition as in Korn and Wilmott [9], it follows that $\hat{\pi}_p \equiv \hat{\pi}$ (see the paragraph between Eqs. (5) and (6) for details). This result shows that the crash hedging strategy remains the same even if the probability of a crash is known. Thus, this result justifies the wording *worst-case scenario* of the above-developed concept. This is due to the fact that the worst-case scenario should be independent of the probability of the worst case and which has been shown above. Let us summarize this result in a proposition.

Proposition 1 *Given that the probability of a crash is positive, the worst-case scenario portfolio problem as it has been defined in Eq. (3) is independent of the probability of the worst-case occurring.*

If the probability of a crash is zero, the worst-case scenario portfolio problem reduces to the classical crash-free portfolio problem.

4 The q-quantile Crash Hedging Strategy

Obviously, the concept of the worst case scenario has the disadvantage that additional information (namely the given probability of a crash and the probability distribution of the crash sizes) is not used. However, if the probability of a crash and the probability of the crash size is known, it is possible to construct the *(lower) q-quantile crash hedging strategy*.

Assume that $p(t) \in [0, 1]$ is the given probability of a crash at time $t \in [0, T]$ and assume that $f(k, t) \in [0, 1]$ is the given density of the distribution function for a crash of size $k \in [k_*, k^*]$ at time t. Moreover, suppose that a function $q : [0, T] \longrightarrow [0, 1]$ is given. With this, define

$$
k_q(t; \pi) := \begin{cases} 0 & \text{if } 1 - p(t) \geq q(t) \\ \inf\left\{ k_q : 1 - p(t) + p(t) \int_{k_*}^{k_q} f(k, t)\, dk \geq q(t) \right\} & \text{if } 1 - p(t) < q(t) \text{ and } \pi \geq 0 \\ \sup\left\{ k_q : 1 - p(t) + p(t) \int_{k_q}^{k^*} f(k, t)\, dk \geq q(t) \right\} & \text{if } 1 - p(t) < q(t) \text{ and } \pi < 0 \end{cases}
$$

for any given portfolio strategy π. This has the following interpretation. The probability that at most a crash of size $k_q(t)$ at time t happens is $q(t)$. Equivalently, the probability that a crash higher than $k_q(t)$ will happen at time t is less than $1 - q(t)$. Obviously, this is a value at risk approach which relaxes the worst-case scenario approach.

Notice that the worst case of a non-negative portfolio strategy is either a crash of size k^* or no crash. On the other hand, the worst case of a negative portfolio strategy is either a crash of size k_* or no crash. Correspondingly, the q-quantile calculates differently for negative portfolio strategies (see the third row) than for the non-negative portfolio strategies (see the second row). Furthermore, denote by

$$
K_q(t) := \begin{cases} \{0\} & \text{if } k_q(t) = 0 \\ \{0\} \cup [k_*, k_q(t)] & \text{if } k_q(t) \neq 0 \text{ and } \pi \geq 0 \\ \{0\} \cup [k_q(\tau), k^*] & \text{if } k_q(t) \neq 0 \text{ and } \pi < 0 \end{cases} .
$$

Definition 2 (i) The problem to solve

$$\sup_{\pi(\cdot)\in A(x)} \inf_{\substack{0\le\tau\le T,\\ k\in K_q(\tau)}} \mathbb{E}\left[\ln\left(X^\pi(T)\right)\right], \tag{16}$$

where the terminal wealth $X^\pi(T)$ in the case of a crash of size k at time τ is given by

$$X^\pi(T) = [1 - \pi(\tau)k] \, X_1^{\pi,\tau,X_0^\pi(\tau)}(T), \tag{17}$$

with $X_1^{\pi,\tau,X_0^\pi(\tau)}(t)$ as above, is called the *(lower) q-quantile scenario portfolio problem.*

(ii) The *value function* to the above problem is defined via

$$w_q(t,x) := \sup_{\pi(\cdot)\in A(t,x)} \inf_{\substack{t\le\tau\le T,\\ k\in K_q(\tau)}} \mathbb{E}\left[\ln\left(X^{\pi,t,x}(T)\right)\right]. \tag{18}$$

(iii) A portfolio strategy $\hat{\pi}_q$ determined via the equation

$$w_q(t,x) = v_1\left(t, x\left(1 - \hat{\pi}_q(t)k_q(t)\right)\right) \qquad \text{for all } t \in [0, T] \text{ with } k_q(t) > 0$$

will be called a *(lower) q-quantile crash hedging strategy.*

(iv) A portfolio strategy $\tilde{\pi}_q$ is a *partial (lower) q-quantile crash hedging strategy,* if it is for any $t \in [0, T]$ either a q-quantile crash hedging strategy or a solution to the q-quantile scenario portfolio problem.

It is straightforward to see that the 1-quantile scenario portfolio problem is equivalent to the worst-case scenario portfolio problem given in Eq. (3). Moreover, the 1-quantile crash hedging strategy is equivalent to the crash hedging strategy in Definition 3.1 in Menkens [16, p. 602].

Remark 1 (i) Clearly, the definition given in Eq. (16) is different from the corresponding definition given in Sect. 3 and it leads only to the same solution in the two extreme cases of either $p = 1$ or $p = 0$.

(ii) Notice that the q-quantile scenario portfolio problem is only a q-quantile concerning the crash. The randomness of the market movement represented in the model by a geometric Brownian motion has been averaged out, namely by taking the expectation—and not the q-quantile.

Define the *support of k_q* to be

$$\text{supp}\left(k_q\right) := \left\{t \in [0, T] : k_q(t) > 0\right\}.$$

Using this, it is possible to show the following.

Theorem 1 *Let us suppose that k_q is continuously differentiable on* $\text{supp}\left(k_q\right)$ *with respect to t.*

(i) *Then there exists a unique (lower) q-quantile crash hedging strategy $\hat{\pi}_q$, which is on* supp (k_q) *given by the solution of the differential equation*

$$\hat{\pi}_q'(t) = \left(\hat{\pi}_q(t) - \frac{1}{k_q(t)}\right)\left[\frac{\sigma_0^2}{2}\left(\hat{\pi}_q(t) - \pi_0^*\right)^2 + \Psi_1 - \Psi_0\right] - \hat{\pi}_q(t)k_q'(t), \quad (19)$$

$$\hat{\pi}_q(T) = 0. \tag{20}$$

For $t \in [0, T] \setminus$ supp (k_q) set $\hat{\pi}_q(t) := \pi_0^*$.

Moreover, if $\Psi_1 \geq r_0$, then the q-quantile crash hedging strategy is bounded by

$$0 \leq \hat{\pi}_q(t) < \frac{1}{k_q(t)} \leq \frac{1}{k_*} \quad \text{for } t \in \text{supp } (k_q).$$

Additionally, if $\Psi_1 \leq \Psi_0$ and $\pi_0^* \geq 0$, the q-quantile crash hedging strategy has another upper bound with $\hat{\pi}_q(t) < \pi_0^* - \sqrt{\frac{2}{\sigma_0^2}(\Psi_0 - \Psi_1)}$.

On the other side, if $\Psi_1 < r_0$ the q-quantile crash hedging strategy is bounded by

$$\pi_0^* - \sqrt{\frac{2}{\sigma_0^2}(\Psi_0 - \Psi_1)} < \hat{\pi}_q(t) < 0 \quad \text{for } t \in [0, T).$$

(ii) *If $\Psi_1 < \Psi_0$ and $\pi_0^* < 0$, there exists a partial q-quantile crash hedging strategy $\tilde{\pi}_q$ at time t (which is different from $\hat{\pi}_q$), if*

$$S_q(t) := T - \frac{\ln\left(1 - \pi_0^* k_q(t)\right)}{\Psi_0 - \Psi_1} > 0 \quad \text{for } t \in \text{supp } (k_q). \tag{21}$$

With this, $\tilde{\pi}_q(t)$ is given by the unique solution of the differential equation

$$\tilde{\pi}_q'(t) = \left(\tilde{\pi}_q(t) - \frac{1}{k_q(t)}\right)\left[\frac{\sigma_0^2}{2}\left(\tilde{\pi}_q(t) - \pi_0^*\right)^2 + \Psi_1 - \Psi_0\right] - \tilde{\pi}_q(t)k_q'(t),$$

$$\tilde{\pi}_q\left(S_q(t)\right) = \pi_0^*.$$

For $S_q(t) \leq 0$ set $\tilde{\pi}_q(t) := \pi_0^*$. This partial crash hedging strategy is bounded by

$$\pi_0^* - \sqrt{\frac{2}{\sigma_0^2}(\Psi_0 - \Psi_1)} < \tilde{\pi}_q(t) \leq \pi_0^* < 0.$$

If k_q is independent of the time t, the optimal portfolio strategy for an investor, who wants to maximize her q-quantile scenario portfolio problem, is given by

$$\bar{\pi}_q(t) := \min\left\{\hat{\pi}_q(t), \tilde{\pi}_q(t), \pi_0^*\right\} \quad \text{for all } t \in [0, T], \tag{22}$$

where $\tilde{\pi}_q$ will be taken into account, if it exists. $\bar{\pi}_q$ will also be called the optimal q-quantile crash hedging strategy.

Remark 2 Let us write $\hat{\pi}_k(t)$ (instead of $\hat{\pi}_q(t)$) to emphasize the dependence on k, whenever needed. It follows from Eqs. (19) and (20) that

$$\hat{\pi}_k'(T) = -\frac{1}{k}[\Psi_1 - r_0] \xrightarrow{k \downarrow 0, k \neq 0} \begin{cases} \infty & \text{if } \Psi_1 < r_0 \\ 0 & \text{if } \Psi_1 = r_0 \\ -\infty & \text{if } \Psi_1 > r_0 \end{cases}. \tag{23}$$

(a) First, observe that this implies that $\hat{\pi}_q(t) \equiv 0$ if $\Psi_1 = r_0$, that is this is the only case where both the optimal q-quantile crash hedging strategy and the optimal crash hedging strategy are constant. That is, everything is invested in the risk-free asset if $\Psi_1 = r_0$.

(b) Second, notice that $\hat{\pi}_{k_1}' < \hat{\pi}_{k_2}'$ for $k_1 < k_2$. Hence, $\hat{\pi}_{k_1} \geq \hat{\pi}_{k_2}$ with strict inequality applying on $[0, T)$. Thus, in particular, $\hat{\pi}_q(t) > \hat{\pi}(t)$ for $t \in [0, T)$ for any q which satisfies $q(t) < 1$ for $t \in [0, T)$.

(c) Third, for the remainder of this remark, let us consider only the case that $\Psi_1 \leq \Psi_0$ and $\pi_0^* \geq 0$ (the other cases follow similarly). In this situation, one has that $\hat{\pi}_k(t) \leq \pi_0^* - \sqrt{\frac{2}{\sigma_0^2}(\Psi_0 - \Psi_1)}$. Thus, it is clear that

$$\psi(t) := \begin{cases} 0 & \text{for } t = T \\ \pi_0^* - \sqrt{\frac{2}{\sigma_0^2}(\Psi_0 - \Psi_1)} & \text{else} \end{cases}$$

is an upper bound for any $\hat{\pi}_k$ with $k > 0$. It follows that

$$\hat{\pi}_k(t) \longrightarrow \psi(t) \quad \text{pointwise for } k \downarrow 0 \text{ with } k \neq 0,$$

because of the convergence (23). Finally, keep in mind that the case $k = 0$ yields π_0^* as the optimal portfolio with $\pi_0^* \neq \psi$. An example is given in Fig. 4.

Proof (of Theorem 1) If $k_q(t)$ is constant in t this theorem follows from Theorem 4.1 in Korn and Wilmott [9, p.181], (for generalizations of this theorem see either Theorem 4.2 in Korn and Menkens [10, p.135] or Theorem 3.1 in Menkens [16, p.603]) by replacing k^* with k_q. To verify the differential equation in the general case, keep in mind that by differentiating the—modified—Equation (A.5) in Korn and Wilmott [9, p.183] (or Eq. (3.1) in Menkens [16, p.602]) with respect to t, $k_q(t)$ has also to be differentiated with respect to t. This leads to the differential equation (19).

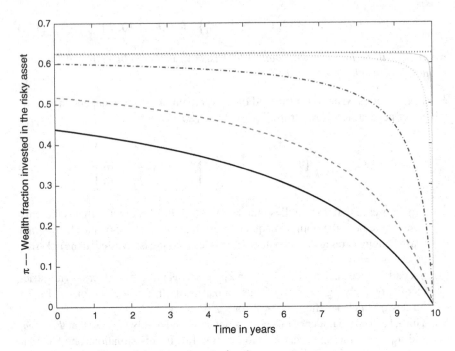

Fig. 4 Example of $k \longrightarrow 0$ for $\Psi_1 = \Psi_0$ and $\pi_0^* \geq 0$

5 Examples

5.1 Uniformly Distributed Crash Sizes

Suppose that the crash time has probability $p(t) = p$ and that the crash size is uniformly distributed on $[k_*, k^*]$, that is

$$f(k, t) = \left\{ \begin{array}{ll} \frac{1}{k^* - k_*} & \text{for } k \in [k_*, k^*] \\ 0 & \text{otherwise} \end{array} \right\}.$$

Using the defining equation for k_q, that is

$$1 - p + p \int_{k_*}^{k_q} \frac{1}{k^* - k_*} \, dk = q$$

this leads to the following equation for k_q

$$k_q = k_* + \frac{q + p - 1}{p} \left[k^* - k_* \right].$$

For $q = 1$, we get the worst case back, that is $k_1 = k^*$, as constructed.

5.2 Conditional Exponential Distributed Crash Sizes

Assume that the crash sizes are exponential distributed on the interval $[k_*, k^*]$. This means that

$$f(k, t) = \begin{cases} \frac{\lambda e^{-\lambda k}}{e^{-\lambda k_*} - e^{-\lambda k^*}} & \text{for } k \in [k_*, k^*] \\ 0 & \text{otherwise} \end{cases}.$$

With this, k_q calculates to

$$k_q = -\frac{1}{\lambda} \ln \left(\frac{1 - q - p}{p} \left[e^{-\lambda k_*} - e^{-\lambda k^*} \right] + e^{-\lambda k_*} \right).$$

Again, $k_q = k^*$ for $q = 1$.

5.3 Conditional Exponential Distributed Crash Sizes with Exponential Distributed Crash Times

Suppose that not only the crash height has a conditional exponential distribution, but also the crash time has a conditional distribution, independent of the crash size, that is

$$p(t) = q + (p - q) \frac{1 - e^{-\theta t}}{1 - e^{-\theta T}}. \tag{24}$$

This means that the probability of a crash happening is moving from q if $t = 0$ to p if $t = T$ in an exponential decreasing way if $q > p$. The defining equation of k_q writes in this case to

$$1 - q - (p - q) \frac{1 - e^{-\theta t}}{1 - e^{-\theta T}} + \left[q + (p - q) \frac{1 - e^{-\theta t}}{1 - e^{-\theta T}} \right] \frac{e^{-\lambda k_*} - e^{-\lambda k_q(t)}}{e^{-\lambda k_*} - e^{-\lambda k^*}} = q.$$

This gives

$$k_q(t) = -\frac{1}{\lambda} \ln \left(e^{-\lambda k^*} + \frac{[1 - q][1 - e^{-\theta T}] \left[e^{-\lambda k_*} - e^{-\lambda k^*} \right]}{q[1 - e^{-\theta T}] + (p - q)[1 - e^{-\theta t}]} \right).$$

Clearly, this is an example where k_q depends on the time t. Its derivative calculates to

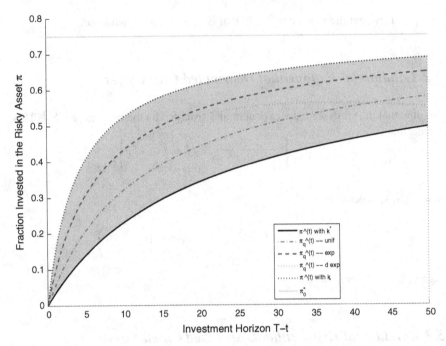

Fig. 5 The range of (optimal) q-quantile crash hedging strategies for $\Psi_1 = \Psi_0$ and $\pi_0^* \geq 0$. This graphic shows $\hat{\pi}_{k^*}$ (*solid line*), $\hat{\pi}_{k_*}$ (*dotted line*), the range of possible optimal q–quantile crash hedging strategies (grey area) if k_q is constant, and π_0^* (*solid straight line*). The *dash–dotted line* is a uniform distributed example (see Sect. 5.1), the *dashed line* is an exponential distributed example (see Sect. 5.2), and the *dotted line* is a time–varying example (see Sect. 5.3).

$$\frac{dk_q}{dt}(t) = -\frac{1}{\lambda} \frac{\dfrac{-(p-q)[1-q]\theta e^{-\theta t}\left[1-e^{-\theta T}\right]}{\left[q[1-e^{-\theta T}]+(p-q)[1-e^{-\theta t}]\right]^2}\left[e^{-\lambda k_*} - e^{-\lambda k^*}\right]}{e^{-\lambda k^*} + \dfrac{[1-q][1-e^{-\theta T}]}{q[1-e^{-\theta T}]+(p-q)[1-e^{-\theta t}]}\left[e^{-\lambda k_*} - e^{-\lambda k^*}\right]}$$

$$= \frac{1}{\lambda\left[q\left[1-e^{-\theta T}\right]+(p-q)\left[1-e^{-\theta t}\right]\right]}$$

$$\times \frac{(p-q)\left[1-q\right]\theta e^{-\theta t}\left[1-e^{-\theta T}\right]\left[e^{-\lambda k_*} - e^{-\lambda k^*}\right]}{(p-q)\left[1-e^{-\theta t}\right]e^{-\lambda k^*} + \left[1-e^{-\theta T}\right]\left[e^{-\lambda k_*}\left[1-q\right] - e^{-\lambda k^*}\left[1-2q\right]\right]}.$$

Figure 5 shows the potential range of the optimal q-quantile crash hedging strategy (the gray shaded area) if $k_q(t) \neq 0$ is constant. Obviously, in the case of $k_q(t) = 0$, one has that $\hat{\pi}_q(t) = \pi_0^*$ (that is the optimal strategy is to invest according to the Merton fraction). Moreover, if $k_q(t)$ is not constant, it can happen that the corresponding q-quantile crash hedging strategy moves outside the given range. However, this usually happens only if the derivative $\frac{dk_q}{dt}$ is sufficiently large—which is not the case for most situations. The parameters used in these figures are $k^* = 0.5$, $k_* = 0.1$,

$\pi_0^* = 0.75$, and $\sigma_0 = 0.25$. Additionally, the examples discussed above are plotted for the choices of $p = 0.1$ and $q = 0.95$. The dashed line is the example of a uniform distribution with $p = 10\%$ and $q = 5\%$. The dash–dotted line is the example of an exponential distribution with the additional parameter $\lambda = 10$ (where the other parameters are as above) and the dotted line is the example of a time varying crash probability given in Eq. (24) with the additional parameter $\theta = 0.1$. Notice that the first two examples lead to similar strategies as in Korn and Wilmott [9], just that k^* is replaced by k_q, which is constant in those two examples. The third example is clearly different from that. Starting with an investment horizon of $T = 50$ years, the optimal strategy is to increase the fraction invested in the risky asset up to an investment horizon of about 30 years. This is due to the fact that the probability of a crash happening is 95 % at $T = 50$ and it is exponentially decreasing to 10 % as the investment horizon is reached.

6 Deterministic Portfolio Strategies

Definition 3 Let π be an admissible portfolio strategy.

$$\pi_d(t) := \mathbb{E}[\pi(t)] \quad \text{for all } t \in [0, T]$$

will be called the *(to π) corresponding deterministic portfolio strategy*.

If $\pi_d = \pi$, then π_d is admissible because π is admissible. If $\pi_d(t) \neq \pi(t)$ for some $t \in [0, T]$, then there exist $\omega_1, \omega_2 \in \Omega$, which depend on t, such that

$$\pi(t, \omega_1) \leq \pi_d(t) \leq \pi(t, \omega_2).$$

Thus, π_d is bounded and therefore admissible.

Definition 4 Let us define

$$k_\pi(t) := k^* \cdot \mathbb{1}_{\{\pi(t) \geq 0\}} + k_* \cdot \mathbb{1}_{\{\pi(t) < 0\}}.$$

Lemma 1 *Let π be an admissible portfolio strategy. Then the corresponding deterministic portfolio strategy to π yields in the initial crash-free market at least the same expected utility of terminal wealth as π. If, additionally $\pi(t) < \frac{1}{k^*}$ holds for all $t \in [0, T]$, then π_d yields in the initial market with a possible crash at least the same worst case expected utility of terminal wealth as π.*

Remark 3 This Lemma is important because often an optimization problem is solved only on the set of deterministic strategies (see e.g., Korn and Wilmott [9], Korn and Menkens [10], or Christiansen [6]) and not on the set of stochastic strategies (which include the deterministic ones). This is done because it is often simpler to solve the optimization problem on the set of deterministic strategies.

Proof (of Lemma 1) Using the Theorem of Fubini, one has for any admissible portfolio strategy π

$$\mathscr{J}_0(t, x, \pi) = \ln(x) + \mathbb{E}\left[\int_t^T \Psi_0 - \frac{\sigma_0^2}{2}\left(\pi(s) - \pi_0^*\right)^2 ds\right]$$

$$= \ln(x) + \int_t^T \Psi_0 - \frac{\sigma_0^2}{2}\mathbb{E}\left[\left(\pi(s) - \pi_0^*\right)^2\right] ds$$

$$= \ln(x) + \int_t^T \Psi_0 - \frac{\sigma_0^2}{2}\left(\mathbb{E}[\pi(s)] - \pi_0^*\right)^2 - \frac{\sigma_0^2}{2}\mathrm{Var}\left(\pi(s)\right) ds$$

$$= \ln(x) + \int_t^T \Psi_0 - \frac{\sigma_0^2}{2}\left(\pi_d(s) - \pi_0^*\right)^2 - \frac{\sigma_0^2}{2}\mathrm{Var}\left(\pi(s)\right) ds$$

$$= \mathscr{J}_0(t, x, \pi_d) - \frac{\sigma_0^2}{2}\int_t^T \mathrm{Var}\left(\pi(s)\right) ds$$

$$\leq \mathscr{J}_0(t, x, \pi_d).$$

This is the case if no crash happens. In the case that a crash has happened, one gets with the definition

$$A_\pi(t) := \ln\left(1 - \mathbb{E}[\pi(t)]k_{\pi_d}(t)\right) - \mathbb{E}[\ln\left(1 - \pi(t)k_\pi(t)\right)]$$

the following

$$v_1(t, x(1 - \pi(t)k_\pi(t))) = \ln(x) + \mathbb{E}[\ln\left(1 - \pi(t)k_\pi(t)\right)] + \Psi_1(T - t)$$

$$= \ln(x) + \ln\left(1 - \pi_d(t)k_{\pi_d}(t)\right) + \Psi_1(T - t) - A_\pi(t)$$

$$= v_1\left(t, x\left(1 - \pi_d(t)k_{\pi_d}(t)\right)\right) - A_\pi(t)$$

$$\leq v_1\left(t, x\left(1 - \pi_d(t)k_{\pi_d}(t)\right)\right),$$

where it has been used for the last inequality that $A_\pi(t) \geq 0$. However, this is Jensen's inequality which holds if $1 - \pi(t)k_\pi(t) \geq 0$. The latter holds for $\pi(t) < \frac{1}{k^*}$, which is the assumption. This proves the assertion.

Remark 4 The condition $\pi(t) < \frac{1}{k^*}$ is natural if a crash of size k^* can happen, because it avoids that the investor can go bankrupt. Since $k^* \leq 1$, the condition means that the investor is not allowed to be too much leveraged.

7 Conclusion

It has been shown that the worst-case scenario approach of Korn and Wilmott [9] will not make use of additional probabilistic information of a crash happening. This is overcome by introducing a q-quantile approach which is a Value at Risk ansatz to the worst-case scenario method. Examples are given; in particular, one extreme example shows that it is possible with the q-quantile approach to obtain optimal portfolio strategies which are first increasing and then decreasing. Finally, it is shown that any stochastic portfolio strategy will give a lower expected utility of terminal wealth (or a lower worst-case scenario bound) than the corresponding deterministic portfolio strategy (defined by taking the expectation of the stochastic portfolio strategy)).

Acknowledgments I like to thank Prof. Ralf Korn for many fruitful discussions, for generating a stimulating working atmosphere, not only when I was his PhD student but also when I visited him. I also benefited from discussions with Christian Ewald, Frank Thomas Seifried, and Mogens Steffensen. Moreover, the feedback from Rudi Zagst and an anonymous referee improved this paper considerably.

Support from DFG through the SPP 1033 *Interagierende stochastische Systeme von hoher Komplexität* and partial support by the Science Foundation Ireland via the *Edgeworth Centre* (Grant No. 07/MI/008) and *FMC2* (Grant No. 08/SRC/FMC1389) is gratefully acknowledged.

Open Access This chapter is distributed under the terms of the Creative Commons Attribution Noncommercial License, which permits any noncommercial use, distribution, and reproduction in any medium, provided the original author(s) and source are credited.

References

1. Aase, K.K.: Optimum portfolio diversification in a general continuous-time model. Stochast. Process. Appl. **18**, 81–98 (1984)
2. Belak, C., Christensen, S., Menkens, O.: Worst-case optimal investment with a random number of crashes. Stat. Probab. **90**, 140–148 (2014)
3. Belak, C., Christensen, S., Menkens, O.: Worst-case portfolio optimization in a market with bubbles. Preprint, available at http://ssrn.com/abstract=2319913 (2013)
4. Belak, C., Menkens, O., Sass, J.: Worst-case portfolio optimization with proportional transaction costs. Preprint, available at http://ssrn.com/abstract=2207905 (2013)
5. Bertsimas, D., Brown, D.B., Caramanis, C.: Theory and applications of robust optimization. SIAM Rev. **53**(3), 464–501 (2011)
6. Christiansen, M.C.: A variational approach for mean-variance-optimal deterministic consumption and investment. In: Glau, K. et al. (eds.) Innovations in Quantitative Risk Management, Springer Proceedings in Mathematics & Statistics, vol. 99 (2015)
7. Desmettre, s., Korn, R., Seifried, F.T.: Worst-case consumption-portfolio optimization, Int. J. Theor. Appl. Financ. (2014)
8. Hua, P., Wilmott, P.: Modelling market crashes: The worst-case scenario. Working Paper Series 1999-MF-24, Oxford Financial Research Centre, London. See http://www.finance.ox.ac.uk/Papers/MathematicalFinance/index.shtml (1999)
9. Korn, R., Wilmott, P.: Optimal portfolios under the threat of a crash. Int. J. Theor. Appl. Financ. **5**(2), 171–187 (2002)

10. Korn, R., Menkens, O.: Worst-case scenario portfolio optimization: a new stochastic control approach. Math. Methods Oper. Res. **62**(1), 123–140 (2005)
11. Korn, R.: Worst-case scenario investment for insurers. Insur.: Math. Econ. **36**(1), 1–11 (2005)
12. Korn, R., Steffensen, M.: On worst-case portfolio optimization. SIAM J. Control Optim. **46**(6), 2013–2030 (2007)
13. Korn, R., Menkens, O., Steffensen, M.: Worst-case-optimal dynamic reinsurance for large claims. Eur. Actuar. J. **2**(1), 21–48 (2012)
14. Mataramvura, S., Øksendal, B.: Risk minimizing portfolios and HJBI equations for stochastic differential games. Stochastics **80**(4), 317–337 (2008)
15. Menkens, O.: Costs and benefits of crash hedging. Preprint, available at http://ssrn.com/abstract=2397233 February 2014
16. Menkens, O.: Crash hedging strategies and worst-case scenario portfolio optimization. Int. J. Theor. Appl. Financ. **9**(4), 597–618 (2006)
17. Merton, R.C.: Lifetime portfolio selection under uncertainty: The continuous-time case. Rev. Econ. Stat. **51**, 247–257 (1969)
18. Mönnig, L.: A worst-case optimization approach to impulse perturbed stochastic control with application to financial risk management. PhD thesis, Technische Universität Dortmund (2012)
19. Øksendal, B., Sulem, A.: Portfolio optimization under model uncertainty and bsde games. Quant. Financ. **11**(11), 1665–1674 (2011)
20. Pelsser, A.: Pricing in incomplete markets. Preprint, available at http://ssrn.com/abstract=1855565 May 2011
21. Rustem, B., Howe, M.: Algorithms for Worst-case Design and Applications to Risk Management. Princton University Press, Princeton (2002)
22. Seifried, F.T.: Optimal investment for worst-case crash scenarios: a martingale approach. Math. Oper. Res. **35**(3), 559–579 (2010)
23. Wald, A.: Statistical decision functions which minimize the maximum risk. Ann. Math. **46**(2), 265–280 (1945)
24. Wald, A.: Statistical Decision Functions. Wiley, New York (1950)

Improving Optimal Terminal Value Replicating Portfolios

Jan Natolski and Ralf Werner

Abstract Currently, several large life insurance companies apply the replicating portfolio technique for valuation and risk management of their liabilities. In [7], the two most common approaches, cash-flow matching and terminal value matching, have been investigated from a theoretical perspective and it has been shown that optimal terminal value replicating portfolios are not suitable to replicate liability cash-flows by construction. Thus, their usage for asset liability management is rather restricted, especially for out-of-sample cash profiles of liabilities. In this paper, we therefore enhance the terminal value approach by an additional linear regression of the corresponding optimal dynamic numéraire strategy to overcome this drawback. We show that terminal value matching together with an approximated dynamic strategy has in-sample and out-of-sample performance very close to the optimal cash-flow matching portfolio and, due to computational advantages, can thus be used as an alternative for cash-flow matching, especially in risk and asset liability management.

1 Introduction

In the last years, market consistent valuation has become the standard approach toward risk management of life insurance policies, see for example [3]. Due to the complexity of life insurance contracts, most academics and practitioners resort to Monte Carlo methods for valuation purposes. However, the difficulty is to find a computationally efficient yet sufficiently accurate algorithm. For instance, contracts may include surrender options, which allow the policy holder every year to cancel the contract and withdraw the value of her account. In this context, [1, 2] and several

J. Natolski · R. Werner (✉)
University of Augsburg, Universitätsstraße 14, 86159 Augsburg, Germany
e-mail: ralf.werner@math.uni-augsburg.de

J. Natolski
e-mail: jan.natolski@math.uni-augsburg.de

© The Author(s) 2015
K. Glau et al. (eds.), *Innovations in Quantitative Risk Management*,
Springer Proceedings in Mathematics & Statistics 99,
DOI 10.1007/978-3-319-09114-3_16

other authors therefore resort to the well-known least squares Monte Carlo approach, which was originally introduced by [6] to price American options. In contrast, [9] first suggested valuation of with-profits guaranteed annuity options, which are typical life insurance products, via static replicating portfolios. To hedge against interest rate risk, a portfolio is built of vanilla swaptions and a remarkably good fit of the market value of annuity options is obtained. The purpose of constructing a replicating portfolio is to approximate the liability cash-flows of an insurance company by a portfolio formed by a finite number of selected financial instruments. If the approximation is accurate, one obtains a good estimate of the market value of liabilities from the fair value of the replicating portfolio. In current literature, two portfolio construction approaches stand out. The first one aims to match liability cash-flows and cash-flows of the replicating portfolio at each time point. The second one is less restrictive as it only demands that accrued terminal values of the cash-flows match well at some final time horizon T.

For risk purposes, insurance companies want to compute the fair value of their assets and liabilities, i.e., the market consistent embedded value (MCEV) under shifted market conditions now or one year in the future. More precisely, having found a replicating portfolio which matches the fair value of liabilities under current market conditions, one performs instantaneous shocks on known parameters (such as volatility, forward rate curve, etc.) and checks if fair values are still matched. This is commonly referred to as a comparison of sensitivities between the fair value of the replicating portfolio and the fair value of liabilities. If sensitivities are similar, it is usually assumed that fair values will be roughly matched one year in the future even if rare events in the 99.5 % quantile take place. For instance, this is the motivation for [4] to put additional constraints in the optimization problem to guarantee that fair values are close to one another under various stress scenarios. Figure 1 illustrates the dependence between initial asset prices and the fair value of liabilities and a replicating portfolio. It can be observed that fair values are close to each other and behave quite similar, but not fully identical.

For the purpose of improving terminal value matching, we start with the setup as given in [7], that is, we consider the cash-flow matching problem and the terminal value matching problem as proposed in [9] and [8], respectively, and relax the requirement of static replication by allowing for dynamic investment strategies in the numéraire asset. We briefly review the theoretical results derived therein, before we investigate in more detail the benefit of our approach based on market scenarios generated by an insurance company: First, we compare the in-sample and out-of-sample performance of the two replicating portfolios. Then, in the main contribution of this article, we take a closer look at the optimal dynamic investment strategy and approximate it by a time-dependent linear combination of the replicating assets. In our particular example, the approximation turns out to be remarkably accurate as in-sample and out-of sample tests will show.

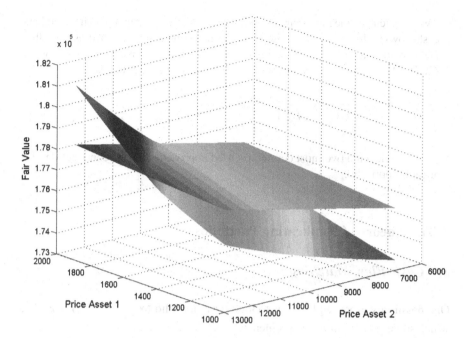

Fig. 1 Fair value of liabilities and of the replicating portfolio depending on initial asset prices

2 The Mathematical Setup

This setup roughly follows that of [3]: Let $\left(\Omega, \mathscr{F}, (\mathscr{F}_t)_{t \in \mathscr{T}}, \mathbb{Q}\right)$ be a filtered probability space[1] in discrete time $\mathscr{T} := \{t = 0, 1 \ldots, T\}$ with risk-neutral measure $\mathbb{Q}$. On this probability space, we introduce a frictionless, arbitrage-free financial market as follows.

- Let $\left(R_t^F\right)_t \in \mathbb{R}^n, t \in \mathscr{T}$ be Markovian financial risk factors (e.g., interest rates).
- Let $\left(R_t^L\right)_t \in \mathbb{R}^l, t \in \mathscr{T}$ be Markovian risk factors, independent of $\left(R_t^F\right)_{t \in \mathscr{T}}$, affecting the liabilities of an insurance company (e.g., mortality rates).
- $\left(R_t^F\right)_{t \in \mathscr{T}}$ and $\left(R_t^L\right)_{t \in \mathscr{T}}$ generate filtrations $\left(\mathscr{F}_t^F\right)_{t \in \mathscr{T}}$ and $\left(\mathscr{F}_t^L\right)_{t \in \mathscr{T}}$ respectively. We assume $\mathscr{F}_t = \mathscr{F}_t^F \vee \mathscr{F}_t^L, \ \forall t \in \mathscr{T}$.
- There are m financial assets, an $\mathbb{R}^{d_1}$-valued $\left(\mathscr{F}_t^F\right)$-adapted process $\left(D_t^F\right)_{t \in \mathscr{T}}$ (e.g., risk factors, book values, moving averages, etc.) and a function $C_i^F : \{1, \ldots, T\} \times \mathbb{R}^{d_1} \mapsto \mathbb{R}$ for each asset i such that $C_i^F(t, D_t^F)$ is the cash payment of asset i at time t. At T, this cash payment represents the remaining value of the asset. D^L and C^L are defined analogously; however, liabilities may also depend on the financial risk factors D^F, i.e., $C^L = C^L(t, D_t^F, D_t^L)$.

---

[1] Similarly to [7] we assume that all technical requirements are fulfilled (square integrability, completeness of filtration, ...).

- $(N_t)_{t \in \mathcal{T}}$ denotes the numéraire (with initial value $N_0 = 1$, paying no intermediate cash-flows) which is used in the dynamic investment strategy. We assume that N_T is paid as a cash-flow at the final time horizon. For convenience, let us write $C_0^F(t, D_t^F)$ for the cash payment of the numéraire at time t, that is

$$
C_0^F\left(t, D_t^F\right) = \begin{cases} 0, & t = 1, \ldots, T-1 \\ N_T, & t = T \end{cases}.
$$

Next, we review the two most commonly used approaches for the construction of a replicating portfolio.

3 The Theory of Replicating Portfolios

3.1 Cash-Flow Matching

One possibility proposed by [9] is to look for a portfolio $(\alpha_{\text{opt}}^0, \ldots, \alpha_{\text{opt}}^m) \in \mathbb{R}^{m+1}$, which solves the optimization problem[2]

$$
\min_{\alpha \in \mathbb{R}^{m+1}} \sum_{t=1}^{T} \left[\mathbb{E}^{\mathbb{Q}} \left(\left[\frac{C^L(t, D_t^F, D_t^L)}{N_t} - \sum_{i=0}^{m} \alpha^i \frac{C^F_i(t, D_t^F)}{N_t} \right]^2 \right) \right]^{\frac{1}{2}}. \quad \text{(RP}_{\text{CF}}\text{)}
$$

The objective function penalizes the difference between two cash payments at each time t. The role of the discounting factor $1/N_t$ is to assign equal weight to mismatches of equal size in terms of their discounted value. An alternative approach is discounted terminal value matching.

3.2 Discounted Terminal Value Matching

The terminal value of a cash-flow is obtained by summing all cash payments accrued to the terminal time T with the risk-free interest rate. By discounted terminal value, we mean the accrued terminal value discounted to the present. In mathematical notation, the discounted accrued liability cash-flow and the discounted accrued cash-flow of a replicating portfolio $\alpha = (\alpha^0, \ldots, \alpha^m) \in \mathbb{R}^{m+1}$ are given by

---

[2] The existence of a minimum has been shown in [7].

$$\tilde{A}^L := \sum_{t=1}^{T} \frac{C^L\left(t, D_t^F, D_t^L\right)}{N_t},$$

$$\tilde{A}^F(\alpha) := \sum_{t=1}^{T} \left[\sum_{i=0}^{m} \alpha^i \frac{C_i^F\left(t, D_t^F\right)}{N_t} \right].$$

The observation that although two cash-flows may have entirely different cash payment profiles, they still have the same fair value, leads to the alternative optimization proble[3]

$$\min_{\alpha \in \mathbb{R}^{m+1}} \left[\mathbb{E}^{\mathbb{Q}} \left(\left[\tilde{A}^L - \tilde{A}^F(\alpha) \right]^2 \right) \right]^{\frac{1}{2}}. \qquad (\mathrm{RP}_{\tilde{\mathrm{TV}}})$$

Originally, this problem was introduced by [8] with the difference that they considered non-discounted terminal values.

4 Equivalence of Cash-Flow Matching and Discounted Terminal Value Matching

Next, we recall the connection between $(\mathrm{RP}_{\mathrm{CF}})$ and $(\mathrm{RP}_{\tilde{\mathrm{TV}}})$ as established in [7]. If the numéraire asset can be bought or sold at any time, problems $(\mathrm{RP}_{\mathrm{CF}})$ and $(\mathrm{RP}_{\tilde{\mathrm{TV}}})$ are practically the same. The brief explanation is that cash-flow mismatches can be laid off by an appropriate strategy in this asset. These mismatches then sum up to the discounted terminal value mismatch and thus problems $(\mathrm{RP}_{\mathrm{CF}})$ and $(\mathrm{RP}_{\tilde{\mathrm{TV}}})$ are intimately linked.

In more detail, suppose that the insurance company is allowed to invest and finance cash-flows from trading the numéraire asset at all times $t = 1, \ldots, T$. Define the following linear space of processes

$$\mathscr{A} = \left\{ (\delta_t)_{t=1,\ldots,T} : \forall t = 1, \ldots, T-1, \ \delta_t \in \mathscr{L}^2(\mathscr{F}_t), \ \sum_{t=1}^{T} \delta_t = 0 \right\}.$$

Any process from this space represents an adapted strategy of investments in or borrowing from the numéraire asset, so δ_t is the number of assets bought or sold short at time t. Here, $\delta_t > 0$ is interpreted as a purchase, which corresponds to a negative cash-flow for the insurer and $\delta_t < 0$ as a sale, which corresponds to a positive cash-flow. The condition $\sum_{t=1}^{T} \delta_t = 0$ ensures that strategies have zero discounted terminal value. Note that strategies from $\mathscr{A}$ are not necessarily predictable. At each time point, the insurer can incorporate all information available at that time to make a decision on the trade δ_t. Only at T is the insurer bound to clear the balance thus making δ_T predictable.

---

[3] For the existence of a minimum, see [7].

The introduction of such strategies turns out to be the key link between problems (RP$_{\mathrm{CF}}$) and (RP$_{\tilde{\mathrm{TV}}}$): The discounted terminal value $\tilde{A}^F(\alpha, \delta)$ corresponding to an investment strategy (α, δ) with $\alpha \in \mathbb{R}^{m+1}, \delta \in \mathscr{A}$ is given by

$$\tilde{A}^F(\alpha, \delta) = \tilde{A}^F(\alpha),$$

where

$$\tilde{A}^F(\alpha, \delta) := \sum_{t=1}^{T} \left[\sum_{i=0}^{m} \alpha^i \frac{C_i^F(t, D_t^F)}{N_t} \right] - \sum_{t=1}^{T} \delta_t.$$

In other words, the discounted terminal value only depends on the initial portfolio $\alpha^0, \alpha^1, \ldots, \alpha^m$ in the assets. Thus, we write $\tilde{A}^F(\alpha)$ instead of $\tilde{A}^F(\alpha, \delta)$. We say that two investment strategies (α, δ) and $(\beta, \hat{\delta})$ with $\alpha, \beta \in \mathbb{R}^{m+1}, \delta, \hat{\delta} \in \mathscr{A}$ are FV-equivalent iff

$$\alpha = \beta.$$

Note that due to the above, initial portfolios of two FV-equivalent investment strategies have equal fair value, as they produce identical discounted terminal values.

Based on the extension from static portfolios to partially dynamic strategies, we define corresponding optimization problems,

$$\inf_{\alpha \in \mathbb{R}^{m+1}, \delta \in \mathscr{A}} \sum_{t=1}^{T} \left[\mathbb{E}^Q \left(\left[\frac{C^L(t, D^F_t, D^L_t)}{N_t} - \left(\sum_{i=0}^{m} \alpha^i \frac{C^F_i(t, D^F_t)}{N_t} - \delta_t \right) \right]^2 \right) \right]^{\frac{1}{2}}, \quad \text{(GRP}_{\mathrm{CF}}\text{)}$$

the generalized cash-flow matching problem and

$$\inf_{\alpha \in \mathbb{R}^{m+1}, \delta \in \mathscr{A}} \left[\mathbb{E}^Q \left(\left[\tilde{A}^L - \tilde{A}^F(\alpha) \right]^2 \right) \right]^{\frac{1}{2}}, \quad \text{(GRP}_{\tilde{\mathrm{TV}}}\text{)}$$

the generalized discounted terminal value matching problem. Based on the following two additional weak assumptions, the main results follow.

Assumption 1 The matrix $\mathbb{E}^Q(Q^F)$ is positive definite, where

$$Q^F := \left(\tilde{A}_i^F \tilde{A}_j^F \right)_{i,j=0,\ldots,m},$$

with discounted terminal value $\tilde{A}_i^F$ of the cash-flow generated by asset i given as

$$\tilde{A}_i^F := \sum_{t=1}^{T} \frac{C_i^F(t, D_t^F)}{N_t}.$$

Assumption 2 Let $\alpha_{\text{opt}} = \left(\alpha_{\text{opt}}^0, \alpha_{\text{opt}}^1, \ldots, \alpha_{\text{opt}}^m \right)$ be the solution to $(\text{RP}_{\tilde{\text{TV}}})$. The cash-flow mismatch $C^L \left(T, D_T^F, D_T^L \right) - \sum_{i=0}^m \alpha_{\text{opt}}^i C_i^F \left(T, D_T^F \right)$ is not $\mathscr{F}_{T-1}$-measurable.

The following properties of the two optimization problems and their connections were derived in [7].

1. **Properties of $(\text{GRP}_{\tilde{\text{TV}}})$ and the relationship to $(\text{RP}_{\tilde{\text{TV}}})$:**

    a. Under Assumption 1, the solution to $(\text{RP}_{\tilde{\text{TV}}})$ exists, is unique and given by $\alpha_{\text{opt}} = \mathbb{E}^{\mathbb{Q}} \left(Q^F \right)^{-1} \mathbb{E}^{\mathbb{Q}} \left(\tilde{A}^F \tilde{A}^L \right)$. The set of solutions to $(\text{GRP}_{\tilde{\text{TV}}})$ is the FV-equivalence class of the solution to $(\text{RP}_{\tilde{\text{TV}}})$.

    b. The optimal value of $(\text{GRP}_{\tilde{\text{TV}}})$ is equal to the optimal value of $(\text{RP}_{\tilde{\text{TV}}})$.

2. **Properties of (GRP_{CF}) and the relationship to $(\text{RP}_{\tilde{\text{TV}}})$, $(\text{GRP}_{\tilde{\text{TV}}})$ and (RP_{CF}):**

    a. Under Assumptions 1 and 2, the solution to (GRP_{CF}) exists and is unique with initial portfolio given by the solution to $(\text{RP}_{\tilde{\text{TV}}})$ and strategy $\delta \in \mathscr{A}$ such that *cash-flows are perfectly matched* at times $t = 1, \ldots, T - 1$.

    b. Under Assumption 1, the set of solutions to $(\text{GRP}_{\tilde{\text{TV}}})$ is the equivalence class of the solution to (GRP_{CF}).

    c. The optimal value of (GRP_{CF}) is smaller than or equal to the optimal value of (RP_{CF}). Under Assumptions 1 and 2, equality is achieved iff for times $t = 1, \ldots, T - 1$ the liability cash-flow is perfectly replicated by the cash-flow of the portfolio solving $(\text{RP}_{\tilde{\text{TV}}})$.

3. **Fair values of $(\text{GRP}_{\tilde{\text{TV}}})$ and (GRP_{CF}):**

    a. The fair value of the solutions to $(\text{GRP}_{\tilde{\text{TV}}})$ and the fair value of the solution to (GRP_{CF}) are equal to the fair value of the liability cash-flow.

The main drawback of the generalized terminal value approach lies in the introduction of the dynamic strategy in the numéraire asset: as the optimal δ_t depend on the liability cash-flow (see Property 2.c above), this strategy is not available out-of-sample to reproduce (unknown!) liability cash-flows. Although the main purpose of replicating portfolios in risk management is fair value replication, asset liability management usually requires cash-flow replication as well.

Therefore, the optimal numéraire strategy has to be estimated based on available information up to time t, which then in turn allows a reproduction of liability cash-flows, even in a terminal value approach. The most simple approach toward this end is a standard linear regression of the optimal δ_t against the information available in time t. Besides the obvious usage of prices of financial instruments as explaining variables, any further available information (e.g., non-traded risk factors like interest rate, etc.) could in theory be used for this purpose.

Starting with the portfolio solving $(\text{RP}_{\tilde{\text{TV}}})$, we compute $(\delta_t)_{t=1,\ldots,T}$ such that cash-flows are perfectly matched in-sample except for T. The idea is to approximate δ_t, $t = 1, \ldots T - 1$ by an ordinary linear regression, that is

$$\hat{\delta}_t(a) := a_t^1 \frac{C_{\text{CA}}^F\left(t, D_t^F\right)}{N_t} + a_t^2 \frac{C_{\text{SP}}^F\left(t, D_t^F\right)}{N_t} + a_t^3 \frac{C_{\text{NK}}^F\left(t, D_t^F\right)}{N_t}, \quad t = 1, \dots, T-1,$$

$$\hat{\delta}_T(a) := -\sum_{t=1}^{T-1} \hat{\delta}_t$$

where $a \in \mathbb{R}^{T-1 \times 3}$ solves the problem

$$\min_{a \in \mathbb{R}^{T-1 \times 3}} \sum_{t=1}^{T} \left[\mathbb{E}^{\mathbb{Q}} \left(\left[\frac{C^L(t, D_t^F, D_t^L)}{N_t} - \left(\sum_{i=0}^{m} \alpha_{\text{opt}}^i \frac{C_i^F(t, D_t^F)}{N_t} - \hat{\delta}_t(a) \right) \right]^2 \right) \right]^{\frac{1}{2}}.$$

In other words, we solve (GRP$_{\text{CF}}$) with $\alpha_{\text{opt}}^0, \dots, \alpha_{\text{opt}}^m$ fixed and optimal for (RP$_{\widetilde{\text{TV}}}$) and δ_t restricted to have the form above. Note that the parameters $(a_t^1, a_t^2, a_t^3)_{t=1,\dots T-1}$ are known to the insurer at present. The hope is that the portfolio obtained from matching discounted terminal values together with dynamic investment strategy $(\hat{\delta}_t)_{t=1,\dots,T}$ will produce at least a similar out-of-sample objective value as the static portfolio obtained from solving (RP$_{\text{CF}}$).

5 Example

Based on financial market scenarios provided by a life insurer, we carry out some numerical analysis to compare the performance of the portfolios solving (RP$_{\text{CF}}$) and (RP$_{\widetilde{\text{TV}}}$). The results above imply that in an in-sample test the terminal value technique will outperform the cash-flow matching technique. On the other hand, it is not clear what happens in an out-of-sample test. This will also depend on the robustness of both methods.

Since scenarios for liability cash-flows were unavailable, we implemented the model proposed by [5]. A policy holder pays an initial premium P_0, which is invested by the insurer in a corresponding portfolio of assets with value process $(A_t)_{t \in \mathbb{N}}$. The value of the contract $(L_t)_{t \in \mathbb{N}}$ now evolves according to the following recursive formula.

$$L_{t+1} = L_t \left(1 + \max\left(r_G, \rho\left(\frac{A_t - L_t}{L_t} - \gamma \right) \right) \right), \quad t = 0, 1, \dots,$$

where r_G is the interest guaranteed to the policy holder, ρ is the level of participation in market value earnings and γ is a target buffer ratio.

To generate liability cash-flows, we assumed that starting in January 1998 the insurance company receives one client every year up to 2012. Each client pays an initial nominal premium of 10.000 Euros. All contracts run 15 years. At maturity, the value of the contract is paid to the policy holder generating a liability cash-flow.

The portfolio in which the premia are invested consists of the Standard and Poors 500 index, the Nikkei 225 index and the cash account. We normed the values of the cash account, the S and P 500 and the Nikkei 225 so that all three have value 1 Euro in year 2012. Every year the portfolio is adjusted such that 80% of the value are invested in the cash account and 10% are invested in each index.

For the construction of the replicating portfolio, we chose the same three assets. Cash-flows are generated by selling or buying assets every year. Since we are constructing a static portfolio, the decision how many assets will be bought or sold each year in the future has to be made in the present. Hence, one may regard the replicating assets as 3×15 call options with strike 0, one option for each index/cash account and each year. We chose the cash account as the numéraire asset in the market.

In order to make the evolution of contract values more sensitive to changes of financial asset prices, we assumed a low guaranteed interest rate of $r_G = 2.0\%$, a high participation ratio $\rho = 0.75$ and a low target buffer ratio $\gamma = 0.05$. From 1,000 scenarios,[4] we chose to use 500 for the construction of the replicating portfolios and the remaining 500 for an out-of-sample performance test. The portfolio is constructed

Table 1 Optimal replicating portfolios (in thousand Euros) for problems (RP$_{CF}$) and (RP$_{TV}$) and their fair value

Year	Cash-flow match			Terminal value match		
	Cash account	S&P	Nikkei	Cash account	S&P	Nikkei
Total initial position	173.010	2.688	0.886	178.091	11.174	−12.047
Fair value	176.6			177.2		
2012	14.089	0	0	0	3.062	−5.530
2013	13.518	0.042	−0.005	0	2.739	−2.969
2014	13.141	0.052	−0.125	0	4.511	−8.693
2015	12.757	0.171	−0.091	0	−1.790	3.402
2016	12.719	0.237	−0.137	0	−0.745	−3.480
2017	12.728	0.261	−0.038	0	−0.093	−0.459
2018	11.685	0.250	0.046	0	−1.600	4.009
2019	11.300	0.304	0.081	0	1.269	−8.352
2020	10.964	0.212	0.065	0	0.123	7.723
2021	10.606	0.196	0.131	0	1.013	2.730
2022	10.308	0.208	0.169	0	2.139	−1.091
2023	10.155	0.269	0.287	0	2.179	−1.842
2024	9.847	0.184	0.191	0	−1.522	4.333
2025	9.645	0.152	0.143	0	0.828	−0.190
2026	9.549	0.150	0.168	178.091	−0.939	−1.637

The sample fair value of liabilities for the first 500 scenarios is 1.76×10^5 Euros

---

[4] As scenarios were provided by a life insurance company, only this restricted number of scenarios was available. Scenario paths for the Nikkei and the S&P indices as well as the cash account

Table 2 Values of the objective function in (RP$_{CF}$) for optimal portfolios to (RP$_{CF}$) and (RP$_{\widetilde{TV}}$) relative to the fair value of liabilities

	In-sample (%)	Out-of-sample (%)
Cash-flow	8.72	9.23
Terminal value	193.2	192.8

in year 2012. Tables 1 and 2 show optimal portfolios and the magnitude of in-sample and out-of-sample mismatches. The numbers in Table 1 show which quantity (in thousands) of each asset should be bought or sold at the end of each particular year and the total initial position in year 2012. For the mismatches in Table 2, we computed the objective value of the cash-flow matching problem for both portfolios in-sample and out-of sample and divided by the fair value of liabilities. Therefore, these numbers can be viewed as a relative error.

It needs to be noted that in the terminal value matching problem, all strategies concerning purchases and sales of the cash account lead to the same objective value. Hence, the terminal position of 178.091 could have been spread in all possible manners over the years 2012–2026 without any difference.

As one may have expected, the replicating portfolio obtained from discounted terminal value matching very badly matches cash payments in particular years since these mismatches are not penalized by the objective function of the discounted terminal value matching problem. Consequently, a replicating portfolio obtained from terminal value matching is of little use to the insurer if cash payments are supposed to match well at each point in time. As already explained, the missing remedy is to employ an approximation of the appropriate dynamic investment strategy in the numéraire asset.

We implemented the linear approximation of the optimal dynamic investment strategy as outlined at the end of Sect. 4 for the same scenarios that were used for the portfolio optimizations (see Fig. 2). Table 3 shows the optimal parameters $(a_t^1, a_t^2, a_t^3)_{t=1,....,T-1}$ and the coefficients of determination R^2.

On first sight, it is striking how large the coefficients of determination (R^2) are (on average above 80 %). However, since the optimal δ_t is a linear combination of discounted financial cash-flows $C_{CA}^F\left(t, D_t^F\right)/N_t$, $C_{SP}^F\left(t, D_t^F\right)/N_t$ and $C_{NK}^F\left(t, D_t^F\right)/N_t$ and discounted liability cash-flow $C^L\left(t, D_t^F, D_t^L\right)/N_t$, this is not too surprising. Actually, if liability cash-flows were known, i.e., available for the regression, a perfect fit (i.e., $R^2 = 100\%$) would be obtainable. In all other cases, the liability cash-flow is approximated by the asset cash-flows rather well.

Analogous to Table 2, Table 4 shows the in-sample and out-of-sample objective function values for the portfolio solving (RP$_{CF}$) and the portfolio solving (RP$_{\widetilde{TV}}$)

(Footnote 4 continued)
were generated with standard models from the Barrie and Hibbert Economic Scenario Generator (see www.barrhibb.com/economic_scenario_generator).

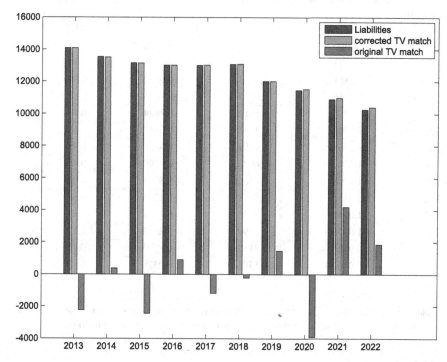

Fig. 2 The bar chart shows cash-flows of liabilities and the optimal terminal value replicating portfolio with and without a dynamic correction in the first ten years

together with dynamic investment strategy $(\hat{\delta}_t)_{t=1,\dots,T}$ relative to the fair value of the liabilities.

Clearly, the dynamic strategy in the replicating assets significantly improves the quality of the cash-flow match. Yet, the optimal portfolio for cash-flow matching still slightly outperforms this dynamic variant due to the reasoning given above.

We also regressed with additional in-the-money call options on the cash-flows, but there was only a negligible improvement in-sample and out-of-sample. Possibly, one may achieve better results with a more sophisticated choice of regressors, but that seems unlikely or at least challenging given the high coefficients of determination. Further, all results obtained above have been tested to be quite stable when changing the number of scenarios or changing the specific choice of liabilities. Of course, a more detailed analysis based on a real-world example could provide further valuable insights.

Table 3 Parameters (in thousands) obtained from linear regression and the coefficients of determination

Year	a^1	a^2	a^3	R^2
2012	1.4089	−2.3976	0.6460	1
2013	1.3634	−1.9468	0.3073	0.96
2014	1.3361	−3.2633	0.9379	0.99
2015	1.2956	1.6990	−0.4615	0.87
2016	1.2954	0.8510	0.3424	0.91
2017	1.2903	0.3433	0.0117	0.18
2018	1.1768	1.4854	−0.4825	0.83
2019	1.1267	−0.6920	0.9790	0.96
2020	1.0790	0.1306	−0.8758	0.94
2021	1.0380	−0.5618	−0.2773	0.69
2022	1.0143	−1.4692	0.1697	0.61
2023	1.0078	−1.4899	0.2622	0.78
2024	0.9846	1.3168	−0.4793	0.93
2025	0.9706	−0.5706	0.0376	0.43

Table 4 Values of the objective function in (RP$_{CF}$) for the optimal portfolio to (RP$_{CF}$) and the optimal portfolio (RP$_{\tilde{TV}}$) with strategy $(\hat{\delta}_t)_{t=1,...,T}$ relative to the fair value of liabilities

	Cash-flow (%)	T.V. w. correction (%)
In-sample	8.72	10.16
Out-of-sample	9.23	11.05

6 Conclusion

Motivated by the theoretical results in [7], we improved the cash-flow matching quality of the optimal terminal value portfolio without deterioration of the terminal value match. This is achieved by the introduction of a deterministic strategy (e.g., in replicating assets or risk factors) which approximates the optimal non-deterministic strategy. It turned out that with the dynamic correction the terminal value matching technique is comparable (but still slightly inferior) to the static cash-flow matching technique in terms of in-sample as well as out-of-sample performance. Due to the high coefficients of determination, a significant improvement by a more selected choice of explaining variables seems unlikely. Taking into account that in contrast to cash-flow matching, terminal value matching has an explicit analytic solution and that the least squares problems involved in the approximation of the dynamic strategy are also numerically negligible, this might thus represent a computationally more efficient alternative to the standard cash-flow matching approach. Further evidence can only be obtained by the careful examination of a real-world scenario.

Acknowledgments The authors would like to thank Pierre Joos, Christoph Winter, and Axel See-mann for very helpful discussions and feedback. We are also grateful for the generous constant financial support by Allianz Deutschland AG. Finally, many thanks go to two anonymous referees of this paper for valuable comments which helped to improve presentation.

Open Access This chapter is distributed under the terms of the Creative Commons Attribution Noncommercial License, which permits any noncommercial use, distribution, and reproduction in any medium, provided the original author(s) and source are credited.

References

1. Andreatta, G., Corradin, S.: Fair value of life liabilities with embedded options: an application to a portfolio of Italian insurance policies. Working Paper, RAS Spa, Pianificazione Redditività di Gruppo
2. Baione, F., De Angelis, P., Fortunati, A.: On a fair value model for participating life insurance policies. Invest. Manag. Financ. Innov. **3**(2), 105–115 (2006)
3. Bauer, D., Bergmann, D., Kiesel, R.: On the risk-neutral valuation of life insurance contracts with numerical methods in view. Astin Bull. **40**, 65–95 (2010)
4. Dubrana, L.: A formalized hybrid portfolio replication technique applied to participating life insurance portfolios. Available at http://www.ludovicdubrana.com// (2013) Accessed 30 Dec 2013
5. Grosen, A., Jørgensen, P.L.: Fair valuation of life insurance liabilities: the impact of interest rate guarantees, surrender options, and bonus policies. Insur.: Math. Econ. **26**(1), 37–57 (2000)
6. Longstaff, F., Schwartz, E.: Valuing american options by simulation: a simple least-squares approach. Rev. Financ. Stud. **14**(1), 113–147 (2001)
7. Natolski, J., Werner, R.: Mathematical analysis of different approaches for replicating portfolios. Euro. Actuar. J. (2014). doi:10.1007/s13385-014-0094-z
8. Oechslin, J., Aubry, O., Aellig, M., Kappeli, A., Bronnimann, D., Tandonnet A., Valois, G.: Replicating embedded options. Life and Pensions Risk, 47–52 (2007)
9. Pelsser, A.: Pricing and hedging guaranteed annuity options via static option replication. Insur.: Math. Econ. **33**(2), 283–296 (2003)

Part IV
Computational Methods for Risk Management

Risk and Computation

Rüdiger U. Seydel

Abstract Computation is based on models and applies algorithms. Both a model and an algorithm can be sources of risks, which will be discussed in this paper. The *risk from the algorithm* stems from erroneous results, the topic of the first part of this paper. We attempt to give a definition of *computational risk*, and propose how to avoid it. Concerning the underlying model, our concern will not be the "model error". Rather, even the reality (or a perfect model) can be subjected to structural changes: Nonlinear relations of underlying laws can trigger sudden or unexpected changes in the dynamical behavior. These phenomena must be analyzed, as far they are revealed by a model. A computational approach to such a *structural risk* will be discussed in the second part. The paper presents some guidelines on how to limit computational risk and assess structural risk.

Keywords Computational risk · Structural risk · Accuracy of algorithms · Bifurcation

Mathematical Subject Classification 91B30 · 91G60 · 65Y20 · 65P30

1 Computational Risk

Early computer codes concentrated on the evaluation of special functions. The emphasis was to deliver *full accuracy* (say, seven correct decimal digits on a 32-bit machine) in minimal time. Many of these algorithms are based on formulas of [1, 6]. Later the interest shifted to more complex algorithms such as solving differential equations, where discretizations are required. Typically, the errors are of the type $C \Delta^p$, where Δ represents a discretization parameter, p denotes the convergence order of the method, and C is a hardly assessable error coefficient. A control of the error is highly complicated, costly, and frequently somewhat vague, and is source of computational risk.

R.U. Seydel (✉)
Mathematisches Institut, Universität zu Köln, Weyertal 86, 50931 Köln, Germany
e-mail: seydel@math.uni-koeln.de

© The Author(s) 2015
K. Glau et al. (eds.), *Innovations in Quantitative Risk Management*,
Springer Proceedings in Mathematics & Statistics 99,
DOI 10.1007/978-3-319-09114-3_17

This first part of the paper discusses how to assess the risk from erroneous results of algorithms. Accuracy properties of algorithms will have to be reconsidered.

1.1 Efficiency of Algorithms

The performance of algorithms can be well compared in a diagram depicting the costs (computing time) over the achieved relative error. In case, the output of an algorithm consists of more than one real number, then we think of the largest of all these errors. Now, for a certain computational task, select and run a set of algorithms, and enter the points representing their performance into the diagram. Schematically, the dots look as in Fig. 1.[1]

For nontrivial computational tasks, there will be hardly a method that is simultaneously both highly accurate and extremely fast; there is always a trade-off. Hence, one will not find algorithms in the lower left corner, below the curve in Fig. 1. This (smoothed) curve is the *efficient frontier*. It can be defined in the Pareto sense as minimizing computing time and maximizing accuracy. Clearly, the aim of researchers is to push the frontier down; the curve is not immutable in time. The smoothed frontier in Fig. 1 may serve as idealized vehicle to define efficiency: Each method on the frontier is efficient.

This notion of efficiency allows to define the "best" algorithm for a certain task almost uniquely. A reasonable computational accuracy must be put into relation to the underlying model error. So, indicate the size of the model error on the horizontal axis, and let a vertical line at that position cut the efficient frontier, which completes the choice of the proper algorithm. Of course, the efficient frontier is a snapshot that compares an artificial selection of algorithms.

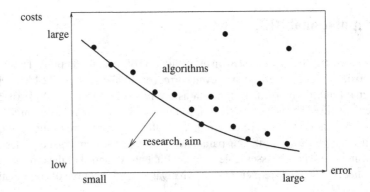

Fig. 1 Costs (computing time) of algorithms over relative error

---

[1] An example of such a diagram for the task of pricing American-style options is, for example, Fig. 4.19 in [14], however, for the root mean square error of a set of 60 problems.

Notice that this error in the final result does not explicitly consider intermediate errors or inconsistencies in the algorithm. For example, errors from solving linear equations, instability caused by propagation of rounding errors, or discretization errors do not enter explicitly. The final lumped error is seen with the eyes of the user.

1.2 Risk of an Algorithm

Computational methods involve parameters on which the accuracy depends. Discretizations are characterized by their fineness[2] M. For example, a binomial method for option pricing may work with $M = 100$ or $M = 50$ time intervals. Let us call the first algorithm B-100, and the second B-50. Here "algorithm" is understood as an implementation where all accuracy parameters (as M) are fixed; B-100 is an algorithm different from B-50.

Now we are prepared to define the computational risk for a given model:

Computational Risk:
The chosen algorithm does not deliver the required accuracy.

For example, when an algorithm provides results with an error of 0.002 where we required 0.001 (three decimal digits), this would be strictly seen as *failure*. Nowadays in practice, it is widespread not to notice such a failure. As a "safety measure" one frequently chooses unnecessarily high values of the fineness M. This makes a failure less likely, but leads to overshooting and a lack of efficiency. As outlined above, we assume that the optimal algorithm is chosen such that it correctly matches the required accuracy.

1.3 Eliminate the Risk

Occasionally, it was suggested to establish algorithms with guaranteed accuracy [9]. Related algorithms are highly involved, expensive and hence used rarely. Although the idea of guaranteed accuracy is not really new, it seems appropriate to be pushed forward for applications in finance. For example, algorithms for option pricing have reached a level of sophistication which may allow to pursue as second step the establishing of *dependable* accuracy information.

In this paper, we propose to unburden algorithms from relevant accuracy and error control. Rather the algorithms should be made as fast as possible, without iterating to convergence. As mentioned above, for each algorithm the mesh fineness M will be fixed. Then the algorithm has fixed costs, and can be regarded as "analytic method". The implementation matters. External fine-tuning is not available, and the computer programs can be regarded as hard-wired.

---

[2] Number of subintervals into which an underlying interval is subdivided by a discretization

Table 1 Fictive entry in an accuracy file

Correct digits	Algorithm
2	A
3	B-50
4	B-100
5	C

Then these "ultimate" versions of algorithms are investigated for their accuracy. We suggest to gather accuracy or error information into a file *separate* from the algorithm. This "file" can be a look-up table, or a set of inequalities for parameters. Typically, the accuracy results will be determined empirically. As an illustration, the accuracy information for a certain task (say, pricing an American-style vanilla put option) and a specific set of parameters (strike, volatility σ, interest rate r, time to maturity T) might look as in Table 1. As application, one chooses the algorithm according to the information file.

1.4 Effort

Certainly, the above suggestion amounts to a big endeavor. In general, original papers do not contain the required accuracy information. Instead, usually, convergence behavior, stability, and intermediate errors are analyzed. Accuracy is mostly tested on a small selection of numerical examples. It will be a challenge to researchers, to provide the additional accuracy information for "any" set of parameters. The best way to organize this is left open. Strong results will establish inequalities for the parameters that guarantee certain accuracy. Weaker results will establish multidimensional tables of discrete values of the parameters, and the application will interpolate the accuracy.

To encourage the work, let us repeat the advantages: Accuracy information and conditions under which algorithms fail will be included in external files. The algorithms will be slimmed down, the production runs will be faster, and the costs on a particular computer are fixed and known in advance. The computational risk will be eliminated.

1.5 Example

As an example, consider the pricing of a vanilla American put at the money, with one year to maturity. We choose an algorithm that implements the analytic interpolation

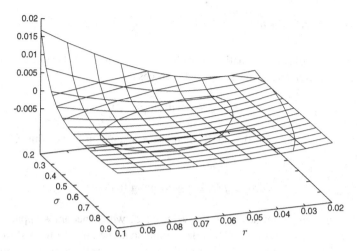

Fig. 2 Relative error. level curves: -0.004, -0.002, 0, $+0.002$

method by Johnson [7].[3] For the specific option problem, the remaining parameters are r and σ. Figure 2 shows the relative error in the calculated price of the option depending on r and σ, and implicitly a *map of accuracies*. For the underlying rectangle of realistic r, σ-values, and the assumed type of option, a result can be summarized as follows:

In case $\sigma > 3r$ holds, the absolute of the relative accuracy is smaller than 0.005 (two and a half digits).

Of course, the accuracy result can be easily refined.

2 Assessing Structural Risk

We now turn to the second topic of the paper, on how to assess structural changes in a model computationally. This is based on *dynamical systems*, in which the dynamical behavior depends on a certain model parameter. Critical threshold values of this parameter will be decisive. Below we shall understand "structural risk" as given by the distance to the next threshold value of the critical parameter. An early paper stressing the role threshold values (bifurcations) can play for a risk analysis is [11]. The approach has been applied successfully in electrical engineering for assessing voltage collapse, see [3]. We begin with recalling some basic facts from dynamical systems.

---

[3] For analytic methods, strong results may be easier to obtain because implementation issues are less relevant.

2.1 Simplest Attractor

The basic mean reversion equation is well-known in finance: This is a stochastic differential equation (SDE) for a stochastic process σ_t

$$d\sigma_t = \alpha(\zeta - \sigma_t)\,dt + \gamma\sigma_t^\delta\,dW_t$$

with constant $\alpha, \zeta, \gamma, \delta > 0$, and W_t denotes a Wiener process. This SDE is of the type

$$d\sigma_t = f(\sigma_t)\,dt + \text{driving force.}$$

The response of σ_t is attracted by the value of ζ, which becomes apparent by a simple stability analysis of the SDE's deterministic kernel, the ordinary differential equation (ODE) $\dot{x} = f(x) = \alpha(\zeta - x)$. The state $x = \zeta$ is the simplest example of an attractor.[4]

For more flexibility, a constant (and unknown) value of ζ can be replaced by a suitable process ζ_t, which in turn is driven by some model equation. This adds a second equation. A simple example of such a system is the *tandem equation*

$$d\sigma_t = \alpha_1(\zeta_t - \sigma_t)\,dt + \gamma\sigma_t\,dW_t$$
$$d\zeta_t = \alpha_2(\sigma_t - \zeta_t)\,dt$$

An ODE stability analysis of its deterministic kernel does not reveal an attractor. Rather the equilibrium is degenerate, the Jacobian matrix is singular. Simulating the tandem system shows two trajectories dancing about each other, but drifting across the phase space erratically. What is needed is some anchoring, which can be provided by an additional nonlinear term.

2.2 Mean-Field Models

We digress for a moment to emphasize that the above tandem is a *mean-field model*. In canonical variables x_1, x_2, it is of the type

$$dx_1 = \alpha_1^* \left[\tfrac{1}{2}(x_1 + x_2) - x_1 \right] dt + \gamma_1 x_1\,dW_t^{(1)}$$
$$dx_2 = \alpha_2^* \left[\tfrac{1}{2}(x_1 + x_2) - x_2 \right] dt + \gamma_2 x_2\,dW_t^{(2)}$$

which generalizes to $x_1, \ldots, x_n$. The reversion is to the mean

---

[4] The equilibrium $x = \zeta$ is stable since $df/dx = -\alpha < 0$; for $t \to \infty$, x approaches ζ.

$$\bar{x} := \frac{1}{n} \sum_i^n x_i \,,$$

and a key element for modeling interaction among agents [4, 5]. More general mean-field models include an additional nonlinear term, and are of the type

$$\dot{x} = \beta * f(x) + \alpha * \text{interaction} + \gamma * \text{ext.forces}\,.$$

Notice that the dimension n is a parameter, and the solution structure thus depends on the number of variables. The parameters α measure the size of cooperation, and γ the strength of external random forces. The nonlinearity $f(x)$ and the balance of the parameters β, α, γ, n control the dynamics.

2.3 Artificial Example

As noted above, a suitable nonlinear term can induce a dynamic control that prevents the trajectories from drifting around erratically. Here we choose a cubic nonlinearity of the Duffing-type $f(x) = x - x^3$, since it represents a classical bistability [13]. For slightly more flexibility, we shift the location of equilibria by a constant s; otherwise, we choose constants artificially. For the purpose of demonstration, our artificial example is the system

$$dx_1 = 0.1(x_1 - s)\left\{1 - (x_1 - s)^2\right\} dt + 0.5\,[x_2 - x_1]\,dt + 0.1x_1 dW_t$$
$$dx_2 = 0.5\,[x_1 - x_2]\,dt$$

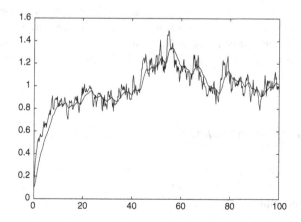

Fig. 3 Artificial example of Sect. 2.3: x_1 and x_2 over time t, for $s = 2$, starting at 0.1

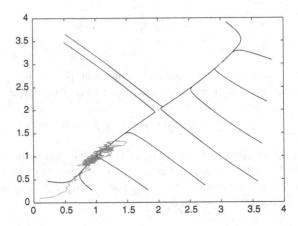

Fig. 4 x_1, x_2-phase plane, with the trajectory of Fig. 3, and 11 trajectories of the unforced system

Clearly, there are three ODE equilibria, namely two stable nodes at $x_1 = x_2 = s \pm 1$ and a saddle at $x_1 = x_2 = s$. For graphical illustrations of the response, see Figs. 3 and 4.

Figure 3 depicts the quick attraction of the trajectories (starting at 0.1) toward the smaller node at $s - 1 = 1$. This dynamical response is shown again in Fig. 4 in the x_1, x_2-phase portrait. As a background, this figure shows 11 trajectories of the deterministic kernel, where the random perturbation is switched off. Starting from 11 initial points in the plane, the trajectories approach one of the two stable nodes. This part of the plane consists of two basins of attraction, separated by a separatrix that includes the saddle. The phase portrait of the deterministic kernel serves as *skeleton* of the dynamics possible for the randomly perturbed system.

Now imagine to increase the strength of the random force (enlarge γ). For sufficiently large γ, the trajectories may jump across the wall of the separatrix. Then the dynamics is attracted by the other node. Obviously, these transitions between the two regimes may happen repeatedly. In this way, one of the stylized facts can be modeled, namely the *volatility clustering* [2][5]. This experiment underlines the modeling power of such nonlinear systems.

2.4 Structure in Phase Spaces

The above gentle reminder on dynamical systems has exhibited the three items *node*, *saddle*, and *separatrix*. There are many more "beasts" in the phase space. The following is an incomplete list:

---

[5] Phases with high and low volatility are separated from each other.

- stationary state,
- periodic behavior,
- chaotic behavior,
- jumps, discontinuities,
- loss or gain of stability.

These qualitative labels stand for the structure of dynamical responses. The structure may change when a parameter is varied. Although a "parameter" is a constant, it may undergo slow variations, or may be manipulated by some external (political) force. Such changes in the "constant" parameter are called *quasi-stationary*. Typically, our parameter is in the role of a control parameter. Some variations in the parameter may have little consequences on the response of the system. But there are *critical threshold* values of the parameter, where the changes in the structure can have dramatic consequences. At these thresholds, small changes in the parameter can trigger essential changes in the state of the system. The mathematical mechanism that explains such qualitative changes is *bifurcation*.[6]

When a system drifts toward a bifurcation, then this must be considered as risk!

Bifurcation is at the heart of systemic risk. Hence there is a need for a tool that signals bifurcations in advance.

2.5 Risk Index

Let λ denote a bifurcation parameter of a dynamical system. For an underlying model, we denote by λ_0 a numerically calculated critical threshold value of λ. At this point, the model error enters, because λ_0 is based on the model. The distance to λ_0 is a measure of structural risk. This is the distance between the current operation point (λ) and the closest bifurcation. To signal the distance, the *risk index*

$$R(\lambda) := \frac{\lambda}{|\lambda - \lambda_0| - \varepsilon}$$

was suggested [12]. (ε is a small number representing several sources of error.) The larger the value of R, the closer the risk is. The index gives risk a quantitative meaning, invariant of the scaling of the model.[7] A *feasible range* of the parameter λ has been defined by

$$\mathscr{F}_c := \{ \lambda \mid R(\lambda) < c \},$$

and its complement is the *risk area of level c*.

---

[6] For an introduction into bifurcation and related numerical methods, see [13].

[7] Essentially, this is a deterministic approach. One may think of incorporating a volatility into R.

2.6 Example

Sometimes, stock prices behave cyclically, and one may ask whether there is an underlying deterministic kernel with periodic structure. In this context, behavioral trading models are of interest. Lux [8] in his model splits traders into chartists and fundamentalists, and models their impact on the price of an asset. The variables are

- $p(t)$ market price of an asset, with fundamental value p^*;
- z proportion of chartists, and
- $x(t)$ their sentiment index, between -1 for pessimistic and $+1$ for optimistic.

The growth $\dot{p}$ will be proportional to zx (impact of chartists) and to $(1-z)(p^*-p)$ (impact of fundamentalists). Combining these two impacts leads to the first of the two equations in the system

$$\dot{p} = \beta\left(z\,x\,\xi_c + (1-z)(p^*-p)\xi_f\right)$$
$$\dot{x} = 2zv_1(\tanh(U_{+-}) - x)\cosh(U_{+-}) + (1-z)(1-x^2)v_2(\sinh(U_{+f}) - \sinh(U_{-f}))$$

The second equation models the sentiment x, with incentive functions U_{+-}, U_{+f}, U_{-f}

$$U_{+-} := \alpha_1 x + \alpha_2\frac{\dot{p}}{v_1}$$
$$U_{+f} := \alpha_3\left(\frac{1}{p}\left(rp^* + \frac{\dot{p}}{v_2}\right) - r - s\left|\frac{p^*-p}{p}\right|_*\right)$$
$$U_{-f} := \alpha_3\left(r - \frac{1}{p}\left(rp^* + \frac{\dot{p}}{v_2}\right) - s\left|\frac{p^*-p}{p}\right|_*\right)$$

$|\ |_*$ is a smoothed version of $|\ |$, and the chosen constants are:

$\beta = 0.5$, $\xi_c = 5$, $\xi_f = 5$, $v_1 = 0.5$, $v_2 = 0.75$,
$\alpha_1 = 1.02$, $\alpha_2 = 0.25$, $\alpha_3 = 1.5$, $r = 0.1$, $s = 0.8$, $p^* = 10$.

This is an ODE system. The original model [8] includes a third equation for the proportion z. Our modified model is simpler in that it takes z as external parameter (our λ). The concern will be the structure of the response of the system as it varies with z.

For the chosen constants, we calculate $\lambda_0 = 0.6914$ as critical threshold value of the parameter z [10]. This is a *Hopf bifurcation*, at which periodic cycles are born out of a stationary state. Accordingly, we have the two regimes

- $z < \lambda_0$: $(p, x) = (p^*, 0)$ stable stationary, and
- $z > \lambda_0$: stable periodic motion (cyclic behavior of the asset price).

At the Hopf point, there is a transition between the regimes. The risk index R signals the critical threshold by large values (Fig. 5). For the chosen constants, the threshold occurs at a proportion of chartists of about 70% of the traders.

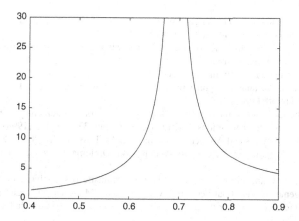

Fig. 5 Risk index R over parameter z. *Left wing* index along the stationary states. *Right wing* index along the periodic states

2.7 Summary

We summarize the second part of the paper. Provided a good model exists,[8] we suggest to begin with calculating the bifurcations/threshold values of parameters. They are the pivoting points of possible trend switching. The distance between the current operation point of the real financial system and the bifurcation point must be observed. Large values of the risk index can be used as indicator, signaling how close the risk is. This can be used as a tool for a stress test.

Acknowledgments The paper has benefited from discussions with Roland C. Seydel.

Open Access This chapter is distributed under the terms of the Creative Commons Attribution Noncommercial License, which permits any noncommercial use, distribution, and reproduction in any medium, provided the original author(s) and source are credited.

References

1. Abramowitz, M., Stegun, I.A. (eds.): Handbook of Mathematical Functions. Dover, New York (1968)
2. Cont, R.: Empirical properties of asset returns: stylized facts and statistical issues. Quant. Finance **1**, 223–236 (2001)
3. Eidiani, M.: A reliable and efficient method for assessing voltage stability in transmission and distribution networks. Int. J. Electr. Power Energy Syst. **33**, 453–456 (2011)

---

[8] Admittedly, quite an assumption! But a lack of a good model is equivalent to a lack of understanding. Of course, models in economics are not perfect. The challenge is to find a model that captures all relevant nonlinearities.

4. Garnier, J., Papanicolaou, G., Yang, T.-W.: Large deviations for a mean field model of systemic risk. SIAM J. Financial Math. **4**, 151–184 (2013)

5. Haldane, A.G.: Rethinking the financial network. Talk in Amsterdam (2009) http://www.bis.org/review/r090505e.pdf?frames=0

6. Hart, J.F.: Computer Approximations. John Wiley, New York (1968)

7. Johnson, H.E.: An analytic approximation for the American put price. J. Financial Quant. Anal. **18**, 141–148 (1983)

8. Lux, T.: The socio-economic dynamics of speculative markets: interacting agents, chaos, and the fat tails of return distributions. J. Econ. Behav. Organ. **33**, 143–165 (1998)

9. Moore, R.E.: Methods and Applications of Interval Analysis. SIAM, Philadelphia (1979)

10. Quecke, S.: Dynamische Systeme am Aktienmarkt. Diploma thesis, University of Cologne (2003)

11. Seydel, R.: Risk and bifurcation. Towards a deterministic risk analysis. In: Risk Analysis and Management in a Global Economy. Volume I: Risk Management in Europe. New Challenges for the Industrial World, pp. 318–339. Monograph Series Institut für Technikfolgenabschätzung (1997)

12. Seydel, R.: A new risk index. ZAMM **84**, 850–855 (2004)

13. Seydel, R.: Practical Bifurcation and Stability Analysis (First Edition 1988), 3rd edn. Springer, New York (2010)

14. Seydel, R.: Tools for Computational Finance, 5th edn. Springer, London (2012)

Extreme Value Importance Sampling for Rare Event Risk Measurement

D.L. McLeish and Zhongxian Men

Abstract We suggest practical and simple methods for Monte Carlo estimation of the (small) probabilities of large losses using importance sampling. We argue that a simple optimal choice of importance sampling distribution is a member of the generalized extreme value distribution and, unlike the common alternatives such as Esscher transform, this family achieves bounded relative error in the tail. Examples of simulating rare event probabilities and conditional tail expectations are given and very large efficiency gains are achieved.

Keywords Rare event simulation · Risk measurement · Relative error · Monte Carlo methods · Importance sampling

1 Introduction

Suppose $\mathbf{Y} = (Y_1, Y_2, \ldots, Y_m)$ is a vector of independent random variables each with cumulative distribution function (cdf) F and probability density function (pdf) f with respect to Lebesgue measure. Suppose we wish to estimate the probability of a large loss, $p_t = P(L(\mathbf{Y}) > t)$ where $L(\mathbf{Y})$ is the loss determined by the realization $\mathbf{Y}$ (usually assumed to be monotonic in its components) and t is some predetermined threshold. There are many different loss functions $L(\mathbf{Y})$ used in rare event simulation, including barrier hitting probabilities of sums or averages of independent random variables, or of processes such as an Ornstein–Uhlenbeck or Feller process. The methods discussed here are designed for problems in which a small number of continuous factors are the primary contributors to large losses. We wish to use *importance sampling* (IS) (see [3], Sect. 4.6 or [12] p. 183): generate independent replications of $\mathbf{Y}$ repeatedly, say n times, from an alternative distribution, say one with pdf $f_{IS}(\mathbf{y})$ and then estimate the above expected value using the IS estimator

D.L. McLeish (✉)
University of Waterlo, 200 University Avenue West, Waterloo, ON N2L 3G1, Canada
e-mail: dlmcleis@uwaterloo.ca

Z. Men
e-mail: zmen@uwaterloo.ca

© The Author(s) 2015   317
K. Glau et al. (eds.), *Innovations in Quantitative Risk Management*,
Springer Proceedings in Mathematics & Statistics 99,
DOI 10.1007/978-3-319-09114-3_18

$E_\theta([I(L(\mathbf{Y}) > t)f(\mathbf{Y})/f_{IS}(\mathbf{Y})])$, where $I(*)$ denotes the indicator function. We use

$$\hat{p}_t = \frac{1}{n}\sum_{i=1}^{n} I(L(\mathbf{Y}_i) > t)\frac{f(\mathbf{Y}_i)}{f_{IS}(\mathbf{Y}_i)}, \quad \text{where} \quad \mathbf{Y}_j \sim f_{IS}(\mathbf{y}). \tag{1}$$

If we denote by E_{IS} the expected value under the IS distribution f_{IS}, and by E the expected value under the original distribution f, then

$$E_{IS}(\hat{p}_t) = E_{IS}\left(I(L(\mathbf{Y}) > t)\frac{f(\mathbf{Y})}{f_{IS}(\mathbf{Y})}\right)$$
$$= p_t$$

confirming that this is an *unbiased* estimator. There is a great deal of the literature on such problems when the event of interest is "rare", i.e., when p_t is very small, and many different approaches depending on the underlying loss function and distribution. We do not attempt a review of the literature in the limited space available. Excellent reviews of the methods and applications are given in Chap. 6 of [1] and Chap. 10 of [9]. Highly efficient methods have been developed for tail estimation in very simple problems, such as when the loss function consists of a sum of independent identically distributed increments. In this paper, we will provide practical tools for simulation of such problems in many examples of common interest. For rare events, the variance or standard error is less suitable as a performance measure than a version scaled by the mean because in estimating very small probabilities such as 0.0001, it is not the absolute size of the error that matters but its size relative to the true value.

Definition 1 The relative error (RE) of the importance sample estimator is the ratio of the estimator's standard deviation to its mean.

Simulation is made more difficult for rare events because crude Monte Carlo fails. As a simple illustration, suppose we wish to estimate a very small probability p_t. To this end, we generate n values of $L(\mathbf{Y}_i)$ and estimate this probability with $\hat{p} = X/n$ where X is the number of times that $L(\mathbf{Y}_i) > t$ and X has a Binomial(n, p_t) distribution. In this case, the relative error is

$$\text{RE} = \frac{\sqrt{\text{Var}(\frac{X}{n})}}{E\left(\frac{X}{n}\right)} = \frac{\sqrt{\frac{1}{n}p_t(1-p_t)}}{p_t} = n^{-1/2}\sqrt{\frac{1-p_t}{p_t}}.$$

For rare events, p_t is small and the relative error is very large. If we wish a normal-based confidence interval for p_t of the form $(\hat{p}_t - 0.1\hat{p}_t, \hat{p}_t + 0.1\hat{p}_t)$ for example, we are essentially stipulating a certain relative error (RE = 0.05102) whatever the value of p_t. In order to achieve a reasonable bound on the relative error, we would need to use sample sizes that were of the order of p_t^{-1}, i.e., larger and larger sample sizes for rarer events.

Returning to the estimator (1), if we take its variance, we obtain

$$n^{-1}\left[E_{\text{IS}}\left(I(L(\mathbf{Y}) > t)\frac{f^2(\mathbf{Y})}{f_{\text{IS}}^2(\mathbf{Y})}\right) - p_t^2\right] = n^{-1}\left[E\left(I(L(\mathbf{Y}) > t)\frac{f(\mathbf{Y})}{f_{\text{IS}}(\mathbf{Y})}\right) - p_t^2\right]$$

the relative error is

$$\text{RE}(f_{\text{IS}}; t, n) = n^{-1/2}\left[p_t^{-2}E\left(I(L(\mathbf{Y}) > t)\frac{f(\mathbf{Y})}{f_{\text{IS}}(\mathbf{Y})}\right) - 1\right]^{1/2}.$$

The relative error decreases linearly in $n^{-1/2}$ but it is the factor

$$p_t^{-2}\left[E\left(I(L(\mathbf{Y}) > t)\frac{f(\mathbf{Y})}{f_{\text{IS}}(\mathbf{Y})}\right)\right]^{1/2},$$

highly sensitive to t when p_t is small, that determines whether an IS distribution is good or bad for a given problem. There is a large literature regarding the use of importance sampling for such problems, much recommending the use of the exponential tilt or *Esscher transform*. The suggestion is to adopt an IS distribution of the form

$$f_{\text{IS}}(y) = \text{constant} \times e^{\theta y} f(y) \qquad (2)$$

and then tune the parameter θ so that the IS estimator is as efficient as possible (see, for example [1, 7, 15]). Chapter 10 of [9] provides a detailed discussion of methods and applications as well as a discussion of the boundedness of relative error. McLeish [11] demonstrates that the IS distribution (2) is suboptimal and unlike the alternatives we explore there, does not typically achieve bounded relative error. We argue for the use of the generalized extreme value (GEV) family of distributions for such problems. A loose paraphrase of the theme of the current paper is "all you really need[1] is GEV". Indeed in Appendix A, we prove a result (Proposition 1) which shows that, under some conditions, there is always an importance sampling estimator whose relative error is bounded in the tail obtained by generating the distance along one principal axis from an extreme value distribution, while leaving the other coordinates unchanged in distribution. We now consider some one-dimensional problems.

2 The One-Dimensional Case

Consider estimating $P(L(Y) > t)$ where the value of t is large, the random variable Y is one-dimensional and $L(y)$ is monotonically increasing. We would like to use an importance sample distribution for which, by adjusting the values of the parameters,

---

[1] For importance sampling estimates of rare events, at least, with apologies to the Beatles.

we can have small relative error for any large t. We seek a parametric family $\{f_\theta; \theta \in \Theta\}$ of importance sample estimators which have bounded relative error as follows:

Definition 2 Suppose $\mathcal{H}$ is the class of non-negative integrable functions $\{I(L(Y) > t);$ for $t > T\}$. We say a parametric family $\{f_\theta; \theta \in \Theta\}$ has **bounded relative error** for the class $\mathcal{H}$ if

$$\sup_{t > T} \inf_{\theta \in \Theta} \mathrm{RE}(f_\theta; t, n) < \infty.$$

A parametric family has bounded relative error for estimating functions in a class $\mathcal{H}$ if, for each $t > T$, there exists a parameter value θ which provides bounded relative error. Indeed, a bound on the relative error of approximately $0.738n^{-1/2}$ can be achieved by importance sampling if we know the tail behavior of the distribution. There are very few circumstances under which the exponential tilt, families of continuous densities of the form

$$f_\theta(x) = \text{constant} \times e^{\theta y} f(y),$$

provides bounded relative error. The literature recommending the exponential tilt usually rests on demonstrating *logarithmic efficiency* (see [1], p. 159 or Sect. 10.1 of [9]), a substantially weaker condition that does not guarantee a bound in the relative error. Although we may design a simulation optimally for a specific criterion such as achieving small relative error in the estimation of $P(L(Y) > t)$, we are more often interested in the nature of the whole tail beyond t. For example, we may be interested in $E[(L(Y) - t) I(L(Y) > t)] = \int_t^\infty P(L(Y) > s)ds$ and this would require that a single simulation be efficient for estimating all parameters $P(L(Y) > s), s > t$. The property of bounded relative error provides some assurance that the family used adapts to the whole tail, rather than a single quantile.

For simplicity, we assume for the present that Y is univariate, has a continuous distribution, and $L(Y)$ is a strictly increasing function of Y. Then

$$P(L(Y) > t) = \int_{L^{-1}(t)}^{\infty} f(y)dy$$

and we can achieve bounded relative error if we use an importance sample distribution drawn from the family

$$f_\theta(y) = \text{constant} \times e^{\theta T(y)} f(y) \tag{3}$$

where $T(y)$ behaves, for large values of y, roughly like a linear function of $\bar{F}(y) = 1 - F(y)$. If $T(y) \sim -\bar{F}(y)$ as $y \to \infty$, the optimal parameter θ is $\theta_t = \frac{k_2}{p_t} \simeq \frac{1.5936}{p_t}$[2] and the limit of the relative error of the IS estimator is $\simeq 0.738n^{-1/2}$ (see Appendix

---

[2] Here $k_2 \simeq 1.5936$ is the unique positive solution to the equation $e^{-k} + \frac{k}{2} = 1$.

A). The simplest and most tractable family of distributions with appropriate tail behavior is the GEV distribution associated with the density $f(y)$.

We now provide an intuitive argument in favor of the use of the GEV family of IS distributions. For a more rigorous justification, see Appendix A.

The choice $T(y) = -\bar{F}(y)$ provides asymptotic bounded relative error [11]. Consider a family of cumulative distribution functions

$$F_\theta(y) = \frac{e^{-\theta \bar{F}(y)} - e^{-\theta}}{1 - e^{-\theta}}.$$

The corresponding probability density function $F_\theta'(y)$ is of the form (3). As $\theta \to \infty$ and $y \to \infty$ in such a way that $\theta \bar{F}(y)$ converges to a nonzero constant, then

$$(F(y))^\theta = (1 - \bar{F}(y))^\theta \sim e^{-\theta \bar{F}(y)}, \tag{4}$$

so that

$$F_\theta(y) \sim (F(y))^\theta. \tag{5}$$

Therefore, $F_\theta(y)$ is asymptotically equivalent to the distribution of the maximum of θ observations from the original target distribution F. This, suitably normalized, converges to a member of the GEV family of distributions. We also show in [11] that the optimal parameter is asymptotic to $\theta = k_2/p_t$ as $p_t \to 0$. Consequently,

$$F_\theta(t) \sim (1 - \bar{F}(t))^\theta \sim (1 - p_t)^{k_2/p_t} \sim e^{-k_2} \simeq 0.203.$$

Thus, when we use the corresponding extreme value importance sample distribution, about 20.3 % of the observations will fall below t and the other 79.7 % will fall above, and this can be used to identify one of the parameters of the IS distribution. Of course, only the observations greater than t are directly relevant to estimating quantities like $P(L > t)$. This leads to the option of conditioning on the event $L > t$ and using the generalized Pareto family (see Appendix B).

The three distinct classes of extreme value distributions and some of their basic properties are outlined in Appendix B. All have simple closed forms for their pdf, cdf, and inverse cdf and can be easily and efficiently generated. In addition to a shape parameter ξ, they have location and scale parameters d, c so the cdf is $H_\xi(\frac{x-d}{c})$, where $H_\xi(x)$ is the cdf for unit scale and 0 location parameter. We say that a given continuous cdf F falls in the maximum domain of attraction of an extreme value cdf $H_\xi(x)$ if there exist sequences c_n, d_n such that

$$F^n(d_n + c_n x) \to H_\xi(x) \quad \text{as } n \to \infty.$$

We will choose an extreme value distribution with parameters that approximate the distribution of the maximum of $\theta_t = k_2/p_t$ random variables from the original density $f(y)$. Further properties of the extreme value distributions and detail on the choice of parameters is given in Appendix B. Proposition 1 in Appendix A shows that if F is in the domain of attraction of H_0, then H_0 provides a family of IS distributions with bounded relative error. It is also unique in the sense that any other such IS distribution has tails essentially equivalent to those of H_0. A similar result can be proved for $\xi \neq 0$. The superiority of the extreme value distributions for importance sampling stems from the bound on relative error, but equally from their ease of simulation, the simple closed form for the pdf and cdf and the maximum stability property, which says that the distribution of the maximum of i.i.d. random variables drawn from this distribution is a member of the same family.

We get a better sense of the extent of the variance reduction using IS if we compare sample sizes required to achieve a certain relative error. If we use crude random sampling in order to estimate p_t using a sample size n_{cr}, the relative error of the crude estimator is

$$\mathrm{RE(crude)} = \frac{\sqrt{\frac{p_t(1-p_t)}{n_{cr}}}}{p_t} = \sqrt{\frac{1-p_t}{n_{cr}\, p_t}}$$

whereas if we use a GEV importance sample of size n_{IS}, the relative error is $\mathrm{RE}(IS) \simeq 0.738 n_{\mathrm{IS}}^{-1/2}$. Equating these, the ratio of the sample sizes required for a fixed relative error is $\frac{n_{cr}}{n_{\mathrm{IS}}} \simeq \frac{1.84}{p_t}$ for p_t small. Indeed, if $p_t = 10^{-4}$, an importance sample estimator based on a sample size 5×10^6 is quite feasible on a small laptop computer, but is roughly equivalent to a crude Monte Carlo estimator of sample size 9.2×10^{10}, possible only on the largest computers.

3 Examples

3.1 Example 1: Simulation Estimators of Quantiles and TailVar for the Normal Distribution

Rarely when we wish to simulate an expected value in the region of the space $[L(Y) > t]$ is this the only quantity of interest. More commonly, we are interested in various functions sensitive to the tail of the distribution. This argues for using an IS estimator with bounded relative error rather than the more common practice of simply conditioning on the region of interest. For a simple example, suppose Y follows a $N(0, 1)$ distribution and we wish to estimate a property of the tail defined by $Y > t$, where t is large. Suppose we simulate from the conditional distribution given $Y > t$, that is from the pdf

$$\frac{1}{1 - \Phi(t)} \phi(y) I(y > t) \tag{6}$$

where ϕ and Φ are the standard normal pdf and cdf respectively. If we wish also to estimate $P(Y > t + s | Y > t) \sim e^{-st - \frac{s^2}{2}}$ for $s > 0$ fixed, sampling from this pdf is highly inefficient, since for n simulations from pdf (6), the RE for estimating $P(Y > t + s | Y > t)$ is approximately $n^{-1/2}\sqrt{e^{st + \frac{s^2}{2}} - 1}$ and this grows extremely rapidly in both t and s. We would need a sample size of around $n = 10^4 e^{st + \frac{s^2}{2}}$ (or about 60 trillion if $s = 3$ and $t = 6$) from the IS density (6) to achieve a RE of 1 %.

Crude Monte Carlo fails here but use of IS with the usual standard exponential tilt or Esscher transform with $T(y) = y$, though very much better, still fails to deliver bounded relative error. In [11], it is shown that the relative error is $\sim (\frac{\pi}{2})^{1/4} \sqrt{t/n} \to \infty$ as $p_t \to 0$. While the IS distribution obtained by the exponential tilt is a very large improvement over crude Monte Carlo and logarithmically efficient, it still results in an unbounded relative error as $p_t \to 0$.

The Normal distribution is in the maximum domain of attraction of the Gumbel distribution ($\xi = 0$) so our arguments suggest that we should use as IS distribution

$$H_0(\frac{y - d}{c}) = \exp(-e^{-(y-d)/c}) \tag{7}$$

with parameters c, d selected to match the distribution function $(\Phi(y))^{k_2/p_t}$ (see Appendix A). Using this Gumbel distribution as an IS distribution permits a very substantial increase in efficiency.

To show how effective this is as an IS distribution, we simulate from the Gumbel distribution with cdf (7). The weights attached to a given IS simulated value of Y are the ratio of the two pdfs, the standard normal and the Gumbel, or

$$w(Y) = c_\theta \phi(Y) \exp(e^{-\frac{Y - d_\theta}{c_\theta}} + \frac{Y - d_\theta}{c_\theta}). \tag{8}$$

For example with $t = 4.7534$, $p_t = 10^{-6}$ and Gumbel parameters $c = 0.20$ and $d = 4.85$, the relative error in 10^6 simulations was $0.729n^{-1/2}$. We can compare this with the exponential tilt, equivalent to using the normal$(t, 1)$ distribution as an IS distribution, whose relative error is $2.32n^{-1/2}$, or with crude Monte Carlo, with relative error around $10^3 n^{-1/2}$.

Suppose that our interest is in estimating the conditional tail expectation or TVaR$_\alpha$ based on simulations. The TVaR$_\alpha$ is defined as $E(Y | Y > t) = \frac{E[YI(Y>t)]}{P(Y>t)}$. We designed the GEV parameters for simulating the numerator, $E[YI(Y > t)]$. If we are interested in estimating TVaR$_{0.0001}$ by simulation, $t = $ VaR$_{0.0001} = 3.719$ the true value is

$$\text{TVaR}_{0.0001} = \frac{\int_{3.719}^{\infty} z e^{-z^2/2} dz}{\int_{3.719}^{\infty} e^{-z^2/2} dz} \simeq 3.9585.$$

We will generate random variables Y_i using the Gumbel (0.282, 3.228) distribution and then attach weights (8) to these observations. The estimate of TVaR$_\alpha$ is then the

Table 1 Relative error of estimators of p_t and of $E[YI(Y > t)]$

	$n^{1/2} \times$ RE $\hat{p}_t$	$n^{1/2} \times$ RE $\hat{E}[YI(Y > t)]$
1. Crude	100	100
2. SN (tilt)	2.05	2.05
3. EV IS	0.73	0.73
4. Cond EV IS	0.47	0.47

average of the values $w(Y_i) \times Y_i$ averaged only over those values that are greater than t.

The Gumbel distribution is supported on the whole real line, while the region of interest is only that portion of the space greater than t so one might generate Y_i from the Gumbel distribution conditional on the event $Y_i > t$ rather than unconditionally. The probability $P(Y_i > t)$ where Y_i is distributed according to the Gumbel(c_θ, d_θ) distribution is $\exp(-e^{-(t-d_\theta)/c_\theta})$ and this is typically around 0.80 indicating that about 20 % of the time the Gumbel random variables fall in the "irrelevant" portion of the sample space $S < t$. Since it is easy to generate from the conditional Gumbel distribution $Y|Y > t$ this was also done for a further improvement in efficiency. This conditional distribution converges to the generalized Pareto family of distributions (see Theorem 2 of Appendix B). In this case, since $\xi = 0$, $P(Y - u \le z|Y > u) \to 1 - e^{-z}$ as $u \to \infty$. Therefore, in order to approximately generate from the conditional distribution of the tail of the Gumbel, we generate the excess from an exponential distribution.

Table 1 provides a summary of the results of these simulations. Several simulation methods for estimating $\mathrm{TVaR}_\alpha = E(Y|Y > t) = \frac{E[YI(Y>t)]}{p_t}$ with $p_t = P(Y > t)$ as well as estimates of p_t are compared. Since TVaR is a ratio, we consider estimates of the denominator and numerator, i.e., p_t and $E[YI(Y > t)]$ separately. The underlying distribution of Y is normal in all cases. The methods investigated are:

1. **Crude Simulation (Crude)** Generate independently Y_i, $i = 1, \ldots, n$ from original (normal) distribution. Estimate p_t using $\frac{1}{n}\sum_{i=1}^{n} I(Y_i > t)$ and estimate $E[YI(Y > t)]$ using $\frac{1}{n}\sum_{i=1}^{n} Y_i I(Y_i > t)$.

2. **Exponential Tilt or Shifted Normal IS (SN)** Generate independently Y_i, $i = 1, \ldots, n$ from $N(t, 1)$ distribution. Estimate p_t using $\frac{1}{n}\sum_{i=1}^{n} w_i I(Y_i > t)$ and estimate $E[YI(Y > t)]$ using $\frac{1}{n}\sum_{i=1}^{n} w_i Y_i I(Y_i > t)$ where w_i are the weights, obtained as the likelihood ratio

$$w_i = w(Y_i) = \frac{\phi(Y_i)}{\phi(Y_i - t)}.$$

Since the exponential tilt, applied to a Normal distribution, results in another Normal distribution with a shifted mean and the same variance, this is an application of the exponential tilt.

3. **Extreme Value IS (EVIS)** Generate independently Y_i, $i = 1, \ldots, n$ from the Gumbel(c, d) distribution. Estimate p_t using $\frac{1}{n}\sum_{i=1}^{n} w_i I(Y_i > t)$ and estimate

$E[YI(Y > t)]$ using $\frac{1}{n}\sum_{i=1}^{n} w_i Y_i I(Y_i > t)$ where w_i are the weights, obtained as the likelihood ratio

$$w_i = w(Y_i) = \frac{\phi(Y_i)}{\frac{1}{c}h_0(\frac{Y_i-d}{c})}$$

where $\frac{1}{c}h_0(\frac{Y_i-d}{c})$ is the corresponding Gumbel pdf.

4. **Conditional Extreme Value IS (Cond EVIS)** Generate independently Y_i, $i = 1,\ldots,n$ from the Gumbel(c,d) distribution conditioned on $Y > t$. Estimate p_t using $\frac{1}{n}\sum_{i=1}^{n} w_i I(Y_i > t) = \frac{1}{n}\sum_{i=1}^{n} w_i$ and estimate $E[YI(Y > t)]$ using $\frac{1}{n}\sum_{i=1}^{n} w_i Y_i I(Y_i > t) = \frac{1}{n}\sum_{i=1}^{n} w_i Y_i$ where w_i are the weights, obtained as the likelihood ratio

$$w_i = w(Y_i) = \frac{\phi(Y_i)}{g(Y_i)}$$

and where

$$g(s) = \frac{\frac{1}{c}h_0(\frac{s-d}{c})}{1 - H_0(\frac{t-d}{c})}, \quad \text{for } s > t$$

is the corresponding conditional Gumbel pdf.

Conditional Normal IS Since we are interested in the tail behavior of the random variable Y given $Y > t$ it would be natural to simulate from the conditional distribution $Y|Y > t$. Unfortunately, this is an infeasible method because it requires advance knowledge of $p_s = P(Y > s)$ for all $s \geq t$.

We indicate in Table 1 the relative error of these various methods in the case $p_t = 0.0001, t = 3.719$. The corresponding parameters of the Gumbel distribution that we used were $c = 0.243$, $d = 3.84$ but the results are quite robust to the values of these parameters. Notice that the efficiency gain of the conditional extreme value simulation, as measured by the ratio of variances, is around $\left(\frac{100}{0.47}\right)^2 \simeq 45,270$ relative to a crude Monte Carlo and around $\left(\frac{2.05}{0.47}\right)^2 \simeq 19$ relative to the exponential tilt.

3.2 Example 2: Simulating a Portfolio Credit Risk Model

We provide a simulation of a credit risk model using importance sampling. The model, once the industry standard, is the normal copula model for portfolio credit risk, introduced in Morgan's CreditMetrics system[3] (see [5]). Under this model, the k'th firm defaults with probability $p_k(Z)$, and this probability depends on m unobserved factors that comprise the vector Z. Losses on a portfolio then take the form $L = \sum_{k=1}^{v} c_k Y_k$ where Y_k, the default indicator, is a Bernoulli random variable with $P(Y_k = 1) = p_k(Z)$, (denoted $Y_k \sim \text{Bern}(p_k(Z))$), p_k are functions of

---

[3] Very popular prior to 2008!

common factors Z, and c_k is the portfolio exposure to the $k'th$ default. Suppose we wish to estimate $P(L > t)$.

3.2.1 One-Factor Case

In the simplest one-factor case, p_k are functions $p_k(Z) = \Phi\left(\dfrac{a_k Z + \Phi^{-1}(\rho_k)}{\sqrt{1-a_k^2}}\right)$ of a common standard normal $N(0, 1)$ random variable Z, the scalars a_k are the factor loadings or weights and ρ_k represents the marginal probability of default (it is easy to see that $E[p_k(Z)] = \rho_k$).

If we wish to simulate an event $P(L > t)$ which has small probability, there are two parallel opportunities for importance sampling, both investigated by [4].

For example, for each given Z, we might replace the Bernoulli distribution of Y_k with a distribution having higher probabilities of default, i.e., replace $p_k(Z)$ by q_k where $q_k \geq p_k(Z)$. The choice of q_k is motivated by an exponential tilt as is argued in [4]. Conditional on the factor Z, the tilted Bernoulli random variables Y_k are such that $E(\sum_{k=1}^{v} c_k Y_k | Z) = L$. We *do not require the use of importance sampling in this second stage* of the simulation so having used an IS distribution for Z, we generate Bernoulli $(p_i(Z))$ random variables Y_i. There are two similar alternatives in the first stage, generate Z from the Gumbel or generate $\tilde{L} = \sum_{k=1}^{v} c_k p_k(Z)$, a proxy for the loss, from the Gumbel distribution. These two alternatives give similar results since the Gumbel is the extreme value distribution corresponding to both Z and $\tilde{L}$, and $\tilde{L}$ is a nondecreasing function of Z. In Table 2, we give the results corresponding to the second of these alternatives, simulating $\tilde{L} = \sum_{k=1}^{v} c_k p_k(Z)$ and then solving for the factor Z. Unlike [4], where a shifted normal IS distribution for Z is used, we use the Gumbel distribution for L motivated by the arguments of Sect. 2. Extreme value importance sampling provides a very substantial variance reduction over crude simulation of course, but also over importance sampling using the exponential tilt. We determine appropriate parameters for the Gumbel extreme value distribution by quantile matching and then draw Z from a Gumbel(c, d) distribution. We use the parameters taken from the numerical example in [4], i.e., $v = 1{,}000$ obligors, exposures c_k are 1, 4, 9, 16, and 25 with 200 at each level of exposure, and the marginal default probabilities $\rho_k = 0.01\,(1 + \sin 16\frac{\pi k}{v})$ so that they range from 0 to 2%. The factor loadings a_k were generated as uniform random variables on the interval $(0,1)$. In summary, the main difference with [4] is our use of the Gumbel distribution for simulating $\tilde{L}$ rather than the shifted normal and the lack of a tilt for Y_k.

The resulting relative errors estimated from 30,000 simulations are shown in Table 2, and evidently there is a significant variance reduction achieved by the choice of the Gumbel distribution. For example, when the threshold t was chosen to be 2,000, there was a decrease in the variance by a factor of approximately $\left(\frac{2.33}{0.69}\right)^2$ or about 11

Table 2 Relative error of estimators: Crude, shifted normal (G & L) and EVIS

t	p_t	n	$n^{1/2}$ RE(Crude)	$n^{1/2}$ RE(EVIS)	$n^{1/2}$ RE(G&L)
1,500	0.0075	30,000	12.01	0.70	2.03
2,000	0.0041	30,000	15.34	0.69	2.33
3,000	0.0022	30,000	22.12	0.70	2.74

and a much more substantial decrease over crude by a factor of around $\left(\frac{15.34}{0.69}\right)^2$ or about 494.

3.2.2 Multifactor Case

In the multifactor case, the event that an obligor k fails is determined by a Bernoulli random variable $Y_k \sim \text{Bern}(p_k)$. The loss function $L = \sum_{k=1}^{\nu} c_k Y_k$ is then a linear function of Y_k and corresponding exposures c_k. We wish to estimate the probability of a large loss: $P(L > t)$. The values p_k are functions

$$p_k(\mathbf{Z}) = \Phi\left(\frac{\mathbf{a}_k \mathbf{Z} + \Phi^{-1}(\rho_k)}{\sqrt{1 - \mathbf{a}_k \mathbf{a}_k^T}}\right)$$

of a number of factors $\mathbf{Z}^T = (Z_1, \ldots, Z_m)$ where the individual factors $Z_i, i = 1, \ldots, m$ are independent standard normal random variables. Here ρ_k is the marginal probability that obligor k fails, i.e., $P(Y_k = 1) = E[p_k(\mathbf{Z})] = \rho_k$ since $\mathbf{a}_k \mathbf{Z}$ is $N(0, \mathbf{a}_k \mathbf{a}_k^T)$ (see [4], p. 1644) and the row vectors $\mathbf{a}_k$ are factor loadings which relate the factors to the specific obligors.

Simulation Model We begin with brief description of the model simulated in [4, p. 1650], that is the basis of our comparison. We assume $\nu = 1,000$ obligors, the marginal probabilities of default $\rho_k = 0.01(1 + \sin(16\pi \frac{k}{\nu}))$, and the exposures $c_k = [\frac{5k}{\nu}]^2, k = 1, \ldots, \nu$. The components of the factor loading vector $\mathbf{a}_k$ were generated as independent $U(0, \frac{1}{\sqrt{m}})$, where m is the number of factors. The simulation described in [4] is a two-stage Monte Carlo IS method. The first stage simulates the latent factors in $\mathbf{Z}$ by IS, where the importance distributions are independent univariate Normal distributions with means obtained by solving equating modes and with variances unchanged equal to 1. Specifically they choose the normal IS distribution having the same mode as

$$P(L > t | \mathbf{Z} = \mathbf{z}) e^{-\mathbf{z}^T \mathbf{z}/2} \simeq \left(1 - \Phi\left(\frac{t - \mu(\mathbf{z})}{\sigma(\mathbf{z})}\right)\right) e^{-\mathbf{z}^T \mathbf{z}/2} \tag{9}$$

because this is approximately proportional to $P(\mathbf{Z} = \mathbf{z}|L > t)$, the ideal IS distribution. In other words, the IS distribution for Z_i is $N(\mu_i, 1), i = 1, \ldots, m$ where the vector of values of μ_i is given by (see [4], Eq. (20))

$$\mu = \max_{\mathbf{z}} P(L > t|\mathbf{Z} = \mathbf{z})e^{-\mathbf{z}^T \mathbf{z}/2} \tag{10}$$

with (see [4], p. 1648)

$$P(L > t|\mathbf{Z} = \mathbf{z}) \simeq 1 - \Phi\left(\frac{t - \mathrm{E}[L|\mathbf{Z} = \mathbf{z}]}{\sqrt{\mathrm{Var}[L|\mathbf{Z} = \mathbf{z}]}}\right) \text{ with } \mathrm{E}[L|\mathbf{Z} = \mathbf{z}] > t. \tag{11}$$

Conditional on the values of the latent factors $\mathbf{Z}$, the second stage of the algorithm in [4] is to twist the Bernoulli random variables Y_k using modified Bernoulli distributions, i.e., with a suitable change in the values of the probabilities $P(Y_k = 1), k = 1, \ldots, \nu$. Our comparisons below are with this two-stage form of the IS algorithm.

Our simulation for this portfolio credit risk problem is a one-stage IS simulation algorithm. If there are m factors in the portfolio credit risk model we simulate $m - 1$ of them $\tilde{Z}_i$ from univariate normal $N(\mu_i, 1), i = 1, \ldots, m-1$ with a different mean, as in [4], but then we simulate an approximation to the total loss, $\tilde{L}$, from a Gumbel distribution, and finally set $\tilde{Z}_m$ equal to the value implied by $\tilde{Z}_1, \ldots, \tilde{Z}_{m-1}$ and $\tilde{L}$. This requires solving an equation

$$\tilde{L}(\tilde{Z}_1, \ldots, \tilde{Z}_{m-1}, \tilde{Z}_m) = \tilde{L} \tag{12}$$

for $\tilde{Z}_m$. The parameters $\mu = (\mu_1, \mu_2, \ldots, \mu_{m-1})$ are obtained from the crude simulation. Having solved (12), we attach weight to this IS point $(\tilde{Z}_1, \ldots, \tilde{Z}_m)$ equal to

$$\omega = \frac{c \times \prod_{i=1}^{m} \phi(\tilde{Z}_i)}{\frac{\partial L}{\partial Z_m} \times \prod_{i=1}^{m-1} \phi(\tilde{Z}_i - \mu_i)} e^{(\tilde{L}-d)/c} \exp\left(e^{-(\tilde{L}-d)/c}\right), \tag{13}$$

and

$$\frac{\partial L}{\partial Z_m} = \sum_{k=1}^{\nu} \frac{a_{k,m}}{\sqrt{1 - \mathbf{a}_k \mathbf{a}_k^T}} \phi\left(\frac{\mathbf{a}_k \tilde{\mathbf{Z}} + \Phi^{-1}(\rho_k)}{\sqrt{1 - \mathbf{a}_k \mathbf{a}_k^T}}\right).$$

We choose the parameters $\mu_i, i = 1, \ldots, m - 1$ for the above IS distributions using estimates of the quantity

$$\mu = E(\mathbf{Z}|L > t) = E\left(\mathbf{Z}\left|\sum_{k=1}^{\nu} c_k p_k(\mathbf{Z}) > t\right.\right) \tag{14}$$

based on the preliminary simulation, with the parameters c, d of the Gumbel obtained from (24).

We summarize our algorithm for the portfolio credit risk problem as follows:

1. Conduct a crude MC simulation and estimate the parameter μ in (14).
2. Estimate parameters c and d of the Gumbel distribution (24) where $E(\tilde{L})$ is estimated by average$(L|L > t)$.
3. Repeat (a)–(d) for independent simulations as $j = 1, \ldots, n$ where n is the sample size of the simulation.

   (a) Generate $\tilde{L}$ from the Gumbel(c, d) distribution.
   (b) Generate $\tilde{Z}_i, i = 1, \ldots, m - 1$ from the univariate normal $N(\mu_i, 1)$ distributions.
   (c) Solve $\tilde{L}(\tilde{Z}_1, \ldots, \tilde{Z}_{m-1}, \tilde{Z}_m) = t$ for $\tilde{Z}_m$ and calculate (13).
   (d) Simulate a loss $L_j = \sum_{k=1}^{\nu} c_k Y_k$ where $Y_k \sim \text{Bern}(p_k(\tilde{Z}))$ with $p_k(\tilde{Z}) = \Phi\left(\frac{a_k \tilde{Z} + \Phi^{-1}(\rho_k)}{\sqrt{1 - a_k a_k^T}}\right)$.

4. Estimate p_t using a weighted average

$$\frac{1}{n} \sum_{j=1}^{n} \omega_j I(L_j > t).$$

5. Estimate the variance of this estimator using n^{-1} times the sample variance of the values $\omega_j I(L_j > t), j = 1, \ldots, n$.

Simulation Results The results in Table 3 were obtained by using crude Monte Carlo, importance sampling using the GEV distribution as the IS distribution, and the IS approach proposed in [4]. In the crude simulations, the sample size is 50,000, while in the later two methods, the sample size is 10,000.

Notice that for a modest number of factors there is a very large reduction in variance over the crude (for example the ratio of relative error corresponding to 2 factors, $t = 2,500$ corresponds to an efficiency gain or variance ratio of nearly 2,400) and a significant improvement over the Glasserman and Li [4] simulation with a variance ratio of approximately 4. This improvement erodes as the number of factors increases, and in fact the method of Glasserman and Li has smaller variance in this case when $m = 10$. In general, ratios of multivariate densities of large dimension tend to be quite "noisy"; although the weights have expected value 1, they often have large variance. A subsequent paper will deal with the large dimensional case.

Table 3 Comparison between crude simulation, EVIS and Glasserman and Li (2005) for the credit risk model

t	p_t	n	$n^{1/2}$ RE (crude)	$n^{1/2}$ RE (EVIS)	$n^{1/2}$ RE (G&L)
2 factors					
1,500	0.0034	50,000	17.1	0.99	1.73
2,000	0.0015	10,000	26.2	0.96	1.82
2,500	0.00038	10,000	51.3	1.05	1.93
3 factors					
1,500	0.00305	50,000	18.94	1.24	1.72
2,000	0.00111	10,000	31.61	1.15	1.82
2,500	0.00042	10,000	49.99	1.35	1.99
5 factors					
1,500	0.00289	50,000	18.87	1.39	1.71
2,000	0.00099	10,000	39.52	1.55	1.81
2,500	0.00035	10,000	55.89	1.57	1.88
10 factors					
1,500	0.00246	50,000	20.83	1.84	1.79
2,000	0.00081	10,000	33.70	2.15	1.89
2,500	0.00029	10,000	57.73	3.06	1.98

4 Conclusion

The family of extreme value distributions are ideally suited to rare event simulation. They provide a very tractable family of distributions and have tails which provide bounded relative error regardless of how rare the event is. Examples of simulating values of risk measures demonstrate a very substantial improvement over crude Monte Carlo and a smaller improvement over competitors such as the exponential tilt. This advantage is considerable for relatively low-dimensional problems, but there may be little or no advantage over an exponential tilt when the dimensionality of the problem increases.

Open Access This chapter is distributed under the terms of the Creative Commons Attribution Noncommercial License, which permits any noncommercial use, distribution, and reproduction in any medium, provided the original author(s) and source are credited.

Appendix A: Assumptions and Results

We suppose without loss of generality that the argument to the loss function is a multivariate normal MNV$(0, I_m)$ random vector $\mathbf{Z}$, since any (possibly dependent) random vector $\mathbf{Y}$ can be generated from such a $\mathbf{Z}$. We begin by assuming that "large" values of $L(\mathbf{Z})$ are determined by the distance of $\mathbf{Z} = (Z_1, Z_2, \ldots, Z_m)$ from the origin in a specific direction, i.e.,

Assumption 1 There exists a direction vector $\mathbf{v} \in \Re^m$ such that, for all fixed vectors $\mathbf{w} \in \Re^m$,

$$\frac{P(L(Z_0 \mathbf{v}) > t)}{P(L(Z_0 \mathbf{v} + \mathbf{w}) > t)} \to 1 \text{ as } t \to \infty \tag{15}$$

where Z_0 is $N(0, 1)$.

We propose an importance sampling distribution generated as follows:

$$\mathbf{Z} = Y\mathbf{v} + (I_m - \mathbf{v}\mathbf{v}')\varepsilon, \quad \text{where } \varepsilon \sim \text{MVN}(\mathbf{0}, \mathbf{I}_m), \tag{16}$$

where Y has the extreme value distribution $H_0(\frac{y-d}{c})$. If we replace the distribution of Y by the standard normal, it is easy to see that (16) gives $\mathbf{Z} \sim \text{MVN}(\mathbf{0}, \mathbf{I}_m)$ so the IS weight function in this case is simply the ratio of the two univariate distributions for Y.

Assumption 2 Suppose that for any fixed $\mathbf{w} \in \Re^m$, there exits y_0 such that $L(y\mathbf{v} + \mathbf{w})$ is an increasing function of y for $y > y_0$.

Proposition 1 *Under assumptions 1 and 2, there is a sequence of importance sampling distributions of the form* (16) *which provides bounded relative error asymptotic to* $cn^{-1/2}$ *as* $p_t \to 0$ *where* $c \simeq 0.738$.

In order to prove this result, we will use the following lemma, a special case of Corollary 1 of [11]:

Lemma 1 *Suppose the random variable Y has a continuous distribution with cdf F_Y. Suppose that $T(y)$ is nondecreasing and for some real number a we have $a + T(y) \sim -\overline{F}_Y(y)$ as $y \to y_F^-$ with $y_F = \sup\{y; F_Y(y) < 1\} \le \infty$. Then the IS estimator for sample size n obtained from density* (3) *with $\theta = \theta_t = \frac{k_2}{p_t}$ has bounded RE asymptotic to* $cn^{-1/2}$ *as* $p_t \to 0$ *where* $c = \frac{1}{k_2}\sqrt{e^{k_2} - 1 - k_2^2} \simeq 0.738$.[4]

Proof of Proposition 1. The condition (15) allows us to solve an asymptotically equivalent univariate problem, i.e., estimate $P(L_1(Y) > t)$ where $L_1(Y) = L(Y\mathbf{v})$, $Y \sim N(0, 1)$. Clearly, the Normal distribution for Y satisfies $F_Y \in \text{MDA}(H_0(x))$ so that there exist sequences c_n, d_n such that $F_Y^n(d_n + c_n x) \to H_0(x)$ as $n \to \infty$ for the GEV H_0. Lemma 1 shows that the importance sampling distribution

$$f_\theta(y) = \text{constant} \times e^{-\frac{k_2}{p_t}\overline{F}_Y(y)} f(y) \tag{17}$$

---

[4] $k_2 \simeq 1.5936$ and $c \simeq 0.738$ are the unique positive solutions to the equations $e^k = \frac{1}{1-\frac{k}{2}} = 1 + k^2(1 + c^2)$.

provides bounded relative error for the estimation of p_t as $t \to \infty$ and $p_t \to 0$. Note that the probability density function of the maximum $n = \frac{k_2}{p_t} + 1$ random variables drawn from the distribution $F_Y(y)$ is given by

$$n \, (F_Y(y))^{n-1} \, f(y) = \text{constant} \times (1 - \overline{F}_Y(y))^{\frac{k_2}{p_t}} f(y) \sim \text{constant} \times e^{-\frac{k_2}{p_t} \overline{F}_Y(y)} f(y). \tag{18}$$

Furthermore, by the local limit or density version of convergence to the extreme value distributions, (see Theorem 2 (b) [2] or [14]), with $y = d_n + c_n x$, and $x = (y - d_n)/c_n$,

$$n F_Y^{n-1}(d_n + c_n x) f(d_n + c_n x) \to H_0'(x) = \frac{1}{c_n} \exp\left(-x - e^{-x}\right) \text{ as } n \to \infty$$

which implies, combining (17) and (18), that

$$n c_n \, (F_Y(y))^{n} \, f(y) \sim \exp\left(-(y - d_n)/c_n - e^{-(y-d_n)/c_n}\right). \tag{19}$$

Therefore, the extreme value distribution provides a bounded relative error importance sampling distribution, equivalent to (17).

Appendix B: Maximum Domain of Attraction and Properties of The Generalized Extreme Value Distributions

Maximum domain of attraction

If there are sequences of real constants c_n and d_n, $n = 1, 2, \ldots$ where $c_n > 0$ for all n, such that

$$F^n(d_n + c_n x) \to H(x) \text{ as } n \to \infty, \tag{20}$$

for some nondegenerate cdf $H(x)$, then we say that F is in the maximum domain of attraction (MDA) of the cdf H and write $F \in \text{MDA}(H)$. The Fisher–Tippett theorem (see Theorem 7.3 of [13]) characterizes the possible limiting distributions H as members of the generalized extreme value distribution (GEV). A cdf is a member of this family if it has cumulative distribution function of the form $H_\xi(\frac{x-d}{c})$ where $c > 0$ and

$$H_0(x) = \exp(-e^{-x}), \quad H_\xi(x) = e^{-(1+\xi x)^{-1/\xi}} \quad \text{for } \xi \neq 0 \text{ and } \xi x > -1. \tag{21}$$

Theorem 1 (Fisher–Tippet, Gnedenko) *If $F \in \text{MDA}(H)$ for some nondegenerate cdf H, then H must take the form (21).*

The properties of the GEV distributions listed in Table 4 are obtained from routine calculations and properties in [13] or [10].

Table 4 Some properties of the generalized extreme value distributions

Property	$\xi = 0$	$\xi \neq 0$ and $\xi x > -1$
cdf $= H_\xi(x)$	$\exp(-e^{-x})$	$\exp(-(1+\xi x)^{-1/\xi})$
pdf $= h_\xi(x)$	$e^{-x}\exp(-e^{-x})$	$(1+\xi x)^{-\frac{1}{\xi}-1}\exp(-(1+\xi x)^{-1/\xi})$
Mode: satisfies $H_\xi(x) = \exp(-1-\xi)$	0	$\frac{(1+\xi)^{-\xi}-1}{\xi}$
$H_\xi(x \vert x > x_{1/2})$	$\frac{\exp(1-e^{-x})}{e-1}$, for $x > x_{1/2}$	$\frac{\exp(-(1+\xi x)^{-1/\xi})}{1-\exp(-1-\xi)}$, for $x > x_{1/2}$
Median	$x_{1/2} = -\ln(\ln 2)$	$\frac{(\ln 2)^{-\xi}-1}{\xi}$
Inverse $H_\xi^{-1}(p)$	$-\ln(-\ln p)$	$\frac{(-\ln p)^{-\xi}-1}{\xi}$
Mean	$\gamma = $ (Euler's constant) $\simeq 0.577216$	$\begin{cases} \frac{\Gamma(1-\xi)-1}{\xi} & \text{if } \xi < 1 \\ \infty & \text{if } \xi \geq 1 \end{cases}$
Variance	$\frac{\pi^2}{6} \simeq 1.645$	$\begin{cases} \frac{\Gamma(1-2\xi)-2\Gamma(1-\xi)+1}{\xi} & \text{if } \xi < \frac{1}{2} \\ \infty & \text{if } \xi \geq \frac{1}{2} \end{cases}$
Random number generator $U \sim U(0,1)$	$-\ln(-\ln U)$	$\frac{(-\ln U)^{-\xi}-1}{\xi}$

Choosing the parameters c and d

The GEV has $H_\xi(\frac{y-d}{c})$ and probability density function $c^{-1}h_\xi(\frac{y-d}{c})$. Other parameters can be easily found in the above table. We wish to choose an extreme value distribution with parameters corresponding to the maximum of a sample of $\theta_t = k_2/p_t$ random variables from the original density $f(y)$. In other words, we wish to find values of d_{θ_t} and c_{θ_t} so that

$$(F(y))^{\theta_t} \simeq H_\xi\left(\frac{y - d_{\theta_t}}{c_{\theta_t}}\right) \tag{22}$$

and this leads to matching t with the quantile corresponding to $e^{-k_2} \simeq 0.203$. In other words, one parameter is determined by the equation

$$\frac{t - d_{\theta_t}}{c_{\theta_t}} = H_\xi^{-1}(e^{-k_2}) = \begin{cases} -\ln(k_2) & \xi = 0 \\ \frac{k_2^{-\xi}-1}{\xi} & \xi \neq 0 \end{cases} \tag{23}$$

Another parameter can be determined using the crude simulation and the values of Y for which $L(Y) > t$. We can match another quantile, for example, the median, the mode, or the sample mean which estimates $E[Y \mid L(Y) > t]$. In the case of standard normally distributed inputs and the Gumbel distribution, matching the conditional expected value $E(L \mid L > t)$ and (23) results approximately in:

$$c = \frac{E(\tilde{L}) - t}{1.0438}, \quad \text{and} \quad d = t + 0.46659c. \tag{24}$$

Here $E(\tilde{L}) = \text{average}(L|L > t)$ based on a preliminary crude simulation of values of L simulated under the original distribution. Of course, one could also use maximum likelihood estimation to determine appropriate parameters for the ID distribution (see [10]) but the specific choice of estimator seemed to have little impact on the quality of the importance sampling provided that the estimated GEV density was sufficiently dispersed.

As an alternative to simulating $\tilde{L}$ from the GEV, we may simulate instead from $\tilde{L}|\tilde{L} > t$, resulting in the generalized Pareto distribution. For a given c.d.f. F, the **conditional excess distribution** is

$$F_u(y) = P(X - u > y | X > u) = \frac{F(u+y) - F(u)}{1 - F(u)}, \quad x \geq 0.$$

Then the conditional excess distribution can be approximated by the so-called generalized Pareto distribution for large values of u (see [13], Theorem 7.20):

Theorem 2 (Pickands, Balkema, de Haan) $F \in \text{MDA}(H_\xi)$ *for some* ξ *if and only if*

$$\lim_{u \to \infty} \sup_x |F_u(x) - G_{\xi, \beta(u)}(x)| \to 0$$

for some positive measurable function $\beta(u)$ *where* $G_{\xi, \beta}$ *is the c.d.f of the* **Generalized Pareto (GP) distribution:**

$$G_{\xi, \beta}(y) = \begin{cases} 1 - (1 + \frac{\xi y}{\beta})^{-1/\xi} & for \quad \begin{array}{l} \xi > 0, \quad y > 0 \ or \\ \xi < 0, \quad 0 < y < \frac{\beta}{-\xi} \end{array} \\ 1 - e^{-y/\beta} & for \quad \xi = 0, \quad y > 0 \end{cases} \quad (25)$$

References

1. Asmussen, S., Glynn, P.W.: Stochastic Simulation: Algorithms and Analysis. Springer, New York (2007)
2. De Haan, L., Resnick, S.I.: Local limit theorems for sample extremes. Ann. Probab. **10**, 396–413 (1982)
3. Glassserman, P.: Monte Carlo Methods in Financial Engineering. Springer, New York (2004)
4. Glasserman, P., Li, J.: Importance sampling for portfolio credit risk. Manag. Sci. **51**(11), 1643–1656 (2005)
5. Gupton, G., Finger, C.C., Bhatia, M.: CreditMetrics, Technical Document of the Morgan Guaranty Trust Company http://www.defaultrisk.com/pp_model_20.htm (1997)
6. Hall, P.: On the rate of convergence of normal extremes. J. Appl. Probab. **16**, 433–439 (1979)
7. Homem-de-Mello, T., Rubinstein, R.Y.: Rare event probability estimation using cross-entropy. In: Proceedings of the 2002 Winter Simulation Conference, pp. 310–319 (2002)
8. Kroese, D., Rubinstein, R.Y.: Simulation and the Monte Carlo Method, 2nd edn. Wiley, New York (2008)
9. Kroese, D., Taimre, T., Botev, Z.I.: Handbook of Monte Carlo Methods. Wiley, New York (2011)

10. Kotz, S., Nadarajah, S.: Extreme Value Distributions: Theory and Applications. Imperial College Press, London (2000)
11. McLeish, D.L.: Bounded relative error importance sampling and rare event simulation. ASTIN Bullet. **40**, 377–398 (2010)
12. McLeish, D.L.: Monte Carlo Simulation and Finance. Wiley, New York (2005)
13. McNeil, A.J., Frey, R., Embrechts, P.: Quantitative Risk Management. Princeton University Press, Princeton (2005)
14. Pickands, J.: Sample sequences of maxima. Ann. Math. Stat. **38**, 1570–1574 (1967)
15. Ridder, A., Rubinstein, R.: Minimum cross entropy method for rare event simulation. Simulation **83**, 769–784 (2007)

A Note on the Numerical Evaluation of the Hartman–Watson Density and Distribution Function

German Bernhart and Jan-Frederik Mai

Abstract The Hartman–Watson distribution is an infinitely divisible probability law on the positive half-axis whose density is difficult to evaluate near zero. We compare three different methods to evaluate this density and show that the straightforward implementation along Yor's explicit formula can be improved significantly by resorting to dedicated Laplace inversion algorithms. In particular, the best method seems to be an approach that is specifically designed for distributions from the Bondesson class, to which the Hartman–Watson distribution belongs. The latter approach can furthermore be extended to yield an efficient Laplace inversion algorithm for evaluating the distribution function of the Hartman–Watson law.

Keywords Hartman–Watson law · Laplace inversion · Infinitely divisible distributions · Bondesson class

Mathematics Subject Classification (2000) 65C50 · 60E07

1 Introduction

In the process of studying the probability distribution of the integral over a geometric Brownian motion, [14] introduced the function

$$
\theta(r, x) := \frac{r \, e^{\frac{\pi^2}{2x}}}{\sqrt{2\,\pi^3\,x}} \int_0^\infty e^{-\frac{y^2}{2x} - r\,\cosh(y)} \, \sinh(y) \, \sin\left(\frac{\pi\,y}{x}\right) \mathrm{d}y, \tag{1}
$$

for $r, x > 0$. Denoting by

G. Bernhart (✉)
Technische Universität München, Parkring 11, 85748 Garching-Hochbrück, Germany
e-mail: german.bernhart@tum.de

J.-F. Mai
XAIA Investment GmbH, Sonnenstraße 19, 80331 München, Germany

© The Author(s) 2015
K. Glau et al. (eds.), *Innovations in Quantitative Risk Management*,
Springer Proceedings in Mathematics & Statistics 99,
DOI 10.1007/978-3-319-09114-3_19

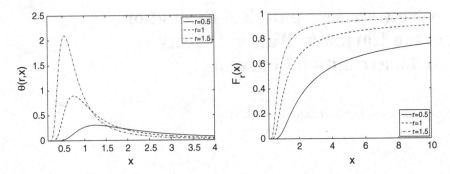

Fig. 1 *Left* The function $\theta(r, x)$ for three different parameters r and values $x \in (0.15, 4)$. *Right* The distribution function $F_r(x)$ for three different parameters r and values $x \in (0.15, 10)$

$$I_\nu(z) := \sum_{m=0}^{\infty} \frac{1}{m! \, \Gamma(m + \nu + 1)} \left(\frac{z}{2}\right)^{2m+\nu} \qquad (2)$$

the modified Bessel function of the first kind, the function $f_r(x) := \theta(r, x)/I_0(r)$, $x > 0$, $r > 0$, is the density of a one-parametric probability law, say μ_r, on the positive half-axis, called the *Hartman–Watson law*. The Hartman–Watson law arises as the first hitting time of certain diffusion processes, see [10], and is of paramount interest in mathematical finance in the context of Asian option pricing, see [2, 6, 14]. It was shown in [7] that this law is infinitely divisible with Laplace transform given by $\varphi_r(u) := I_{\sqrt{2u}}(r)/I_0(r)$, $u \geq 0$. Moreover, it follows from a result in [10] that μ_r is not only infinitely divisible, but even within the so-called Bondesson class, which is a large subfamily of infinitely divisible laws that is introduced in and named after [5]. Notice in particular that it follows from this fact together with ([13, Theorem 6.2, p. 49]) that the function $\Psi_r(u) := -\log\left(I_{\sqrt{2u}}(r)/I_0(r)\right)$, $u \geq 0$, is a so-called *complete Bernstein function*, which allows for a holomorphic extension to the sliced complex plane $\mathbb{C} \setminus (-\infty, 0)$. We will make use of this observation in Sect. 5.

It is well-known that the numerical evaluation of the density of the Hartman–Watson law near zero is a challenging task because the integrand in the formula for $\theta(r, x)$ is highly oscillating. The following sections discuss several methods to evaluate the function $\theta(r, x)$ accurately. Figure 1 visualizes the function $\theta(r, x)$ for three different parameters r and values $x \in (0.15, 4)$, where all numerical computation routines discussed in the present note yield exactly the same result.

When looking at Fig. 1, mathematical intuition suggests that the approximation $\theta(0.5, x) \approx 0$ for $x < 0.15$ might be a pragmatic—and numerically efficient—implementation close to zero. Nevertheless, [9] considers the numerical evaluation close to zero and obtains significant errors, see Sect. 3. Moreover, [6] studies the asymptotic behavior of $f_r(x)$ as $x \downarrow 0$ and [2] study the behavior of the distribution function F_r of μ_r as the argument tends to zero. We like to mention that the right tail of the Hartman–Watson distribution μ_r becomes extremely heavy as $r \downarrow 0$. For instance, the distribution function $F_r(x) = \int_0^x f_r(t)\,dt$ is still significantly smaller than 1 for $x = 10$ and different r, see Fig. 1.

The remaining article is organized as follows. Section 2 illustrates the occurrence of the Hartman–Watson distribution, in particular, in mathematical finance. Section 3 discusses the direct implementation of Formula (1). Section 4 proposes the use of the Gaver–Stehfest Laplace inversion technique. Section 5 proposes a complex Laplace inversion algorithm to numerically evaluate f_r and F_r. Finally, Sect. 6 concludes.

2 Occurrence of the Hartman–Watson Law

The most prominent occurrence of the Hartman–Watson distribution is probably in directional statistics (see [8]): if $\mathbf{W}_t$ denotes a two-dimensional Brownian motion on the unit circle and $\tau \sim \mu_r$ is independent thereof, then $\mathbf{W}_\tau$ has the same law as $(\cos(X), \sin(X))$, where X follows the so-called von Mises distribution with parameter r, which has density given by

$$f_X(x) = \frac{1}{2\pi I_0(r)}\, e^{r\,\cos(x)}, \quad -\pi < x < \pi.$$

The von Mises distribution is the most prominent law for an angle in the field of directional statistics, because it constitutes a tractable approximation to the "wrapped normal distribution" (i.e., the law of $Y \bmod 2\pi$ when Y is normal), which is difficult to work with.

The importance of the Hartman–Watson distribution in the context of mathematical finance originates from the fact that

$$\frac{e^{-\frac{x^2}{2t}}}{\sqrt{2\pi t}}\, \mathbb{P}(A_t \in du \mid W_t = x) = \frac{1}{u}\, e^{-\frac{1+e^{2x}}{2u}}\, \theta\!\left(\frac{e^x}{u}, t\right),$$

where W_t denotes standard Brownian motion and $A_t = \int_0^t e^{2\,W_s}\,ds$ an associated integrated geometric Brownian motion, see [14]. The process A_t, and hence the Hartman–Watson distribution, naturally enters the scene when Asian stock derivatives, i.e., derivatives with "averaging periods," are considered in the Black–Scholes world, see, e.g., [2, 9]. Another example, which is mathematically based on the exactly same reasoning, has recently been given in [3]: when the Black–Scholes model is enhanced by the introduction of stochastic repo margins, this leads to a convexity adjustment for all kinds of stock derivatives which involves the density of the Hartman–Watson distribution.

Let us furthermore briefly sketch a potential third application, which uses a stochastic representation for the Hartman–Watson law. Consider a diffusion process $\{X_t\}_{t\geq 0}$ satisfying the SDE

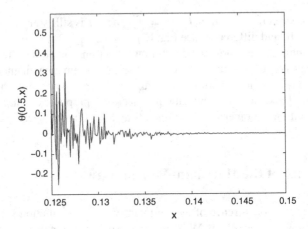

Fig. 2 Evaluation of Formula (1) for $r = 0.5$ and $x \in [0.125, 0.15]$ in MATLAB applying the built-in adaptive quadrature routine quadgk, which can handle infinite integration domains

$$dX_t = X_t \left(\left(\frac{1}{2} + X_t \, \frac{I_1(X_t)}{I_0(X_t)} \right) dt + dW_t \right), \quad X_0 = r > 0.$$

This explodes with probability one, as can be seen from Feller's test for explosion (the drift increases rapidly), i.e., there exists a stopping time $\tau \in (0, \infty)$ such that paths of $\{X_t\}$ are well defined on $[0, \tau)$ and $\lim_{t \uparrow \tau} X_t = \infty$ almost surely. Such explosive diffusions are used to model fatigue failures in solid materials. X_t describes the evolution of the length of the longest crack and τ is the time point of ultimate damage. Kent [10] shows that $\tau \sim \mu_r$. We may rewrite τ as the first hitting time of zero of the stochastic process $Y_t := 1/X_t$, starting at $Y_0 = 1/r > 0$. Observing the stock price $S_0 > 0$ of a highly distressed company facing bankruptcy, it might now make sense to model the evolution of this company's stock price until default as $S_t := Y_t$ setting $r := 1/S_0$. The time of bankruptcy is defined as the first time the stock price hits zero, which has a Hartman–Watson law. A similar model, assuming S_t to follow a CEV process that is allowed to diffuse to zero, is applied in [1].

3 Straightforward Implementation Based on Formula (1)

Regarding the exact numerical evaluation of the Hartman–Watson density, the article [9] shows that a straightforward numerical implementation of the Formula (1) for $r = 0.5$ and $x \in [0.125, 0.15]$ yields significant numerical errors. In particular, Fig. 2 in [9] shows that one ends up with negative density values. We come to the same conclusion, see Fig. 2.

4 Evaluation via Gaver–Stehfest Laplace Inversion

We apply the Gaver–Stehfest algorithm in order to obtain $\theta(r, \cdot)$ from its Laplace transform $I_{\sqrt{2}\cdot}(r)$ via Laplace inversion for fixed values of r. For a rigorous proof and a good explanation of this method, see [11]. In particular, it is not difficult to observe from Yor's expression (1) that ([11] Theorem 1(iii)) applies, which justifies the approximation

$$\theta(r, x) \approx \frac{\log(2)}{x} \sum_{k=1}^{2n} a_k(n) \, I_{\sqrt{2k \log(2)/x}}(r), \tag{3}$$

for $n \in \mathbb{N}$ large enough, where for $j = 1, \ldots, 2n$ we have

$$a_k(n) = \frac{(-1)^{k+n}}{n!} \sum_{j=\lfloor (k+1)/2 \rfloor}^{\min\{k,n\}} j^{n+1} \binom{n}{j} \binom{2j}{j} \binom{j}{k-j}.$$

The Gaver–Stehfest algorithm has the nice feature that only evaluations of the Laplace transform on the positive half-axis are required. In particular, the required modified Bessel function $I_{\sqrt{2u}}(r)$ is efficient and easy to compute for $u > 0$. In MATLAB, it is available as the built-in function `besseli`. The drawback of the Gaver–Stehfest algorithm is that it requires high-precision arithmetic because the involved constants $a_k(n)$ are alternating and become huge and difficult to evaluate. For practical implementations, this prevents the use of large n, which would theoretically be desirable due to the convergence result of [11]. Nevertheless, our empirical investigation shows that $n = 10$ is still feasible on a standard PC without further precision arithmetic considerations and yields reasonable results for the considered parameterization. However, for larger values of r, the algorithm is less stable as can be seen at the end of Sect. 5.

The obtained values of $\theta(r, x)$ are visualized in Fig. 3. Comparing them to the brute force implementation in Fig. 2, the error for small x becomes significantly smaller.

5 Evaluation via a Complex Laplace Inversion Method for the Bondesson Class

As already mentioned in the introduction, the Hartman–Watson law is in the Bondesson class which allows to apply a Laplace inversion algorithm specifically derived for such distributions in [4]. Furthermore, this method has the advantage that it immediately implies as a corollary a similar formula for the distribution function F_r. To be precise, we have the formula

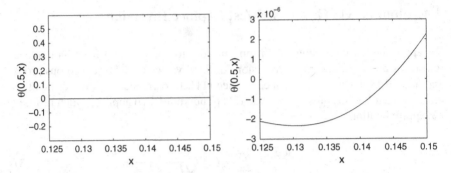

Fig. 3 Evaluation of $\theta(r, x)$ for $r = 0.5$ and $x \in [0.125, 0.15]$ in MATLAB applying the Gaver–Stehfest approximation (3) with $n = 10$. *Left* The y-axis is precisely the same as in Fig. 2 for comparability. *Right* The y-axis is made finer to visualize smaller errors (scale 10^{-6})

$$\theta(r, x) = \frac{M e^{x a}}{\pi} \int_0^1 \mathrm{Im}\left(\bar{e}^{x M \log(v) (b i - a)} I_{\sqrt{2 (a - M \log(v) (b i - a))}}(r) (b i - a)\right) \frac{dv}{v}$$

$$(4)$$

with arbitrary parameters $a, b > 0$ and $M > 2/(a x)$, and this integral is a proper Riemannian integral, since the integrand vanishes for $v \downarrow 0$, see [4]. Regarding the choice of the parameters, [4] have shown that $a = 1/x$, $M = 3$ is usually a good choice and we will use these parameters. Concerning the remaining parameter b, we choose $b = a$. For the evaluation of the distribution function F_r, it is also shown in [4] that

$$F_r(x) = \frac{M e^{x a}}{\pi} \int_0^1 \mathrm{Im}\left(\bar{e}^{x M \log(v) (b i - a)} \frac{\varphi_r(a - M \log(v) (b i - a))}{a - M \log(v) (b i - a)} (b i - a)\right) \frac{dv}{v}$$

with the same parameter restrictions as above. One particular challenge with this method is that the modified Bessel function I_ν needs to be evaluated for complex ν. A straightforward implementation sufficient for our needs is achieved by using the partial sums related to the representation in Eq. (2). It has the advantage that error bounds can be computed, as for $r > 0$ and $S_n^\nu(r) := \sum_{m=0}^n \frac{1}{m! \Gamma(m + \nu + 1)} (\frac{r}{2})^{2m + \nu}$, one can compute

$$\left| S_n^\nu(r) - I_\nu(r) \right| \leq \left(\frac{r}{2}\right)^{\mathrm{Re}(\nu)} \sum_{m=n+1}^\infty \frac{1}{m! |\Gamma(m + \nu + 1)|} \left(\frac{r}{2}\right)^{2m}.$$

Using the Gamma functional equation $\Gamma(z + 1) = \Gamma(z) z$, it is easy to see that $|\Gamma(z + 1)| \geq |\Gamma(z)|$ for $|z| \geq 1$. Thus, for $n \geq -\mathrm{Re}(\nu) - 1$, the sequence $\{|\Gamma(m + \nu + 1)|\}_{m=n+1, n+2, \ldots}$ is increasing, yielding

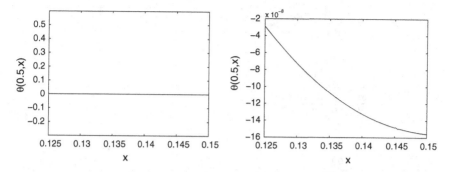

Fig. 4 Evaluation of $\theta(r, x)$ for $r = 0.5$ and $x \in [0.125, 0.15]$ in MATLAB applying the Laplace inversion formula (4) with $a = b = 1/x$ and $M = 3$. The modified Bessel function is implemented with accuracy 10^{-6}. *Left* The y-axis is precisely the same as in Fig. 2 for comparability. *Right* The y-axis is made finer to visualize smaller errors (scale 10^{-8})

$$\left| S_n^\nu(r) - I_\nu(r) \right| \le \frac{\left(\frac{r}{2}\right)^{\mathrm{Re}(\nu)}}{|\Gamma(n + \nu + 2)|} \sum_{m=n+1}^{\infty} \frac{1}{m!} \left(\frac{r^2}{4}\right)^m,$$

where the series term is the residual of the Taylor expansion of $\exp(-r^2/4)$, which allows for a closed-form estimate. Consequently, one is able to choose n such that the modified Bessel function is approximated up to a given accuracy. Using the Gamma functional equation, one has to compute the complex Gamma function only once which further increases efficiency. The complex Gamma function is computed using the Lanczos approximation, see [12].[1]

Figure 4 shows the resulting values of $\theta(r, x)$, where the modified Bessel function is approximated with accuracy 10^{-6}. Formula (4) is evaluated in MATLAB applying the built-in adaptive quadrature routine quadgk. Comparing the results to the Gaver–Stehfest inversion, the error for small x is again significantly reduced and the results can be even improved by further increasing the accuracy of the modified Bessel function.

A second comparison of the presented methods is included for a larger value of r. The Laplace inversion method for the Bondesson class represents the most stable and accurate algorithm as can be seen in Fig. 5, which visualizes the values of $\theta(r, x)$ for small x and $r = 3$. Whereas the straightforward implementation based on Formula (1) fails due to numerical problems and the choice $n = 10$ is not ideal for the Gaver–Stehfest Laplace inversion, the Bondesson method yields stable results.

---

[1] We use the implementation of P. Godfrey published on http://www.mathworks.com/matlabcentral/fileexchange/3572-gamma.

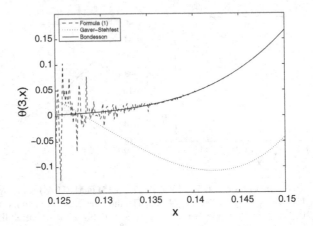

Fig. 5 Evaluation of $\theta(r, x)$ for $r = 3$ and $x \in [0.125, 0.15]$ using the three presented methods with the same specifications as before, i.e., the Gaver–Stehfest approximation (3) with $n = 10$ and the Laplace inversion formula (4) with $a = b = 1/x$ and $M = 3$

6 Conclusion

We compared three different methods to numerically evaluate the density of the Hartman–Watson law. We found that Laplace inversion algorithms significantly outperform direct implementation of Yor's formula (1). Moreover, a dedicated algorithm for distributions of the Bondesson class was proposed to numerically evaluate the distribution function of the Hartman–Watson law efficiently.

Open Access This chapter is distributed under the terms of the Creative Commons Attribution Noncommercial License, which permits any noncommercial use, distribution, and reproduction in any medium, provided the original author(s) and source are credited.

References

1. Atlan, M., Leblanc, B.: Hybrid equity-credit modelling. Risk Magazine August: 61–66 (2005)
2. Barrieu, P., Rouault, A., Yor, M.: A study of the Hartman-Watson distribution motivated by numerical problems related to the pricing of Asian options. J. Appl. Probab. **41**, 1049–1058 (2004)
3. Bernhart, G., Mai, J.-F.: On convexity adjustments for stock derivatives due to stochastic repo margins. Working paper (2014)
4. Bernhart, G., Mai, J.-F., Schenk, S., Scherer M.: The density for distributions of the Bondesson class. J. Comput. Financ. (2014, to appear)
5. Bondesson, L.: Classes of infinitely divisible distributions and densities: Zeitschrift für Wahrscheinlichkeitstheorie und verwandte Gebiete **57**, 39–71 (1981)
6. Gerhold, S.: The Hartman–Watson distribution revisited: asymptotics for pricing Asian options. J. Appl. Probab. **48**(3), 892–899 (2011)

7. Hartman, P.: Completely monotone families of solutions of nth order linear differential equations and infinitely divisible distributions. Ann. Scuola Norm. Sup. Pisa **4**(3), 267–287 (1976)
8. Hartman, P., Watson, G.: "Normal" distribution functions on spheres and the modified Bessel functions. Ann. Probab. **5**, 582–585 (1974)
9. Ishiyama, K.: Methods for evaluating density functions of exponential functionals represented as integrals of geometric Brownian motion. Methodol. Comput. Appl. Probab. **7**, 271–283 (2005)
10. Kent, J.T.: The spectral decomposition of a diffusion hitting time. Ann. Probab. **10**(1), 207–219 (1982)
11. Kuznetsov, A.: On the convergence of the Gaver–Stehfest algorithm. SIAM J. Numer. Anal. (2013, forthcoming)
12. Lanczos, C.: A precision approximation of the gamma function. J. Soc. Ind. Appl. Math. Ser. B Numer. Anal. **1**, 86–96 (1964)
13. Schilling, R.L., Song, R., Vondracek Z. Bernstein Functions. Studies in Mathematics, Vol. 37, de Gruyter, Berlin. (2010)
14. Yor, M.: On some exponential functionals of Brownian motion. Adv. Appl. Probab. **24**(3), 509–531 (1992)

Computation of Copulas by Fourier Methods

Antonis Papapantoleon

Abstract We provide an integral representation for the (implied) copulas of dependent random variables in terms of their moment generating functions. The proof uses ideas from Fourier methods for option pricing. This representation can be used for a large class of models from mathematical finance, including Lévy and affine processes. As an application, we compute the implied copula of the NIG Lévy process which exhibits notable time-dependence.

1 Introduction

Copulas provide a complete characterization of the dependence structure between random variables and link in a very elegant way the joint distribution with the marginal distributions via Sklar's theorem. However, they are a rather static concept and do not blend well with stochastic processes which can be used to describe the random evolution of dependent quantities, e.g., the evolution of several stock prices. Therefore, other methods to create dependence in stochastic models have been developed. Multivariate stochastic processes spring immediately to mind, for example, Lévy or affine processes (cf. e.g., [3, 4, 17] or [15]), while in mathematical finance models using time changes or linear mixture models have been developed; see, e.g., [5, 10, 12, 13] or [11], to mention just a small part of the existing literature. In these approaches, however the copula is typically not known explicitly. Another very interesting approach is due to [9], who introduced Lévy copulas to characterize the dependence structure of Lévy processes.

In this note, we provide a new representation for the (implied) copula of a multidimensional random variable in terms of its moment generating function. The derivation of the main result borrows ideas from Fourier methods for option pricing, and the motivation stems from the knowledge of the moment generating function in most of the aforementioned models. This paper is organized as follows: in Sect. 2 we provide the representation of the copula in terms of the moment generating function;

A. Papapantoleon (✉)
Institute of Mathematics, TU Berlin, Straße des 17. Juni 136, 10623 Berlin, Germany
e-mail: papapan@math-tu-berlin.de

© The Author(s) 2015   347
K. Glau et al. (eds.), *Innovations in Quantitative Risk Management*,
Springer Proceedings in Mathematics & Statistics 99,
DOI 10.1007/978-3-319-09114-3_20

the results are proved for random variables for simplicity, while stochastic processes are considered as a corollary. In Sect. 3, we provide two examples to showcase how this method can be applied, for example, in performing sensitivity analysis of the copula with respect to the parameters of the model. Finally, Sect. 4 concludes with some remarks.

2 Copulas via Fourier Transform Methods

Let $\mathbb{R}^n$ denote the n-dimensional Euclidean space, $\langle \cdot, \cdot \rangle$ the Euclidean scalar product and $\mathbb{R}^n_-$ the negative orthant, i.e., $\mathbb{R}^n_- = \{x \in \mathbb{R}^n : x_i < 0 \; \forall i\}$. We consider a random variable $X = (X_1, \ldots, X_n)^\top \in \mathbb{R}^n$ defined on a probability space $(\Omega, \mathscr{F}, \mathbf{P})$. We denote by F the cumulative distribution function (cdf) of X and by f its probability density function (pdf). Let C denote the copula of X and c its copula density function. Analogously, let F_i and f_i denote the cdf and pdf respectively of the marginal X_i, for all $i \in \{1, \ldots, n\}$. In addition, we denote by F_i^{-1} the generalized inverse of F_i, i.e., $F_i^{-1}(u) = \inf\{v \in \mathbb{R} : F_i(v) \geq u\}$.

We denote by M_X the (extended) moment generating function of X:

$$M_X(u) = \mathbf{E}[e^{\langle u, X \rangle}], \tag{1}$$

for all $u \in \mathbb{C}^n$ such that $M_X(u)$ exists. Let us also define the set

$$\mathscr{I} = \{R \in \mathbb{R}^n : M_X(R) < \infty \text{ and } M_X(R + \mathrm{i}\cdot) \in L^1(\mathbb{R}^n)\}.$$

In the sequel, we will assume that the following condition is in force.

Assumption $(\mathbb{D})$. $\mathscr{R} := \mathscr{I} \cap \mathbb{R}^n_- \neq \emptyset$.

Remark 1 The integrability of the moment generating function required by Assumption $(\mathbb{D})$ has the following implications:

(a) the distribution function F is absolutely continuous with respect to the Lebesgue measure;
(b) the density function f is bounded and continuous;
(c) the marginal distribution functions F_i are also absolutely continuous.

See [17, Proposition 2.5] for (a) and (b) and [8, Theorem 12.2] for (c).

Theorem 1 *Let X be a random variable that satisfies Assumption $(\mathbb{D})$. The copula of X is provided by*

$$C(u) = \frac{1}{(-2\pi)^n} \int_{\mathbb{R}^n} M_X(R + \mathrm{i}v) \frac{e^{-\langle R + \mathrm{i}v, x \rangle}}{\prod_{i=1}^n (R_i + \mathrm{i}v_i)} \mathrm{d}v \Big|_{x_i = F_i^{-1}(u_i)}, \tag{2}$$

where $u \in [0, 1]^n$ and $R \in \mathscr{R}$.

Proof Assumption ($\mathbb{D}$) implies that $F_1, \ldots, F_n$ are continuous and we know from Sklar's theorem that the copula of X is unique and provided by

$$C(u_1, \ldots, u_n) = F\left(F_1^{-1}(u_1), \ldots, F_n^{-1}(u_n)\right); \tag{3}$$

see, e.g., [14, Theorem 5.3] for a proof in this setting and [16] for an elegant proof in the general case.

We will evaluate the joint cdf F using the methodology of Fourier methods for option pricing. That is, we will think of the cdf as the "price" of a digital option on several fictitious assets. Let us define the function

$$g(y) = 1_{\{y_1 \le x_1, \ldots, y_n \le x_n\}}(y), \quad x, y \in \mathbb{R}^n, \tag{4}$$

and denote by $\widehat{g}$ its Fourier transform. Then we have that

$$
\begin{aligned}
F(x) &= \mathbf{P}(X_1 \le x_1, \ldots, X_n \le x_n) \\
&= \mathbf{E}\left[1_{\{X_1 \le x_1, \ldots, X_n \le x_n\}}\right] = \mathbf{E}[g(X)] \\
&= \frac{1}{(2\pi)^n} \int_{\mathbb{R}^n} M_X(R + \mathrm{i}v)\widehat{g}(\mathrm{i}R - v)\mathrm{d}v,
\end{aligned}
\tag{5}
$$

where we have applied Theorem 3.2 in [6]. The prerequisites of this theorem are satisfied due to Assumption ($\mathbb{D}$) and because $g_R \in L^1(\mathbb{R}^n)$, where $g_R(x) := \mathrm{e}^{-\langle R, x \rangle} g(x)$ for $R \in \mathbb{R}^n_-$.

Finally, the statement follows from (3) and (5) once we have computed the Fourier transform of g. We have for $R_i < 0, i \in \{1, \ldots, n\}$,

$$
\begin{aligned}
\widehat{g}(\mathrm{i}R - v) &= \int_{\mathbb{R}^n} \mathrm{e}^{\mathrm{i}\langle \mathrm{i}R - v, y \rangle} g(y)\mathrm{d}y \\
&= \int_{\mathbb{R}^n} \mathrm{e}^{\mathrm{i}\langle \mathrm{i}R - v, y \rangle} 1_{\{y_1 \le x_1, \ldots, y_n \le x_n\}}\mathrm{d}y \\
&= \prod_{i=1}^{n} \int_{-\infty}^{x_i} \mathrm{e}^{(-R_i - \mathrm{i}v_i)y_i}\mathrm{d}y_i \\
&= (-1)^n \prod_{i=1}^{n} \frac{\mathrm{e}^{-(R_i + \mathrm{i}v_i)x_i}}{R_i + \mathrm{i}v_i},
\end{aligned}
\tag{6}
$$

which concludes the proof.

Remark 2 If the moment generating function of the marginals is known, the inverse function can be easily computed numerically. We have that

$$F_i^{-1}(u) = \inf\{v \in \mathbb{R} : F_i(v) \geq u\}$$
$$= \inf\{v \in \mathbb{R} : \mathbf{E}\big[1_{\{X_i \leq v\}}\big] \geq u\},$$

where the expectation can be computed using (5) again, while a root finding algorithm provides the infimum (using the continuity of F_i).

We can also compute the copula density function using Fourier methods, which resembles the computation of Greeks in option pricing.

Lemma 1 *Let X be a random variable that satisfies Assumption* $(\mathbb{D})$ *and assume further that the marginal distribution functions $F_1, \ldots, F_n$ are strictly increasing and continuously differentiable. Then, the copula density function c of X is provided by*

$$c(u) = \frac{1}{(2\pi)^n \prod_{i=1}^n f_i(x_i)} \int\limits_{\mathbb{R}^n} M_X(R + iv) \, e^{-\langle R+iv, x\rangle} dv \bigg|_{x_i = F_i^{-1}(u_i)}, \qquad (7)$$

where $u \in (0, 1)^n$ and $R \in \mathscr{R}$.

Proof The distribution functions F and $F_1, \ldots, F_n$ are absolutely continuous hence the copula density exists, cf. [14, p. 197]. Let $u \in (0, 1)^n$, then we have that $x_i = F_i^{-1}(u_i)$ is finite for every $i \in \{1, \ldots, n\}$, hence $e^{-\langle R, x\rangle}$ is bounded. Using Assumption $(\mathbb{D})$ we get that the function $M_X(R+iv)e^{-\langle R+iv, x\rangle}$ is integrable and we can interchange differentiation and integration. Then we have that

$$c(u) = \frac{\partial^n}{\partial u_1 \ldots \partial u_n} C(u_1, \ldots, u_n)$$

$$= \frac{\partial^n}{\partial u_1 \ldots \partial u_n} \frac{1}{(-2\pi)^n} \int\limits_{\mathbb{R}^n} M_X(R + iv) \frac{e^{-\langle R+iv, x\rangle}}{\prod_{i=1}^n (R_i + iv_i)} \bigg|_{x_i = F_i^{-1}(u_i)} dv$$

$$= \frac{1}{(-2\pi)^n} \int\limits_{\mathbb{R}^n} \frac{M_X(R + iv)}{\prod_{i=1}^n (R_i + iv_i)} \frac{\partial^n}{\partial u_1 \ldots \partial u_n} e^{-\langle R+iv, x\rangle} \bigg|_{x_i = F_i^{-1}(u_i)} dv. \qquad (8)$$

Now, since the marginal distribution functions are continuously differentiable, using the chain rule and the inverse function theorem we get that

$$\frac{\partial^n}{\partial u_1 \ldots \partial u_n} \left(e^{-\langle R+iv, x\rangle} \bigg|_{x_i = F_i^{-1}(u_i)} \right)$$

$$= (-1)^n \prod_{i=1}^n (R_i + iv_i) e^{-\langle R+iv, x\rangle} \frac{1}{\prod_{i=1}^n f_i(x_i)} \bigg|_{x_i = F_i^{-1}(u_i)}, \qquad (9)$$

which combined with (8) yields the required result.

A natural application of these representations is for the calculation of the copula of a random variable X_t from a multidimensional stochastic process $X =$

$(X_t)_{t\geq 0}$. There are many examples of stochastic processes where the corresponding characteristic functions are known explicitly. Prominent examples are Lévy processes, self-similar additive ("Sato") processes and affine processes.

Corollary 1 *Let* $X = (X_t)_{t\geq 0}$ *be an* $\mathbb{R}^n$-*valued stochastic process on a filtered probability space* $(\Omega, \mathscr{F}, (\mathscr{F}_t)_{t\geq 0}, \mathbf{P})$. *Assume that the random variable* X_t, $t \geq 0$, *satisfies Assumption* $(\mathbb{D})$. *Then, the copula of* X_t *is provided by*

$$C_t(u) = \frac{1}{(-2\pi)^n} \int\limits_{\mathbb{R}^n} M_{X_t}(R + iv) \frac{e^{-\langle R+iv,x\rangle}}{\prod_{i=1}^n (R_i + iv_i)} dv \bigg|_{x_i = F_{X_t^i}^{-1}(u_i)}, \tag{10}$$

where $u \in [0,1]^n$ *and* $R \in \mathscr{R}$. *An analogous statement holds for the copula density function* c_t *of* X_t.

3 Examples

We will demonstrate the applicability and flexibility of Fourier methods for the computation of copulas using two examples. First, we consider a 2D normal random variable and next a 2D normal inverse Gaussian (NIG) Lévy process. Although the copula of the normal random variable is the well-known Gaussian copula, little was known about the copula of the NIG distribution until recently; see Theorem 5.13 in [18] for a special case. Hammerstein [19, Chap. 2] has now provided a general characterization of the (implied) copula of the multidimensional NIG distribution using properties of normal mean-variance mixtures.

Example 1 The first example is simply a "sanity check" for the proposed method. We consider the two-dimensional Gaussian distribution and compute the corresponding copula for correlation values equal to $\rho = \{-1, 0, 1\}$; see Fig. 1 for the resulting contour plots. Of course, the copula of this example is the Gaussian copula, which for correlation coefficients equal to $\{-1, 0, 1\}$ corresponds to the *countermonotonicity* copula, the *independence* copula and the *comonotonicity* copula respectively. This is also evident from Fig. 1.

Example 2 Let $X = (X_t)_{t\geq 0}$ be a two-dimensional NIG Lévy process, i.e.

$$X_t = (X_t^1, X_t^2) \sim \text{NIG}_2(\alpha, \beta, \delta t, \mu t, \Delta), \quad t \geq 0. \tag{11}$$

The parameters satisfy: $\alpha, \delta > 0$, $\beta, \mu \in \mathbb{R}^2$, and $\Delta \in \mathbb{R}^{2\times 2}$ is a symmetric, positive definite matrix (w.l.o.g. we can assume $\det(\Delta) = 1$). Moreover, $\alpha^2 > \langle \beta, \Delta\beta \rangle$. The moment generating function of X_1, for $u \in \mathbb{R}^2$ with $\alpha^2 - \langle \beta + u, \Delta(\beta + u) \rangle \geq 0$, is

$$M_{X_1}(u) = \exp\left(\langle u, \mu \rangle + \delta\left(\sqrt{\alpha^2 - \langle \beta, \Delta\beta \rangle} - \sqrt{\alpha^2 - \langle \beta + u, \Delta(\beta + u) \rangle}\right)\right),$$

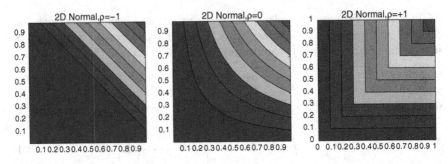

Fig. 1 Contour plots of copulas for Example 1

cf. [1]. The marginals are also NIG distributed and we have that $X_t^i \sim \text{NIG}(\hat{\alpha}^i, \hat{\beta}^i, \hat{\delta}^i t,$ $\hat{\mu}^i t)$, where

$$\hat{\alpha}^i = \sqrt{\frac{\alpha^2 - \beta_j^2(\delta_{jj} - \delta_{ij}^2 \delta_{ii}^{-1})}{\delta_{ii}}}, \ \hat{\beta}^i = \beta_i + \beta_j \delta_{ij}^2 \delta_{ii}^{-1}, \ \hat{\delta}^i = \delta\sqrt{\delta_{ii}}, \ \hat{\mu}^i = \mu_i,$$

for $i = \{1, 2\}$ and $j = \{2, 1\}$; cf. e.g., [2, Theorem 1]. Assumption ($\mathbb{D}$) is satisfied for $R \in \mathbb{R}_-^2$ such that $\alpha^2 - \langle \beta + R, \Delta(\beta + R) \rangle \geq 0$; see Appendix B in [6]. Hence $\mathcal{R} \neq \emptyset$.

Therefore, we can apply Theorem 1 to compute the copula of the NIG distribution. The parameters used in the numerical example are similar to [6, pp. 233–234]: $\alpha = 10.20$, $\beta = \left(\begin{smallmatrix} -3.80 \\ -2.50 \end{smallmatrix}\right)$, $\delta = 0.150$, $\mu \equiv 0$, and two matrices $\Delta^+ = \left(\begin{smallmatrix} 1 & 0 \\ 0 & 1 \end{smallmatrix}\right)$ and $\Delta^- = \left(\begin{smallmatrix} 1 & -1 \\ -1 & 2 \end{smallmatrix}\right)$, which lead to positive and negative correlation. The correlation coefficients are $\rho_+ = 0.1015$ and $\rho_- = -0.687$ respectively.

The contour plots are exhibited in Figs. 2 and 3 and show clearly the influence of the different mixing matrices Δ^+ and Δ^- to the dependence structure. Moreover, we can also observe that time has a significant effect on the dependence structure of the multidimensional NIG Lévy process. This is an interesting observation, since the correlation matrix is invariant over time (which is true for any Lévy process).

4 Final Remarks

We will not elaborate on the speed of Fourier methods compared with Monte Carlo methods in the multidimensional case; the interested reader is referred to [7] for a careful analysis. Moreover, [20] provides recommendations on the efficient implementation of Fourier integrals using sparse grids in order to deal with the "curse of dimensionality." Let us point out though that the computation of the copula function will be much quicker than the computation of the copula density, since the integrand in (2) decays much faster than the one in (7). One should think of the analogy to

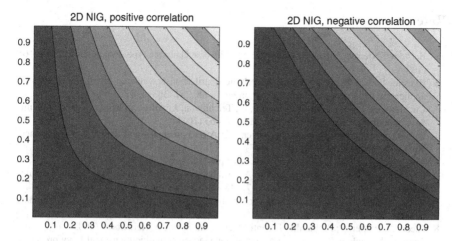

Fig. 2 Contour plots of copulas for NIG, $t = 1$

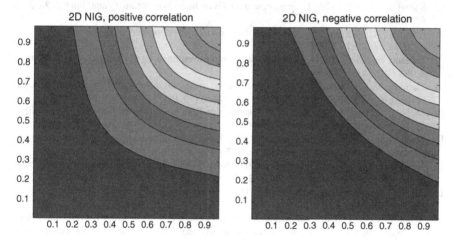

Fig. 3 Contour plots of copulas for NIG, $t = \frac{1}{2}$

option prices and option Greeks again. Finally, it seems tempting to use these formulas for the computation of tail dependence coefficients. However, due to numerical instabilities at the limits, they did not yield any meaningful results.

Acknowledgments The author thanks E. A. von Hammerstein for helpful comments and suggestions.

Open Access This chapter is distributed under the terms of the Creative Commons Attribution Noncommercial License, which permits any noncommercial use, distribution, and reproduction in any medium, provided the original author(s) and source are credited.

References

1. Barndorff-Nielsen, O.E.: Processes of normal inverse Gaussian type. Financ. Stoch. **2**, 41–68 (1998)
2. Blæsild, P.: The two-dimensional hyperbolic distribution and related distributions, with an application to Johannsen's bean data. Biometrika **68**, 251–263 (1981)
3. Cuchiero, C., Filipović, D., Mayerhofer, E., Teichmann, J.: Affine processes on positive semi-definite matrices. Ann. Appl. Probab. **21**, 397–463 (2011)
4. Duffie, D., Filipović, D., Schachermayer, W.: Affine processes and applications in finance. Ann. Appl. Probab. **13**, 984–1053 (2003)
5. Eberlein, E., Madan, D.: On correlating Lévy processes. J. Risk **13**, 3–16 (2010)
6. Eberlein, E., Glau, K., Papapantoleon, A.: Analysis of Fourier transform valuation formulas and applications. Appl. Math. Financ. **17**, 211–240 (2010)
7. Hurd, T.R., Zhou, Z.: A Fourier transform method for spread option pricing. SIAM J. Financ. Math. **1**, 142–157 (2010)
8. Jacod, J., Protter, P.: Probability Essentials, 2nd edn. Springer, Heidelberg (2003)
9. Kallsen, J., Tankov, P.: Characterization of dependence of multidimensional Lévy processes using Lévy copulas. J. Multivar. Anal. **97**, 1551–1572 (2006)
10. Kawai, R.: A multivariate Lévy process model with linear correlation. Quant. Financ. **9**, 597–606 (2009)
11. Khanna, A., Madan D.: Non Gaussian models of dependence in returns. Preprint, SSRN/1540875, (2009)
12. Luciano, E., Schoutens, W.: A multivariate jump-driven financial asset model. Quant. Financ. **6**, 385–402 (2006)
13. Luciano, E., Semeraro, P.: A generalized normal mean-variance mixture for return processes in finance. Int. J. Theor. Appl. Financ. **13**, 415–440 (2010)
14. McNeil, A., Frey, R., Embrechts, P.: Quantitative Risk Management: Concepts, Techniques, Tools. Princeton University Press, (2005)
15. Muhle-Karbe, J., Pfaffel, O., Stelzer, R.: Option pricing in multivariate stochastic volatility models of OU type. SIAM J. Financ. Math. **3**, 66–94 (2012)
16. Rüschendorf, L.: On the distributional transform, Sklar's theorem, and the empirical copula process. J. Stat. Plan. Infer. **139**, 3921–3927 (2009)
17. Sato, K.: Lévy Processes and Infinitely Divisible Distributions. Cambridge University, Cambridge (1999)
18. Schmidt, V.: Dependencies of extreme events in finance. Ph.D. thesis, University of Ulm (2003)
19. Hammerstein, E.A.v.: Generalized hyperbolic distributions: theory and applications to CDO pricing. Ph.D. thesis, University of Freiburg (2011)
20. Villiger S.: Basket option pricing on sparse grids using fast Fourier transforms. Master's thesis, ETH Zürich (2007)

Part V
Dependence Modelling

Goodness-of-fit Tests for Archimedean Copulas in High Dimensions

Christian Hering and Marius Hofert

Abstract A goodness-of-fit transformation for Archimedean copulas is presented from which a test can be derived. In a large-scale simulation study it is shown that the test performs well according to the error probability of the first kind and the power under several alternatives, especially in high dimensions where this test is (still) easy to apply. The test is compared to commonly applied tests for Archimedean copulas. However, these are usually numerically demanding (according to precision and runtime), especially when the dimension is large. The transformation underlying the newly proposed test was originally used for sampling random variates from Archimedean copulas. Its correctness is proven under weaker assumptions. It may be interpreted as an analogon to Rosenblatt's transformation which is linked to the conditional distribution method for sampling random variates. Furthermore, the suggested goodness-of-fit test complements a commonly used goodness-of-fit test based on the Kendall distribution function in the sense that it utilizes all other components of the transformation except the Kendall distribution function. Finally, a graphical test based on the proposed transformation is presented.

Keywords Archimedean copulas · Goodness-of-fit tests · High dimensions · Rosenblatt transformation · Transformation method for sampling

AMS 2000 subject classifications 62F03 · 62H15

C. Hering (✉)
Institute of Number Theory and Probability Theory, Ulm University,
Helmholtzstraße 18, 89081 Ulm, Germany
e-mail: christian.hering@uni-ulm.de

M. Hofert
Department of Mathematics, Technische Universität München, 85748 Garching, Germany
e-mail: marius.hofert@tum.de

© The Author(s) 2015
K. Glau et al. (eds.), *Innovations in Quantitative Risk Management*,
Springer Proceedings in Mathematics & Statistics 99,
DOI 10.1007/978-3-319-09114-3_21

1 Introduction

From risk R management practice, there is an increasing interest in copula theory
and applications in high dimensions. One of the reasons is that vectors of risk factor
changes are typically high-dimensional and have to be adequately modeled; see
[23, Chap. 2]. In high dimensions, the inherent model risk can be substantial. It
is, thus, of interest to test whether an estimated or assumed (dependence) model
is appropriate. One of our goals is, therefore, to present and explore goodness-of-
fit tests in high dimensions for a widely used class of copulas in practice, namely
Archimedean copulas. We also investigate the influence of the dimension on the
conducted goodness-of-fit tests and address the problems that arise specifically in
high dimensions.

It is clear that especially in high dimensions, the exchangeability of Archimedean
copulas becomes an increasingly strong assumption for certain applications. This
point of criticism applies equally well to all exchangeable copula models including
the well-known homogeneous Gaussian or t copulas. However, note that these models
are indeed applied in banks and insurance companies, typically in high dimensions, in
order to introduce (tail-) dependence to joint models for risks as opposed to assuming
(tail) independence. We therefore believe that it is important to investigate such
models in high dimensions.

Archimedean copulas are copulas which admit the functional form

$$C(\boldsymbol{u}) = \psi(\psi^{-1}(u_1) + \cdots + \psi^{-1}(u_d)), \ \boldsymbol{u} \in [0, 1]^d, \tag{1}$$

for an (*Archimedean*) *generator* ψ, i.e., a continuous, decreasing function $\psi :
[0, \infty] \to [0, 1]$ which satisfies $\psi(0) = 1$, $\psi(\infty) = \lim_{t \to \infty} \psi(t) = 0$, and which
is strictly decreasing on $[0, \inf\{t : \psi(t) = 0\}]$. A necessary and sufficient condition
under which (1) is indeed a proper copula is that ψ is d-*monotone*, i.e., ψ is continuous on $[0, \infty]$, admits derivatives up to the order $d - 2$ satisfying $(-1)^k \psi^{(k)}(t) \geq 0$
for all $k \in \{0, \ldots, d - 2\}$, $t \in (0, \infty)$, and $(-1)^{d-2} \psi^{(d-2)}(t)$ is decreasing and
convex on $(0, \infty)$, see [20] or [22]. For reasons why Archimedean copulas are used
in practice, see [9] or [19].

Goodness-of-fit techniques for copulas only more recently gained interest, see,
e.g., [5, 6, 8, 11–14], and references therein. Although usually presented in a d-
dimensional setting, only some of the publications actually try to apply goodness-
of-fit tests in more than two dimensions, including [5, 26] up to dimension $d = 5$
and [4] up to dimension $d = 8$. The common deficiency of goodness-of-fit tests for
copulas in general, but also for the class of Archimedean copulas, is their limited
applicability when the dimension becomes large. This is mainly due to the lack of
a simple or at least numerically accessible form as the dimension becomes large.
Furthermore, parameter estimation usually becomes much more demanding in high
dimensions; see [19].

As a general goodness-of-fit test, the transformation of [25] is well known.
It is important to note that the inverse of this transformation leads to a popular

sampling algorithm, the conditional distribution method, see, e.g., [10]. In other words, for a bijective transformation which converts d independent and identically distributed ("i.i.d.") standard uniform random variables to a d-dimensional random vector distributed according to some copula C, the corresponding inverse transformation may be applied to obtain d i.i.d. standard uniform random variables from a d-dimensional random vector following the copula C. In this work, we suggest this idea for goodness-of-fit testing based on a transformation originally proposed by [29] for sampling Archimedean copulas. With the recent work of [22] we obtain a more elegant proof of the correctness of this transformation under weaker assumptions. We then apply the first $d - 1$ components to build a general goodness-of-fit test for d-dimensional Archimedean copulas. This complements goodness-of-fit tests based on the dth component, the Kendall distribution function, see, e.g., [13, 26], or [14]. Our proposed test can be interpreted as an Archimedean analogon to goodness-of-fit tests based on Rosenblatt's transformation for copulas in general as it establishes a link between a sampling algorithm and a goodness-of-fit test. The appealing property of tests based on the inverse of the transformation of [29] for Archimedean copulas is that they are easily applied in any dimension, whereas tests based on Rosenblatt's transformation, as well as tests based on the Kendall distribution function are typically numerically challenging. The transformation can also be conveniently used for graphical goodness-of-fit testing as recently advocated by [16].

This paper is organized as follows. In Sect. 2, commonly used goodness-of-fit tests for copulas in general are recalled. In Sect. 3, the new goodness-of-fit test for Archimedean copulas is presented. Section 4 contains details about the conducted simulation study. The results are presented in Sect. 5 and the graphical goodness-of-fit test is detailed in Sect. 6. Finally, Sect. 7 concludes.

2 Goodness-of-fit Tests for Copulas

Let $X = (X_1, \ldots, X_d)$, $d \geq 2$, denote a random vector with distribution function H and continuous marginals $F_1, \ldots, F_d$. In a copula model for X, one would like to know whether C is well represented by a parametric family $\mathscr{C}_0 = \{C(\cdot\,; \boldsymbol{\theta}) : \boldsymbol{\theta} \in \Theta\}$ where Θ is an open subset of $\mathbb{R}^p$, $p \in \mathbb{N}$. In other words, one would like to test the null hypothesis

$$H_0 : C \in \mathscr{C}_0 \tag{2}$$

based on realizations of independent copies X_i, $i \in \{1, \ldots, n\}$, of X. For testing H_0, the (usually unknown) marginal distributions are treated as nuisance parameters and are replaced by their slightly scaled empirical counterparts, the *pseudo-observations* $U_i = (U_{i1}, \ldots, U_{id})$, $i \in \{1, \ldots, n\}$, with

$$U_{ij} = \frac{n}{n+1} \hat{F}_{nj}(X_{ij}), \; i \in \{1, \ldots, n\}, \; j \in \{1, \ldots, d\}, \tag{3}$$

where $\hat{F}_{nj}(x) = \frac{1}{n} \sum_{k=1}^{n} \mathbb{1}_{\{X_{kj} \leq x\}}$ denotes the empirical distribution function of the jth data column (the data matrix consisting of the entries X_{ij}, $i \in \{1, \ldots, n\}$, $j \in \{1, \ldots, d\}$), see [14]. Following the latter approach one ends up with rank-based pseudo-observations which are interpreted as observations of C (besides the known issues of this interpretation, see Remark 1 below) and are, therefore, used for estimating θ and testing H_0.

In order to conduct a goodness-of-fit test, the pseudo-observations U_i, $i \in \{1, \ldots, n\}$, are usually first transformed to some variables U_i', $i \in \{1, \ldots, n\}$, so that the distribution of the latter is known and sufficiently simple to test under the null hypothesis. For Rosenblatt's transformation (see Sect. 2.1), U_i', $i \in \{1, \ldots, n\}$, is also d-dimensional, for tests based on the Kendall distribution function (described in Sect. 2.2), it is one-dimensional, and for the goodness-of-fit approach we propose in Sect. 3, it is $(d - 1)$-dimensional. If not already one-dimensional, after such a transformation, U_i', $i \in \{1, \ldots, n\}$, is usually mapped to one-dimensional quantities Y_i, $i \in \{1, \ldots, n\}$, such that the corresponding distribution F_Y is again known under the null hypothesis. So indeed, instead of (2), one usually considers some adjusted hypothesis $H_0^* : F_Y \in \mathscr{F}_0$ under which a goodness-of-fit test can easily be carried out in a one-dimensional setting. For mapping the variates to a one-dimensional setting, different approaches exist, see Sect. 2.2. Note that if H_0^* is rejected, so is H_0.

Remark 1 As, e.g., [8] describe, there are two problems with the approach described above. First, the pseudo-observations U_i, $i \in \{1, \ldots, d\}$, are neither realizations of perfectly independent random vectors nor are the components perfectly following univariate standard uniform distributions. This affects the null distribution of the test statistic under consideration. All copula goodness-of-fit approaches suffer from these effects since observations from the underlying copula are never directly observed in practice. A solution may be a bootstrap to access the exact null distribution. Particularly in high dimensions, it is often time-consuming, especially for goodness-of-fit tests suggested in the copula literature so far. Second, using estimated copula parameters additionally affects the null distribution.

2.1 Rosenblatt's Transformation and a Corresponding Test

The transformation introduced by [25] is a standard approach for obtaining realizations of standard uniform random vectors U_i', $i \in \{1, \ldots, n\}$, given random vectors U_i, $i \in \{1, \ldots, n\}$, from an absolutely continuous copula C which can then be tested directly or further mapped to one-dimensional variates for testing purposes. Consider a representative d-dimensional random vector $U \sim C$. To obtain $U' \sim \mathrm{U}[0, 1]^d$ (i.e., a random vector with independent components, each uniformly distributed on $[0, 1]$), [25] proposed the transformation $R : U \to U'$, given by

$$U_1' = U_1,$$
$$U_2' = C_2(U_2 \mid U_1),$$
$$\vdots$$
$$U_d' = C_d(U_d \mid U_1, \ldots, U_{d-1}),$$

where for $j \in \{2, \ldots, d\}$, $C_j(u_j \mid u_1, \ldots, u_{j-1})$ denotes the conditional distribution function of U_j given $U_1 = u_1, \ldots, U_{j-1} = u_{j-1}$. We denote this method for constructing goodness-of-fit tests by "R" in what follows.

Remark 2 Note that the inverse transformation R^{-1} of Rosenblatt's transformation leads to the *conditional distribution method* for sampling copulas, see, e.g., [10]. This link brings rise to the general idea of using sampling algorithms based on one-to-one transformations to construct goodness-of-fit tests. This is done in Sect. 3 to construct a goodness-of-fit test for Archimedean copulas based on a transformation originally proposed by [29] for sampling random variates.

To find the quantities $C_j(u_j \mid u_1, \ldots, u_{j-1})$, $j \in \{2, \ldots, d\}$, for a specific copula C (under weak conditions), the following connection between conditional distributions and partial derivatives is usually applied; see [27, p.20]. Assuming C admits continuous partial derivatives with respect to the first $d - 1$ arguments, one has

$$C_j(u_j \mid u_1, \ldots, u_{j-1}) = \frac{D_{j-1,\ldots,1} C^{(1,\ldots,j)}(u_1, \ldots, u_j)}{D_{j-1,\ldots,1} C^{(1,\ldots,j-1)}(u_1, \ldots, u_{j-1})}, \quad j \in \{2, \ldots, d\}, \tag{4}$$

where $C^{(1,\ldots,k)}$ denotes the k-dimensional marginal copula of C corresponding to the first k arguments and $D_{j-1,\ldots,1}$ denotes the mixed partial derivative of order $j - 1$ with respect to the first $j - 1$ arguments. For a d-dimensional Archimedean copula C with $(d - 1)$-times continuously differentiable generator ψ, one has

$$C_j(u_j \mid u_1, \ldots, u_{j-1}) = \frac{\psi^{(j-1)}\left(\sum_{k=1}^{j} \psi^{-1}(u_k)\right)}{\psi^{(j-1)}\left(\sum_{k=1}^{j-1} \psi^{-1}(u_k)\right)}, \quad j \in \{2, \ldots, d\}. \tag{5}$$

The problem when applying (4) or (5) in high dimensions is that it is usually quite difficult to access the derivatives involved, the price which one has to pay for such a general transformation. Furthermore, numerically evaluating the derivatives is often time-consuming and prone to errors.

Genest et al. [14] propose a test statistic based on the empirical distribution function of the random vectors U_i', $i \in \{1, \ldots, d\}$. As an overall result, the authors recommend to use a distance between the distribution under H_0, assumed to be standard uniform on $[0, 1]^d$, and the empirical distribution, namely

$$S_{n,d}^B = n \int\limits_{[0,1]^d} (D_n(\mathbf{u}) - \Pi(\mathbf{u}))^2 \, d\mathbf{u},$$

where $\Pi(\mathbf{u}) = \prod_{j=1}^d u_j$ denotes the independence copula and $D_n(\mathbf{u}) = \frac{1}{n}\sum_{i=1}^n \mathbb{1}_{\{U_i' \le u\}}$ the empirical distribution function based on the random vectors U_i', $i \in \{1, \ldots, d\}$. We refer to this transformation as "$S_{n,d}^B$" in what follows.

2.2 Tests in a One-Dimensional Setting

In order to apply a goodness-of-fit test in a one-dimensional setting one has to summarize the d-dimensional pseudo-observations U_i or U_i' via one-dimensional quantities Y_i, $i \in \{1, \ldots, n\}$, for which the distribution is known under the null hypothesis. In what follows, some popular mappings achieving this task are described.

N_d: Under H_0, the one-dimensional quantities $Y_i = F_{\chi_d^2}(\sum_{j=1}^d \Phi^{-1}(U_{ij}')^2)$, $i \in \{1, \ldots, n\}$, should be i.i.d. according to a standard uniform distribution, where $F_{\chi_d^2}$ denotes the distribution function of a χ^2 distribution with d degrees of freedom and Φ^{-1} denotes the quantile function of the standard normal distribution. This transformation can be found, e.g., in [8] and is denoted by "N_d" in what follows.

K_C: For a copula C let K_C denote the *Kendall distribution function*, i.e., $K_C(t) = \mathbb{P}(C(U) \le t)$, $t \in [0, 1]$, where $U \sim C$, see [3] or [22]. Under H_0 and if K_C is continuous, the random variables $Y_i = K_C(C(U_i))$ should be i.i.d. according to a standard uniform distribution. This approach for goodness-of-fit testing will be referred to as "K_C". Note that in this case, no multidimensional transformation of the data is performed beforehand.

K_Π: One can also consider the random vectors U_i', $i \in \{1, \ldots, n\}$, in conjunction with the independence copula, i.e., define $\tilde{Y}_i = \prod_{j=1}^d U_{ij}'$, where $\tilde{Y}_i$ has distribution function $K_\Pi(t) = t \sum_{k=0}^{d-1} \frac{1}{k!}(-\log t)^k$. Under H_0, the sample $Y_i = K_\Pi(\tilde{Y}_i)$, $i \in \{1, \ldots, n\}$, should indicate a uniform distribution on the unit interval. This approach is referred to as "K_Π".

In the approaches N_d, K_C, and K_Π we have to test the hypothesis that realizations of the random variables Y_i, $i \in \{1, \ldots, n\}$, follow a uniform distribution on the unit interval. This may be achieved in several ways, the following two approaches are applied in what follows.

χ^2: Pearson's χ^2 test, see [24, p. 391], shortly referred to as "χ^2".

AD: The so-called Anderson-Darling test, a specifically weighted Cramér-von Mises test, see [1, 2]. This method is referred to as "AD".

3 A Goodness-of-fit Test for Archimedean Copulas

The goodness-of-fit test we now present is based on the following transformation from [29] for generating random variates from Archimedean copulas. Note that we present a rather short proof of this interesting result, under weaker assumptions.

Theorem 1 (The main transformation) *Let* $U \sim C, d \geq 2$, *where* C *is an Archimedean copula with d-monotone generator* ψ *and continuous Kendall distribution function* K_C. *Then* $U' \sim U[0, 1]^d$, *where*

$$U'_j = \left(\frac{\sum_{k=1}^{j} \psi^{-1}(U_k)}{\sum_{k=1}^{j+1} \psi^{-1}(U_k)} \right)^j, \quad j \in \{1, \ldots, d-1\}, \quad U'_d = K_C(C(U)). \quad (6)$$

Proof As shown in [22], $(\psi^{-1}(U_1), \ldots, \psi^{-1}(U_d))$ has an ℓ_1-norm symmetric distribution with survival copula C and radial distribution $F_R = \mathcal{W}_d^{-1}[\psi]$, where $\mathcal{W}_d[\cdot]$ denotes the Williamson d-transform. Hence, $(\psi^{-1}(U_1), \ldots, \psi^{-1}(U_d)) \stackrel{d}{=} RS$, where $R \sim F_R$ and $S \sim U(\{x \in \mathbb{R}_+^d \mid ||x||_1 = 1\})$ are independent. For $Z_{(0)} = 0$, $Z_{(d)} = 1$, and $(Z_1, \ldots, Z_{d-1}) \sim U[0, 1]^{d-1}$, it follows from [7, p. 207] that $S_j \stackrel{d}{=} Z_{(j)} - Z_{(j-1)}, j \in \{1, \ldots, d\}$, independent of R. This implies that $\psi^{-1}(U_j) \stackrel{d}{=} R(Z_{(j)} - Z_{(j-1)}), j \in \{1, \ldots, d\}$, and hence that U' is in distribution equal to $W = ((Z_{(1)}/Z_{(2)})^1, \ldots, (Z_{(d-1)}/Z_{(d)})^{d-1}, K_C(\psi(R)))$. Since K_C is continuous and $\psi(R) \sim K_C$, $K_C(\psi(R))$ is uniformly distributed in $[0, 1]$. Furthermore, as a function in R, $K_C(\psi(R))$ is independent of $(W_1, \ldots, W_{d-1})$. It therefore suffices to show that $(W_1, \ldots, W_{d-1}) \sim U[0, 1]^{d-1}$, a proof of which can be found in [7, p. 212]. $\square$

The transformation $T : U \to U'$ given in (6) can be interpreted as an analogon to Rosenblatt's transformation R specifically for Archimedean copulas. Both T and R uniquely map d random variables to d random variables and can therefore be used in both directions, for generating random variates and goodness-of-fit tests; the latter approach for T is proposed in this paper. The advantage of this approach for obtaining the random variables (or their realizations in form of given data) $U'_i \sim U[0, 1]^d, i \in \{1, \ldots, n\}$, from $U_i \sim C, i \in \{1, \ldots, n\}$, in comparison to Rosenblatt's transformation lies in the fact that it is typically much easier to compute the quantities in (6) than accessing the derivatives in (5). One can then proceed as for Rosenblatt's transformation and use any of the transformations listed in Sect. 2.2 to transform U'_i, $i \in \{1, \ldots, n\}$, to the one-dimensional quantities $Y_i, i \in \{1, \ldots, n\}$, for testing H_0^*. A test involving the transformation T to obtain the random vectors $U'_i \sim U[0, 1]^d$, $i \in \{1, \ldots, n\}$, is referred to as approach "T_d" in what follows.

Note that evaluating the transformation T might only pose difficulties for the last component U'_d, the Kendall distribution function K_C, whereas computing U'_j, $j \in \{1, \ldots, d-1\}$, is easily achieved for any Archimedean copula with explicit generator inverse. Furthermore, for large d, evaluation of K_C often gets more and more complicated from a numerical point of view (see [18] for the derivatives involved), except for specific cases such as Clayton's family where all involved derivatives of ψ are directly accessible, see, e.g., [29], and therefore K_C can be computed directly via[1] $K_C(t) = \sum_{k=0}^{d-1}(0 - \psi^{-1}(t))^k \psi^{(k)}(\psi^{-1}(t))/k!$, see, e.g., [3] or [22]. Moreover, note that applying T_d for obtaining the transformed data U'_i, $i \in \{1, \ldots, n\}$, requires n-times the evaluation of the Kendall distribution function K_C, which can be computationally intensive, especially in simulation studies involving bootstrap procedures. With the informational loss inherent in the goodness-of-fit tests following the approaches addressed in Sect. 2.2 in mind, one may therefore suggest to omit the last component T_d of T and only consider $T_1, \ldots, T_{d-1}$, i.e., using the data $(U'_{i1}, \ldots, U'_{id-1})$, $i \in \{1, \ldots, n\}$, for testing purposes if d is large. This leads to fast goodness-of-fit tests for Archimedean copulas in high dimensions. A goodness-of-fit test based on omitting the last component of the transformation T is referred to as approach "T_{d-1}" in what follows.

4 A Large-Scale Simulation Study

4.1 The Experimental Design

In our experimental design, focus is put on two features, the error probability of the first kind, i.e., if a test maintains its nominal level, and the power under several alternatives. To distinguish between the different approaches we use either pairs or triples, e.g., the approach "(T_{d-1}, N_{d-1}, AD)" denotes a goodness-of-fit test based on first applying our proposed transformation T without the last component, then using the approach based on the χ^2_{d-1} distribution to transform the data to a one-dimensional setup, and then applying the Anderson-Darling statistic to test H_0^*; similarly, "$(T_{d-1}, S^B_{n,d-1})$" denotes a goodness-of-fit test which uses the approach $S^B_{n,d-1}$ for reducing the dimension and testing H_0^*.

In the conducted Monte Carlo simulation,[2] the following ten different goodness-of-fit approaches are tested:

---

[1] It also follows from this formula that K_C converges pointwise to the unit jump at zero for $d \to \infty$.

[2] All computations were conducted on a compute node (part of the bwGRiD Cluster Ulm) which consists of eight cores (two four-core Intel Xeon E5440 Harpertown CPUs with 2.83 GHz and 6 MB second level cache) and 16 GB memory. The algorithms are implemented in C/C++ and compiled using GCC 4.2.4 with option O2 for code optimization. Moreover, we use the algorithms of the Numerical Algorithms Group, the GNU Scientific Library 1.12, and the OpenMaple interface of Maple 12. For generating uniform random variates an implementation of the Mersenne Twister by [28] is used. For the Anderson-Darling test, the procedures suggested in [21] are used.

$$(T_{d-1}, N_{d-1}, \chi^2), \ (T_{d-1}, N_{d-1}, AD), \ (T_{d-1}, S_{n,d-1}^B), \ (K_C, \chi^2), \ (K_C, AD),$$

$$(T_d, N_d, AD), \ (T_d, K_\Pi, AD), \ (T_d, S_{n,d}^B), \ (R, N_d, AD), \ (R, S_{n,d}^B). \quad (7)$$

Similar to [14], we investigate samples of size $n = 150$ and parameters of the copulas such that Kendall's tau equals $\tau = 0.25$. We work in $d = 5$ and $d = 20$ dimensions for comparing the goodness-of-fit tests given in (7). For every scenario, we simulate the corresponding Archimedean copulas of Ali-Mikhail-Haq ("A"), Clayton ("C"), Frank ("F"), Gumbel ("G"), and Joe ("J"), see, e.g., [15], as well as the Gaussian ("Ga") and t copula with four degrees of freedom ("t_4"); note that we use one-parameter copulas ($p = 1$) in our study only for simplicity. Whenever computationally feasible, $N = 1{,}000$ replications are used for computing the empirical level and power. In some cases, see Sect. 5, less than 1,000 replications had to be used. For all tests, the significance level is fixed at $\alpha = 5\,\%$. For the univariate χ^2-tests, ten cells were used.

Concerning the use of Maple, we proceed as follows. For computing the first $d - 1$ components $T_1, \ldots, T_{d-1}$ of the transformation T involved in the first three and the sixth to eighth approach listed in (7), Maple is only used if working under double precision in C/C++ leads to errors. With errors, nonfloat values including nan, -inf, and inf, as well as float values less than zero or greater than one are meant. For computing the component T_d, Maple is used to generate C/C++ code. To decrease runtime, the function is then hard coded in C/C++, except for Clayton's family where an explicit form of all derivatives and hence K_C is known, see [29]. The same holds for computing K_C for the approaches (K_C, χ^2) and (K_C, AD). For the approaches involving Rosenblatt's transform, a computation in C/C++ is possible for Clayton's family in a direct manner, whereas again Maple's code generator is used for all other copula families to obtain the derivatives of the generator. If there are numerical errors from this approach we use Maple with a high precision for the computation. If Rosenblatt's transformation produces errors even after computations in Maple, we disregard the corresponding goodness-of-fit test and use the remaining test results of the simulation for computing the empirical level and power.

Due to its well-known properties, we use the maximum likelihood estimator ("MLE") to estimate the copula parameters, based on the pseudo-observations of the simulated random vectors $U_i \sim C, i \in \{1, \ldots, n\}$. Besides building the pseudo-observations, note that parameter estimation may also affect the null distribution. This is generally addressed by using a bootstrap procedure for accessing the correct null distribution, see Sect. 4.2 below. Note that a bootstrap can be quite time-consuming in high dimensions, even parameter estimation already turns out to be computationally demanding. For the bootstrap versions of the goodness-of-fit approaches involving the generator derivatives, we were required to hard code the derivatives in order to decrease runtime. Note that such effort is not needed for applying our proposed goodness-of-fit test (T_{d-1}, N_{d-1}, AD), since it is not required to access the generator derivatives.

4.2 The Parametric Bootstrap

For our proposed approach (T_{d-1}, N_{d-1}, AD) it is not clear whether the bootstrap procedure is valid from a theoretical point of view; see, e.g., [8] and [14]. However, empirical results, presented in Sect. 5, indicate the validity of this approach, described as follows.

1. Given the data X_i, $i \in \{1, \ldots, n\}$, build the pseudo-observations U_i, $i \in \{1, \ldots, n\}$ as given in (3) and estimate the unknown copula parameter vector $\boldsymbol{\theta}$ by its MLE $\hat{\boldsymbol{\theta}}_n$.

2. Based on U_i, $i \in \{1, \ldots, n\}$, the given Archimedean family, and the parameter estimate $\hat{\boldsymbol{\theta}}_n$, compute the first $d-1$ components U'_{ij}, $i \in \{1, \ldots, n\}$, $j \in \{1, \ldots, d-1\}$, of the transformation T as in Eq. (6) and the one-dimensional quantities $Y_i = \sum_{j=1}^{d-1} (\Phi^{-1}(U'_{ij}))^2$, $i \in \{1, \ldots, n\}$. Compute the Anderson-Darling test statistic $A_n = -n - \frac{1}{n} \sum_{i=1}^{n} (2i-1)[\log(F_{\chi^2_{d-1}}(Y_{(i)})) + \log(1 - F_{\chi^2_{d-1}}(Y_{(n-i+1)}))]$.

3. Choose the number M of bootstrap replications. For each $k \in \{1, \ldots, M\}$ do:

   a. Generate a random sample of size n from the given Archimedean copula with parameter $\hat{\boldsymbol{\theta}}_n$ and compute the corresponding vectors of componentwise scaled ranks (i.e., the pseudo-observations) $U^*_{i,k}$, $i \in \{1, \ldots, n\}$. Then, estimate the unknown parameter vector $\boldsymbol{\theta}$ by $\hat{\boldsymbol{\theta}}^*_{n,k}$.

   b. Based on $U^*_{i,k}$, $i \in \{1, \ldots, n\}$, the given Archimedean family, and the parameter estimate $\hat{\boldsymbol{\theta}}^*_{n,k}$, compute the first $d-1$ components $U'^*_{ij,k}$, $i \in \{1, \ldots, n\}$, $j \in \{1, \ldots, d-1\}$, of the transformation T as in Eq. (6) and $Y^*_{i,k} = \sum_{j=1}^{d-1} (\Phi^{-1}(U'^*_{ij,k}))^2$, $i \in \{1, \ldots, n\}$. Compute the Anderson-Darling test statistic $A^*_{n,k} = -n - \frac{1}{n} \sum_{i=1}^{n} (2i-1)[\log(F_{\chi^2_{d-1}}(Y^*_{(i),k})) + \log(1 - F_{\chi^2_{d-1}}(Y^*_{(n-i+1),k}))]$.

4. An approximate p-value for (T_{d-1}, N_{d-1}, AD) is given by $\frac{1}{M} \sum_{k=1}^{M} \mathbb{1}_{\{A^*_{n,k} > A_n\}}$.

The bootstrap procedures for the other approaches can be obtained similarly. For the bootstrap procedure using Rosenblatt's transformation see, e.g., [14]. For our simulation studies, we used $M = 1,000$ bootstrap replications. Note that, together with the number $N = 1,000$ of test replications, simulation studies are quite time-consuming, especially if parameters need to be estimated and especially if high dimensions are involved.

Applying the MLE in high dimensions is numerically challenging and time-consuming; see also [19]. Although our proposed goodness-of-fit test can be applied in the case $d = 100$, it is not easy to use the bootstrap described above in such high dimensions. We therefore, for $d = 100$, investigate only the error probability of the first kind similar to the case A addressed in [8]. For this, we generate $N = 1,000$ 100-dimensional samples of size $n = 150$ with parameter chosen such that Kendall's tau equals $\tau = 0.25$ and compute for each generated data set the p-value of the test

(T_{d-1}, N_{d-1}, AD) as before, however, this time with the known copula parameter. Finally, the number of rejections among the 1,000 conducted goodness-of-fit tests according to the five percent level is reported. The results are given at the end of Sect. 5.

5 Results

We first present selected results obtained from the large-scale simulation study conducted for the 10 different goodness-of-fit approaches listed in (7). These results summarize the main characteristics found in the simulation study. As an overall result, we found that the empirical power against all investigated alternatives increases if the dimension gets large. As expected, so does runtime.

We start by discussing the methods that show a comparably weak performance in the conducted simulation study. We start with the results that are based on the test statistics $S_{n,d-1}^B$ or $S_{n,d}^B$ to reduce the dimension. Although keeping the error probability of the first kind, the goodness-of-fit tests $(T_{d-1}, S_{n,d-1}^B)$, $(T_d, S_{n,d}^B)$, and $(R, S_{n,d}^B)$ show a comparably weak performance against the investigated alternatives, at least in our test setup as described in Sect. 4.1. For example, for $n = 150$, $d = 5$, and $\tau = 0.25$, the method $(T_d, S_{n,d}^B)$ leads to an empirical power of 5.2 % for testing Clayton's copula when the simulated copula is Ali-Mikhail-Haq's, 11.5 % for testing the Gaussian copula on Frank copula data, 7.7 % for testing Ali-Mikhail-Haq's copula on data from Frank's copula, and 6.4 % for testing Gumbel's copula on data from Joe's copula. Similarly for the methods $(T_{d-1}, S_{n,d-1}^B)$ and $(R, S_{n,d}^B)$. We therefore do not further report on the methods involving $S_{n,d-1}^B$ or $S_{n,d}^B$ in what follows. The method (T_d, K_Π, AD) also shows a rather weak performance for both investigated dimensions and is therefore omitted. Since the cases of (K_C, χ^2) and (K_C, AD) as well as the approaches (T_{d-1}, N_{d-1}, AD) and $(T_{d-1}, N_{d-1}, \chi^2)$ do not significantly differ, we only report the results based on the Anderson-Darling tests.

Now consider the goodness-of-fit testing approaches (T_{d-1}, N_{d-1}, AD), (K_C, AD), and (T_d, N_d, AD). Recall that (T_{d-1}, N_{d-1}, AD) is based on the first $d - 1$ components of the transformation T addressed in Eq. (6), (K_C, AD) applies only the last component of T, and (T_d, N_d, AD) applies the whole transformation T in d dimensions, where all three approaches use the Anderson-Darling test for testing H_0^*. The test results for the three goodness-of-fit tests with $n = 150$, $\tau = 0.25$, and $d \in \{5, 20\}$ are reported in Tables 1, 2, and 3, respectively. As mentioned above, we use a bootstrap procedure to obtain approximate p-values and test the hypothesis based on those p-values. We use $N = 1,000$ repetitions wherever possible. In all cases involving Joe's copula as H_0 copula only about 650 repetitions could be finished. As Tables 1 and 2 reveal, in many cases, (T_{d-1}, N_{d-1}, AD) shows a larger empirical power than (K_C, AD) (for both d), but the differences in either direction can be large (consider the case of the t_4 copula when the true one is Clayton (both d) and the case of the Frank copula when the true is one is Clayton (both d)). Overall,

Table 1 Empirical power in % for (T_{d-1}, N_{d-1}, AD) based on $N = 1{,}000$ replications with $n = 150$, $\tau = 0.25$, and $d = 5$ (*left*), respectively $d = 20$ (*right*)

	True copula, $d = 5$							True copula, $d = 20$						
H_0	A	C	F	G	J	Ga	t_4	A	C	F	G	J	Ga	t_4
A	**4.8**	10.5	68.5	97.8	100.0	34.2	94.0	**5.2**	4.8	98.1	97.8	100.0	47.2	100.0
C	35.4	**4.7**	92.8	99.6	100.0	84.2	100.0	95.3	**6.1**	100.0	100.0	100.0	100.0	100.0
F	2.9	10.5	**5.3**	58.5	94.8	15.8	99.4	0.3	12.8	**5.4**	63.5	100.0	77.6	100.0
G	24.5	56.6	8.9	**5.2**	10.3	17.0	99.3	99.4	100.0	24.9	**5.2**	77.0	100.0	100.0
J	71.7	92.9	41.1	13.7	**4.9**	76.4	100.0	98.6	98.4	84.4	6.9	**5.2**	100.0	100.0

Table 2 Empirical power in % for (K_C, AD) based on $N = 1{,}000$ replications with $n = 150$, $\tau = 0.25$, and $d = 5$ (*left*), respectively $d = 20$ (*right*)

	True copula, $d = 5$							True copula, $d = 20$						
H_0	A	C	F	G	J	Ga	t_4	A	C	F	G	J	Ga	t_4
A	**6.1**	33.7	13.5	38.3	83.6	11.5	44.4	**4.2**	16.8	0.0	1.7	8.9	59.5	82.4
C	30.6	**5.1**	95.5	86.9	99.3	28.8	7.7	65.9	**5.6**	100.0	99.8	100.0	45.5	4.1
F	41.4	97.6	**4.0**	63.7	59.5	48.1	88.9	90.0	100.0	**5.2**	99.9	100.0	98.5	100.0
G	12.0	24.3	41.1	**4.9**	5.4	6.9	16.3	9.5	56.8	93.0	**6.5**	60.7	1.3	8.3
J	70.1	50.5	70.5	3.0	**5.5**	29.0	12.8	100.0	100.0	99.8	1.8	**6.7**	100.0	100.0

Table 3 Empirical power in % for (T_d, N_d, AD) based on $N = 1{,}000$ replications with $n = 150$, $\tau = 0.25$, and $d = 5$ (*left*), respectively $d = 20$ (*right*)

	True copula, $d = 5$							True copula, $d = 20$						
H_0	A	C	F	G	J	Ga	t_4	A	C	F	G	J	Ga	t_4
A	**4.2**	8.4	36.4	83.1	99.7	21.6	98.4	**5.3**	16.2	98.0	96.6	100.0	68.8	100.0
C	6.9	**4.7**	16.9	65.9	90.2	25.3	100.0	86.3	**5.3**	99.7	99.9	100.0	100.0	100.0
F	4.4	3.1	**4.9**	16.7	46.1	9.1	99.2	0.4	5.6	**5.0**	30.8	100.0	25.9	100.0
G	3.8	5.8	1.8	**5.0**	15.8	3.7	98.7	94.7	100.0	8.2	**7.1**	85.3	98.6	100.0
J	11.1	17.5	6.4	4.8	**4.8**	10.8	99.7	100.0	100.0	74.8	3.5	**5.3**	98.7	100.0

when the true copula is the t_4 copula, (T_{d-1}, N_{d-1}, AD) performs well. Given the comparably numerically simple form of (T_{d-1}, N_{d-1}, AD), this method can be quite useful. Interestingly, by comparing Table 1 with Table 3, we see that if the transformation T with all d components is applied, there is actually a loss in power for the majority of families tested (the cause of this behavior remains an open question). Note that in Table 2 for the case where the Ali-Mikhail-Haq copula is tested, the power decreases in comparison to the five-dimensional case. This might be due to numerical difficulties occurring when K_C is evaluated in this case, since the same behavior is visible for the method (K_C, χ^2).

Table 4 shows the empirical power of the method (R, N_d, AD). In comparison to our proposed goodness-of-fit approach (T_{d-1}, N_{d-1}, AD), the approach (R, N_d, AD) overall performs worse. For $d = 5$, there are only two cases where

Table 4 Empirical power in % for (R, N_d, AD) based on $N = 1{,}000$ replications with $n = 150$, $\tau = 0.25$, and $d = 5$ (*left*), respectively $d = 20$ (*right*)

H_0	True copula, $d = 5$							True copula, $d = 20$						
	A	C	F	G	J	Ga	t_4	A	C	F	G	J	Ga	t_4
A	**4.5**	8.9	46.9	79.1	98.8	11.0	94.2	*	*	*	*	*	*	*
C	11.7	**5.0**	17.7	53.5	68.8	10.4	99.7	93.4	5.3	100.0	100.0	100.0	100.0	100.0
F	3.4	2.6	**5.5**	15.8	61.6	5.7	99.5	–	–	–	–	–	–	–
G	4.9	4.0	1.2	**3.0**	14.5	1.2	97.9	–	–	–	–	–	–	–
J	21.1	21.8	9.5	4.3	**3.6**	7.2	99.7	–	–	–	–	–	–	–

(R, N_d, AD) performs better than (T_{d-1}, N_{d-1}, AD) which are testing the Ali-Mikhail-Haq copula when the true copula is t_4 and testing Joe's copula when the true one is Gumbel. In the high-dimensional case $d = 20$, only results for the Clayton copula are obtained. In this case the actual number of repetitions for calculating the empirical power is approximately 500. For the cases when testing the Ali-Mikhail-Haq, Gumbel, Frank, or Joe copula, no reliable results were obtained since only about 20 repetitions could be run in the runtime provided by the grid. This is due to the high-order derivatives involved in this transformation, which slow down computations considerably; see [19] for more details.

Another aspect, especially in a high-dimensional setup is numerical precision. In going from the low- to the high-dimensional case we faced several problems during our computations. For example, the approach (R, N_d, AD) shows difficulties in testing the H_0 copula of Ali-Mikhail-Haq for $d = 20$. Even after applying Maple (with `Digits` set to 15; default is 10), the goodness-of-fit tests indicated numerical problems. The numerical issues appearing in the testing approaches (K_C, AD) and (T_d, N_d, AD) when evaluating the Kendall distribution function were already mentioned earlier, e.g., in Sect. 4.1. In principal, one could be tempted to choose a (much) higher precision than standard `double` in order to obtain more reliable testing results. However, note that this significantly increases runtime. Under such a setup, applying a bootstrap procedure would not be possible anymore. In high dimensions, only the approaches (T_{d-1}, N_{d-1}, AD) and $(T_{d-1}, N_{d-1}, \chi^2)$ can be applied without facing computational difficulties according to precision and runtime.

Concerning the case $d = 100$, we checked if the error probability of the first kind according to the 5%-level is kept. As results of the procedure described in the end of Sect. 4.2, we obtained 4.6, 4.2, 5.0, 5.5, and 4.9% for the families of Ali-Mikhail-Haq, Clayton, Frank, Gumbel, and Joe, respectively.

6 A Graphical Goodness-of-fit Test

A plot often provides more information than a single p-value, e.g., it can be used to determine where deviations from uniformity are located; see [16] who advocate graphical goodness-of-fit tests in higher dimensions. We now briefly apply the trans-

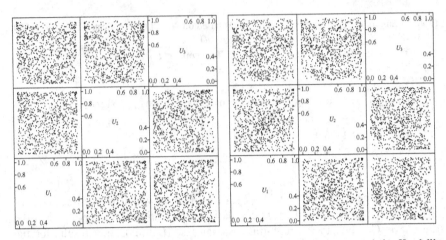

Fig. 1 Data from a Gaussian (*left*) and t_4 (*right*) copula with parameter chosen such that Kendall's tau equals 0.5, transformed with a Gumbel copula with parameter such that Kendall's tau equals 0.5. The deviations from uniformity are small but visible, especially in the corners of the different panels

formation $T : U \to U'$ addressed in Theorem 1 to graphically check how well the transformed variates indeed follow a uniform distribution. Figures 1, 2, and 3 show scatter-plot matrices of 1,000 generated three-dimensional vectors of random variates which are transformed with T under various assumed models (the captions are self-explanatory). Since K_C is easily computed in three dimensions, we also use this last component of T.

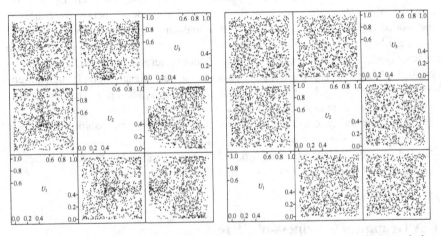

Fig. 2 Data from a Clayton (*left*) and Gumbel (*right*) copula with parameter chosen such that Kendall's tau equals 0.5, transformed with a Gumbel copula with parameter such that Kendall's tau equals 0.5. The deviation from uniformity for the Clayton data is clearly visible. Since the Gumbel data is transformed with the correct family and parameter, the resulting variates are indeed uniformly distributed in the unit hypercube

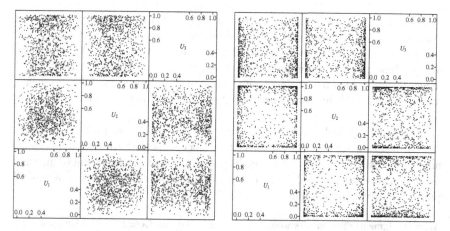

Fig. 3 Data from a Gumbel copula with parameter chosen such that Kendall's tau equals 0.5, transformed with a Gumbel copula with parameter such that Kendall's tau equals 0.2 (*left*) and 0.8 (*right*), respectively. Deviations from uniformity are easily visible

7 Conclusion and Discussion

Goodness-of-fit tests for Archimedean copulas, also suited to high dimensions were presented. The proposed tests are based on a transformation T whose inverse is known for generating random variates. The tests can, therefore, be viewed as analogs to tests based on Rosenblatt's transformation, whose inverse is also used for sampling (known as the conditional distribution method). The suggested goodness-of-fit tests proceed in two steps. In the first step, the first $d - 1$ components of T are applied. They provide a fast and simple transformation from d to $d - 1$ dimensions. This complements known goodness-of-fit tests using only the dth component of T, the Kendall distribution function, but which require the knowledge of the generator derivatives. In a second step, the $d - 1$ components are mapped to one-dimensional quantities, which simplifies testing. This second step is common to many goodness-of-fit tests and hence any such test can be applied.

The power of the proposed testing approach was compared to other known goodness-of-fit tests in a large-scale simulation study. In this study, goodness-of-fit tests in comparably high dimensions were investigated. The computational effort (precision, runtime) involved in applying commonly known testing procedures turned out to be tremendous. The results obtained from these tests in higher dimensions have to be handled with care: Numerical issues for the methods for which not all repetitions could be run without problems might have introduced a bias. To apply commonly known goodness-of-fit tests in higher dimensions requires (much) more work in the future, especially on the numerical side. Computational tools which systematically check for numerical inaccuracies and which are implemented on the paradigm of defensive programming might provide a solution here; see [17] for a first work in this direction.

In contrast, our proposed approach is easily applied in any dimension and its evaluation requires only small numerical precision. Due to the short runtimes, it could also be investigated with a bootstrap procedure, showing good performance in high dimensions. Furthermore, it easily extends to the multiparameter case. To reduce the effect of non-robustness with respect to the permutation of the arguments, one could randomize the data dimensions as is done for Rosenblatt's transformation, see [4].

Finally, a graphical goodness-of fit test is outlined. This is a rather promising field of research for high-dimensional data, since, especially in high dimensions, none of the existing models fits perfectly, and so a graphical assessment of the parts (or dimensions) of the model which fit well and those which do not is in general preferable to a single p-value.

Acknowledgments The authors would like to thank Christian Genest (McGill University) for valuable feedback on this paper. The authors would also like to thank Christian Mosch and the Communication and Information Center of Ulm University for providing computational power via the bwGRiD Cluster Ulm and assistance in using it.

Open Access This chapter is distributed under the terms of the Creative Commons Attribution Noncommercial License, which permits any noncommercial use, distribution, and reproduction in any medium, provided the original author(s) and source are credited.

References

1. Anderson, T.W., Darling, D.A.: Asymptotic theory of certain goodness-of-fit criteria based on stochastic processes. Ann. Math. Stat. **23**(2), 193–212 (1952)
2. Anderson, T.W., Darling, D.A.: A test of goodness of fit. J. Am. Stat. Assoc. **49**, 765–769 (1954)
3. Barbe, P., Genest, C., Ghoudi, K., Rémillard, B.: On Kendall's process. J. Multivar. Anal. **58**, 197–229 (1996)
4. Berg, D.: Copula goodness-of-fit testing: an overview and power comparison. Eur. J. Financ. **15**(7—-8), 675–701 (2009). http://www.informaworld.com/10.1080/13518470802697428
5. Berg, D., Bakken, H.: A copula goodness-of-fit approach based on the conditional probability integral transformation (2007). http://www.danielberg.no/publications/Btest.pdf
6. Breymann, W., Dias, A., Embrechts, P.: Dependence structures for multivariate high-frequency data in finance. Quant. Financ. **3**, 1–14 (2003)
7. Devroye, L.: Non-Uniform Random Variate Generation. Springer, Heidelberg (1986)
8. Dobrić, J., Schmid, F.: A goodness of fit test for copulas based on Rosenblatt's transformation. Comput. Stat. Data Anal. **51**, 4633–4642 (2007)
9. Embrechts, P., Hofert, M.: Comments on: inference in multivariate archimedean copula models. TEST **20**(2), 263–270 (2011). doi:http://dx.doi.org/10.1007/s11749-011-0252-4
10. Embrechts, P., Lindskog, F., McNeil, A.J.: Modelling dependence with copulas and applications to risk management. In: Rachev, S. (ed.) Handbook of Heavy Tailed Distributions in Finance, pp. 329–384. Elsevier, Amsterdam (2003)
11. Fermanian, J.D.: Goodness of fit tests for copulas. J. Multivar. Anal. **95**(1), 119–152 (2005)
12. Genest, C., Rivest, L.P.: Statistical inference procedures for bivariate Archimedean copulas. J. Am. Stat. Assoc. **88**(423), 1034–1043 (1993)

13. Genest, C., Quessy, J.F., Rémillard, B.: Goodness-of-fit procedures for copula models based on the probability integral transformation. Scand. J. Stat. **33**, 337–366 (2006)
14. Genest, C., Rémillard, B., Beaudoin, D.: Goodness-of-fit tests for copulas: a review and a power study. Insur. Math. Econ. **44**, 199–213 (2009)
15. Hofert, M.: Efficiently sampling nested Archimedean copulas. Comput. Stat. Data Anal. **55**, 57–70 (2011). doi: http://dx.doi.org/10.1016/j.csda.2010.04.025
16. Hofert, M., Mächler, M.: A graphical goodness-of-fit test for dependence models in higher dimensions. J. Comput. Graph. Stat. (2013). doi: http://dx.doi.org/10.1080/10618600.2013.812518
17. Hofert, M., Mächler, M.: Parallel and other simulations in R made easy: An end-to-end study (2014)
18. Hofert, M., Mächler, M., McNeil, A.J.: Likelihood inference for Archimedean copulas in high dimensions under known margins. J. Multivar. Anal. **110**, 133–150 (2012). doi: http://dx.doi.org/10.1016/j.jmva.2012.02.019
19. Hofert, M., Mächler, M., McNeil, A.J.: Archimedean copulas in high dimensions: estimators and numerical challenges motivated by financial applications. J. de la Société Française de Statistique **154**(1), 25–63 (2013)
20. Malov, S.V.: On finite-dimensional Archimedean copulas. In: Balakrishnan, N., Ibragimov, I., Nevzorov, V. (eds.) Asymptotic Methods in Probability and Statistics with Applications, pp. 19–35. Birkhäuser, Boston (2001)
21. Marsaglia, G., Marsaglia, J.C.W.: Evaluating the Anderson Darling distribution. J. Stat. Softw. **9**(2), 1–5 (2004)
22. McNeil, A.J., Nešlehová, J.: Multivariate Archimedean copulas, d-monotone functions and l_1-norm symmetric distributions. Ann. Stat. **37**(5b), 3059–3097 (2009)
23. McNeil, A.J., Frey, R., Embrechts, P.: Quantitative Risk Management: Concepts, Techniques, Tools. Princeton University, Princeton (2005)
24. Rao, C.R.: Linear Statistical Inference and its Applications. Wiley-Interscience, Hoboken (2001)
25. Rosenblatt, M.: Remarks on a multivariate transformation. Ann. Math. Stat. **23**(3), 470–472 (1952)
26. Savu, C., Trede, M.: Goodness-of-fit tests for parametric families of Archimedean copulas. Quant. Financ. **8**(2), 109–116 (2008)
27. Schmitz, V.: Copulas and stochastic processes. Ph.D. thesis, Rheinisch-Westfälische Technische Hochschule Aachen
28. Wagner, R.: Mersenne twister random number generator (2003).http://www-personal.umich.edu/wagnerr/MersenneTwister.html
29. Wu, F., Valdez, E.A., Sherris, M.: Simulating exchangeable multivariate Archimedean copulas and its applications. Commun. Stat. Simul. Comput. **36**(5), 1019–1034 (2007)

Duality in Risk Aggregation

Raphael Hauser, Sergey Shahverdyan and Paul Embrechts

Abstract A fundamental problem in risk management is the robust aggregation of different sources of risk in a situation where little or no data are available to infer information about their dependencies. A popular approach to solving this problem is to formulate an optimization problem under which one maximizes a risk measure over all multivariate distributions that are consistent with the available data. In several special cases of such models, there exist dual problems that are easier to solve or approximate, yielding robust bounds on the aggregated risk. In this chapter, we formulate a general optimization problem, which can be seen as a doubly infinite linear programming problem, and we show that the associated dual generalizes several well-known special cases and extends to new risk management models we propose.

1 Introduction

An important problem in quantitative risk management is to aggregate several individually studied types of risks into an overall position. Mathematically, this translates into studying the worst-case distribution tails of $\Psi(X)$, where $\Psi : \mathbb{R}^n \to \mathbb{R}$ is a given function that represents the risk (or undesirability) of an outcome, and where X is a

Associate Professor in Numerical Mathematics, and Tanaka Fellow in Applied Mathematics at Pembroke College, Oxford. This author was supported through grant EP/H02686X/1 from the Engineering and Physical Sciences Research Council of the UK.
This author was supported through grant EP/H02686X/1 from the Engineering and Physical Sciences Research Council of the UK.

R. Hauser (✉) · S. Shahverdyan
Mathematical Institute, Oxford University, Andrew Wiles Building, Radcliffe Observatory Quarter, Woodstock Road, Oxford OX2 6GG, UK
e-mail: hauser@maths.ox.ac.uk

S. Shahverdyan
e-mail: sergey.shahverdyan@maths.ox.ac.uk

P. Embrechts
Department of Mathematics, ETH Zurich, 8092 Zurich, Switzerland
e-mail: embrechts@math.ethz.ch

© The Author(s) 2015
K. Glau et al. (eds.), *Innovations in Quantitative Risk Management*,
Springer Proceedings in Mathematics & Statistics 99,
DOI 10.1007/978-3-319-09114-3_22

random vector that takes values in $\mathbb{R}^n$ and whose distribution is only partially known. For example, one may only have information about the marginals of X and possibly partial information about some of the moments.

To solve such problems, duality is often exploited, as the dual may be easier to approach numerically or analytically [2–5, 14]. Being able to formulate a dual is also important in cases where the primal is approachable algorithmically, as solving the primal and dual problems jointly provides an approximation guarantee throughout the run of a solve: if the duality gap (the difference between the primal and dual objective values) falls below a chosen threshold relative to the primal objective, the algorithm can be stopped with a guarantee of approximating the optimum to a fixed precision that depends on the chosen threshold. This is a well-known technique in convex optimization, see, e.g., [1].

Although for some special cases of the marginal problem analytic solutions and powerful numerical heuristics exist [6, 12, 13, 18, 19], these techniques do not apply when additional constraints are imposed to force the probability measures over which we maximize the risk to conform with empirical observations: In a typical case, the bulk of the empirical data may be contained in a region D that can be approximated by an ellipsoid or the union of several (disjoint or overlapping) polyhedra. For a probability measure μ to be considered a reasonable explanation of the true distribution of (multidimensional) losses, one would require the probability mass contained in D to lie in an empirically estimated confidence region, that is, $\ell \leq \mu(D) \leq u$ for some estimated bounds $\ell < u$. In such a situation, the derivation of robust risk aggregation bounds via dual problems remains a powerful and interesting approach.

In this chapter, we formulate a general optimization problem, which can be seen as a doubly infinite linear programming problem, and we show that the associated dual generalizes several well known special cases. We then apply this duality framework to a new class of risk management models we propose in Sect. 4.

2 A General Duality Relation

Let $(\Phi, \mathfrak{F})$, $(\Gamma, \mathfrak{G})$ and $(\Sigma, \mathfrak{S})$ be complete measure spaces, and let $A : \Gamma \times \Phi \to \mathbb{R}$, $a : \Gamma \to \mathbb{R}$, $B : \Sigma \times \Phi \to \mathbb{R}$, $b : \Sigma \to \mathbb{R}$, and $c : \Phi \to \mathbb{R}$ be bounded measurable functions on these spaces and the corresponding product spaces. Let $\mathcal{M}_{\mathfrak{F}}$, $\mathcal{M}_{\mathfrak{G}}$ and $\mathcal{M}_{\mathfrak{S}}$ be the set of signed measures with finite variation on $(\Phi, \mathfrak{F})$, $(\Gamma, \mathfrak{G})$, and $(\Sigma, \mathfrak{S})$ respectively. We now consider the following pair of optimization problems over $\mathcal{M}_{\mathfrak{F}}$ and $\mathcal{M}_{\mathfrak{G}} \times \mathcal{M}_{\mathfrak{S}}$, respectively,

$$\text{(P)} \quad \sup_{\mathscr{F} \in \mathcal{M}_{\mathfrak{F}}} \int_{\Phi} c(x) \, \mathrm{d}\mathscr{F}(x)$$

$$\text{s.t.} \quad \int_{\Phi} A(y, x) \, \mathrm{d}\mathscr{F}(x) \leq a(y), \quad (y \in \Gamma),$$

$$\int_{\varPhi} B(z, x) \, d\mathscr{F}(x) = b(z), \quad (z \in \varSigma),$$

$$\mathscr{F} \geq 0,$$

and

(D) $\quad \inf_{(\mathscr{G}, \mathscr{S}) \in \mathscr{M}_{\mathfrak{G}} \times \mathscr{M}_{\mathfrak{S}}} \int_{\varGamma} a(y) \, d\mathscr{G}(y) + \int_{\varSigma} b(z) \, d\mathscr{S}(z),$

$$\text{s.t.} \int_{\varGamma} A(y, x) \, d\mathscr{G}(y) + \int_{\varSigma} B(z, x) \, d\mathscr{S}(z) \geq c(x), \quad (x \in \varPhi),$$

$$\mathscr{G} \geq 0.$$

We claim that the infinite-programming problems (P) and (D) are duals of each other.

Theorem 1 (Weak Duality) *For every (P)-feasible measure $\mathscr{F}$ and every (D)-feasible pair $(\mathscr{G}, \mathscr{S})$ we have*

$$\int_{\varPhi} c(x) \, d\mathscr{F}(x) \leq \int_{\varGamma} a(y) \, d\mathscr{G}(y) + \int_{\varSigma} b(z) \, d\mathscr{S}(z).$$

Proof Using Fubini's Theorem, we have

$$\int_{\varPhi} c(x) \, d\mathscr{F}(x) \leq \int_{\varGamma \times \varPhi} A(y, x) \, d(\mathscr{G} \times \mathscr{F})(y, x) + \int_{\varSigma \times \varPhi} B(z, x) \, d(\mathscr{S} \times \mathscr{F})(z, x)$$

$$\leq \int_{\varGamma} a(y) \, d\mathscr{G}(y) + \int_{\varSigma} b(z) \, d\mathscr{S}(z).$$

In various special cases, such as those discussed in Sect. 3, strong duality is known to hold subject to regularity assumptions, that is, the optimal values of (P) and (D) coincide. Another special case under which strong duality applies is when the measures $\mathscr{F}$, $\mathscr{G}$, and $\mathscr{S}$ have densities in appropriate Hilbert spaces, see the forthcoming DPhil thesis of the second author [17].

We remark that the quantifiers in the constraints can be weakened if the set of allowable measures is restricted. For example, if $\mathscr{G}$ is restricted to lie in a set of measures that are absolutely continuous with respect to a fixed measure $\mathscr{G}_0 \in \mathscr{M}_{\mathfrak{G}}$, then the quantifier $(y \in \varGamma)$ can be weakened to $(\mathscr{G}_0\text{-almost all } y \in \varGamma)$.

3 Classical Examples

Our general duality relation of Theorem 1 generalizes many classical duality results, of which we now point out a few examples. Let $p(x_1, \ldots, x_k)$ be a function of k arguments. Then we write

$$1_{\{x:p(x)\geq 0\}} := 1_{\{y:p(y)\geq 0\}}(x) = \begin{cases} 1 & \text{if } p(x) \geq 0, \\ 0 & \text{otherwise.} \end{cases}$$

In other words, we write the argument x of the indicator function directly into the set $\{y : p(y) \geq 0\}$ that defines the function, rather than using a separate set of variables y. This abuse of notation will make it easier to identify which inequality is satisfied by the arguments where the function $1_{\{y:p(y)\geq 0\}}(x)$ takes the value 1.

We start with the Moment Problem studied by Bertsimas and Popescu [2], who considered generalized Chebychev inequalities of the form

$$(\text{P'}) \quad \sup_X \; P[r(X) \geq 0]$$

$$\text{s.t.} \; E_\mu[X_1^{k_1}, \ldots, X_n^{k_n}] = b_k, \quad (k \in J),$$

$$X \text{ a random vector taking values in } \mathbb{R}^n,$$

where $r : \mathbb{R}^n \to \mathbb{R}$ is a multivariate polynomial and $J \subset \mathbb{N}^n$ is a finite sets of multi-indices. In other words, some moments of X are known. By choosing $\Phi = \mathbb{R}^n$, $\Gamma = \emptyset$, $\Sigma = J \cup \{0\}$,

$$B(k, x) = x_1^{k_1}, \ldots, x_n^{k_n}, \quad b(k) = b_k, \quad (k \in J),$$
$$B(0, x) = 1_{\mathbb{R}^n}, \quad b(0) = 1,$$

and $c(x) = 1_{\{x:r(x)\geq 0\}}$, where we made use of the abuse of notation discussed above, problem (P') becomes a special case of the primal problem considered in Sect. 2,

$$(\text{P}) \quad \sup_{\mathscr{F}} \int_{\mathbb{R}^n} 1_{\{x:r(x)\geq 0\}} \, d\mathscr{F}(x)$$

$$\text{s.t.} \int_{\mathbb{R}^n} x_1^{k_1}, \ldots, x_n^{k_n} \, d\mathscr{F}(x) = b_k, \quad (k \in J),$$

$$\int_{\mathbb{R}^n} 1 \, d\mathscr{F}(x) = 1,$$

$$\mathscr{F} \geq 0.$$

Our dual

$$\text{(D)} \quad \inf_{(z,z_0)\in\mathbb{R}^{|J|+1}} \sum_{k\in J} z_k b_k + z_0$$

$$\text{s.t.} \sum_{k\in J} z_k x_1^{k_1}, \ldots, x_n^{k_n} + z_0 \geq 1_{\{x:r(x)\geq 0\}}, \quad (x \in \mathbb{R}^n)$$

is easily seen to be identical with the dual (D') identified by Bertsimas and Popescu,

$$\text{(D')} \quad \inf_{(z,z_0)\in\mathbb{R}^{|J|+1}} \sum_{k\in J} z_k b_k + z_0$$

$$\text{s.t.} \ \forall x \in \mathbb{R}^n, r(x) \geq 0 \Rightarrow \sum_{k\in J} z_k x_1^{k_1}, \ldots, x_n^{k_n} + z_0 - 1 \geq 0,$$

$$\forall x \in \mathbb{R}^n, \sum_{k\in J} z_k x_1^{k_1}, \ldots, x_n^{k_n} + z_0 \geq 0.$$

Note that since Γ, Σ are finite, the constraints of (D') are polynomial copositivity constraints. The numerical solution of semi-infinite programming problems of this type can be approached via a nested hierarchy of semidefinite programming relaxations that yield better and better approximations to (D'). The highest level problem within this hierarchy is guaranteed to solve (D') exactly, although the corresponding SDP is of exponential size in the dimension n, in the degree of the polymomial r, and in $\max_{k\in J}(\sum_i k_i)$. For further details see [2, 7, 10], and Sect. 4.6 below.

Next, we consider the Marginal Problem studied by Rüschendorf [15, 16] and Ramachandran and Rüschendorf [14],

$$\text{(P')} \quad \sup_{\mathscr{F}\in\mathscr{M}_{F_1,\ldots,F_n}} \int_{\mathbb{R}^n} h(x)\,d\mathscr{F}(x),$$

where $\mathscr{M}_{F_1,\ldots,F_n}$ is the set of probability measures on $\mathbb{R}^n$ whose marginals have the cdfs F_i ($i = 1,\ldots,n$). Problem (P') can easily be seen as a special case of the framework of Sect. 2 by setting $c(x) = h(x)$, $\Phi = \mathbb{R}^n$, $\Gamma = \emptyset$, $\Sigma = \mathbb{N}_n \times \mathbb{R}$, $B(i,z,x) = 1_{\{y:\, y_i \leq z\}}$ (using the abuse of notation discussed earlier), and $b_i(z) = F_i(z)$ ($i \in \mathbb{N}_n$, $z \in \mathbb{R}$),

$$\text{(P)} \quad \sup_{\mathscr{F}} \int_{\mathbb{R}^n} h(x)\,d\mathscr{F}(x)$$

$$\text{s.t.} \int_{\mathbb{R}} 1_{\{x_i \leq z\}}\,d\mathscr{F}(x) = F_i(z), \quad (z \in \mathbb{R}, i \in \mathbb{N}_n)$$

$$\mathscr{F} \geq 0.$$

Taking the dual, we find

$$\text{(D)}\quad \inf_{\mathscr{S}_1,\ldots,\mathscr{S}_n}\ \sum_{i=1}^{n}\int_{\mathbb{R}} F_i(z)\,d\mathscr{S}_i(z)$$

$$\text{s.t.}\ \sum_{i=1}^{n}\int_{\mathbb{R}} 1_{\{x_i\le z\}}\,d\mathscr{S}_i(z)\ge h(x),\quad (x\in\mathbb{R}^n).$$

The signed measures $\mathscr{S}_i$ being of finite variation, the functions $S_i(z)=\mathscr{S}((-\infty,z])$ and the limits $s_i=\lim_{z\to\infty}S_i(z)=\mathscr{S}((-\infty,+\infty))$ are well defined and finite. Furthermore, using $\lim_{z\to-\infty}F_i(z)=0$ and $\lim_{z\to+\infty}F_i(z)=1$, we have

$$\sum_{i=1}^{n}\int_{\mathbb{R}} F_i(z)\,d\mathscr{S}(z)=\sum_{i=1}^{n}\left(F_i(z)S_i(z)\big|_{-\infty}^{+\infty}-\int_{\mathbb{R}} S_i(z)\,d F_i(z)\right)$$

$$=\sum_{i=1}^{n}s_i-\sum_{i=1}^{n}\int_{\mathbb{R}} S_i(z)\,d F_i(z)$$

$$=\sum_{i=1}^{n}\int_{\mathbb{R}} (s_i-S_i(z))\,d F_i(z),$$

and likewise,

$$\sum_{i=1}^{n}\int_{\mathbb{R}} 1_{\{x_i\le z\}}\,d\mathscr{S}_i(z)=\sum_{i=1}^{n}\int_{x_i}^{+\infty} 1\,d\mathscr{S}_i(z)=\sum_{i=1}^{n}(s_i-S_i(x_i)).$$

Writing $h_i(z)=s_i-S_i(z)$, (D) is, therefore, equivalent to

$$\text{(D')}\quad \inf_{h_1,\ldots,h_n}\ \sum_{i=1}^{n}\int_{\mathbb{R}} h_i(z)\,d F_i(z)$$

$$\text{s.t.}\ \sum_{i=0}^{n} h_i(x_i)\ge h(x),\quad (x\in\mathbb{R}^n).$$

This is the dual identified by Ramachandran and Rüschendorf [14]. Due to the general form of the functions h_i, the infinite programming problem (D') is not directly usable in numerical computations. However, for specific $h(x)$, (D')-feasible functions $(h_1,\ldots,h_n)$ can sometimes be constructed explicitly, yielding an upper bound on the optimal objective function value of (P') by virtue of Theorem 1. Embrechts and Puccetti [3–5] used this approach to derive quantile bounds on $X_1+\cdots+X_n$, where

X is a random vector with known marginals but unknown joint distribution. In this case, the relevant primal objective function is defined by $h(x) = 1_{\{x:\,e^T x \geq t\}}$, where $t \in \mathbb{R}$ is a fixed level. More generally, $h(x) = 1_{\{x:\,\Psi(x) \geq t\}}$ can be chosen, where Ψ is a relevant risk aggregation function, or $h(x)$ can model any risk measure of choice.

Our next example is the Marginal Problem with Copula Bounds, an extension to the marginal problem mentioned in [3]. The copula defined by the probability measure $\mathscr{F}$ with marginals F_i is the function

$$\mathscr{C}_{\mathscr{F}} : [0,1]^n \to [0,1],$$

$$u \mapsto F\left(F_1^{-1}(u_1), \ldots, F_n^{-1}(u_n)\right).$$

A copula is any function $\mathscr{C} : [0,1]^n \to [0,1]$ that satisfies $\mathscr{C} = \mathscr{C}_{\mathscr{F}}$ for some probability measure $\mathscr{F}$ on $\mathbb{R}^n$. Equivalently, a copula is the multivariate cdf of any probability measure on the unit cube $[0,1]^n$ with uniform marginals. In quantitative risk management, using the model

$$\sup_{\mathscr{F} \in \mathscr{M}_{F_1,\ldots,F_n}} \int_{\mathbb{R}^n} h(x)\,\mathrm{d}\mathscr{F}(x)$$

to bound the worst-case risk for a random vector X with marginal distributions F_i can be overly conservative, as no dependence structure between the coordinates of X_i is assumed given at all. The structure that determines this dependence being the copula $\mathscr{C}_{\mathscr{F}}$, where $\mathscr{F}$ is the multivariate distribution of X, Embrechts and Puccetti [3] suggest problems of the form

$$(\text{P'}) \qquad \sup_{\mu \in \mathscr{M}_{F_1,\ldots,F_n}} \int_{\mathbb{R}^n} h(x)\,\mathrm{d}\mu(x),$$

$$\text{s.t. } \mathscr{C}_{\mathrm{lo}} \leq \mathscr{C}_{\mathscr{F}} \leq \mathscr{C}_{\mathrm{up}},$$

as a natural framework to study the situation in which partial dependence information is available. In problem (P'), $\mathscr{C}_{\mathrm{lo}}$ and $\mathscr{C}_{\mathrm{up}}$ are given copulas, and inequality between copulas is defined by pointwise inequality,

$$\mathscr{C}_{\mathrm{lo}}(u) \leq \mathscr{C}_{\mathscr{F}}(u) \quad (u \in [0,1]^n).$$

Once again, (P') is a special case of the general framework studied in Sect. 2, as it is equivalent to write

(P) $\displaystyle\sup_{\mathscr{F}} \int_{\mathbb{R}^n} h(x)\,\mathrm{d}\mathscr{F}(x)$

s.t. $\displaystyle\int_{\mathbb{R}^n} 1_{\{x \le (F_1^{-1}(u_1),\dots,F_n^{-1}(u_n))\}}(u,x)\,\mathrm{d}\mathscr{F}(x) \le \mathscr{C}_{\mathrm{up}}(u), \quad (u \in [0,1]^n),$

$\displaystyle\int_{\mathbb{R}^n} -1_{\{x \le (F_1^{-1}(u_1),\dots,F_n^{-1}(u_n))\}}(u,x)\,\mathrm{d}\mathscr{F}(x) \le -\mathscr{C}_{\mathrm{lo}}(u), \quad (u \in [0,1]^n),$

$\displaystyle\int_{\mathbb{R}^n} 1_{\{x_i \le z\}}(z,x)\,\mathrm{d}\mathscr{F}(x) = F_i(z), \quad (i \in \mathbb{N}_n,\ z \in \mathbb{R}),$

$\mathscr{F} \ge 0.$

The dual of this problem is given by

(D) $\displaystyle\inf_{\mathscr{G}_{\mathrm{up}},\mathscr{G}_{\mathrm{lo}},\mathscr{S}_1,\dots,\mathscr{S}_n} \int_{[0,1]^n} \mathscr{C}_{\mathrm{up}}(u)\,\mathrm{d}\mathscr{G}_{\mathrm{up}}(u) - \int_{[0,1]^n} \mathscr{C}_{\mathrm{lo}}(u)\,\mathrm{d}\mathscr{G}_{\mathrm{lo}}(u) + \sum_{i=1}^n \int_{\mathbb{R}} F_i(z)\,\mathrm{d}\mathscr{S}_i(z)$

s.t. $\displaystyle\int_{[0,1]^n} 1_{\{x \le (F_1^{-1}(u_1),\dots,F_n^{-1}(u_n))\}}(u,x)\,\mathrm{d}\mathscr{G}_{\mathrm{up}}(u)$

$\displaystyle - \int_{[0,1]^n} 1_{\{x \le (F_1^{-1}(u_1),\dots,F_n^{-1}(u_n))\}}(u,x)\,\mathrm{d}\mathscr{G}_{\mathrm{lo}}(u)$

$\displaystyle + \sum_{i=1}^n \int_{\mathbb{R}} 1_{\{x_i \le z\}}\,\mathrm{d}\mathscr{S}_i(z) \ge h(x), \quad (x \in \mathbb{R}^n),$

$\mathscr{G}_{\mathrm{lo}}, \mathscr{G}_{\mathrm{up}} \ge 0.$

Using the notation s_i, S_i introduced in Sect. 3, this problem can be written as

$\displaystyle\inf_{\mathscr{G}_{\mathrm{up}},\mathscr{G}_{\mathrm{lo}},\mathscr{S}_1,\dots,\mathscr{S}_n} \int_{[0,1]^n} \mathscr{C}_{\mathrm{up}}(u)\,\mathrm{d}\mathscr{G}_{\mathrm{up}}(u) - \int_{[0,1]^n} \mathscr{C}_{\mathrm{lo}}(u)\,\mathrm{d}\mathscr{G}_{\mathrm{lo}}(u) + \sum_{i=1}^n \int_{\mathbb{R}} (s_i - S_i(z))\,\mathrm{d}F_i(z)$

s.t. $\displaystyle\mathscr{G}_{\mathrm{up}}(\mathscr{B}(x)) - \mathscr{G}_{\mathrm{lo}}(\mathscr{B}(x)) + \sum_{i=1}^n (s_i - S_i(x_i)) \ge h(x), \quad (x \in \mathbb{R}^n),$

$\mathscr{G}_{\mathrm{up}}, \mathscr{G}_{\mathrm{lo}} \ge 0,$

where $\mathscr{B}(x) = \{u \in [0,1]^n : u \ge (F_1(x_1),\dots,F_n(x_n))\}$. To the best of our knowledge, this dual has not been identified before.

Due to the high dimensionality of the space of variables and constraints both in the primal and dual, the marginal problem with copula bounds is difficult to solve numerically, even for very coarse discrete approximations.

4 Robust Risk Aggregation via Bounds on Integrals

In quantitative risk management, distributions are often estimated within a parametric family from the available data. For example, the tails of marginal distributions may be estimated via extreme value theory, or a Gaussian copula may be fitted to the multivariate distribution of all risks under consideration, to model their dependencies. The choice of a parametric family introduces *model uncertainty*, while fitting a distribution from this family via statistical estimation introduces *parameter uncertainty*. In both cases, a more robust alternative would be to study models in which the available data is only used to estimate upper and lower bounds on finitely many integrals of the form

$$\int_\Phi \phi(x) \, \mathrm{d}\mathscr{F}(x), \tag{1}$$

where $\phi(x)$ is a suitable test function. A suitable way of estimating upper and lower bounds on such integrals from sample data x_i ($i \in \mathbb{N}_k$) is to estimate confidence bounds via bootstrapping.

4.1 Motivation

To motivate the use of constraints in the form of bounds on integrals (1), we offer the following explanations: First of all, discretized marginal constraints are of this form with piecewise constant test functions, as the requirement that $F_i(\xi_k) - F_i(\xi_{k-1}) = b_k$ ($k = 1, \ldots, \ell$) for a fixed set of discretization points $\xi_0 < \cdots < \xi_\ell$ can be expressed as

$$\int_\Phi 1_{\{\xi_k \leq x_i \leq \xi_{k-1}\}} \, \mathrm{d}\mathscr{F}(x) = b_k, \quad (k = 1, \ldots, \ell). \tag{2}$$

It is, furthermore, quite natural to relax each of these equality constraints to two inequality constraints

$$b_{k,i}^\ell \leq \int_\Phi 1_{\{\xi_k \leq x_i \leq \xi_{k-1}\}} \, \mathrm{d}\mathscr{F}(x) \leq b_{k,i}^u$$

when b_k is estimated from data.

More generally, constraints of the form $\mathrm{P}[X \in S_j] \leq b_j^u$ for some measurable $S_j \subseteq \mathbb{R}^n$ of interest can be written as

$$\int_{\Phi} 1_{S_j}(x)\, \mathrm{d}\, \mathscr{F}(x) \le b_j^u.$$

A collection of ℓ constraints of this form can be relaxed by replacing them by a convex combination

$$\int_{\Phi} \sum_{j=1}^{\ell} w_j 1_{S_j}(x)\, \mathrm{d}\, \mathscr{F}(x) \le \sum_{j=1}^{\ell} w_j b_j^u,$$

where the weights $w_j > 0$ satisfy $\sum_j w_j = 1$ and express the relative importance of each constituent constraint. Nonnegative test functions thus have a natural interpretation as importance densities in sums-of-constraints relaxations. This allows one to put higher focus on getting the probability mass right in regions where it particularly matters (e.g., values of X that account for the bulk of the profits of a financial institution), while maximzing the risk in the tails without having to resort to too fine a discretization.

While this suggests to use a piecewise approximation of a prior estimate of the density of X as a test function, the results are robust under mis-specification of this prior, for as long as $\phi(x)$ is nonconstant, constraints that involve the integral (1) tend to force the probability weight of X into the regions where the sample points are denser. To illustrate this, consider a univariate random variable with density $f(x) = 2/3(1+x)$ on $x \in [0, 1]$ and test function $\phi(x) = 1 + ax$ with $a \in [-1, 1]$. Then $\int_0^1 \phi(x)f(x)\, \mathrm{d}\, x = 1 + 5a/9$. The most dispersed probability measure on $[0, 1]$ that satisfies

$$\int_0^1 \phi(x)\, \mathrm{d}\, \mathscr{F}(x) = 1 + \frac{5a}{9} \tag{3}$$

has an atom of weight 4/9 at 0 and an atom of weight 5/9 at 1 independently of a, as long as $a \ne 0$. The constraint (3) thus forces more probability mass into the right half of the interval $[0, 1]$, where the unknown (true) density $f(x)$ has more mass and produces more sample points.

As a second illustration, take the density $f(x) = 3x^2$ and the same linear test function as above. This time we find $\int_0^1 \phi(x)f(x)\, \mathrm{d}\, x = 1 + 3a/4$, and the most dispersed probability measure on $[0, 1]$ that satisfies

$$\int_0^1 \phi(x)\, \mathrm{d}\, \mathscr{F}(x) = 1 + \frac{3a}{4}$$

has an atom of weight 3/4 at 0 and an atom of weight 1/4 at 1 independently of $a \neq 0$, with similar conclusions as above, except that the effect is even stronger, correctly reflecting the qualitative features of the density $f(x)$.

4.2 General Setup and Duality

Let Φ be decomposed into a partition $\Phi = \bigcup_{i=1}^{k} \Xi_i$ of polyhedra Ξ_i with nonempty interior, chosen as regions in which a reasonable number of data points are available to estimate integrals of the form (1).

Each polyhedron has a primal description in terms of generators,

$$\Xi_i = \mathrm{conv}(q_1^i, \ldots, q_{n_i}^i) + \mathrm{cone}(r_1^i, \ldots, r_{o_i}^i)$$

where $\mathrm{conv}(q_1^i, \ldots, q_{n_i}^i)$ is the polytope with vertices $q_n^i \in \mathbb{R}^n$, and

$$\mathrm{cone}(r_1^i, \ldots, r_{o_i}^i) = \left[\sum_{m=1}^{o_i} \xi_m r_m^i : \xi_m \geq 0 \ (m \in \mathbb{N}_{o_i}) \right]$$

is the polyhedral cone with recession directions $r_m^i \in \mathbb{R}^n$. Each polyhedron also has a dual description in terms of linear inequalities,

$$\Xi_i = \bigcap_{j=1}^{k_i} \left\{ x \in \mathbb{R}^n : \langle f_j^i, x \rangle \geq \ell_j^i \right\},$$

for some vectors $f_j^i \in \mathbb{R}^n$ and bounds $\ell_j^i \in \mathbb{R}$. The main case of interest is where Ξ_i is either a finite or infinite box in $\mathbb{R}^n$ with faces parallel to the coordinate axes, or an intersection of such a box with a linear half-space, in which case it is easy to pass between the primal and dual descriptions. Note however that the dual description is preferrable, as the description of a box in $\mathbb{R}^n$ requires only $2n$ linear inequalities, while the primal description requires 2^n extreme vertices.

Let us now consider the problem

$$\text{(P)} \quad \sup_{\mathscr{F} \in \mathscr{M}_{\mathfrak{F}}} \int_{\Phi} h(x) \, d\mathscr{F}(x)$$

$$\text{s.t.} \int_{\Phi} \phi_s(x) \, d\mathscr{F}(x) \leq a_s, \quad (s = 1, \ldots, M),$$

$$\text{s.t.} \int_{\Phi} \psi_t(x) \, d\mathscr{F}(x) = b_t, \quad (t = 1, \ldots, N),$$

$$\int_{\Phi} 1 \, d\,\mathscr{F}(x) = 1,$$

$$\mathscr{F} \geq 0,$$

where the test functions ψ_t are piecewise linear on the partition $\Phi = \bigcup_{i=1}^{k} \Xi_i$, and where $-h(x)$ and the test functions ϕ_s are piecewise linear on the infinite polyhedra of the partition, and either jointly linear, concave, or convex on the finite polyhedra (i.e., polytopes) of the partition. The dual of (P) is

(D) $\displaystyle \inf_{(y,z)\in\mathbb{R}^{M+N+1}} \sum_{s=1}^{M} a_s y_s + \sum_{t=1}^{N} b_t z_t + z_0,$

s.t. $\displaystyle \sum_{s=1}^{M} y_s \phi_s(x) + \sum_{t=1}^{N} z_t \psi_t(x) + z_0 \, 1_\Phi(x) - h(x) \geq 0, \quad (x \in \Phi),$

$$\tag{4}$$

$$y \geq 0.$$

We remark that (P) is a semi-infinite programming problem with infinitely many variables and finitely many constraints, while (D) is a semi-infinite programming problem with finitely many variables and infinitely many constraints. However, the constraint (4) of (D) can be rewritten as copositivity requirements over the polyhedra Ξ_i,

$$\sum_{s=1}^{M} y_s \phi_s(x) + \sum_{t=1}^{N} z_t \psi_t(x) + z_0 \, 1_\Phi(x) - h(x) \geq 0, \quad (x \in \Xi_i), \quad (i = 1, \ldots, k).$$

Next we will see how these copositivity constraints can be handled numerically, often by relaxing all but finitely many constraints. Nesterov's first-order method can be adapted to solve the resulting problems, see [8, 9, 17].

In what follows, we will use the notation

$$\varphi_{y,z}(x) = \sum_{s=1}^{M} y_s \phi_s(x) + \sum_{t=1}^{N} z_t \psi_t(x) + z_0 - h(x).$$

4.3 Piecewise Linear Test Functions

The first case we discuss is when $\phi_s|_{\Xi_i}$ and $h|_{\Xi_i}$ are jointly linear. Since we furthermore assumed that the functions $\psi_t|_{\Xi_i}$ are linear, there exist vectors $v_s^i \in \mathbb{R}^n$, $w_t^i \in \mathbb{R}^n$, $g^i \in \mathbb{R}^n$ and constants $c_s^i \in \mathbb{R}$, $d_t^i \in \mathbb{R}$ and $e^i \in \mathbb{R}$ such that

$$\phi_s|_{\varXi_i}(x) = \langle v_s^i, x \rangle + c_s^i,$$
$$\psi_t|_{\varXi_i}(x) = \langle w_t^i, x \rangle + d_t^i,$$
$$h|_{\varXi_i}(x) = \langle g^i, x \rangle + e^i.$$

The copositivity condition

$$\sum_{s=1}^{M} y_s \phi_s(x) + \sum_{t=1}^{N} z_t \psi_t(x) + z_0 \, 1_\Phi(x) - h(x) \geq 0, \quad (x \in \varXi_i)$$

can then be written as

$$\langle f_j^i, x \rangle \geq \ell_j^i, \quad (j = 1, \dots, k_i) \implies$$

$$\left\langle \sum_{s=1}^{M} y_s v_s^i + \sum_{t=1}^{N} z_t w_t^i - g^i, \, x \right\rangle \geq e^i - \sum_{s=1}^{M} y_s c_s^i - \sum_{t=1}^{N} z_t d_t^i - z_0.$$

By Farkas' Lemma, this is equivalent to the constraints

$$\sum_{s=1}^{M} y_s v_s^i + \sum_{t=1}^{N} z_t w_t^i - g^i = \sum_{j=1}^{k_i} \lambda_j^i f_j^i, \tag{5}$$

$$e^i - \sum_{s=1}^{M} y_s c_s^i - \sum_{t=1}^{N} z_t d_t^i - z_0 \leq \sum_{j=1}^{k_i} \lambda_j^i \ell_j^i, \tag{6}$$

$$\lambda_j^i \geq 0, \quad (j = 1, \dots, k_i), \tag{7}$$

where λ_j^i are additional auxiliary decision variables.

Thus, if all test functions are linear on all polyhedral pieces $\varXi_i$, then the dual (D) can be solved as a linear programming problem with $M + N + 1 + \sum_{i=1}^{k} k_i$ variables and $k(n + 1)$ linear constraints, plus bound constraints on y and the λ_j^i. More generally, if some but not all polyhedra correspond to jointly linear test function pieces, then jointly linear pieces can be treated as discussed above, while other pieces can be treated as discussed below.

Let us briefly comment on numerical implementations, further details of which are described in the second author's thesis [17]: An important case of the above described framework corresponds to a discretized marginal problem in which $\phi_s(x)$ are piecewise constant functions chosen as follows for $s = (i, j), (\iota = 1, \dots, n; j = 1, \dots, m)$: Introduce $m + 1$ breakpoints $\xi_0^\iota < \xi_1^\iota < \cdots < \xi_m^\iota$ along each coordinate axis ι, and consider the infinite slabs

$$S_{\iota, j} = \left\{ x \in \mathbb{R}^n : \xi_{j-1}^\iota \leq x_\iota \leq \xi_j^\iota \right\}, \quad (j = 1, \dots, m).$$

Then choose $\phi_{i,j}(x) = 1_{S_{i,j}}(x)$, the indicator function of slab $S_{i,j}$. We remark that this approach corresponds to discretizing the constraints of the Marginal Problem described in Sect. 3, but not to discretizing the probability measures over which we maximize the aggregated risk.

While the number of test functions is nm and thus linear in the problem dimension, the number of polyhedra to consider is exponentially large, as all intersections of the form

$$\Xi_{i,\mathbf{j}} = \bigcap_{\iota=1}^{n} S_{\iota,j_{\iota}}$$

for the m^n possible choices of $\mathbf{j} \in \mathbb{N}_m^n$ have to be treated separately. In addition, in VaR applications $h(x)$ is taken as the indicator function of an affine half-space $\{x : \sum x_{\iota} \geq \tau\}$ for a suitably chosen threshold τ, and for CVaR applications $h(x)$ is chosen as the piecewise linear function $h(x) = \max(0, \sum x_{\iota} - \tau)$. Thus, polyhedra $\Xi_{i,\mathbf{j}}$ that meet the affine hyperplane $\{x : \sum x_{\iota} = \tau\}$ are further sliced into two separate polyhedra. A straightforward application of the above described LP framework would thus lead to an LP with exponentially many constraints and variables. Note however that the constraints (5)–(7) now read

$$g^i = \sum_{j=1}^{k_i} \lambda_j^i f_j^i, \tag{8}$$

$$e^i - \sum_{s=1}^{M} y_s c_s^i - z_0 \leq \sum_{j=1}^{k_i} \lambda_j^i \ell_j^i, \tag{9}$$

$$\lambda_j^i \geq 0, \quad (j = 1, \ldots, k_i), \tag{10}$$

as $v_s^i = 0$ and no test functions $\psi_t(x)$ were used, with $g^i = [\,1 \ldots 1\,]^{\mathsf{T}}$ when $\Xi_i \subseteq \{x : \sum x_{\iota} \geq \tau\}$ and $g^i = 0$ otherwise. That is, the vector that appears in the left-hand side of Constraint (8) is fixed by the polyhedron Ξ_i alone and does not depend on the decision variables y, z_0. Since z_0 is to be chosen as small as possible in an optimal solution of (D), the constraint (9) has to be made as slack as possible. Therefore, the optimal values of λ_j^i are also fixed by the polyhedron Ξ_i alone and are identifiable by solving the small-scale LP

$$(\lambda_j^i)^* = \arg\max_{\lambda} \sum_{j=1}^{k_i} \lambda_j^i \ell_j^i$$

$$\text{s.t.} - g^i = \sum_{j=1}^{k_i} \lambda_j^i f_j^i,$$

$$\lambda_j^i \geq 0, \quad (j = 1, \ldots, k_i).$$

In other words, when the polyhedron Ξ_i is considered for the first time, the variables $(\lambda_j^i)^*$ can be determined once and for all, after which the constraints (8)–(10) can be replaced by

$$e^i - \sum_s y_s c_s^i - z_0 \leq C_i,$$

where $C_i = \sum_{j=1}^{k_i} (\lambda_j^i)^* \ell_j^i$, and where the sum on the left-hand side only extends over the n indices s that correspond to test functions that are nonzero on Ξ_i. Thus, only the $nm + 1$ decision variables (y, z_0) are needed to solve (D). Furthermore, the exponentially many constraints correspond to an extremely sparse constraint matrix, making the dual of (D) an ideal candidate to apply the simplex algorithm with delayed column generation. A similar approach is possible for the situation where ϕ_s is of the form

$$\phi_s(x) = 1_{S_s}(x) \times (\langle v_s, x \rangle + c_s),$$

for all $s = (\iota, j)$. The advantage of using test functions of this form is that fewer breakpoints $\xi_{\iota, j}$ are needed to constrain the distribution appropriately.

4.4 Piecewise Convex Test Functions

When $\phi_s|_{\Xi_i}$ and $-h|_{\Xi_i}$ are jointly convex, then $\varphi_{y,z}(x)$ is convex. The copositivity constraint

$$\sum_{s=1}^{M} y_s \phi_s(x) + \sum_{t=1}^{N} z_t \psi_t(x) + z_0 1_\Phi(x) - h(x) \geq 0, \quad (x \in \Xi_i)$$

can then be written as

$$\langle f_j^i, x \rangle \geq \ell_j^i, \quad (j = 1, \ldots, k_i) \implies \varphi_{y,z}(x) \geq 0,$$

and by Farkas' Theorem (see ,e.g., [11]), this condition is equivalent to

$$\varphi_{y,z}(x) + \sum_{j=1}^{k_i} \lambda_j^i \left(\ell_j^i - \langle f_j^i, x \rangle \right) \geq 0, \quad (x \in \mathbb{R}^n), \tag{11}$$

$$\lambda_j^i \geq 0, \quad (j = 1, \ldots, k_i),$$

where λ_j^i are once again auxiliary decision variables. While (11) does not reduce to finitely many constraints, the validity of this condition can be checked numerically by globally minimizing the convex function $\varphi_{y,z}(x) + \sum_{j=1}^{k_i} \lambda_j^i \left(\ell_j^i - \langle f_j^i, x \rangle \right)$. The constraint (11) can then be enforced explicitly if a line-search method is used to solve the dual (D).

4.5 Piecewise Concave Test Functions

When $\phi_s|_{\Xi_i}$ and $-h|_{\Xi_i}$ are jointly concave but not linear, then $\varphi_{y,z}(x)$ is concave and $\Xi_i = \mathrm{conv}(q_1^i, \ldots, q_{n_i}^i)$ is a polytope. The copositivity constraint

$$\sum_{s=1}^{M} y_s \phi_s(x) + \sum_{t=1}^{N} z_t \psi_t(x) + z_0 1_\Phi(x) - h(x) \geq 0, \quad (x \in \Xi_i) \qquad (12)$$

can then be written as

$$\varphi_{y,z}(q_j^i) \geq 0, \quad (j = 1, \ldots, n_i).$$

Thus, (12) can be replaced by n_i linear inequality constraints on the decision variables y_s and z_t.

4.6 Piecewise Polynomial Test Functions

Another case that can be treated via finitely many constraints is when $\phi_s|_{\Xi_i}$, $\psi_t|_{\Xi_i}$, and $h|_{\Xi_i}$ are jointly polynomial. The approach of Lasserre [7] and Parrilo [10] can be applied to turn the copositivity constraint

$$\langle f_j^i, x \rangle \geq \ell_j^i, \quad (j = 1, \ldots, k_i) \implies \varphi_{y,z}(x) \geq 0,$$

into finitely many linear matrix inequalities. However, this approach is generally limited to low-dimensional applications.

5 Conclusions

Our analysis shows that a wide range of duality relations in use in quantitative risk management can be understood from the single perspective of a generalized duality relation discussed in Sect. 2. An interesting class of special cases is provided by formulating a finite number of constraints in the form of bounds on integrals.

The duals of such models are semi-inifinite optimization problems that can often be reformulated as finite optimization problems, by making use of standard results on copositivity.

Acknowledgments Part of this research was conducted while the first author was visiting the FIM at ETH Zurich during sabbatical leave from Oxford. He thanks for the support and the wonderful research environment he encountered there. The research of the first two authors was supported through grant EP/H02686X/1 from the Engineering and Physical Sciences Research Council of the UK.

Open Access This chapter is distributed under the terms of the Creative Commons Attribution Noncommercial License, which permits any noncommercial use, distribution, and reproduction in any medium, provided the original author(s) and source are credited.

References

1. Borwein, J. and Lewis, A.: Convex Analysis and Nonlinear Optimization: Theory and Examples (CMS Books in Mathematics)
2. Bertsimas, D., Popescu, I.: Optimal inequalities in probability theory: a convex optimization approach. SIAM J. Optim. **15**(3), 780–804 (2005)
3. Embrechts, P., Puccetti, G.: Bounds for functions of dependent risks. Financ. Stoch. **10**, 341–352 (2006)
4. Embrechts, P., Puccetti, G.: Bounds for functions of multivariate risks. J. Multivar. Analy. **97**, 526–547 (2006)
5. Embrechts, P., Puccetti, G.: Aggregating risk capital, with an application to operational risk. Geneva Risk Insur. Rev. **31**, 71–90 (2006)
6. Embrechts, P., Puccetti, G., Rüschendorf, L.: Model uncertainty and VaR aggregation. J. Bank. Financ. **37**(8), 2750–2764 (2013)
7. Lasserre, J.B.: Global optimization with polynomials and the problem of moments. SIAM J. Optim. **11**(3), 796–817 (2001)
8. Nesterov, Y.: A method of solving a convex programming problem with convergence rate $O(1/k^2)$. Soviet Math. Dokl. **27**(2), 372–376 (1983)
9. Nesterov, Y.: Smooth minimization of non-smooth functions. Math. Program., Ser. A **103**, 127–152 (2005)
10. Parrilo, P.A.: Semidefinite programming relaxations for semialgebraic problems. Math. Program., Ser. B **96**, 293–320 (2003)
11. Pólik, I., Terlaky, T.: A survey of the S-Lemma. SIAM Rev. **49**, 371–418 (2007)
12. Puccetti, G., Rüschendorf, L.: Computation of sharp bounds on the distribution of a function of dependent risks. J. Comp. Appl. Math. **236**(7), 1833–1840 (2012)
13. Puccetti, G., Rüschendorf, L.: Sharp bounds for sums of dependent risks. J. Appl. Probab. **50**(1), 42–53 (2013)
14. Ramachandran, D., Rüschendorf, L.: A general duality theorem for marginal problems. Probab. Theory Relat. Fields **101**(3), 311–319 (1995)
15. Rüschendorf, L.: Random variables with maximum sums. Adv. Appl. Probab. **14**(3), 623–632 (1982)
16. Rüschendorf, L.: Construction of multivariate distributions with given marginals. Ann. Inst. Stat. Math. **37**(2), 225–233 (1985)
17. Shahverdyan, S.: Optimisation Methods in Risk Management. DPhil thesis, Oxford Mathematical Institute (2014)

18. Wang, R., Peng, L., Yang, J.: Bounds for the sum of dependent risks and worst value-at-risk with monotone marginal densities. Financ. Stoch. **17**(2), 395–417 (2013)
19. Wang, B., Wang, R.: The complete mixability and convex minimization problems with monotone marginal densities. J. Multivar. Anal. **102**(10), 1344–1360 (2011)

Some Consequences of the Markov Kernel Perspective of Copulas

Wolfgang Trutschnig and Juan Fernández Sánchez

Abstract The objective of this paper is twofold: After recalling the one-to-one correspondence between two-dimensional copulas and Markov kernels having the Lebesgue measure λ on $[0, 1]$ as fixed point, we first give a quick survey over some consequences of this interrelation. In particular, we sketch how Markov kernels can be used for the construction of strong metrics that strictly distinguish extreme kinds of statistical dependence, and show how the translation of various well-known copula-related concepts to the Markov kernel setting opens the door to some surprising mathematical aspects of copulas. Secondly, we concentrate on the fact that iterates of the star product of a copula A with itself are Cesáro convergent to an idempotent copula $\hat{A}$ with respect to any of the strong metrics mentioned before and prove that $\hat{A}$ must have a very simple form if the Markov operator T_A associated with A is quasi-constrictive in the sense of Lasota.

Keywords Copula · Doubly stochastic measure · Markov kernel · Markov operator

1 Introduction

In 1996, Olsen et al. (see [23]) proved the existence of an isomorphism between the family $\mathscr{C}$ of two-dimensional copulas (endowed with the so-called star product) and the family $\mathscr{M}$ of all Markov operators (with the standard composition as binary operation). Using disintegration (see [29]) allows to express the aforementioned Markov operators in terms of Markov kernels, resulting in a one-to-one correspondence of $\mathscr{C}$ with the family $\mathscr{K}$ of all Markov kernels having the Lebesgue

W. Trutschnig (✉)
Department for Mathematics, University of Salzburg, Hellbrunnerstrasse 34,
Salzburg 5020, Austria
e-mail: wolfgang@trutschnig.net

J. Fernández Sánchez
Grupo de Investigación de Análisis Matemático, Universidad de Almería,
La Cañada de San Urbano, Almería, Spain
e-mail: juanfernandez@ual.es

© The Author(s) 2015
K. Glau et al. (eds.), *Innovations in Quantitative Risk Management*,
Springer Proceedings in Mathematics & Statistics 99,
DOI 10.1007/978-3-319-09114-3_23

measure λ on $[0, 1]$ as fixed point. Identifying every copula with its Markov kernel allows to define new metrics D_1, D_2, D_∞ which, contrary to the uniform one, strictly separate independence from complete dependence (full predictability). Additionally, the 'translation' of various copula-related concepts from $\mathscr{C}$ to $\mathscr{M}$ and $\mathscr{K}$ has proved useful in so far that it allowed both, for alternative simple proofs of already known properties as well as for new and interesting results. Section 3 of this paper is a quick incomplete survey over some useful consequences of this translation. In particular, we mention the fact that for each copula $A \in \mathscr{C}$, the iterates of the star product of A with itself are Cesáro converge to an idempotent copula $\hat{A}$ w.r.t. each of the three metrics mentioned before, i.e., we have

$$\lim_{n \to \infty} D_1(s_{*n}(A), \hat{A}) = 0$$

whereby $s_{*n}(A) = \frac{1}{n} \sum_{i=1}^{n} A^{*i}$ for every $n \in \mathbb{N}$. Section 4 contains some new unpublished results and proves that the idempotent limit copula $\hat{A}$ must have a very simple (ordinal-sum-like) form if the Markov operator T_A corresponding to A is quasi-constrictive in the sense of Lasota ([1, 15, 18]).

2 Notation and Preliminaries

As already mentioned before, $\mathscr{C}$ will denote the family of all (two-dimensional) *copulas*, d_∞ will denote the uniform metric on $\mathscr{C}$. For properties of copulas, we refer to [8, 22, 26]. For every $A \in \mathscr{C}$, μ_A will denote the corresponding *doubly stochastic measure*, $\mathscr{P}_\mathscr{C}$, the class of all these doubly stochastic measures. Since copulas are the restriction of two-dimensional distribution functions with $\mathscr{U}(0, 1)$-marginals to $[0, 1]^2$, the Lebesgue decomposition of every element in $\mathscr{P}_\mathscr{C}$ has no discrete component. The Lebesgue measure on $[0, 1]$ and $[0, 1]^2$ will be denoted by λ and λ_2, respectively. For every metric space (Ω, d), the Borel σ-field on Ω will be denoted by $\mathscr{B}(\Omega)$. A *Markov kernel* from $\mathbb{R}$ to $\mathscr{B}(\mathbb{R})$ is a mapping $K : \mathbb{R} \times \mathscr{B}(\mathbb{R}) \to [0, 1]$ such that $x \mapsto K(x, B)$ is measurable for every fixed $B \in \mathscr{B}(\mathbb{R})$ and $B \mapsto K(x, B)$ is a probability measure for every fixed $x \in \mathbb{R}$. Suppose that X, Y are real-valued random variables on a probability space $(\Omega, \mathscr{A}, \mathscr{P})$, then a Markov kernel $K : \mathbb{R} \times \mathscr{B}(\mathbb{R}) \to [0, 1]$ is called *regular conditional distribution of Y given X* if for every $B \in \mathscr{B}(\mathbb{R})$

$$K(X(\omega), B) = \mathbb{E}(\mathbf{1}_B \circ Y | X)(\omega) \tag{1}$$

holds $\mathscr{P}$-a.s. It is well known that for each pair (X, Y) of real-valued random variables a regular conditional distribution $K(\cdot, \cdot)$ of Y given X exists, that $K(\cdot, \cdot)$ is unique $\mathscr{P}^X$-a.s. (i.e., unique for $\mathscr{P}^X$-almost all $x \in \mathbb{R}$) and that $K(\cdot, \cdot)$ only depends on $\mathscr{P}^{X \otimes Y}$. Hence, given $A \in \mathscr{C}$ we will denote (a version of) the regular conditional distribution of Y given X by $K_A(\cdot, \cdot)$ and refer to $K_A(\cdot, \cdot)$ simply as *regular conditional distribution of A* or as *the Markov kernel of A*. Note that for every $A \in \mathscr{C}$, its conditional regular distribution $K_A(\cdot, \cdot)$, and every Borel set $G \in \mathscr{B}([0, 1]^2)$ we have

$$\int\limits_{[0,1]} K_A(x, G_x)\, d\lambda(x) = \mu_A(G), \tag{2}$$

whereby $G_x := \{y \in [0, 1] : (x, y) \in G\}$ for every $x \in [0, 1]$. Hence, as special case,

$$\int\limits_{[0,1]} K_A(x, F)\, d\lambda(x) = \lambda(F) \tag{3}$$

for every $F \in \mathscr{B}([0, 1])$. On the other hand, every Markov kernel $K : [0, 1] \times \mathscr{B}([0, 1]) \rightarrow [0, 1]$ fulfilling (3) induces a unique element $\mu \in \mathscr{P}_{\mathscr{C}}([0, 1]^2)$ via (2). For more details and properties of conditional expectation, regular conditional distributions, and disintegration see [13, 14].

$\mathscr{T}$ will denote the family of all λ-preserving transformations $h : [0, 1] \rightarrow [0, 1]$ (see [34]), $\mathscr{T}_p$ the subset of all bijective $h \in \mathscr{T}$. A copula $A \in \mathscr{C}$ will be called *completely dependent* if and only if there exists $h \in \mathscr{T}$ such that $K(x, E) := \mathbf{1}_E(hx)$ is a regular conditional distribution of A (see [17, 29] for equivalent definitions and main properties). For every $h \in \mathscr{T}$, the corresponding completely dependent copula will be denoted by C_h, the class of all completely dependent copulas by $\mathscr{C}_d$.

A linear operator T on $L^1([0, 1]) := L^1([0, 1], \mathscr{B}([0, 1]), \lambda)$ is called *Markov operator* ([3, 23] if it fulfills the following three properties:

1. T is positive, i.e., $T(f) \geq 0$ whenever $f \geq 0$
2. $T(\mathbf{1}_{[0,1]}) = \mathbf{1}_{[0,1]}$
3. $\int_{[0,1]}(Tf)(x)d\lambda(x) = \int_{[0,1]} f(x)d\lambda(x)$

As mentioned in the introduction $\mathscr{M}$ will denote the class of all Markov operators on $L^1([0, 1])$. It is straightforward to see that the operator norm of T is one, i.e., $\|T\| := \sup\{\|Tf\|_1 : \|f\|_1 \leq 1\} = 1$ holds. According to [23] *there is a one-to-one correspondence between $\mathscr{C}$ and $\mathscr{M}$*—in fact, the mappings $\Phi : \mathscr{C} \rightarrow \mathscr{M}$ and $\Psi : \mathscr{M} \rightarrow \mathscr{C}$, defined by

$$\Phi(A)(f)(x) := (T_A f)(x) := \frac{d}{dx} \int\limits_{[0,1]} A_{,2}(x, t)f(t)d\lambda(t),$$

$$\tag{4}$$

$$\Psi(T)(x, y) := A_T(x, y) := \int\limits_{[0,x]} (T\mathbf{1}_{[0,y]})(t)d\lambda(t)$$

for every $f \in L^1([0, 1])$ and $(x, y) \in [0, 1]^2$ ($A_{,2}$ denoting the partial derivative of A w.r.t. y), fulfill $\Psi \circ \Phi = id_{\mathscr{C}}$ and $\Phi \circ \Psi = id_{\mathscr{M}}$. Note that in case of $f := \mathbf{1}_{[0,y]}$ we have $(T_A \mathbf{1}_{[0,y]})(x) = A_{,1}(x, y)$ λ-a.s. According to [29] the first equality in (4) can be simplified to

$$(T_A f)(x) = \mathbb{E}(f \circ Y | X = x) = \int_{[0,1]} f(y) K_A(x, dy) \qquad \lambda\text{-a.s.} \qquad (5)$$

It is not difficult to show that the uniform metric d_∞ is a metrization of the weak operator topology on $\mathcal{M}$ (see [23]).

3 Some Consequences of the Markov Kernel Approach

In this section, we give a quick survey showing the usefulness of the Markov kernel perspective of two-dimensional copulas.

3.1 Strong Metrics on $\mathcal{C}$

Expressing copulas in terms of their corresponding Markov kernels, the metrics D_1, D_2, D_∞ on $\mathcal{C}$ can be defined as follows:

$$D_1(A, B) := \int_{[0,1]} \int_{[0,1]} \left| K_A(x, [0, y]) - K_B(x, [0, y]) \right| d\lambda(x) \, d\lambda(y) \qquad (6)$$

$$D_2^2(A, B) := \int_{[0,1]} \int_{[0,1]} \left| K_A(x, [0, y]) - K_B(x, [0, y]) \right|^2 d\lambda(x) \, d\lambda(y) \qquad (7)$$

$$D_\infty(A, B) := \sup_{y \in [0,1]} \int_{[0,1]} \left| K_A(x, [0, y]) - K_B(x, [0, y]) \right|^2 d\lambda(x) \qquad (8)$$

The following two theorems state the most important properties of the metrics D_1, D_2 and D_∞.

Theorem 1 ([29]) *Suppose that $A, A_1, A_2, \ldots$ are copulas and let $T, T_1, T_2, \ldots$ denote the corresponding Markov operators. Then the following four conditions are equivalent:*

(a) $\lim_{n \to \infty} D_1(A_n, A) = 0$
(b) $\lim_{n \to \infty} D_\infty(A_n, A) = 0$
(c) $\lim_{n \to \infty} \| T_n f - T f \|_1 = 0$ *for every $f \in L^1([0, 1])$*
(d) $\lim_{n \to \infty} D_2(A_n, A) = 0$

As a consequence, each of the three metrics D_1, D_2 and D_∞ is a metrization of the strong operator topology on $\mathcal{M}$.

Theorem 2 ([29]) *The metric space* $(\mathscr{C}, D_1)$ *is complete and separable. The same holds for* $(\mathscr{C}, D_2)$ *and* $(\mathscr{C}, D_\infty)$. *The topology induced on* $\mathscr{C}$ *by* D_1 *is strictly finer than the one induced by* d_∞.

Remark 3 The idea of constructing metrics via conditioning to the first coordinate can be easily extended to the family $\mathscr{C}^m$ of all m-dimensional copulas for arbitrary $m \geq 3$. For instance, the multivariate version of D_1 on $\mathscr{C}^m$ can be defined by

$$D_1(A, B) = \int_{[0,1]^{m-1}} \int_{[0,1]} |K_A(x, [\mathbf{0}, \mathbf{y}]) - K_B(x, [\mathbf{0}, \mathbf{y}])| d\lambda(x) d\lambda^{m-1}(\mathbf{y}),$$

whereby $[\mathbf{0}, \mathbf{y}] = \times_{i=1}^{m-1}[0, y_i]$ and $K_A(K_B)$ denotes the Markov kernel (regular conditional distribution) of $\mathbf{Y}$ given X for $(X, \mathbf{Y}) \sim A(B)$. As shown in [11], the resulting metric spaces $(\mathscr{C}^m, D_1)$, $(\mathscr{C}^m, D_2)$, $(\mathscr{C}^m, D_\infty)$ are again complete and separable.

3.2 Induced Dependence Measures

The main motivation for the consideration of conditioning-based metrics like D_1 was the need for a metric that, contrary to d_∞, is capable of distinguishing extreme types of statistical dependence, i.e., independence and complete dependence. For the uniform metric d_∞, it is straightforward to construct sequences $(C_{h_n})_{n \in \mathbb{N}}$ of completely dependent copulas (in fact, even sequences of shuffles of M, see [9, 22]) fulfilling $\lim_{n \to \infty} d_\infty(C_{h_n}, \Pi) = 0$—for D_1, however, the following result holds:

Theorem 4 ([29]) *For every* $A \in \mathscr{C}$ *we have* $D_1(A, \Pi) \leq 1/3$. *Furthermore, equality* $D_1(A, \Pi) = 1/3$ *holds if and only if* $A \in \mathscr{C}_d$.

As a straightforward consequence, we may define $\tau_1 : \mathscr{C} \to [0, 1]$ by

$$\tau_1(A) := 3D_1(A, \Pi). \tag{9}$$

This dependence measure τ_1 exhibits the seemingly natural properties that (i) exactly members of the family $\mathscr{C}_d$ (describing complete dependence) are assigned maximum dependence (equal to one) and (ii) Π is the only copula with minimum dependence (equal to zero). Note that (i) means that $\tau_1(A)$ is maximal if and only if A describes the situation of full predictability, i.e., asset Y is a deterministic function of asset X. In particular, all shuffles of M have maximum dependence. Dependence measures based on the metric D_2 may be constructed analogously.

Example 5 For the Farlie-Gumbel-Morgenstern family $(G_\theta) \in [-1, 1]$ of copulas (see [22]), given by

$$G_\theta(x, y) = xy + \theta xy(1 - x)(1 - y), \tag{10}$$

it is straightforward to show that $\tau_1(G_\theta) = \frac{|\theta|}{4}$ holds for every $\theta \in [-1, 1]$ (for details see [29]).

Example 6 For the Marshall-Olkin family $(M_{\alpha,\beta})_{(\alpha,\beta)\in[0,1]^2}$ of copulas (see [22]), given by

$$M_{\alpha,\beta}(x, y) = \begin{cases} x^{1-\alpha} y & \text{if } x^\alpha \geq y^\beta \\ xy^{1-\beta} & \text{if } x^\alpha \leq y^\beta. \end{cases} \tag{11}$$

it can be shown that

$$\zeta_1(M_{\alpha,\beta}) = 3\alpha (1 - \alpha)^z + \frac{6}{\beta} \frac{1 - (1 - \alpha)^z}{z} - \frac{6}{\beta} \frac{1 - (1 - \alpha)^{z+1}}{z + 1} \tag{12}$$

holds, whereby $z = \frac{1}{\alpha} + \frac{2}{\beta} - 1$ (for details again see [29]).

Remark 7 The dependence measure τ_1 is nonmutual, i.e., we do not necessarily have $\tau_1(A) = \tau_1(A^t)$, whereby A^t denotes the transpose of A (i.e., $A^t(x, y) = A(y, x)$). This reflects the fact that the dependence structure of random variables might be strongly asymmetric, see [29] for examples as well as [27] for a measure of mutual dependence.

Remark 8 Since most properties of D_1 in dimension two also hold in the general m-dimensional setting it might seem natural to simply consider $\tau_1(A) := aD_1(A, \Pi)$ as dependence measure on $\mathscr{C}^m$ (a being a normalizing constant). It is, however, straightforward to see that this yields no reasonable notion of a dependence quantification in so far that we would also have $\tau_1(A) > 0$ for copulas A describing independence of X and $\mathbf{Y} = (Y_1, \ldots, Y_{m-1})$. For a possible way to overcome this problem and assign copulas describing the situation in which each component of a portfolio $(Y_1, \ldots, Y_{m-1})$ is a deterministic function of another asset X maximum dependence we refer to [11].

Remark 9 It is straightforward to verify that for samples $(X_1, Y_1), \ldots, (X_n, Y_n)$ from $A \in \mathscr{C}$ the empirical copula $\hat{E}_n$ (see [22, 28]) cannot converge to A w.r.t. D_1 unless we have $A \in \mathscr{C}_d$. Using Bernstein or checkerboard aggregations (smoothing the empirical copula) might make it possible to construct D_1-consistent estimators of $\tau_1(A)$. Convergence rates of these aggregations and other related questions are future work.

3.3 The IFS Construction of (Very) Singular Copulas

Using Iterated Function Systems, one can construct copulas exhibiting surprisingly irregular analytic behavior. The aim of this section is to sketch the construction and then state two main results. For general background on Iterated Function Systems with Probabilities (IFSP, for short), we refer to [16]. The IFSP construction of two-dimensional copulas with fractal support goes back to [12] (also see [2]), for the generalization to the multivariate setting we refer to [30].

Definition 10 ([12]) A $n \times m$-matrix $\tau = (t_{ij})_{i=1,\ldots,n,\, j=1,\ldots,m}$ is called *transformation matrix* if it fulfills the following four conditions: (i) $\max(n, m) \geq 2$, (ii) all entries are non-negative, (iii) $\sum_{i,j} t_{ij} = 1$, and (iv) no row or column has all entries 0. $\mathfrak{T}$ will denote the family of all transformations matrices.

Given $\tau \in \mathfrak{T}$ define the vectors $(a_j)_{j=0}^m$, $(b_i)_{i=0}^n$ of cumulative column and row sums by $a_0 = b_0 = 0$ and

$$a_j = \sum_{j_0 \leq j} \sum_{i=1}^n t_{ij_0} \quad j \in \{1, \ldots, m\}, \qquad b_i = \sum_{i_0 \leq i} \sum_{j=1}^m t_{i_0 j} \quad i \in \{1, \ldots, n\}. \quad (13)$$

Since τ is a transformation matrix both $(a_j)_{j=0}^m$ and $(b_i)_{i=0}^n$ are strictly increasing and $R_{ji} := [a_{j-1}, a_j] \times [b_{i-1}, b_i]$ is a compact rectangle with nonempty interior for all $j \in \{1, \ldots, m\}$ and $i \in \{1, \ldots, n\}$. Set $\tilde{I} := \{(i, j) : t_{ij} > 0\}$ and consider the IFSP $\{[0, 1]^2, (f_{ji})_{(i,j) \in \tilde{I}}, (t_{ij})_{(i,j) \in \tilde{I}}\}$, whereby the affine contraction $f_{ji} : [0, 1]^2 \to R_{ji}$ is given by

$$f_{ji}(x, y) = \left(a_{j-1} + x(a_j - a_{j-1}),\, b_{i-1} + y(b_i - b_{i-1})\right). \quad (14)$$

$Z_\tau^* \in \mathcal{K}([0, 1]^2)$ will denote the attractor of the IFSP (see [16]). The induced operator $\mathcal{V}_\tau$ on $\mathcal{P}([0, 1]^2)$ is defined by

$$\mathcal{V}_\tau(\mu) := \sum_{j=1}^m \sum_{i=1}^n t_{ij}\, \mu^{f_{ji}} = \sum_{(i,j) \in \tilde{I}} t_{ij}\, \mu^{f_{ji}}. \quad (15)$$

It is straightforward to see that $\mathcal{V}_\tau$ maps $\mathcal{P}_{\mathscr{C}}$ into itself so we may view $\mathcal{V}_\tau$ also as operator on $\mathscr{C}$. According to [12] there is exactly one copula $A_\tau^* \in \mathscr{C}$, to which we will refer to as *invariant copula*, such that $\mathcal{V}_\tau(\mu_{A_\tau^*}) = \mu_{A_\tau^*}$ holds. The IFSP construction also converges w.r.t. D_1—the following result holds:

Theorem 11 ([29]) *Let $\tau \in \mathfrak{T}$ be a transformation matrix. Then V_τ is a contraction on the metric space $(\mathscr{C}, D_1)$ and there exists a unique copula A_τ^* such that $V_\tau A_\tau^* = A_\tau^*$ and for every $B \in \mathscr{C}$ we have $\lim_{n \to \infty} D_1(V_\tau^n B, A_\tau^*) = 0$.*

Example 12 Figure 1 depicts the density of $\mathcal{V}_\tau^n(\Pi)$ for $n \in \{1, 2, 3, 5\}$, whereby τ is given by

$$\tau = \begin{pmatrix} \frac{1}{6} & 0 & \frac{1}{6} \\ 0 & \frac{1}{3} & 0 \\ \frac{1}{6} & 0 & \frac{1}{6} \end{pmatrix}.$$

Moreover (again see [12]) the support $Supp(\mu_{A_\tau^*})$ of $\mu_{A_\tau^*}$ fulfills $\lambda_2(Supp(\mu_{A_\tau^*})) = 0$ if τ contains at least one zero. Hence, in this case, $\mu_{A_\tau^*}$ is singular w.r.t. the Lebesgue measure λ_2, we write $\mu_{A_\tau^*} \perp \lambda_2$. On the other hand, if τ contains no zeros we may

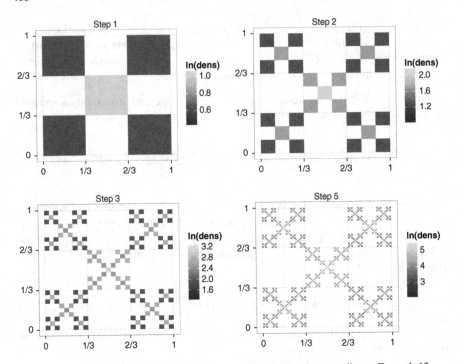

Fig. 1 Image plot of the density of $\mathscr{V}_\tau^n(\Pi)$ for $n \in \{1, 2, 3, 5\}$ and τ according to Example 12

still have $\mu_{A_\tau^\star} \perp \lambda_2$ although in this case $\mu_{A_\tau^\star}$ has full support $[0, 1]^2$. In fact, an even stronger and quite surprising singularity result holds—letting $\hat{\mathfrak{T}}$ denote the family of all transformation matrices τ (i) containing no zeros, (ii) fulfilling that the row sums and column sums through every t_{ij} are identical, and (iii) $\mu_{A_\tau^\star} \neq \lambda_2$ we have the following striking result:

Theorem 13 ([33]) *Suppose that $\tau \in \hat{\mathfrak{T}}$. Then the corresponding invariant copula $A_\tau^\star$ is singular w.r.t. λ_2 and has full support $[0, 1]^2$. Moreover, for λ-almost every $x \in [0, 1]$ the conditional distribution function $y \mapsto F_x^{A_\tau^\star}(y) = K_{A_\tau^\star}(x, [0, y])$ is continuous, strictly increasing and has derivative zero λ-almost everywhere.*

3.4 The Star Product of Copulas

Given $A, B \in \mathscr{C}$ the *star product* $A * B \in \mathscr{C}$ is defined by (see [3, 23])

$$(A * B)(x, y) := \int_{[0,1]} A_{,2}(x, t) B_{,1}(t, y) d\lambda(t) \tag{16}$$

and fulfills $T_{A*B} = \Phi_{A*B} = \Phi(A) \circ \Phi(B) = T_A \circ T_B$, so the mapping Φ in equation (4) actually is an isomorphism. A copula $A \in \mathscr{C}$ is called *idempotent* if $A * A = A$ holds, the family of all idempotent copulas will be denoted by $\mathscr{C}^{ip}$. For a complete characterization of idempotent copulas we refer to [4] (also see [26]). The star product can easily be translated to the Markov kernel setting—the following result holds:

Lemma 14 ([30]) *Suppose that $A, B \in \mathscr{C}$ and let K_A, K_B denote Markov kernels of A and B. Then the Markov kernel $K_A \circ K_B$, defined by*

$$(K_A \circ K_B)(x, F) := \int_{[0,1]} K_B(y, F) K_A(x, dy), \tag{17}$$

is a Markov kernel of $A * B$. Furthermore $\mathscr{C}^{ip}$ is closed in $(\mathscr{C}, D_1)$.

Remark 15 Let $A \in \mathscr{C}$ be arbitrary. If $(X_n)_{n \in \mathbb{N}}$ is a stationary Markov process on $[0, 1]$ with (stationary) transition probability $K_A(\cdot, \cdot)$ and $X_1 \sim \mathscr{U}(0, 1)$ then $(X_n, X_{n+1}) \sim A$ for every $n \in \mathbb{N}$ and Lemma 14 implies that $(X_1, X_{n+1}) \sim A * A * \cdots * A =: A^{*n}$, i.e., the n-step transition probability of the process is given by the Markov kernel of A^{*n}.

Remark 16 In case the copulas A, B are absolutely continuous with densities k_A and k_B it is straightforward to verify that $A * B$ is absolutely continuous with density k_{A*B} given by

$$k_{A*B}(x, y) = \int_{[0,1]} k_A(x, z) k_B(z, y) d\lambda(z). \tag{18}$$

Since the star product of copulas is a natural generalization of the multiplication of doubly stochastic matrices and doubly stochastic idempotent matrices are fully characterizable (see [10, 25]) the following result underlines how much more complex the family of idempotent copulas is (also see [12] for the original result without idempotence).

Theorem 17 ([30]) *For every $s \in (1, 2)$ there exists a transformation matrix $\tau_s \in \mathfrak{T}$ such that:*

1. *The invariant copula $A^\star_{\tau_s}$ is idempotent.*
2. *The Hausdorff dimension of the support of $A^\star_{\tau_s}$ is s.*

Example 18 For the transformation matrix τ from Example 12 the invariant copula $A^\star_\tau$ is idempotent and its support has Hausdorff dimension $\ln 5 / \ln 3$. Hence, setting $A := A^\star_\tau$ and considering the Markov process outlined in Remark 15 we have $(X_i, X_{i+n}) \sim A$ for all $i, n \in \mathbb{N}$. The same holds if we take $A := \mathscr{V}^j_\tau(\Pi)$ for arbitrary $j \in \mathbb{N}$ since this A is idempotent too.

We conclude this section with a general result that will be used later on and which, essentially, follows from Von Neumanns mean ergodic theorem for Hilbert spaces (see [24]) since Markov operators have operator norm one. For every copula $A \in \mathscr{C}$ and every $n \in \mathbb{N}$ as in the Introduction we set

$$s_{*n}(A) = \frac{1}{n} \sum_{i=1}^{n} A^{*i}. \qquad (19)$$

Theorem 19 ([32]) *For every copula A there exists a copula $\hat{A}$ such that*

$$\lim_{n \to \infty} D_1 \big(s_{*n}(A), \hat{A} \big) = 0. \qquad (20)$$

This copula $\hat{A}$ is idempotent, symmetric, and fulfills $\hat{A} * A = A * \hat{A} = \hat{A}$.

As nice by-product, Theorem 19 also offers a very simple proof of the fact that idempotent copulas are necessarily symmetric (originally proved in [4]).

4 Copulas Whose Corresponding Markov Operator Is Quasi-constrictive

Studying asymptotic properties of Markov operators quasi-constrictiveness is a very important concept. To the best of the authors' knowledge, there is no natural/simple characterization of copulas whose Markov operator is quasi-constrictive. The objective of this section, however, is to show that the D_1-limit $\hat{A}$ of $s_{*n}(A)$ has a very simple form if T_A is quasi-constrictive. We start with a definition of quasi-constrictiveness in the general setting. In general, T is a Markov operator on $L^1(\Omega, \mathscr{A}, \mu)$ if the conditions (M1)-(M3) from Sect. 2 with [0, 1] replaced by Ω, $\mathscr{B}([0, 1])$ replaced by $\mathscr{A}$, and λ replaced by μ hold.

Definition 20 ([1, 15, 18]) Suppose that $(\Omega, \mathscr{A}, \mu)$ is a finite measure space and let $\mathscr{D}(\Omega, \mathscr{A}, \mu)$ denote the family of all probability densities w.r.t. μ. Then a Markov operator $T : L^1(\Omega, \mathscr{A}, \mu) \to L^1(\Omega, \mathscr{A}, \mu)$ is called *quasi-constrictive* if there exist constants $\delta > 0$ and $\kappa < 1$ such that for every probability density $f \in \mathscr{D}(\Omega, \mathscr{A}, \mu)$ the following inequality is fulfilled:

$$\limsup_{n \to \infty} \int_E T^n f(x) d\mu(x) \le \kappa \quad \text{for every } E \in \mathscr{A} \text{ with } \mu(E) \le \delta \qquad (21)$$

Komornik and Lasota (see [15]) have shown in 1987 that quasi-constrictivity is equivalent to *asymptotic periodicity*—in particular they proved the following *spectral decomposition theorem*: For every quasi-constrictive Markov operator T there exist an integer $r \geq 1$, densities $g_1, \ldots, g_r \in \mathscr{D}(\Omega, \mathscr{A}, \mu)$ with pairwise disjoint support, essentially bounded non-negative functions $h_1, \ldots, h_r \in L^\infty(\Omega, \mathscr{A}, \mu)$ and a permutation σ of $\{1, \ldots, r\}$ such that for every $f \in L^1(\Omega, \mathscr{A}, \mu)$

$$T^n f(x) = \sum_{i=1}^{r} \left(\int_\Omega f h_i d\mu \right) g_{\sigma^n(i)}(x) + R_n f(x) \quad \text{with} \quad \lim_{n \to \infty} \|R_n f\|_1 = 0 \quad (22)$$

holds. Furthermore (see again [1, 15, 18]), in case of $\mu(\Omega) = 1$ there exists a measurable partition $(E_i)_{i=1}^{r}$ of Ω in sets with positive measure such that g_j and σ in (22) fulfill

$$g_j = \frac{1}{\mu(E_j)} 1_{E_j} \quad \text{and} \quad \mu(E_j) = \mu(E_{\sigma^n(j)}) \quad (23)$$

for every $j \in \{1, \ldots, r\}$ and every $n \in \mathbb{N}$.

Example 21 For every absolutely continuous copula A with density k_A fulfilling $k_A \leq M$ the corresponding Markov operator is quasi-constrictive. This directly follows from the fact that

$$T_A f(x) = \int_{[0,1]} f(y) K_A(x, dy) = \int_{[0,1]} f(y) k_A(x, y) dy \leq M$$

holds for every $f \in \mathscr{D}([0, 1]) := \mathscr{D}([0, 1], \mathscr{B}([0, 1]), \lambda)$.

Example 22 There are absolutely continuous copulas A whose corresponding Markov operator is not quasi-constrictive—one example is the idempotent ordinal-sum-like copula O with unbounded density k_O defined by

$$k_O(x, y) := \sum_{n=1}^{\infty} 2^n 1_{[1-2^{1-n}, 1-2^{-n})}(x, y)$$

for all $x, y \in [0, 1]$ (straightforward to verify).

Before returning to the copula setting we prove a first proposition to the spectral decomposition that holds for general Markov operators on $L^1(\Omega, \mathscr{A}, \mu)$ with $(\Omega, \mathscr{A}, \mu)$ being a probability space.

Lemma 23 *Suppose that $(\Omega, \mathscr{A}, \mu)$ is a probability space and that $T : L^1(\Omega, \mathscr{A}, \mu)$ $\to L^1(\Omega, \mathscr{A}, \mu)$ is a quasi-constrictive Markov operator. Then there exists $r \geq 1$, a measurable partition $(E_i)_{i=1}^{r}$ of Ω in sets with positive measure, densities $h'_1, \ldots, h'_r \in L^\infty(\Omega, \mathscr{A}, \mu) \cap \mathscr{D}(\Omega, \mathscr{A}, \mu)$ and a permutation σ of $\{1, \ldots, r\}$ such that we have $\sum_{i=1}^{r} \mu(E_i) h'_i = 1$ as well as*

$$T^n f(x) = \sum_{i=1}^{r} \left(\int_\Omega f h'_i d\mu \right) 1_{E_{\sigma^n(i)}}(x) + R_n f(x) \quad \text{with} \quad \lim_{n \to \infty} \|R_n f\|_1 = 0 \quad (24)$$

for every $f \in L^1(\Omega, \mathscr{A}, \mu)$ and every $n \in \mathbb{N}$.

Proof Using (22) and (23) it follows that

$$R_n \mathbf{1}_\Omega(x) = 1 - T^n \mathbf{1}_\Omega(x) = \sum_{i=1}^{r} \mathbf{1}_{E_{\sigma^n(i)}}(x) - \sum_{i=1}^{r} \frac{\|h_i\|_1}{\mu(E_i)} \mathbf{1}_{E_{\sigma^n(i)}}(x)$$

$$= \sum_{i=1}^{r} \left(1 - \frac{\|h_i\|_1}{\mu(E_i)}\right) \mathbf{1}_{E_{\sigma^n(i)}}(x)$$

for every $x \in \Omega$, which implies

$$0 = \lim_{n \to \infty} \|R_n \mathbf{1}_\Omega\|_1 = \lim_{n \to \infty} \sum_{i=1}^{r} \left|1 - \frac{\|h_i\|_1}{\mu(E_i)}\right| \mu(E_{\sigma^n(i)}) = \sum_{i=1}^{r} \left|1 - \frac{\|h_i\|_1}{\mu(E_i)}\right| \mu(E_i).$$

Since $\mu(E_i) > 0$ for every $i \in \{1, \ldots, r\}$ this shows that $h_i' := \frac{h_i}{\mu(E_i)} \in L^\infty(\Omega, \mathscr{A}, \mu) \cap \mathscr{D}(\Omega, \mathscr{A}, \mu)$ for every $i \in \{1, \ldots, r\}$. Furthermore we have $\lim_{n \to \infty} \|R_n h_i'\|_1 = 0$ for every fixed i, from which

$$1 = \int_\Omega T^n h_i'(x) d\mu(x) = \lim_{n \to \infty} \int_\Omega \sum_{j=1}^{r} \left(\int_\Omega h_i'(z) h_j'(z) d\mu(z)\right) \mathbf{1}_{E_{\sigma^n(j)}}(x) d\mu(x)$$

$$= \sum_{j=1}^{r} \left(\int_\Omega h_i'(z) h_j'(z) d\mu(z)\right) \mu(E_j)$$

follows. Multiplying both sides with $\mu(E_i)$, summing over $i \in \{1, \ldots, r\}$ yields

$$1 = \int_\Omega \underbrace{\sum_{i=1}^{r} h_i'(z) \mu(E_i) \sum_{j=1}^{r} h_j'(z) \mu(E_j)}_{:=g(z)} d\mu(z)$$

so $g \in \mathscr{D}(\Omega, \mathscr{A}, \mu)$ and at the same time $g^2 \in \mathscr{D}(\Omega, \mathscr{A}, \mu)$. Using Cauchy Schwarz inequality it follows that $g(x) = 1$ for μ-almost every $x \in \Omega$. □

Lemma 24 *Suppose that A is a copula whose corresponding Markov operator T_A is quasi-constrictive. Then there exists $r \geq 1$, a measurable partition $(E_i)_{i=1}^{r}$ of $[0, 1]$ in sets with positive measure, and pairwise different densities $h_1, \ldots, h_r \in L^\infty([0, 1]) \cap \mathscr{D}([0, 1])$ such that the limit copula $\hat{A}$ of $s_{*n}(A)$ is absolutely continuous with density $k_{\hat{A}}$, defined by*

$$k_{\hat{A}}(x, y) = \sum_{i=1}^{r} h_i(y) \mathbf{1}_{E_i}(x) \tag{25}$$

for all $x, y \in [0, 1]$.

Proof Fix an arbitrary $f \in L^1([0, 1])$. Then, using Lemma 23, we have

$$\frac{1}{n}\sum_{j=1}^{n} T_A^j f(x) = \frac{1}{n}\sum_{j=1}^{n}\sum_{i=1}^{r} \int_{[0,1]} f h'_{\sigma^{-j}(i)} d\lambda \mathbf{1}_{E_i}(x) + \frac{1}{n}\sum_{j=1}^{n} R_j f(x)$$

$$= \sum_{i=1}^{r} \mathbf{1}_{E_i}(x) \int_{[0,1]} f(z) \underbrace{\frac{1}{n}\sum_{j=1}^{n} h'_{\sigma^{-j}(i)}(z)\, d\lambda(z)}_{:=g_n^i(z)} + \frac{1}{n}\sum_{j=1}^{n} R_j f(x)$$

for every $x \in [0, 1]$ and every $n \in \mathbb{N}$. Since σ is a permutation $j \to h'_{\sigma^{-j}(i)}(z)$ is periodic for every z and every i, so, for every $i \in \{1, \dots, r\}$, there exists a function h_i such that

$$\lim_{n\to\infty} \frac{1}{n}\sum_{j=1}^{n} h'_{\sigma^{-j}(i)}(z) = h_i(z)$$

for every $z \in [0, 1]$ and every $i \in \{1, \dots, r\}$. Obviously $h_i \in L^\infty([0, 1])$ and, using Lebesgue's theorem on dominated convergence, h_i is also a density, so we have $h_1, \dots, h_r \in L^\infty([0, 1]) \cap \mathscr{D}([0, 1])$. Finally, using Theorem 19 and the fact that $\lim_{n\to\infty} \|R_n f\|_1 = 0$ for every $f \in L^1([0, 1])$, it follows immediately that

$$T_{\hat{A}} f(x) = \int_{[0,1]} f(y) \sum_{i=1}^{r} h_i(y) \mathbf{1}_{E_i}(x) d\lambda(y).$$

This completes the proof since mutually different densities can easily be achieved by building unions from elements in the partition $(E_i)_{i=1}^{r}$ if necessary. $\square$

Using the fact that $\hat{A}$ is idempotent we get the following stronger result:

Lemma 25 *The density $k_{\hat{A}}$ of $\hat{A}$ in Lemma 24 has the form*

$$k_{\hat{A}}(x, y) = \sum_{i,j=1}^{r} m_{i,j} \mathbf{1}_{E_i \times E_j}(x, y),$$

i.e., it is constant on all rectangles $E_i \times E_j$.

Proof According to Theorem 19 the copula $\hat{A}$ is idempotent so $\hat{A}$ is symmetric. Consequently the set

$$\Delta := \{(x, y) \in [0, 1]^2 : k_{\hat{A}}(x, y) = k_{\hat{A}}(y, x)\} \in \mathscr{B}([0, 1]^2)$$

has full measure $\lambda_2(\Delta) = 1$. Using Lemma 24 we have

$$\sum_{i=1}^{r} h_i(y)\mathbf{1}_{E_i}(x) = \sum_{i=1}^{r} h_i(x)\mathbf{1}_{E_i}(y)$$

for every $(x, y) \in \Delta$. Fix arbitrary $i, j \in \{1, \ldots, r\}$. Then we can find $x \in E_i$ such that $\lambda(\Delta_x) = 1$ holds, whereby $\Delta_x = \{y \in [0, 1] : (x, y) \in \Delta\}$. For such x we have $h_i(y) = h_j(x)$ for λ-almost every $y \in E_j$, which firstly implies that h_j is, up to a set of measure zero, constant on E_j and, secondly, that $k_{\hat{A}}$ is constant on $E_i \times E_j$ outside a set of λ_2-measure zero. Since we may modify the density on a set of λ_2-measure zero we can assume that $k_{\hat{A}}$ is of the desired form

$$k_{\hat{A}}(x, y) = \sum_{i,j=1}^{r} m_{i,j}\mathbf{1}_{E_i \times E_j}(x, y),$$

with $M = (m_{i,j})_{i,j=1}^{r}$ being a non-negative, symmetric matrix fulfilling

(a) $\sum_{i,j=1}^{r} m_{i,j}\lambda(E_i)\lambda(E_j) = 1$
(b) $\sum_{j=1}^{r} m_{i,j}\lambda(E_j) = 1$ for every $i \in \{1, \ldots, r\}$
(c) $\sum_{i=1}^{r} m_{i,j}\lambda(E_i) = 1$ for every $j \in \{1, \ldots, r\}$
(d) $\sum_{i=1}^{r} |m_{i,j} - m_{i,l}| > 0$ whenever $j \neq l$. $\qquad\square$

Before proceeding with the final result it is convenient to take a look at the matrix $H = (H_{i,j})_{i,j=1}^{r}$ defined by

$$H_{i,j} := m_{i,j}\lambda(E_j) = \int_{E_j} h_i(z)d\lambda(z) \qquad (26)$$

for all $i, j \in \{1, \ldots, r\}$. According to (a) in the proof of Lemma 25 H is stochastic. Furthermore, idempotence of $\hat{A}$ and Remark 16 imply $k_{\hat{A}} * k_{\hat{A}} = k_{\hat{A}}$, hence

$$\sum_{i=1}^{r} h_i(y)\mathbf{1}_{E_i}(x) = k_{\hat{A}}(x, y) = k_{\hat{A}} * k_{\hat{A}}(x, y)$$

$$= \int_{[0,1]} \sum_{i=1}^{r} h_i(z)\mathbf{1}_{E_i}(x) \sum_{j=1}^{r} h_j(y)\mathbf{1}_{E_j}(z)\, d\lambda(z)$$

$$= \sum_{i,j=1}^{r} \mathbf{1}_{E_i}(x)h_j(y) \int_{E_j} h_i(z)d\lambda(z) = \sum_{i,j=1}^{r} \mathbf{1}_{E_i}(x)h_j(y)H_{i,j}.$$

From this is follows immediately that $h_i(y) = \sum_{j=1}^{r} H_{i,j}h_j(y)$ is fulfilled for every $y \in [0, 1]$ and $i \in \{1, \ldots, r\}$, so, integrating both sides over E_l, we have $H_{i,l} =$

$\sum_{j=1}^{r} H_{i,j} H_{j,l}$, which shows that H is idempotent. Having this, the proof of the following main result of this section will be straightforward.

Theorem 26 *Suppose that A is a copula whose corresponding Markov operator T_A is quasi-constrictive. Then there exist $r \geq 1$ and a measurable partition $(E_i)_{i=1}^{r}$ of $[0, 1]$ in sets with positive measure, such that the limit copula $\hat{A}$ of $s_{*n}(A)$ is absolutely continuous with density $k_{\hat{A}}$ given by*

$$k_{\hat{A}}(x, y) = \sum_{i=1}^{r} \frac{1}{\lambda(E_i)} 1_{E_i \times E_i}(x, y) \tag{27}$$

for all $x, y \in [0, 1]$. In other words, the limit copula $\hat{A}$ has an ordinal-sum-of-Π-like structure.

Proof Since H is an idempotent stochastic matrix and since H can not have any column consisting purely of zeros, up to a permutation, H must have the form (see [5, 21]).

$$\begin{pmatrix} Q_1 & 0 & \cdots & 0 \\ 0 & Q_2 & \cdots & 0 \\ \vdots & \vdots & \ddots & \vdots \\ 0 & 0 & \cdots & Q_s \end{pmatrix}, \tag{28}$$

whereby each Q_i is a strictly positive $r_i \times r_i$-matrix with identical rows and s is the range of H. We will show that $r_i = 1$ for every $i \in \{1, \ldots, s\}$. Suppose, on the contrary, that $r_l \geq 2$ for some l. Then there would be indices $I_l := \{i_1, \ldots, i_{r_l}\} \subseteq \{1, \ldots, r\}$ and $a_1, \ldots, a_{r_l} \in (0, 1)^{r_l}$ with $\sum_{i=1}^{r_l} a_i = 1$ such that Q_l would have the form

$$Q_l = \begin{pmatrix} a_1 & a_2 & \cdots & a_{r_l} \\ a_1 & a_2 & \cdots & a_{r_l} \\ \vdots & \vdots & \ddots & \vdots \\ a_1 & a_2 & \cdots & a_{r_l} \end{pmatrix} = \begin{pmatrix} H_{i_1,i_1} & H_{i_1,i_2} & \cdots & H_{i_1,i_{r_l}} \\ H_{i_2,i_1} & H_{i_2,i_2} & \cdots & H_{i_2,i_{r_l}} \\ \vdots & \vdots & \ddots & \vdots \\ H_{i_{r_l},i_1} & H_{i_{r_l},i_2} & \cdots & H_{i_{r_l},i_{r_l}} \end{pmatrix}. \tag{29}$$

It follows immediately that

$$H_{i_1,i_v} = m_{i_1,i_1}\lambda(E_{i_v}) = H_{i_2,i_v} = m_{i_2,i_v}\lambda(E_{i_v}) = \cdots = H_{i_{r_l},i_v} = m_{i_{r_l},i_v}\lambda(E_{i_v}),$$

so $m_{i_j,i_v} = m_{i_1,i_v}$ for every $j \in \{1, \ldots, r_l\}$ and arbitrary $v \in \{1, \ldots, r_l\}$. Having this symmetry of M implies that all entries of Q_l are identical, which contradicts the fact that the conditional densities are not identical, i.e., the fact that

$$\sum_{j \in I_l} |m_{j,i_1} - m_{j,i_2}| = \sum_{j=1}^{r} |m_{j,i_1} - m_{j,i_2}| > 0$$

whenever $i_1 \neq i_2$. Consequently $r_i = 1$ for every $i \in \{1, \ldots, s\}$ and $k_{\hat{A}}$ has the desired form. $\square$

Remark 27 Consider again the transformation matrix τ from Example 12. Then $\mathscr{V}_\tau^1(\Pi), \mathscr{V}_\tau^2(\Pi), \ldots$ are examples of the ordinal-sum-of-Π-like copulas mentioned in the last theorem.

Acknowledgments The second author acknowledges the support of the Ministerio de Ciencia e Innovación (Spain) under research project MTM2011-22394.

Open Access This chapter is distributed under the terms of the Creative Commons Attribution Noncommercial License, which permits any noncommercial use, distribution, and reproduction in any medium, provided the original author(s) and source are credited.

References

1. Bartoszek, W.: The work of professor Andrzej Lasota on asymptotic stability and recent progress. Opusc. Math. **28**(4), 395–413 (2008)
2. Cuculescu, I., Theodorescu, R.: Copulas: diagonals, tracks. Rev. Roum. Math Pure A. **46**, 731–742 (2001)
3. Darsow, W.F., Nguyen, B., Olsen, E.T.: Copulas and Markov processes. Ill. J. Math. **36**(4), 600–642 (1992)
4. Darsow, W.F., Olsen, E.T.: Characterization of idempotent 2-copulas. Note Mat. **30**(1), 147–177 (2010)
5. Doob, J.: Topics in the theory of Markov chains. Trans. Am. Math. Soc. **52**, 37–64 (1942)
6. Durante, F., Klement, E.P., Quesada-Molina, J., Sarkoci, J.: Remarks on two product-like constructions for copulas. Kybernetika **43**(2), 235–244 (2007)
7. Durante, F., Sarkoci, P., Sempi, C.: Shuffles of copulas. J. Math. Anal. Appl. **352**, 914–921 (2009)
8. Durante, F., Sempi, C.: Copula theory: an introduction. In: Jaworski, P., Durante, F., Härdle, W., Rychlik, T. (eds.) Copula Theory and Its Applications. Lecture Notes in Statistics-Proceedings, vol. 198, pp. 1–31. Springer, Berlin (2010)
9. Durante, F., Fernńdez-Sánchez, J.: Multivariate shuffles and approximation of copulas. Stat. Probab. Lett. **80**, 1827–1834 (2010)
10. Farahat, H.K.: The semigroup of doubly-stochastic matrices. Proc. Glasg. Math. Ass. **7**, 178–183 (1966)
11. Fernández-Sánchez, J., Trutschnig, W.: Conditioning based metrics on the space of multivariate copulas, their interrelation with uniform and levelwise convergence and Iterated Function Systems. to appear in J. Theor. Probab. doi:10.1007/s10959-014-0541-4
12. Fredricks, G.A., Nelsen, R.B., Rodríguez-Lallena, J.A.: Copulas with fractal supports. Insur. Math. Econ. **37**, 42–48 (2005)
13. Kallenberg, O.: Foundations of Modern Probability. Springer, New York (1997)
14. Klenke, A.: Probability Theory—A Comprehensive Course. Springer, Berlin (2007)
15. Komornik, J., Lasota, A.: Asymptotic decomposition of Markov operators. Bull. Pol. Acad. Sci. Math. **35**, 321327 (1987)
16. Kunze, H., La Torre, D., Mendivil, F., Vrscay, E.R.: Fractal Based Methods in Analysis. Springer, New York (2012)
17. Lancaster, H.O.: Correlation and complete dependence of random variables. Ann. Math. Stat. **34**, 1315–1321 (1963)

18. Lasota, A., Mackey, M.C.: Chaos, Fractals and Noise—Stochastic Aspects of Dynamics. Springer, New York (1994)
19. Li, X., Mikusinski, P., Taylor, M.D.: Strong approximation of copulas. J. Math. Anal. Appl. **255**, 608–623 (1998)
20. Mikusínski, P., Sherwood, H., Taylor, M.D.: Shuffles of min. Stochastica **13**, 61–74 (1992)
21. Mukherjea, A.: Completely simple semigroups of matrices. Semigroup Forum **33**, 405–429 (1986)
22. Nelsen, R.B.: An Introduction to Copulas. Springer, New York (2006)
23. Olsen, E.T., Darsow, W.F., Nguyen, B.: Copulas and Markov operators. In: Proceedings of the Conference on Distributions with Fixed Marginals and Related Topics. IMS Lecture Notes. Monograph Series, vol. 28, pp. 244–259 (1996)
24. Parry, W.: Topics in Ergodic Theory. Cambridge University Press, Cambridge (1981)
25. Schwarz, S.: A note on the structure of the semigroup of doubly-stochastic matrices. Math. Slovaca **17**(4), 308–316 (1967)
26. Sempi, C.: Conditional expectations and idempotent copulae. In: Cuadras, C.M., et al. (eds.) Distributions with given Marginals and Statistical Modelling, pp. 223–228. Kluwer, Netherlands (2002)
27. Siburg, K.F., Stoimenov, P.A.: A measure of mutual complete dependence. Metrika **71**, 239–251 (2010)
28. Swanepoel, J., Allison, J.: Some new results on the empirical copula estimator with applications. Stat. Probab. Lett. **83**, 1731–1739 (2013)
29. Trutschnig, W.: On a strong metric on the space of copulas and its induced dependence measure. J. Math. Anal. Appl. **384**, 690–705 (2011)
30. Trutschnig, W., Fernández-Sánchez, J.: Idempotent and multivariate copulas with fractal support. J. Stat. Plan. Inference **142**, 3086–3096 (2012)
31. Trutschnig, W., Fernández-Sánchez, J.: Some results on shuffles of two-dimensional copulas. J. Stat. Plan. Inference **143**, 251–260 (2013)
32. Trutschnig, W.: On Cesáro convergence of iterates of the star product of copulas. Stat. Probab. Lett. **83**, 357–365 (2013)
33. Trutschnig, W., Fernández-Sánchez, J.: Copulas with continuous, strictly increasing singular conditional distribution functions. J. Math. Anal. Appl. **410**, 1014–1027 (2014)
34. Walters, P.: An Introduction to Ergodic Theory. Springer, New York (2000)

Copula Representations for Invariant Dependence Functions

Jayme Pinto and Nikolai Kolev

Abstract Our main goal is to characterize in terms of copulas the linear Sibuya bivariate lack of memory property recently introduced in [12]. As a particular case, one can obtain nonaging copulas considered in the literature.

1 Introduction and Preliminaries

Let X_i be non-negative continuous random variables with survival functions $S_{X_i}(x_i) = P(X_i > x_i)$ and densities $f_{X_i}(x_i), i = 1, 2$. Denote by $S(x_1, x_2) = P(X_1 > x_1, X_2 > x_2)$, the joint survival function of the random vector (X_1, X_2). Following [13], any bivariate survival function can be decomposed as a product of marginal survival functions and a dependence function $\Omega(x_1, x_2)$ via

$$S(x_1, x_2) = S_{X_1}(x_1)S_{X_2}(x_2)\Omega(x_1, x_2) \quad \text{for all} \quad x_1, x_2 \geq 0. \tag{1}$$

The function $\Omega(x_1, x_2)$ represents the free-of-margin influence contribution to the genuine dependence advocated by $S(x_1, x_2)$. A family of Sibuya copulas is introduced in [6], where the authors are motivated by a particular dynamic default model.

Our analysis is based on the following relation

$$S(x_1 + t, x_2 + t) = S(x_1, x_2)S(t, t)B(x_1, x_2; t), \quad t > 0 \tag{2}$$

where $B(x_1, x_2; t)$ is an appropriate "aging" function satisfying the boundary conditions $B(x_1, x_2; 0) = B(0, 0; t) = 1$. In fact, incorporating a time component in the arguments, we replace the product of marginal survival functions in (1) by the product of joint survival functions with nonoverlapping arguments.

J. Pinto (✉) · N. Kolev
Department of Statistics, University of São Paulo, São Paulo, Brazil
e-mail: jaymeaugusto@gmail.com

N. Kolev
e-mail: kolev.ime@gmail.com

© The Author(s) 2015
K. Glau et al. (eds.), *Innovations in Quantitative Risk Management*,
Springer Proceedings in Mathematics & Statistics 99,
DOI 10.1007/978-3-319-09114-3_24

411

In the simplest case, when $B(x_1, x_2; t) = 1$ in (2), one gets the functional equation

$$S(x_1 + t, x_2 + t) = S(x_1, x_2)S(t, t) \qquad (3)$$

for all $x_1, x_2 \geq 0$ and $t > 0$. Bivariate continuous distributions satisfying (3) possess the classical bivariate lack of memory property (BLMP).

The only solution of (3) with exponential marginals is *the Marshall–Olkin bivariate exponential distribution* introduced in [9]. However, there do exist distributions having BLMP with nonexponential marginals. Various solutions of functional equation (3) are presented in [7] where the marginals may have any kind of failure rates: increasing, decreasing, bathtub, etc. It is well-known that BLMP preserves the distribution of (X_1, X_2) and its residual lifetime vector

$$\mathbf{X}_t = (X_{1t}, X_{2t}) = [(X_1 - t, X_2 - t) \mid X_1 > t, X_2 > t]$$

independent of $t \geq 0$, i.e., $(X_1, X_2) \overset{d}{=} \mathbf{X}_t$ implying $X_i \overset{d}{=} X_{it}$, $i = 1, 2$ for all $t \geq 0$.

Remark 1 The vectors (X_1, X_2) and $\mathbf{X}_t$ should necessarily have the same survival copula, which is unique under continuity of X_i, $i = 1, 2$. Therefore, BLMP implies that the corresponding survival copulas are time invariant (nonaging).

The joint survival function of $\mathbf{X}_t$ is given by $S_{\mathbf{X}_t}(x_1, x_2) = S(x_1+t, x_2+t)/S(t, t)$. Its marginal survival functions are $S_{X_{1t}}(x_1) = S(x_1 + t, t)/S(t, t)$ and $S_{X_{2t}}(x_2) = S(t, x_2+t)/S(t, t)$. Applying the Sibuya form representation (1) with respect to the residual lifetime vector $\mathbf{X}_t$ we have

$$S_{\mathbf{X}_t}(x_1, x_2) = S_{X_{1t}}(x_1)S_{X_{2t}}(x_2)\Omega_t(x_1, x_2), \qquad (4)$$

where $\Omega_t(x_1, x_2)$ is the dependence function of $\mathbf{X}_t$.

We will consider a class of continuous bivariate distributions preserving $\Omega_t(x_1, x_2)$ independent of $t \geq 0$, i.e., imposing condition $\Omega_t(x_1, x_2) = \Omega(x_1, x_2)$, where $\Omega(x_1, x_2)$ is the dependence function of (X_1, X_2) from (1). Such a class with memoryless dependence function has been recently introduced in [12] as follows.

Definition 1 The nonnegative continuous bivariate distribution (X_1, X_2) possesses *linear Sibuya BLMP* (to be abbreviated LS-BLMP) if

$$\frac{S_{\mathbf{X}_t}(x_1, x_2)}{S_{X_{1t}}(x_1)S_{X_{2t}}(x_2)} = \frac{S(x_1, x_2)}{S_{X_1}(x_1)S_{X_2}(x_2)} \qquad (5)$$

for all $x_1, x_2, t \geq 0$ and

$$S_{X_{it}}(x_i) = S_{X_i}(x_i)\exp\{-a_i x_i t\} \quad \text{for} \quad a_i \geq 0, \ i = 1, 2. \qquad (6)$$

Observe that BLMP distributions satisfy (5). This means that the class of bivariate continuous distributions with LS-BLMP includes those possessing BLMP.

Let us assume that the partial derivatives of $S(x_1, x_2)$ exist and are continuous. Denote by $r_i(x_1, x_2) = -\partial \ln S(x_1, x_2)/\partial x_i$ the conditional failure rates, $i = 1, 2$. In [12] it is introduced a class $\mathscr{L}(\mathbf{x}; \mathbf{a})$ of nonnegative bivariate continuous distributions that satisfy the relation

$$r(x_1, x_2) = r_1(x_1, x_2) + r_2(x_1, x_2) = a_0 + a_1 x_1 + a_2 x_2 \quad \text{for} \quad a_0, a_1, a_2 \geq 0 \quad (7)$$

for all $x_1, x_2 \geq 0$, where $\mathbf{x} = (x_1, x_2)$ and $\mathbf{a} = (a_0, a_1, a_2)$ is the parameter vector.

When the survival function $S(x_1, x_2)$ is differentiable, the sum $r_1(x_1, x_2) + r_2(x_1, x_2)$ has the following interpretation in terms of directional derivatives: it establishes the performance of $-\ln[S(x_1, x_2)]$ along the lines parallel to $\{x_1 = x_2\}$, i.e., with $45°$ inclination.

Managing a portfolio means observing and controlling its value changes over time to achieve a desired outcome. The vector $(r_1(x_1, x_2), r_2(x_1, x_2))$ of partial derivatives of $-\ln[S(x_1, x_2)]$ is its gradient. With the gradient at hand, the risk manager can evaluate the incremental impact of changes to the portfolio.

The Marshall–Olkin bivariate exponential distribution is a widely used model in risk management and possesses BLMP, see Chap. 3 in [8]. The class $\mathscr{L}(\mathbf{x}; \mathbf{a})$ transforms into BLMP when $a_1 = a_2 = 0$ in (7) and $r_1(x_1, x_2) + r_2(x_1, x_2) = a_0$.

The sum in (7) may serve as a complementary risk measure. For example, the portfolio can be considered "risky" if $r_1(x_1, x_2) + r_2(x_1, x_2) > a_0 + a_1 x_1 + a_2 x_2$, where parameters a_0, a_1 and a_2 are preliminary fixed by an expert.

The joint survival function corresponding to (7) is given by

$$S(x_1, x_2) = \begin{cases} S_{X_1}(x_1 - x_2) \exp\left\{-a_0 x_2 - a_1 x_1 x_2 - \frac{a_2 - a_1}{2} x_2^2\right\}, & \text{if } x_1 \geq x_2 \geq 0; \\ S_{X_2}(x_2 - x_1) \exp\left\{-a_0 x_1 - a_2 x_1 x_2 - \frac{a_1 - a_2}{2} x_1^2\right\}, & \text{if } x_2 \geq x_1 \geq 0. \end{cases}$$

Remark 2 The joint survival function $S(x_1, x_2)$ in the previous expression is proper only for certain marginals $S_{X_1}(x_1)$ and $S_{X_2}(x_2)$. Their choice will determine the range of possible values for the non-negative parameters a_0, a_1 and a_2, see Theorem 5.2.14 and Proposition 5.2.17 in [12]. The nonnegative parameter a_0 plays an important role in the class $\mathscr{L}(\mathbf{x}; \mathbf{a})$. If $a_0 = f_{X_1}(0) + f_{X_2}(0)$, the joint survival function $S(x_1, x_2)$ is absolutely continuous and if $a_0 < f_{X_1}(0) + f_{X_2}(0)$, the distribution exhibits a singular component.

It happens that the class $\mathscr{L}(\mathbf{x}; \mathbf{a})$ specified by (7) can be characterized by the LS-BLMP defined by (5) and (6). The class $\mathscr{L}(\mathbf{x}; \mathbf{a})$ contains continuous bivariate distributions that are symmetric or asymmetric, positive quadrant dependent or negative quadrant dependent, absolutely continuous or exhibit a singular component. In addition, $\mathscr{L}(\mathbf{x}; \mathbf{a})$ can be equivalently represented by relation (2) when $B(x_1, x_2; t) = \exp\{-a_1 x_1 t - a_2 x_2 t\}$, i.e., by

$$\frac{S(x_1 + t, x_2 + t)}{S(t, t)} = S(x_1, x_2) \exp\{-a_1 x_1 t - a_2 x_2 t\}. \tag{8}$$

In Sect. 2, we will characterize the class $\mathscr{L}(\mathbf{x}; \mathbf{a})$ (or equivalently LS-BLMP) in copula terms using the functional equation (8) as base. Recall that the time invariance (nonaging) phenomena of the dependence function $\Omega(x_1, x_2)$ concerns the preservation of the dependence function $\Omega_t(x_1, x_2)$ given in Sibuya form (4). This justifies our suggestion to the corresponding copula be named "Sibuya-type copula." In Sect. 3, we discuss bivariate survival functions with nonaging survival copulas and obtain known relations as particular cases of our findings.

2 Copula Representations of the Class $\mathscr{L}(\mathbf{x}; \mathbf{a})$

Let the vector (X_1, X_2) be a member of the class $\mathscr{L}(\mathbf{x}; \mathbf{a})$. Hence, the survival function of the corresponding residual lifetime vector $\mathbf{X}_t$ is given by (8). Denote by C and C_t, the survival copulas of (X_1, X_2) and $\mathbf{X}_t$, respectively. First, we will find a relation between the survival copulas C and C_t. As a second step, we will obtain a characterizing functional equation for the survival copula C_t that joins the corresponding marginals in both sides of (8).

Theorem 1 *Let (X_1, X_2) belong to the class $\mathscr{L}(\mathbf{x}; \mathbf{a})$. The survival copulas of $\mathbf{X}_t$ and (X_1, X_2) are connected by*

$$C_t(u, v) = C\left(\exp\{-H_1\big(G_{1t}^{-1}(-\ln u)\big)\}, \exp\{-H_2\big(G_{2t}^{-1}(-\ln v)\big)\}\right)$$
$$\times \exp\{-a_1 t G_{1t}^{-1}(-\ln u) - a_2 t G_{2t}^{-1}(-\ln v)\}, \tag{9}$$

where $u, v \in (0, 1]$, $H_i(x_i) = -\ln[S_{X_i}(x_i)]$ and $G_{it}(x_i) = H_i(x_i) + a_i x_i t$, $i = 1, 2$.

Proof The marginals of $\mathbf{X}_t$ have survival functions specified by (6). Using Sklar's theorem, relation (8) can be rewritten in terms of the survival copulas C_t and C as follows

$$C_t\big(S_{X_1}(x_1) \exp\{-a_1 x_1 t\}, S_{X_2}(x_2) \exp\{-a_2 x_2 t\}\big)$$
$$= C\big(S_{X_1}(x_1), S_{X_2}(x_2)\big) \exp\{-a_1 x_1 t - a_2 x_2 t\}. \tag{10}$$

Let $u = S_{X_1}(x_1) \exp\{-a_1 x_1 t\}$ and $v = S_{X_2}(x_2) \exp\{-a_2 x_2 t\}$. From the relations $S_{X_i}(x_i) = \exp\{-H_i(x_i)\}$ and $G_{it}(x_i) = H_i(x_i) + a_i x_i t$, $i = 1, 2$, we get $x_1 = G_{1t}^{-1}(-\ln u)$ and $x_2 = G_{2t}^{-1}(-\ln v)$. Using these Eqs. in (10) we obtain (9). $\square$

Relation (9) shows that the survival copulas of (X_1, X_2) and $\mathbf{X}_t$ do not coincide in general. The time invariance (nonaging) in the class $\mathscr{L}(\mathbf{x}; \mathbf{a})$ (being equivalent to LS-BLMP) is related to the memoryless dependence function Ω_t of the residual lifetime vector $\mathbf{X}_t$, see relation (5). For comparison only, recall that the time invariance for BLMP distributions is concerned with the joint distribution of $\mathbf{X}_t$.

Substituting $a_1 = a_2 = 0$ in (9), we get $C_t(u, v) = C(u, v)$ for all $t \geq 0$, i.e., the survival copula C_t is time invariant, see Remark 1. The conclusion is same if X_1 and X_2 are independent, i.e., $C(u, v) = uv$. Thus, we have the following result.

Corollary 1 *Under conditions of Theorem 1 if*
  (i) $a_1 = a_2 = 0$ *or*
  (ii) X_1 *is independent of* X_2,
  then $C_t(u, v) = C(u, v)$ *for all* $u, v \in (0, 1]$ *and* $t \geq 0$.

The next example illustrates the relations established.

Example 1 Let the vector (X_1, X_2) belong to $\mathscr{L}(\mathbf{x}; \mathbf{a})$. Suppose that the marginals are exponentially distributed, i.e., $S_{X_i}(x) = \exp\{-\lambda_i x_i\}$, $\lambda_i > 0$, $i = 1, 2$. Therefore, $G_{it}(x) = \lambda_i x + a_i xt$ and $G_{it}^{-1}(u) = u/(\lambda_i + a_i t)$, $i = 1, 2$. From (9) we obtain

$$C_t(u, v) = C\left(\exp\left\{\frac{\lambda_1 \ln u}{\lambda_1 + a_1 t}\right\}, \exp\left\{\frac{\lambda_2 \ln v}{\lambda_2 + a_2 t}\right\}\right) \exp\left\{\frac{a_1 t \ln u}{\lambda_1 + a_1 t} + \frac{a_2 t \ln v}{\lambda_2 + a_2 t}\right\},$$

which can be simplified to

$$C_t(u, v) = C\left(u^{\frac{\lambda_1}{\lambda_1 + a_1 t}}, v^{\frac{\lambda_2}{\lambda_2 + a_2 t}}\right) u^{\frac{a_1 t}{\lambda_1 + a_1 t}} v^{\frac{a_2 t}{\lambda_2 + a_2 t}}. \tag{11}$$

Relation (11) gives a general expression for the survival copula $C_t(u, v)$ corresponding to $\mathbf{X}_t$ for all members of the class $\mathscr{L}(\mathbf{x}; \mathbf{a})$ with exponential marginals.

Assume further that (X_1, X_2) follows Gumbel's type I exponential distribution with survival function

$$S(x_1, x_2) = \exp\{-\lambda_1 x_1 - \lambda_2 x_2 - \theta \lambda_1 \lambda_2 x_1 x_2\}, \quad \theta \in [0, 1], \lambda_1, \lambda_2 > 0,$$

see [5]. This distribution is a member of the class $\mathscr{L}(\mathbf{x}; \mathbf{a})$ and the constants in (7) are specified by $a_0 = \lambda_1 + \lambda_2$ and $a_1 = a_2 = \theta \lambda_1 \lambda_2$. The corresponding survival copula is $C(u, v) = uv \exp\{-\theta \ln u \ln v\}$. Substituting $C(u, v)$ in (11) we obtain $C_t(u, v) = uv \exp\{-\theta \ln u \ln v/[(1 + \theta \lambda_2 t)(1 + \theta \lambda_1 t)]\}$. Therefore, the survival copula $C_t(u, v)$ depends on t as well.

When $t = 0$ in (11) we recover the survival copula $C(u, v)$ of (X_1, X_2) and letting $t \to \infty$, we obtain the independence copula $C_\infty(u, v) = uv$. Notice that the independence of X_1 and X_2 is equivalent to the condition $a_1 = a_2 = 0$.

Now, our interest is to find a characterizing functional equation involving the survival copula C_t of $\mathbf{X}_t$ for the absolutely continuous members of the class $\mathscr{L}(\mathbf{x}; \mathbf{a})$.

Theorem 2 *Let the survival copula C_t of $\mathbf{X}_t$ be differentiable in its arguments. The absolutely continuous random vector (X_1, X_2) belongs to the class $\mathscr{L}(\mathbf{x}; \mathbf{a})$, if and only if there exist non-negative constants a_1 and a_2, such that*

$$C_t \left(\frac{S(x_1 + t, t)}{S(t,t)}, \frac{S(t, x_2 + t)}{S(t,t)} \right) = C_t \left(S_{X_1}(x_1) \exp\{-a_1 x_1 t\}, S_{X_2}(x_2) \exp\{-a_2 x_2 t\} \right),$$

(12)

for all $x_1, x_2, t \geq 0$.

Proof Let us assume that the functional equation (12) is satisfied. We will show that (7) is fulfilled. Taking the derivative in both sides of (12) with respect to t we obtain

$$C_t^1 \left(\tfrac{S(x_1+t,t)}{S(t,t)}, \tfrac{S(t,x_2+t)}{S(t,t)} \right) \tfrac{[S^1(x_1+t,t)+S^2(x_1+t,t)]S(t,t)-S(x_1+t,t)[S^1(t,t)+S^2(t,t)]}{[S(t,t)]^2}$$
$$+ C_t^2 \left(\tfrac{S(x_1+t,t)}{S(t,t)}, \tfrac{S(t,x_2+t)}{S(t,t)} \right) \tfrac{[S^1(t,x_2+t)+S^2(t,x_2+t)]S(t,t)-S(t,x_2+t)[S^1(t,t)+S^2(t,t)]}{[S(t,t)]^2}$$
$$= C_t^1 \left(S_{X_1}(x_1) \exp\{-a_1 x_1 t\}, S_{X_2}(x_2) \exp\{-a_2 x_2 t\} \right) \left(-a_1 x_1 S_{X_1}(x_1) \exp\{-a_1 x_1 t\} \right)$$
$$+ C_t^2 \left(S_{X_1}(x_1) \exp\{-a_1 x_1 t\}, S_{X_2}(x_2) \exp\{-a_2 x_2 t\} \right) \left(-a_2 x_2 S_{X_2}(x_2) \exp\{-a_2 x_2 t\} \right),$$

where the superscripts ^{1} and ^{2} denote the partial derivatives with respect to the first and second arguments of the corresponding functions. Letting $x_1 = 0$ in the last equation we have

$$C_t^2 \left(1, \tfrac{S(t,x_2+t)}{S(t,t)} \right) \tfrac{[S^1(t,x_2+t)+S^2(t,x_2+t)]S(t,t)-S(t,x_2+t)[S^1(t,t)+S^2(t,t)]}{[S(t,t)]^2}$$
$$= C_t^2 \left(1, S_{X_2}(x_2) \exp\{-a_2 x_2 t\} \right) \left(-a_2 x_2 S_{X_2}(x_2) \exp\{-a_2 x_2 t\} \right).$$

When $x_i = 0$ in (12) we get relations (6) in Definition 1, $i = 1, 2$ and therefore

$$\frac{[S^1(t, x_2 + t) + S^2(t, x_2 + t)]S(t, t) - S(t, x_2 + t)[S^1(t, t) + S^2(t, t)]}{[S(t, t)]^2} = -a_2 x_2 S_{X_2}(x_2) \exp\{-a_2 x_2 t\}.$$

Since $r(t, x_2 + t) = -[S^1(t, x_2 + t) + S^2(t, x_2 + t)]/S(t, x_2 + t)$ and $r(t, t) = [S^1(t, t) + S^2(t, t)]/S(t, t)$ we get

$$-\frac{S(t, x_2 + t)}{S(t, t)}[r(t, x_2 + t) - r(t, t)] = -a_2 x_2 S_{X_2}(x_2) \exp\{-a_2 x_2 t\},$$

which is equivalent to

$$r(t, x_2 + t) = r(t, t) + a_2 x_2.$$

(13)

Analogously we obtain the equation

$$r(x_1 + t, t) = r(t, t) + a_1 x_1.$$

(14)

Now, we will represent $r(t, t)$ as a function of a_0, a_1, a_2 and t. Taking the partial derivative of (12) with respect to x_1 we have

$$C_t^1 \left(\frac{S(x_1 + t, t)}{S(t,t)}, \frac{S(t, x_2 + t)}{S(t,t)} \right) \frac{S^1(x_1 + t, t)}{S(t,t)} = C_t^1 \left(S_{X_1}(x_1) \exp\{-a_1 x_1 t\}, S_{X_2}(x_2) \exp\{-a_2 x_2 t\} \right)$$
$$\times \left(-f_{X_1}(x_1) \exp\{-a_1 x_1 t\} - a_1 t S_{X_1}(x_1) \exp\{-a_1 x_1 t\} \right).$$

Applying (6) in the last equation we obtain

$$\frac{S^1(x_1 + t, t)}{S(t, t)} = -f_{X_1}(x_1)\exp\{-a_1 x_1 t\} - a_1 t S_{X_1}(x_1)\exp\{-a_1 x_1 t\}$$

and putting $x_1 = 0$ we have $r_1(t, t) = f_{X_1}(0) + a_1 t$. Similarly we get $r_2(t, t) = f_{X_2}(0) + a_2 t$. The sum of last two equations gives

$$r(t, t) = r_1(t, t) + r_2(t, t) = [f_{X_1}(0) + f_{X_2}(0)] + a_1 t + a_2 t.$$

Let $t = 0$ in last relation to get $f_{X_1}(0) + f_{X_2}(0) = a_0 \geq 0$. Thus,

$$r(t, t) = a_0 + a_1 t + a_2 t.$$

Taking into account (13) and (14), we conclude that $r(x_1, x_2) = a_0 + a_1 x_1 + a_2 x_2$. Therefore, we obtain the relation (7) which defines the class $\mathscr{L}(\mathbf{x}; \mathbf{a})$. In addition, the corresponding bivariate distributions are absolutely continuous because of equation $f_{X_1}(0) + f_{X_2}(0) = a_0$, see Remark 2.

Conversely, assume that the random vector (X_1, X_2) belonging to the class $\mathscr{L}(\mathbf{x}; \mathbf{a})$ is absolutely continuous. Therefore (8), being equivalent to (5) and (6), is valid. In addition, relations (6) show that the marginal distributions in both sides of (8) coincide. Applying Sklar's theorem to (8), we obtain the functional equation (12). □

Since the dependence function Ω_t satisfies the Sibuya form (4), we refer to the survival copula C_t characterized by functional equation (12) as *Sibuya-type copula*.

Example 2 Let us consider the absolutely continuous joint survival function

$$S(x_1, x_2) = \begin{cases} \exp\left\{-\left[\lambda_1 x_1 + \lambda_2 x_2 + \lambda_1 \lambda_2 x_2(\theta_1 x_1 + \frac{\theta_2 - \theta_1}{2} x_2)\right]\right\}, & \text{if } x_1 \geq x_2 \geq 0; \\ \exp\left\{-\left[\lambda_1 x_1 + \lambda_2 x_2 + \lambda_1 \lambda_2 x_1(\theta_2 x_2 + \frac{\theta_1 - \theta_2}{2} x_1)\right]\right\}, & \text{if } x_2 \geq x_1 \geq 0, \end{cases}$$

where $\theta_i \in (0, 1]$, and $\lambda_i > 0$, $i = 1, 2$. This distribution was obtained in [12] and can be named *Generalized Gumbel's bivariate exponential distribution* with parameters λ_i and θ_i, $i = 1, 2$. If $\theta_1 = \theta_2 = \theta$, we get the Gumbel distribution considered in Example 1. The marginal survival functions are $S_{X_i}(x_i) = \exp\{-\lambda_i x_i\}$, $i = 1, 2$.

The survival function of the residual lifetime vector $\mathbf{X}_t$ is given by (8). After some algebra, we get the corresponding survival copula

$$C_t(u, v) = \begin{cases} uv \exp\left\{-\dfrac{\theta_1}{\gamma_1(t)\gamma_2(t)} \ln u \ln v\right\} \exp\left\{-\dfrac{\lambda_1(\theta_2-\theta_1)}{2\lambda_2\gamma_1^2(t)}(\ln v)^2\right\}, \\ \quad \text{if } u^{-\lambda_2\gamma_1(t)} \geq v^{-\lambda_1\gamma_2(t)}; \\ uv \exp\left\{-\dfrac{\theta_2}{\gamma_1(t)\gamma_2(t)} \ln u \ln v\right\} \exp\left\{-\dfrac{\lambda_2(\theta_1-\theta_2)}{2\lambda_1\gamma_2^2(t)}(\ln u)^2\right\}, \\ \quad \text{if } u^{-\lambda_2\gamma_1(t)} < v^{-\lambda_1\gamma_2(t)}, \end{cases}$$

where $\gamma_1(t) = 1 + \lambda_1\theta_2 t$, $\gamma_2(t) = 1 + \lambda_2\theta_1 t$ and $u, v \in (0, 1]$. Fix $a_i = \lambda_1\lambda_2\theta_i$, $i = 1, 2$ in (12) to verify that

$$C_t(\exp\{-\lambda_1 x_1 - \lambda_1\lambda_2\theta_1 x_1 t\}, \exp\{-\lambda_2 x_2 - \lambda_1\lambda_2\theta_2 x_2 t\}) = \frac{S(x_1 + t, x_2 + t)}{S(t, t)},$$

for all $t \geq 0$. Therefore, the generalized Gumbel's bivariate exponential distribution is member of the class $\mathscr{L}(\mathbf{x}; \mathbf{a})$.

3 Bivariate Survival Functions with Nonaging Survival Copulas

In this section, we will consider nonaging survival copulas $C(u, v)$ instead of memoryless dependence functions $\Omega_t(x_1, x_2)$.

Let us denote by $\mathscr{A}$ the class of continuous bivariate survival functions $S(x_1, x_2)$, such that (X_1, X_2) and $\mathbf{X}_t$ have the same survival copula $C(u, v)$. Therefore, the functional equation

$$\frac{C\big(S_{X_1}(x_1 + t), S_{X_2}(x_2 + t)\big)}{C\big(S_{X_1}(t), S_{X_2}(t)\big)} = C\left(\frac{C\big(S_{X_1}(x_1 + t), S_{X_2}(t)\big)}{C\big(S_{X_1}(t), S_{X_2}(t)\big)}, \frac{C\big(S_{X_1}(t), S_{X_2}(x_2 + t)\big)}{C\big(S_{X_1}(t), S_{X_2}(t)\big)}\right) \quad (15)$$

has to be satisfied for all $x_1, x_2 \geq 0$ and $t \geq 0$. We will assume further that the survival copula C is time invariant (or nonaging) if it corresponds to a member of the class $\mathscr{A}$.

Taking into account the conclusion in Remark 1, all bivariate survival functions possessing BLMP belong to $\mathscr{A}$. It happens that this time invariance property is not restricted to BLMP survival functions. For instance, it is well-known that the Clayton bivariate survival function given by

$$S(x_1, x_2) = \left[S_{X_1}^{-\theta}(x_1) + S_{X_2}^{-\theta}(x_2) - 1\right]^{-1/\theta}, \quad \theta \in (0, \infty),$$

has time invariant survival copula. One can find other members of the class $\mathscr{A}$ in Examples 3 and 4.

Let $\mathscr{D}(t) = \{(u,v) \in (0,1] \,|\, u = S_{X_1}(t), v = S_{X_2}(t), t > 0\}$ be a curve on the unit square parameterized by $t > 0$. In such a case, from (15) we may obtain nonaging survival copulas whenever C is invariant on the curve $\mathscr{D}(t)$. In particular, if $X_1 \overset{d}{=} X_2$, we have invariance of the survival copula along the main diagonal of the unit square.

Example 3 [Invariance on the main diagonal] The Cuadras-Augé survival copula

$$C_\alpha(u,v) = [\min(uv)]^\alpha [uv]^{1-\alpha}, \quad \alpha \in [0,1]$$

is invariant on the main diagonal of the unit square, see [2]. Let us initially consider equally distributed marginals $S_{X_1}(x) = S_{X_2}(x) = S_X(x)$. If $S_X(x)$ is exponentially distributed, then $S(x_1, x_2) = C_\alpha(S_{X_1}(x_1), S_{X_2}(x_2))$ is a particular case of the Marshall–Olkin's bivariate exponential distribution, see [9], possessing BLMP and, consequently, belonging to the class $\mathscr{A}$. Now, let X be gamma distributed random variable. In this case, BLMP does not hold true but the corresponding joint survival function still belongs to $\mathscr{A}$.

In a third scenario, where X_1 and X_2 do not share the same distribution but are joined by the Cuadras-Augé survival copula, $S(x_1, x_2)$ neither possesses BLMP nor belongs to $\mathscr{A}$.

Example 4 [Invariance along a curve] The Marshall–Olkin survival copula

$$C_{\alpha,\beta}(u,v) = \min(u^{1-\alpha}v, uv^{1-\beta}), \quad \alpha, \beta \in (0,1)$$

is invariant on the curve $\{(u,v) = (t^\alpha, t^\beta), t \in (0,1)\}$, see [2]. Notice that when $\alpha = \beta$ we obtain the Cuadras-Augé survival copula from Example 3.

Let us consider a baseline survival function $S_X(x)$ and substitute $S_{X_1}(x) = [S_X(x)]^\alpha$ and $S_{X_2}(x) = [S_X(x)]^\beta$. Then, the corresponding joint survival function $S(x_1, x_2) = C_{\alpha,\beta}(S_{X_1}(x_1), S_{X_2}(x_2))$ belongs to $\mathscr{A}$. In particular, if the marginals are exponentially distributed, not necessarily sharing the same parameter, then $S(x_1, x_2)$ possesses BLMP. But choosing X_1 exponentially distributed and X_2 beta distributed, say the corresponding joint survival function is not a member of the class $\mathscr{A}$.

The cases considered in the last two examples depend on the choice of the marginal survival functions. A general invariance property can be obtained when we consider the Clayton survival copula. In such a case, for any marginals we have time invariant survival copulas. We refer the reader to Sect. 4 in [2] for more details on time invariant copulas.

In fact, the Clayton survival copula is the only absolutely continuous copula that is preserved even under bivariate truncation, see [11]. The absolutely continuous assumption is relaxed in Theorem 4.1 in [3]. In [10], it is given a characterization of the survival functions which simultaneously have Clayton survival copula and possess BLMP, see their Theorem 3.2.

In the next statement, we establish a necessary condition to an absolutely continuous bivariate survival function be a member of the class $\mathscr{A}$.

Theorem 3 *Let $S(x_1, x_2)$ be an absolutely continuous survival function belonging to the class $\mathscr{A}$. Then, its survival copula satisfies the functional equation*

$$C(u, v) = \left[u - \frac{f_{X_2}(0)C^2(u, 1)}{a_0} \right] C^1(u, v) + \left[v - \frac{f_{X_1}(0)C^1(1, v)}{a_0} \right] C^2(u, v),$$

$$(16)$$

for all $u, v \in [0, 1]$ and $a_0 > 0$, where C^1 and C^2 denote the partial derivatives of C with respect to the first and second arguments, respectively.

Proof Take the derivative in (15) with respect to t and substitute $t = 0$ to get (16). $\qquad \square$

The knowledge of the first partial derivatives of the survival copula $C(u, v)$ is sufficient to recover the distribution of $\min(U, V)$, where U and V are uniformly distributed with survival copula $C(u, v)$. Really, $P(\min(U, V) > t) = C(t, t)$ for $t \in [0, 1]$. Now, substitute $u = v = t$ in (16) to get the corresponding equation (and main diagonal copula).

Finally, we show two known functional equations which are particular cases of (16). Under assumptions of Theorem 3, let $f_{X_1}(0) = f_{X_2}(0)$. Then

$$C(u, v) = \left[u - \frac{C^2(u, 1)}{2} \right] C^1(u, v) + \left[v - \frac{C^1(1, v)}{2} \right] C^2(u, v).$$

The same equation is obtained in Proposition 3 (ii) in [1] under the condition that X_1 and X_2 are uniformly distributed on the unit square, i.e., $f_{X_1}(0) = f_{X_2}(0) = 1$.

Further, assume that $C(u, v)$ is exchangeable. Thus, $C^2(u, 1) = C^1(1, u)$, $C^2(u, v) = C^1(v, u)$ and the last equation transforms into

$$C(u, v) = \left[u - \frac{C^1(1, u)}{2} \right] C^1(u, v) + \left[v - \frac{C^1(1, v)}{2} \right] C^1(v, u),$$

see Proposition 3 on page 18 in [4].

4 Conclusions

The time invariance of the residual lifetime vector $\mathbf{X}_t$ of (X_1, X_2) is characterized by BLMP in [9]. It tells us that the joint distributions of $\mathbf{X}_t$ and (X_1, X_2) coincide independently of t, i.e., the BLMP holds. In this paper, we consider a more general concept, namely time invariance of the dependence functions of $\mathbf{X}_t$ and (X_1, X_2), given by (4) and (1), respectively.

We offer copula representations for the time invariance property related to bivariate survival functions of the residual lifetime vector $\mathbf{X}_t$. While in Sect. 2, the nonaging phenomena is associated with the dependence function $\Omega_t(x_1, x_2)$, in Sect. 3 our interest is on the survival copula $C_t(u, v)$ of $\mathbf{X}_t$.

We are thankful to the referee and editor for their comments.

Open Access This chapter is distributed under the terms of the Creative Commons Attribution Noncommercial License, which permits any noncommercial use, distribution, and reproduction in any medium, provided the original author(s) and source are credited.

References

1. Charpentier, A.: Tail distribution and dependence measures. Working paper (2003)
2. Charpentier, A., Juri, A.: Limiting dependence structures for tail events, with applications to credit derivatives. J. Appl. Probab. **43**, 563–586 (2006)
3. Durante, F., Jaworski, P.: Invariant dependence structure under univariate truncation. Stat: J. Theoret. Appl. Stat. **46**, 263–277 (2012)
4. Gourieroux, C., Monfort, A.: Age and term structure in duration models. Working paper (2003)
5. Gumbel, E.: Bivariate exponential distributions. J. Am. Stat. Assoc. **55**, 698–707 (1960)
6. Hofert, M., Vrins, F.: Sibuya copulas. J. Multivar. Anal. **114**, 318–337 (2013)
7. Kulkarni, H.: Characterizations and modelling of multivariate lack of memory property. Metrika **64**, 167–180 (2006)
8. Mai, J.-F., Scherer, M.: Simulating Copulas. Imperial College Press, London (2012)
9. Marshall, A., Olkin, I.: A multivariate exponential distribution. J. Am. Stat. Assoc. **62**, 30–41 (1967)
10. Mulero, J., Pellerey, F.: Bivariate aging properties under Archimedean dependence structures. Commun. Stat: Theor. Methods. **39**, 3108–3121 (2010)
11. Oakes, D.: On the preservation of copula structure under truncation. The Canadian J. Stat. **33**, 465–468 (2005)
12. Pinto, J.: Deepening the notions of dependence and aging in bivariate probability distributions. PhD Thesis, University of Sao Paulo (2014)
13. Sibuya, M.: Bivariate extreme statistics I. Ann. Inst. Stat. Math. **11**, 195–210 (1960)

Nonparametric Copula Density Estimation Using a Petrov–Galerkin Projection

Dana Uhlig and Roman Unger

Abstract Nonparametrical copula density estimation is a meaningful tool for analyzing the dependence structure of a random vector from given samples. Usually kernel estimators or penalized maximum likelihood estimators are considered. We propose solving the Volterra integral equation

$$\int_0^{u_1} \cdots \int_0^{u_d} c(s_1, \ldots, s_d)ds_1 \cdots ds_d = C(u_1, \ldots, u_d)$$

to find the copula density $c(u_1, \ldots, u_d) = \frac{\partial^d C}{\partial u_1 \cdots \partial u_d}$ of the given copula C. In the statistical framework, the copula C is not available and we replace it by the empirical copula of the pseudo samples, which converges to the unobservable copula C for large samples. Hence, we can treat the copula density estimation from given samples as an inverse problem and consider the instability of the inverse operator, which has an important impact if the input data of the operator equation are noisy. The well-known curse of high dimensions usually results in huge nonsparse linear equations after discretizing the operator equation. We present a Petrov–Galerkin projection for the numerical computation of the linear integral equation. A special choice of test and ansatz functions leads to a very special structure of the linear equations, such that we are able to estimate the copula density also in higher dimensions.

1 Copula Density Estimation as an Inverse Problem

A copula is a multivariate distribution function of a d-dimensional random vector with uniformly distributed margins. Sklar's theorem ensures that any joint multivariate distribution F of a d-dimensional vector $\mathbf{X} = (X_1, \ldots, X_d)^T$ with margins F_j

D. Uhlig (✉) · R. Unger
Department of Mathematics, Technische Universität Chemnitz, 09107 Chemnitz, Germany
e-mail: dana.uhlig@mathematik.tu-chemnitz.de

R. Unger
e-mail: roman.unger@mathematik.tu-chemnitz.de

© The Author(s) 2015
K. Glau et al. (eds.), *Innovations in Quantitative Risk Management*,
Springer Proceedings in Mathematics & Statistics 99,
DOI 10.1007/978-3-319-09114-3_25

$(j = 1, \ldots, d)$ can be expressed as

$$F(x_1, \ldots, x_d) = C(F_1(x_1), \ldots, F_d(x_d)) \quad \forall \mathbf{x} = (x_1, \ldots, x_d)^T \in \mathbb{R}^d$$

where the copula is unique on $range(F_1) \times \cdots \times range(F_d)$, that is for continuous margins $F_1, \ldots, F_d$ the copula C is unique on the whole domain. Consequently, the copula contains the complete dependence structure of the random vector **X**. For a detailed introduction to copulas and their properties see, for example, [8, 9, Chap. 5] or [10]. In risk management, knowledge of the dependence is of paramount importance.

If the copula is sufficiently smooth, the copula density

$$c(u_1, \ldots, u_d) = \frac{\partial^d C}{\partial u_1 \cdots \partial u_d} \tag{1}$$

exists and then the density gives us the dependence structure in a more convenient way, because usually the graphs of the copulas look very similar and there are only small differences in the slope. For this reason the reconstruction of the copula density is a vibrant field of research in finance and many other scientific fields. Particularly in practical tasks, the dependence structure of more than two random variables is of special interest as the dimension d is large. In the nonparametric statistical estimation, usually kernel estimators are used, but they have often problems with the boundary bias. There are also spline- or wavelet-based approximation methods, but most of them are only discussed in the two-dimensional case. Likewise, in [12], the authors discuss a penalized nonparametrical maximum likelihood method in the two-dimensional case. A detailed survey of literature about nonparametrical copula density estimation can be found in [6]. However, most of the nonparametrical methods are faced with the curse of dimensionality such that the numerical computations are only for sufficiently low dimensions possible. Actually, many authors discuss only the two-dimensional case in non-parametrical copula density estimation.

In this paper we develop an alternative approach based on the theory of inverse problems. The copula density (1) exists only for absolutely continuous copulas. Obviously, the copula is not observable for a sample $\mathbf{X}_1, \mathbf{X}_2, \ldots, \mathbf{X}_T$ in the statistical framework, but we can approximate it with the empirical copula

$$\hat{C}(\mathbf{u}) = \frac{1}{T} \sum_{j=1}^{T} \mathbf{1}_{\{\hat{\mathbf{U}}_j \leq \mathbf{u}\}} = \frac{1}{T} \sum_{j=1}^{T} \prod_{k=1}^{d} \mathbf{1}_{\{\hat{U}_{kj} \leq u_k\}} \tag{2}$$

of the margin transformed pseudo samples $\hat{\mathbf{U}}_1, \hat{\mathbf{U}}_2, \ldots, \hat{\mathbf{U}}_T$ with $\hat{U}_{kj} = \hat{F}_k(X_{kj})$ where

$$\hat{F}_k(x) = \frac{1}{T} \sum_{j=1}^{T} \mathbf{1}_{\{X_{kj} \leq x\}}$$

denotes the empirical margins. It is well-known that the empirical copula uniformly converges to the copula (see [2])

$$\max_{\mathbf{u}\in[0,1]^d} \left| C(\mathbf{u}) - \hat{C}(\mathbf{u}) \right| = \mathscr{O}\left(\frac{(\log\log T)^{\frac{1}{2}}}{T^{\frac{1}{2}}} \right) \quad \text{a.s. for } T \to \infty \tag{3}$$

Therefore, we treat the empirical copula as a noisy representation of the unobservable copula $C^\delta = \hat{C}$. The estimation problem of the density is faced with differentiating the empirical copula, which is obviously not smooth. However, for each density it yields the integral equation

$$\int_0^{u_1} \cdots \int_0^{u_d} c(s_1,\ldots,s_d)\mathrm{d}s_1 \cdots \mathrm{d}s_d = C(u_1,\ldots,u_d) \quad \forall \mathbf{u} = (u_1,\ldots,u_d)^T \in \Omega = [0,1]^d \tag{4}$$

which can be seen as a weak formulation of Eq. (1). In the following, we therefore consider the linear Volterra integral operator $A \in \mathscr{L}\left(L^1(\Omega), L^2(\Omega)\right)$ and solve the linear operator equation

$$Ac = C \tag{5}$$

to find the copula density c. In the following, we assume attainability which means $C \in \mathscr{R}(A)$, hence we only consider copulas $C \in L^2(\Omega)$ which have a solution $c \in L^1(\Omega)$

The injective Volterra integral operator is well-studied in the inverse problem literature. Even in the one-dimensional case, this is an ill-posed operator resulting from the noncontinuity of the inverse A^{-1}, which is the differential operator. Hence, solving Eq. (1) leads to numerical instabilities if the right-hand side of (5) has only a small data error. Because the solution is sensitive to small data errors, regularization methods to overcome the instability are discussed in the inverse problem literature. For a detailed introduction to regularization see, for example, [4, 13].

In Sect. 2 we discuss a discretization of the integral equation (4) and in Sect. 3, we illustrate the numerical instability if we use the empirical copula instead of the exact one and discuss regularization methods for the discretized problem.

The basics to the numerical implementation of the problem and especially the details of the Kronecker multiplication are presented in the authors working paper [14] and a discussion that the Petrov–Galerkin projection is not a simple counting algorithm is done in [15]. This paper gives a summary of the proposed method for effective computation of the right-hand side for larger dimensions and discusses in more detail the analytical aspects of the inverse problem and reasons for the existence of the Kronecker structure.

2 Numerical Approximation

We discuss the numerical computation of the copula density $c \in X = L^1(\Omega)$ from a given copula $C \in Y = L^2(\Omega)$, which is in principle a numerical differentiation and in higher dimensions, a very hard problem (see [1]). Moreover, in practical applications, the measured data C^δ have some noise δ with $\|C - C^\delta\|_Y \leq \delta$ and very often the function is not smooth enough that is $C^\delta \notin C^1(\Omega)$ even $C \in C^1(\Omega)$, which leads to numerical instabilities making a usual numerical differentiation impossible.

For the sake of convenience, we write

$$\int_0^{\mathbf{u}} c(s)ds = C(\mathbf{u}) \quad \forall \mathbf{u} = (u_1, \ldots, u_d)^T \in \Omega = [0, 1]^d$$

for Eq. (4) as a short form. We propose applying a Petrov–Galerkin projection (see [5]) for some discretization size h and consider the finite dimensional approximation

$$c_h(s) = \sum_{j=1}^N c_j \phi_j(s), \tag{6}$$

where $\Phi = \{\phi_1, \phi_2, \ldots, \phi_N\}$ is a basis of the ansatz space V_h. The vector of coefficients $\mathbf{c} = (c_1, \ldots, c_N)^T \in \mathbb{R}^N$ is chosen such that

$$\int_\Omega \int_0^{\mathbf{u}} c_h(s)ds\psi(\mathbf{u})d\mathbf{u} = \int_\Omega C(\mathbf{u})\psi(\mathbf{u})d\mathbf{u} \quad \forall \psi \in \tilde{V}_h. \tag{7}$$

It is sufficient to fulfill Eq. (7) for N linear independent test functions $\psi_i \in \tilde{V}_h$. This yields the system of linear equations

$$K\mathbf{c} = \mathbf{C} \tag{8}$$

with right-hand side

$$C_i = \int_\Omega C(\mathbf{u})\psi_i(\mathbf{u})d\mathbf{u}, \quad i = 1, \ldots, N \tag{9}$$

and the $N \times N$ matrix K with

$$K_{ij} = \int_\Omega \int_0^{\mathbf{u}} \phi_j(s)ds\psi_i(\mathbf{u})d\mathbf{u}.$$

If the exact copula is replaced by the empirical copula, we obtain a noisy representation $\mathbf{C}^\delta$ with

$$C_i^\delta = \int_\Omega \hat{C}(\mathbf{u}) \psi_i(\mathbf{u}) d\mathbf{u}, \quad i = 1, \dots, N \tag{10}$$

of the exact right-hand side $\mathbf{C}$. A typical phenomenon of ill-posed inverse problems is that the numerically computed solution based on noisy data (10) will be high oszillating without choosing a proper regularization. This problem is not caused by the numerical approximation, but rather by the discontinuity of the inverse operator. In Section 3 this will be illustrated. Figure 3 shows the reconstructed density of the Student copula for exact data (9), whereas Fig. 5 shows it for different noise levels.

In principle, we can choose arbitrary ansatz functions $\phi_j \in V_h$ and test functions $\psi_i \in \tilde{V}_h$. However, having the curse of high dimensions in mind, we choose very simple ansatz functions such that the matrix K gets a very special structure allowing us to solve (8) and compute the approximated copula density also for higher dimensional copulas. Obviously, the approximated density (6) is not smooth and in order to obtain a smoother approximated copula C_h with

$$C_h(\mathbf{u}) = \int_0^{\mathbf{u}} c_h(\mathbf{s}) d\mathbf{s}$$

we choose the test functions as integrated ansatz functions, such that the approximated copula

$$C_h(\mathbf{u}) = \sum_{j=1}^N c_j \psi_j(\mathbf{u})$$

is smoother than the approximated density.

We discretize the domain Ω by splitting each one-dimensional interval $[0, 1]$ in n equal subintervals of length $h = \frac{1}{n}$. Hence, we obtain $N = n^d$ equal-sized hypercubes and call these elements $e_1, \dots, e_N$. We number the elements in a specific order, illustrated in Fig. 1 such that if we look at the $(d + 1)$-dimensional problem, the first n^d elements of the new problem have the same number and location as the elements of the d-dimensional problem.

We set $N = n^d$ and choose the ansatz functions

$$\phi_j(\mathbf{u}) = \begin{cases} 1 & \mathbf{u} \in e_j \\ 0 & \text{otherwise} \end{cases} \tag{11}$$

and the test functions ψ_i as the integrated ansatz functions

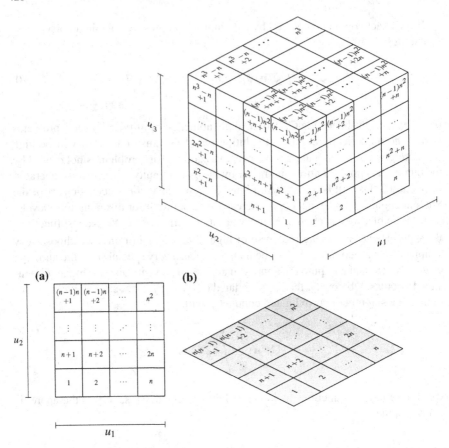

Fig. 1 Discretization of the domain $\Omega = [0, 1]^d$. **a** $d = 2$. **b** $d = 3$

$$\psi_i(\mathbf{u}) = \int_0^\mathbf{u} \phi_i(\mathbf{s})\mathrm{d}s. \tag{12}$$

In contrast to finite element discretizations, the system matrix K is not sparse and the system size $N = n^d$ grows exponentially with the dimension d. A straightforward assembling and solving of the linear system (8) becomes impossible for usual discretizations n. Even in the three-dimensional case, the matrix storage of the system matrix for $n = 80$ needs approximately one terabyte, even when exploiting symmetry, and computing times for assembling and solving such systems will become enormous.

The choices (11) and (12) yield a structure of the $N \times N$ system matrix K, illustrated in Fig. 2, allowing us to solve (8) also for $d > 2$. The matrixplot shows that the $n \times n$ system matrix of the one-dimensional case is equivalent to the upper left $n \times n$ corners of the two- and three-dimensional matrices. Moreover, the other parts of the system matrices are scaled replications of the one-dimensional $n \times n$

(a)   **(b)**

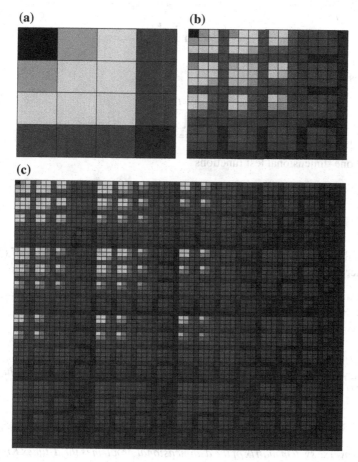

(c)

Fig. 2 Matrixplots of the system matrix K for $n = 4$ and different dimensions d. **a** System matrix for $d = 1$. **b** System matrix for $d = 2$. **c** System matrix for $d = 3$

system matrix. This effect is based by a Kronecker factorization of the d-dimensional system matrix into d one-dimensional matrices of the one-dimensional problem.

One important reason for this structure is that the chosen ansatz functions decomposed into a product of one-dimensional ansatz functions. In order to illustrate this, we consider the lowest corner $\mathbf{b}^i$ of the ith element and define the one-dimensional function

$$\phi_i^k = \mathbf{1}_{\{[b_k^i, b_k^i + h]\}}$$

This yields

$$\phi_i(\mathbf{u}) = \prod_{k=1}^{d} \phi_i^k(u_k) \tag{13}$$

as well as

$$\psi_i(\mathbf{u}) = \prod_{k=1}^{d} \psi_i^k(u_k) \tag{14}$$

with the one-dimensional test functions

$$\psi_i^k(u) = \int_0^u \phi_i^k(s)\mathrm{d}s.$$

We only formulate the main result allowing us to compute solutions of (8) also for higher dimensions d. Details and proofs can be found in the working paper [14].

Theorem 1 *The system matrix for the $(d+1)$-dimensional case can be extracted from the one and d-dimensional system matrices.*

$$^{(d+1)}K = {}^{(1)}K \otimes {}^{(d)}K$$

Corollary 1 *The system matrix $^{(d)}K$ is the d-fold Kronecker product of the $n \times n$ matrix $^{(1)}K$*

$$^{(d)}K = {}^{(1)}K \otimes {}^{(1)}K \otimes \cdots \otimes {}^{(1)}K \tag{15}$$

and the inverse system matrix of the d-dimensional problem is the d-times Kronecker product of the one-dimensional inverse system matrix

$$^{(d)}K^{-1} = {}^{(1)}K^{-1} \otimes {}^{(1)}K^{-1} \otimes \cdots \otimes {}^{(1)}K^{-1}.$$

Following Corollary 1 we only have to assemble the one-dimensional system matrix $^{(1)}K$ of dimension $n \times n$, compute its inverse $^{(1)}K^{-1}$ and have to perform the Kronecker factorization for computing the solution $\mathbf{c} = {}^{(d)}K^{-1}\mathbf{C}$ of (8). Details of the algorithm and an effective Kronecker multiplication are written in [14]. Using effective parallelization methods, the running time can be accelerated. Actually, the computation of the right-hand side (9) is the crucial part and much more expensive than solving the linear system, because we have to evaluate $N = n^d$ different d-dimensional integrals over the whole domain Ω. Note that for our special choice of ansatz functions (6) we have

$$C_i = \int_\Omega C(\mathbf{u})\psi_i(\mathbf{u})\mathrm{d}\mathbf{u} = \sum_{l=i}^{N} \mathbf{1}_{\{b^l \geq b^i\}} \int_{e_l} C(\mathbf{u})\psi_i(\mathbf{u})\mathrm{d}\mathbf{u}, \tag{16}$$

which also reduces the numerical effort. In higher dimensions, the number of elements e_i with zero values grows, such that using Eq. (16) instead of (9) improves the running times.

In the most practical relevant case, where the components of the right-hand side (10) are evaluated over the empirical copula (2), the numerical effort can be radically reduced, because the d-dimensional integral

$$C_i^\delta = \int_\Omega \hat{C}(\mathbf{u})\psi_i(\mathbf{u})d\mathbf{u} = \frac{1}{T}\sum_{j=1}^{T}\int_\Omega \prod_{k=1}^{d} \mathbf{1}_{\{\hat{U}_{kj}\leq u_k\}}\psi_i^k(u_k)d\mathbf{u} = \frac{1}{T}\sum_{j=1}^{T}\prod_{k=1}^{d} I_{ij}^k \quad (17)$$

degenerates in a product of d one-dimensional integrals

$$I_{ij}^k = \int_0^1 \mathbf{1}_{\{\hat{U}_{kj}\leq s\}}\psi_i^k(s)ds = \begin{cases} h(1-b_k^i) - \frac{1}{2}h^2, & \hat{U}_{kj} < b_k^i \\ h(1-b_k^i) - \frac{1}{2}h^2 - \frac{1}{2}(\hat{U}_{kj}-b_k^i)^2, & b_k^i \leq \hat{U}_{kj} \leq b_k^i + h \\ h(1-\hat{U}_{kj}), & \hat{U}_{kj} > b_k^i + h \end{cases}$$

using Eqs. (13) and (14). In this case, the numerical effort is of order $\mathcal{O}(NTd)$ which is an extreme improvement to $\mathcal{O}\left(N3^dT + \frac{N^2+N}{2}3^d\right)$, if the d-dimensional integrals (10) are numerically computed by a usual 3^d-points Gauss formula. We want to point out that the computation of the right-hand side (10) for the empirical copula based on formula (17) is still possible for $d = 9$, whereas the computational effort for computing (16) for an arbitrary given copula C is exorbitant, even if the discretization size n is moderately chosen. The numerical effort is illustrated in Table 1.

Note that contrary to what might be expected, the vector $\mathbf{c} = (c_1, \ldots, c_N)^T$ does not count the number of samples in the elements, even though the approximated solution c_h is a piecewise constant function on the elements and the Petrov–Galerkin projection is not simple counting (for more details see [15]).

Table 1 Computing times using (16) for the independence copula

d	n	N	s_{rhs}	t_{rhs} (s)	t_{rhs} using (16)	t_{solve} (s)	$\|c - c_h\|_{L^1(\Omega)}$
2	30	900	1	0.2	<1 s	0.0005	2.5e − 10
2	60	3,600	1	2.2	<1 s	0.003	4.9e − 9
2	100	10,000	3	6.7	3 s	0.01	2.8e − 8
3	30	27,000	10	60.7	18 s	0.01	4.8e − 7
3	60	216,000	30	1,440	379 s	0.13	3.4e − 5
3	100	1,000,000	30	32,163	8,031 s	1.04	7.1e − 4
4	30	810,000	30	72,989	10,876 s	0.29	1.2e − 3
5	30	24,300,000	30		≈112 days		
6	30	729,000,000	30		≈270 years		

2.1 Examples

In order to illustrate the computing times and approximation quality, we use the independent copula

$$C(\mathbf{u}) = \prod_{k=1}^{d} u_k$$

which has the exact solution $c(u) = 1$. Please note that for this example, we used the exact copula as right-hand side without generating samples. So, there is no data noise and hence $\delta = 0$, which allows us to separate the approximation error and the ill-posedness resulting from the uncontinuity of the inverse operator C^{-1}.

Many authors (see, for example, [11]) look at the integrated square error, which is the squared L^2-norm of the difference between the copula density and its approximation. For the independent copula, the integrated square error can easily be computed

$$\mathrm{ISE}(c, c_h) = \|c - c_h\|_{L^2(\Omega)}^2 = \frac{1}{N} \left\| \mathbf{c} - (1, 1, \ldots, 1)^T \right\|_{l^2}^2 .$$

Actually, this error measure is unsuitable, because the natural space for densities is L^1 instead of L^2 (see [3]) and so we measure the difference in the L^1-norm, which also can be easily computed for the independence copula

$$\|c - c_h\|_{L^1(\Omega)} = \int_{\Omega} |c(\mathbf{u}) - c_h(\mathbf{u})| \, d\mathbf{u} = \frac{1}{N} \left\| \mathbf{c} - (1, 1, \ldots, 1)^T \right\|_{l^1} .$$

In Table 1, we give the following quantities for different discretization steps n in dimension 1 and dimension d: the system size $N = n^d$, the computing times t_{rhs} for assembling the right-hand side, t_{solve} for solving the system, s_{rhs} as the number of computing slaves and the L^1-approximation errors. For the computation of the right-hand side, a parallel OpenMPI implementation was used with s_{rhs} computing slaves. For solving the system with the Kronecker factorization, a sequential C++ implementation is used. The exact computation of an ordinary right-hand side without using the product structure gets still impossible for $d \geq 5$ and the times are estimated computing times. In summary, the example of the independence copula shows that for exact data of the right-hand side, the approximation error is suitable but grows with decreasing discretization size $h = \frac{1}{n}$. We want to point out that this is typical phenomenon of inverse problems, called "regularization by discretization".

If we consider the more practical relevant case, that the empirical copula, generated by T independent samples of the independence copula, is used, we are faced with data noise $\delta > 0$ and ill-posedness. Table 2 shows that the computation based on (17) is still possible for $d \approx 10$. However, the approximation error increase with the dimension d, which is a direct consequence of the ill-posedness, because the

Table 2 Computing times using (17) for $T = 100,000$ samples

d	n	N	s_{rhs}	t_{rhs}	t_{solve} (s)	$\|c - c_h\|_{L^1(\Omega)}$
2	30	900	1	1.3 s	0.0002	8.89e − 2
2	60	3,600	1	5.3 s	0.002	1.76e − 1
2	100	10,000	3	5 s	0.014	2.96e − 1
3	30	27,000	10	7.2 s	0.013	5.17e − 1
3	60	216,000	30	19 s	0.11	1.45e + 0
3	100	1,000,000	30	86 s	0.95	2.71e + 0
4	30	810,000	30	97 s	0.25	2.72e + 0
5	30	24,300,000	30	3,607 s	8.49	1.82e + 2
6	10	1,000,000	30	197 s	0.14	3.50e + 0
7	10	10,000,000	30	2,371 s	1.68	3.54e + 1
8	10	100,000,000	30	26,329 s	18.2	7.28e + 3
9	10	1,000,000,000	30	303,239 s	253	9.63e + 5
10	10	10,000,000,000	30	≈40 days	2,025	

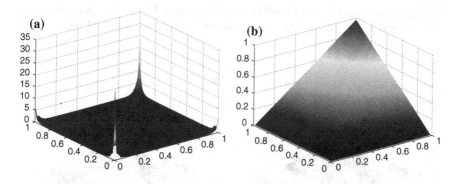

Fig. 3 Student copula, $\rho = 0.5$, $\nu = 1$, $n = 50$, **a** reconstructed density c, **b** copula C

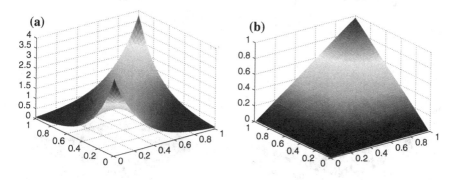

Fig. 4 Frank copula, $\theta = 4$, $n = 50$, **a** reconstructed density c, **b** copula C

condition number of the system matrix K is the condition number of the one-dimensional system matrix $^{(1)}K$ to the power of d.

Naturally, our proposed method works not only for the rather simple independence copula, it also works quite well for all typical copula families. The approximation error for noise free right-hand sides can be neglected. Figures 3 and 4 show the reconstructed densities for the Student and Frank copula, using exact data for the right-hand side. However, ill-posedness is expected when empirical copulas are used. In [14], numerical results for other copula families, like the Gaussian, Gumbel, or Clayton copula, can also be found. However, ill-posedness is expected when empirical copulas are used and we are faced with data noise, which we discuss in the next section.

3 Ill-Posedness and Regularization

Note that in real problems, the copula C is not known and we only have noisy data (10) instead of (9). In order to illustrate the expected numerical instabilities, we have simulated T samples for each two-dimensional copula and present the nonparametric reconstructed densities using the Petrov–Galerkin projection with grid size $n = 50$.

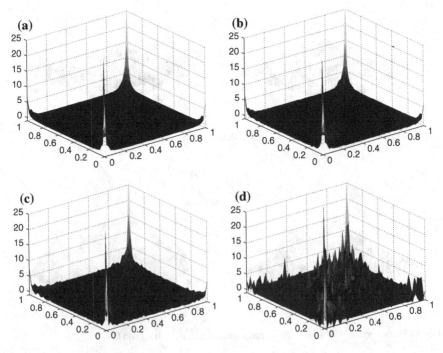

Fig. 5 Student copula density, $\rho = 0.5$, $\nu = 1$, $n = 50$. **a** $T = 1{,}000{,}000$. **b** $T = 100{,}000$. **c** $T = 10{,}000$. **d** $T = 1{,}000$

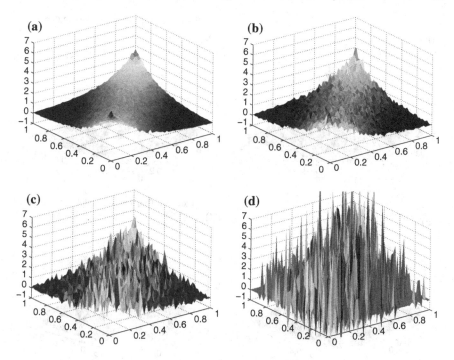

Fig. 6 Frank copula density, $\theta = 4$, $n = 50$. **a** $T = 1,000,000$. **b** $T = 100,000$. **c** $T = 10,000$. **d** $T = 1,000$

A typical problem of ill-posed inverse problems is, that the numerical instability decreases if the grid size n decreases, which can also be seen in Table 1. Therefore, we fix the grid size $n = 50$ and look at the influence of sample size T.

Because of (3), the data noise δ increases if T decreases. Figures 5 and 6 show the expected ill-posedness appearing for decreasing sample size T. Of course, this instabilities also occur for the other copula families, but we restrict our illustration here to these two examples. More examples can be found in [14].

To overcome the ill-posedness, an appropriate regularization for the discretized problem (8) is required. Figures 7 and 8 show the reconstructed copula densities for $T = 1,000$ and $T = 10,000$ samples using the well-known Tikhonov regularization. There is no regularization, if the regularization parameter $\alpha = 0$ is chosen. The left-hand side of the figures shows the unregularized solutions. The choice of the regularization parameter $\alpha = 10^{-8}$ is very naive and arbitrary and serves only as demonstration how the instability can be handled. A better parameter choice should improve the reconstructed densities. It is further work to discuss an appropriate parameter choice rule for Tikhonov regularization as well as other regularization methods.

In order to avoid the complete assembling of the system matrix K leading to high-dimensional systems for $d > 2$, we are interested in regularization methods

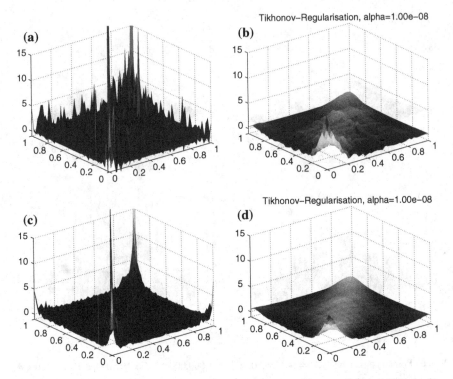

Fig. 7 Regularized Student copula density, $\rho = 0.5$, $\nu = 1$, $n = 50$. **a** $\alpha = 0$, $T = 1{,}000$ samples. **b** $\alpha = 10^{-8}$, $T = 1{,}000$ samples. **c** $\alpha = 0$, $T = 10{,}000$ samples. **d** $\alpha = 10^{-8}$, $T = 10{,}000$ samples

using the special structure (15). In particular, all regularization methods based on the singular value or eigenvalue decomposition of K can be easily handled because the eigenvalue decomposition of the one-dimensional matrix[(1)] $K = V \Lambda V^T$ leads to the eigenvalue decomposition of the system matrix

$$K = (V \otimes \cdots \otimes V)(\Lambda \otimes \cdots \otimes \Lambda)\left(V^T \otimes \cdots \otimes V^T\right).$$

A typical property of Tikhonov regularization is that true peaks in the density will be smoothed. This effect appears in particular for the Student copula density. Hence, the reconstruction quality should be improved, if other regularization methods are used. In the inverse problem theory, it is well-known that Tikhonov regularization accompanies L^2-norm penalization of the regularized solutions. Therefore, L^1 penalties or total variation penalties (see [7]) seem more suitable.

Furthermore, the approximated copula

$$C_h(\mathbf{u}) = \int_0^{\mathbf{u}} c_h(\mathbf{s})\mathrm{d}\mathbf{s} = \sum_{j=1}^{N} c_j \psi_j(\mathbf{u})$$

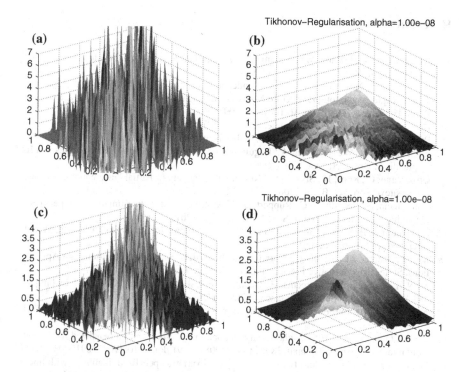

Fig. 8 Regularized Frank copula density, $\theta = 4, n = 50$. **a** $\alpha = 0, T = 1,000$ samples. **b** $\alpha = 10^{-8}$, $T = 1,000$ samples. **c** $\alpha = 0, T = 10,000$ samples. **d** $\alpha = 10^{-8}, T = 10,000$ samples

should yield the typical properties of copulas. For example, the requirement

$$C_h(1, \ldots, 1) \overset{!}{=} 1$$

yields the condition $\sum_{j=1}^{N} c_j = 1$ and the requirements

$$C_h(1, \ldots, 1, u_k, 1, \ldots, 1) \overset{!}{=} u_k \quad k = 1, \ldots, d$$

lead to additional conditions on the vector **c**, which all together can be used to build problem specific regularization methods.

Open Access This chapter is distributed under the terms of the Creative Commons Attribution Noncommercial License, which permits any noncommercial use, distribution, and reproduction in any medium, provided the original author(s) and source are credited.

References

1. Anderssen, R.S., Hegland, M.: For numerical differentiation, dimensionality can be a blessing!. Math. Comput. **68**(227), 1121–1141 (1999)
2. Deheuvels, P.: Non parametric tests of independence. In: Raoult J.P. (ed.) Statistique non Paramétrique Asymptotique. Lecture Notes in Mathematics, vol. 821, pp. 95–107. Springer, Berlin Heidelberg (1980). doi:10.1007/BFb0097426
3. Devroye, L., Györfi, L.: Nonparametric Density Estimation: the L1 View. Wiley Series in Probability and Mathematical Statistics. Wiley, New York (1985)
4. Engl, H., Hanke, M., Neubauer, A.: Regularization of Inverse Problems. Mathematics and Its Applications. Springer, New York (1996)
5. Grossmann, C., Roos, H., Stynes, M.: Numerical Treatment of Partial Differential Equations. Universitext. Springer, Berlin (2007)
6. Kauermann, G., Schellhase, C., Ruppert, D.: Flexible copula density estimation with penalized hierarchical b-splines. Scand. J. Stat. **40**(4), 685–705 (2013)
7. Koenker, R., Mizera, I.: Density estimation by total variation regularization. Adv. Stat. Model. Inference pp. 613–634 (2006)
8. Mai, J., Scherer, M.: Simulating Copulas: Stoch. Models. Sampling Algorithms and Applications. Series in Quantitative Finance. Imperial College Press, London (2012)
9. McNeil, A., Frey, R., Embrechts, P.: Quantitative Risk Management: Concepts. Techniques and Tools. Princeton Series in Finance. Princeton University Press, USA (2010)
10. Nelsen, R.B.: An Introduction to Copulas (Springer Series in Statistics). Springer, New York (2006)
11. Qu, L., Qian, Y., Xie, H.: Copula density estimation by total variation penalized likelihood. Commun. Stat.—Simul. Comput. **38**(9), 1891–1908 (2009). doi:10.1080/03610910903168587
12. Qu, L., Yin, W.: Copula density estimation by total variation penalized likelihood with linear equality constraints. Computat. Stat. Data Anal. **56**(2), 384–398 (2012). doi: http://dx.doi.org/10.1016/j.csda.2011.07.016
13. Schuster, T., Kaltenbacher, B., Hofmann, B., Kazimierski, K.: Regularization Methods in Banach Spaces. Radon Series on Computational and Applied Mathematics. Walter De Gruyter In., Berlin (2012)
14. Uhlig, D., Unger, R.: A Petrov-Galerkin projection for copula density estimation. Technical report, TU Chemnitz, Department of Mathematics (2013). http://www.tu-chemnitz.de/mathematik/preprint/2013/PREPRINT.php?year=2013&num=07
15. Uhlig, D., Unger, R.: The Petrov-Galerkin projection for copula density estimation isn't counting. Technical report, TU Chemnitz, Department of Mathematics (2014). http://www.tu-chemnitz.de/mathematik/preprint/2014/PREPRINT.php?year=2014&num=03

Printed in the United States
By Bookmasters